AF700792

Classical and Modern Direction-of-Arrival Estimation

Classical and Modern Direction-of-Arrival Estimation

Edited by

T. Engin Tuncer

Benjamin Friedlander

ELSEVIER

AMSTERDAM • BOSTON • HEIDELBERG • LONDON
NEW YORK • OXFORD • PARIS • SAN DIEGO
SAN FRANCISCO • SINGAPORE • SYDNEY • TOKYO

Academic Press is an imprint of Elsevier

Academic Press is an imprint of Elsevier
30 Corporate Drive, Suite 400
Burlington, MA 01803, USA

This book is printed on acid-free paper. ♾

Copyright © 2009 by Elsevier Inc. All rights reserved.

Designations used by companies to distinguish their products are often claimed as trademarks or registered trademarks. In all instances in which Academic Press is aware of a claim, the product names appear in initial capital or all capital letters. Readers, however, should contact the appropriate companies for more complete information regarding trademarks and registration.

No part of this publication may be reproduced, stored in a retrieval system, or transmitted in any form or by any means, electronic, mechanical, photocopying, scanning, or otherwise, without prior written permission of the publisher.

Permissions may be sought directly from Elsevier's Science & Technology Rights Department in Oxford, UK: phone: (+44) 1865 843830, fax: (+44) 1865 853333, e-mail: permissions@elsevier.com. You may also complete your request on-line via the Elsevier homepage (*http://elsevier.com*), by selecting "Support & Contact" then "Copyright and Permission" and then "Obtaining Permissions."

Library of Congress Cataloging-in-Publication Data
Application submitted.

ISBN-13: 978-0-12-374524-8

For information on all Academic Press publications
visit our Web site at *www.elsevierdirect.com*

Typeset by: diacriTech, India

Printed in the United States of America
09 10 11 12 13 5 4 3 2 1

Working together to grow
libraries in developing countries

www.elsevier.com | www.bookaid.org | www.sabre.org

ELSEVIER BOOK AID International Sabre Foundation

Contents

Preface

The need for direction-of-arrival (DOA) estimation arises in many engineering applications including wireless communications, radar, radio astronomy, sonar, navigation, tracking of various objects, and rescue and other emergency assistance devices. In its modern version, DOA estimation is usually studied as part of the more general field of array processing. Much of the work in this field, especially in earlier days, focused on radio direction finding—that is, estimating the direction of electromagnetic waves impinging on one or more antennas.

The problem of acoustic direction estimation was also studied extensively, mostly in the context of sonar. In fact much of the development of what is now called "modern DOA estimation" was done for sonar applications where the relatively small bandwidth of the signals to be processed made the computational requirements of advanced algorithms feasible with the technology that existed then. As processing power kept increasing, it became possible to apply advanced techniques to the more demanding wider bandwidth communications and radar signals.

While DOA estimation is now a mature field with a solid theoretical basis and a large number of practical applications, it is still an evolving and quite active field of research. This book attempts to provide a snapshot of the most recent work on this ubiquitous problem and, at the same time, to provide a brief review of the more classical work on direction finding.

The book contains ten chapters. Chapter 1 by Friedlander lays out the fundamentals of the DOA problem. Starting with a discussion of how it all originated, it blends both classical and modern techniques. Chapter 2 by Demmel is a good reference for practicing engineers. It presents the techniques currently used in commercial direction-finding (DF) systems. Chapter 3 by Viberg, Lanne, and Lundgren presents a critical topic for sensor arrays, namely calibration. Different techniques for calibration are presented and numerically compared in this chapter. Chapter 4 by Tuncer, Yasar, and Friedlander discusses narrowband and wideband processing for DOA estimation. The advantages of array interpolation and processing gain achieved by wideband processing are outlined. Chapter 5 by Rübsamen and Gershman presents techniques for search-free DOA estimation for different arrays. Such techniques allow one to use fast algorithms for some unconventional array structures.

Chapter 6 by Amin and Zhang introduces a new dimension to the DOA problem—namely, spatial time-frequency distributions. Direction-of-arrival estimation can be improved as a result of signal-to-noise ratio (SNR) enhancement and source discrimination in the time-frequency domain. Chapter 7 by Abramovich, Johnson, and Mestre discusses an interesting problem in the threshold region. The expected likelihood approach is used as a mechanism to assess the quality of estimates for the low sample case. Chapter 8 by Chevalier, Ferréol, and Albera presents the advantages of higher-order statistics compared to second-order statistics for direction of arrival. It has been shown that a virtual increase of the array aperture by the introduction of virtual sensors can be used to improve resolution and modeling errors. Chapter 9 by Chen and Yao discusses the localization problem in sensor networks. Both maximum likelihood formulation and performance bounds are presented for source localization. Chapter 10 by Amar and Weiss advocates direct position determination for source localization. It has been shown that superior results can be obtained at low SNR even when there are modeling errors.

Acknowledgments

We wish to thank all the authors for their high-quality contributions. This book is a reflection of what is believed to be the renewed interest of people in array signal processing and, more specifically, DOA estimation. It is hoped that this book will be a catalyst for a much wider interest and further collaboration.

List of Contributors

Yuri I. Abramovich (Chapter 7)

Yuri I. Abramovich (an IEEE Fellow) received the Dipl.Eng. (Hons.) degree in radio electronics in 1967 and the Cand.Sci. degree (Ph.D. equivalent) in theoretical radio techniques in 1971, both from the Odessa Polytechnic University, Odessa (Ukraine), USSR. In 1981, he received the D.Sc. degree in radar and navigation from the Leningrad Institute for Avionics, Leningrad (Russia), USSR. From 1968 to 1994, he was with the Odessa State Polytechnic University as a research fellow, professor, and ultimately as Vice-Chancellor of Science and Research. From 1994 to 2006, he was at the Cooperative Research Centre for Sensor Signal and Information Processing (CSSIP), Adelaide, Australia. Since 2000, he has been with the Australian Defence Science and Technology Organisation (DSTO) in Adelaide as Principal Research Scientist, seconded to CSSIP until its closure. His research interests are in signal processing (particularly spatio-temporal adaptive processing, beamforming, signal detection and estimation), its application to radar (particularly over-the-horizon radar), electronic warfare, and communication. He is currently an Associate Editor of *IEEE Transactions on Aerospace and Electronic Systems* and previously served as Associate Editor of *IEEE Transactions on Signal Processing* from 2002 to 2005.

Laurent Albera (Chapter 8)

Laurent Albera received the DEA degree in automatic and signal processing in 2001 from the University of Science of Orsay in Paris. From 2001 to 2004, he was committed for a study within the framework of a research contract Convention Industrielle de Formation par la Recherche (CIFRE) between Thales Communications (formerly Thomson-CSF), the French National Centre for Scientific Research (CNRS), and the University of Nice Sophia-Antipolis. In 2003, he received his Ph.D. degree in sciences from the University of Nice Sophia-Antipolis, France. He is now an assistant professor at the University of Rennes 1 and is affiliated with the research group Laboratoire Traitement du Signal et de l'Image (LTSI), UMR Inserm 642. His research interests include multilinear algebra and high-order statistics in order to perform ICA and source localization.

Alon Amar (Chapter 10)

Alon Amar received the B.Sc. degree in electrical engineering from the Technion–Israel Institute of Technology in 1997, and the M.Sc. and Ph.D. degrees in electrical engineering from Tel Aviv University in Israel in 2003 and 2009, respectively. He is currently a postdoctoral researcher at the Circuits and Systems group at the Faculty of Electrical Engineering, Mathematics and Computer Science of Delft University of Technology in The Netherlands. His main research interests are statistical array processing, source localization, and distributed signal processing.

Moeness Amin (Chapter 6)

Moeness G. Amin received his Ph.D. degree in 1984 from the University of Colorado in Boulder. He has been on the faculty of Villanova University since 1985, where is now a professor in the Department of Electrical and Computer Engineering and the Director of the Center for Advanced Communications. Dr. Amin is the recipient of the 2009 Technical Achievement Award from the European Association of Signal Processing. He is a Fellow of the Institute of Electrical and Electronics Engineers, a Fellow of the International Society of Optical Engineering, recipient of the IEEE Third Millennium Medal, Distinguished Lecturer of the IEEE Signal Processing Society for 2003 and 2004, a member of the Franklin Institute Committee on Science and the Arts, and recipient of the 1997 Villanova University Outstanding Faculty Research Award. Dr. Amin has more than 400 publications in the areas of wireless communications, time-frequency analysis, smart antennas, interference cancellation, GPS technologies, over-the-horizon radar, and radar imaging. He was a Guest Editor of the *Journal of Franklin Institute* September-08 Special Issue on Advances in Indoor Radar Imaging and was a Guest Editor of the *IEEE Transactions on Geoscience and Remote Sensing* May-09 Special Issue on Remote Sensing of Building Interior.

Chiao-En Chen (Chapter 9)

Chiao-En Chen received both the B.S. and M.S. degrees in electrical engineering from National Taiwan University (NTU) in Taipei in 1998 and 2000, respectively. From 2000 to 2002, he served in the R.O.C. army as a missile operations officer and from 2002 to 2003, he was a research assistant in the Department of Physics, NTU. He received his Ph.D. degree in electrical engineering at University of California-Los Angeles (UCLA) in 2008. Since 2008, he has been an assistant professor in the Department of Electrical Engineering and the Department of Communications Engineering, National Chung Cheng University, Chiayi, Taiwan. His research interests include array signal processing, source localization, and MIMO communication systems.

Pascal Chevalier (Chapter 8)

Pascal Chevalier received the M.Sc. degree from Ecole Nationale Supérieure des Techniques Avancées (ENSTA) and the Ph.D. degree from South-Paris University in France in 1985 and 1991, respectively. Since 1991 he has been with Thomson-CSF/RGS (now Thalès-Communications) where he has shared industrial activities (studies, experimentations, expertises, management), teaching activities both in French engineering schools (Supelec, ENST, ENSTA, ESIEE) and French universities (Cergy-Pontoise), and research activities. Since 2000, Dr. Chevalier has also been acting as Technical Manager and Architect of the array processing subsystem as part of a national program of military satellite telecommunications. From 2003 to the present, he has been a Thalès Expert.

Dr. Chevalier's present research interests are in array processing techniques, either blind or informed, second-order or higher-order, spatial- or spatio-temporal, time-invariant or time-varying especially for cyclostationary signals; linear or nonlinear and particularly widely linear for noncircular signals for applications such as TDMA, CDMA, and MIMO radiocommunications networks; satellite telecommunications, spectrum monitoring, and passive listening in HVUHF band. He was a member of the THOMSON-CSF Technical and Scientifical Council between 1995 and 1998. He coreceived the 2003 "Science and Defense" Award from the French Ministry of Defence for its work as a whole about array processing for military radiocommunications. He is author or coauthor of about 25 patents and 100 papers (journal, conferences, and chapters of books). Dr. Chevalier is presently an emeritus member of the Societé des Electriciens et des Electroniciens (SEE).

Franz Demmel (Chapter 2)

Franz Demmel is currently head of R&D Direction Finders at Rohde & Schwarz, Division Radiomonitoring and Radiolocation. He joined the company in 1970 during his education in radio frequency (RF) and communication engineering. After receiving his diploma he worked in the fields of self-matching antennas and active antennas. From 1979 to 2000 he headed a team for antenna development with a focus on wideband antennas, DF arrays, and adaptive antennas. Since 1995 he has been head of R&D Direction Finders; from 2000 to 2007 he was also responsible for development of receivers and has also been in charge of Basic R&D since 2007. Dr. Demmel is lecturer of CCG and was a member of several ITG working groups and conference program committees. His main interests address all aspects of multichannel signal processing.

Anne Ferréol (Chapter 8)

Anne Ferréol was born in Lyon, France. She received the M.Sc. degree from ICPI-Lyon, the Mastère degree from Ecole Nationale Supérieure des Télécommunications (ENST), the Ph.D. degree from Ecole Normale Supérieur de Cachan in France in 1988, 1989, and 2005, respectively. Since 1989 she has been working in the Array Processing Department at THALES-Communications.

Ms. Ferréol co-received the 2003 "Science and Defense" Award from the French Ministry of Defence for its work as a whole about array processing for military radiocommunications. She is a member of the European Network of Excellence NEWCOM.

Benjamin Friedlander (Chapters 1 and 4)

Professor Friedlander is an internationally known expert in the areas of signal processing, array processing, and communications. He has published extensively in the areas of detection, estimation, and filtering; algorithm development; performance analysis; and the application of statistical signal processing to various engineering problems, including digital communications, wireless communications, antenna and sensor arrays, radar and sonar systems, adaptive filtering, and time-frequency analysis.

He received the B.Sc. and the M.Sc. degrees in electrical engineering from the Technion–Israel Institute of Technology, a Ph.D. degree in electrical engineering and a M.Sc. degree in statistics from Stanford University. Before joining the University of California at Santa Cruz, he held various positions in industry; he is currently a professor of electrical engineering. He is the recipient of the 1983 ASSP Senior Award, the 1985 Award for the Best Paper of the Year from the European Association for Signal Processing (EURASIP), the 1989 Technical Achievement Award of the Signal Processing Society, and the IEEE Third Millennium Medal.

Alex B. Gershman (Chapter 5)

Alex B. Gershman received his Diploma and Ph.D. degrees in radiophysics and electronics from the Nizhny Novgorod State University in Russia in 1984 and 1990, respectively. He is currently with the Darmstadt University of Technology in Germany as a professor of Communication Systems. His research interests span the areas of signal processing and communications with the primary emphasis on array processing; beamforming; statistical signal processing, and multi-antenna, multiuser, and cooperative communications.

Dr. Gershman is a Fellow of the IEEE. He has received several awards including the 2004 IEEE Signal Processing Society (SPS) Best Paper Award, the IEEE Aerospace and Electronic Systems Society (AESS) Barry Carlton Award, and the 2001 Wolfgang Paul Award from the Alexander von Humboldt Foundation. He was Editor-in-Chief for *IEEE Signal Processing Letters* (2006–2008) and Chair of the IEEE SPS Sensor Array and Multichannel Technical Committee (2006–2008). Dr. Gershman is currently a member of the IEEE SPS Board of Governors. He is on the editorial boards of the *IEEE Signal Processing Magazine*, the *IEEE Transactions on Signal Processing*, and several other technical journals.

Ben A. Johnson (Chapter 7)

Ben A. Johnson received a B.Sc. degree (cum laude) in physics from Washington State University in 1984, an M.Sc. degree in digital signal processing from University of Southern California in 1988, and a Ph.D. from the University of South Australia in 2009; during that time (his) research has focused on application of spatio-temporal adaptive processing in HF radar. From 1984 to 1989, he was at Hughes Aircraft Company (now Raytheon) as a systems engineer in airborne radar. From 1989 to 1998, he was with Sensis Corporation as a senior radar engineer in ground-based surveillance systems. Since 1998, he has been with Lockheed Martin Australia Electronic Systems Pty. Ltd. on the Jindalee over-the-horizon Operational Radar Network (JORN), first as a senior test engineer and now as technical director.

Maria Lanne (Chapter 3)

Dr. Maria Lanne received the M.Sc. degree in electrical engineering from Chalmers University of Technology, Gothenburg, Sweden, in 1998. Since 1998, she has been with the Department of Microwave and Antennas at Saab Microwave Systems (previously Ericsson Microwave Systems AB) in Gothenburg. From 2003 to 2007 she was an industrial Ph.D. student in the Signal Processing Group at the Department of Signals and Systems, Chalmers University of Technology. Her research interests include array antennas, array calibration, and the interaction between electromagnetic properties and array signal processing.

Astrid Lundgren (Chapter 3)

Dr. Astrid Lundgren received her M.Sc. in chemical engineering in 2002 and a Ph.D. in electrical engineering in 2007, both from Chalmers University of Engineering. Her research interests are in mathematical modeling and system identification, with a focus on local models of nonlinear systems. Since 2008 she has been employed at Volvo Technology Corporation, working in the areas of sensor fusion and hybrid vehicle control.

Xavier Mestre (Chapter 7)

Xavier Mestre (Barcelona, 1974) received the M.Sc. and Ph.D. in electrical engineering from the Technical University of Catalonia (UPC) in 1997 and 2002, respectively. During the pursuit of his Ph.D., he was recipient of a 1998–2001 Ph.D. scholarship (granted by the Catalan government) and was awarded the 2002 Rosina Ribalta second prize for the best doctoral thesis project within the areas of information technologies and communications by the Epson Iberica Foundation. From January 1998 to December 2002, he was with UPC's Communications Signal Processing Group, where he worked as a research assistant.

In January 2003, Dr. Mestre joined the Centre Tecnològic de Telecomunicacions de Catalunya (CTTC), where he currently holds a position as a Senior Research Associate in the area of Radio Communications. He has actively participated in several European projects and contracts with the local industry. Since 2004 he has been coordinator of the Radio Communications Research Area at CTTC. His current research interests include array processing, parametric estimation, random matrix theory, and multicarrier modulations.

Michael Rübsamen (Chapter 5)

Michael Rübsamen received the Dipl.-Ing. degree in electrical engineering from RWTH Aachen University, Germany, in 2006. From October 2004 until July 2005 he held a research scholarship from Bell-Labs, Crawford Hill, where he worked on signal processing algorithms for fiber optic communication systems. In June 2006 he joined the Communication Systems Group at Darmstadt University of Technology in Germany, where he is working towardthe Dr.-Ing. degree in the field of sensor array signal processing. His research interests are focused on direction-of-arrival estimation, robust adaptive beamforming, and parameter estimation. In 2002 he received the Philips Student Award and in 2003 the Henry Ford II Student Award as well as a grant from the German National Academic Foundation (Studienstiftung des deutschen Volkes). He was a student contest finalist at ICASSP 2008 and served as local organization cochair for the IEEE Sensor Array and Multichannel Signal Processing (SAM) Workshop in Darmstadt, Germany, in July 2008.

T. Engin Tuncer (Chapter 4)

T. Engin Tuncer received his B.Sc. and M.Sc. degrees in electrical and electronics engineering from Middle East Technical University, Ankara, Turkey, in 1987 and 1989, respectively, and the Ph.D. degree in electrical engineering from Boston University in 1993. He was an instructor in the Biomedical Engineering Department of Boston University between September 1990 and June 1992. He was an adjunct professor in the Electrical, Computer, and Systems Engineering Department during 1993. From September 1993 to June 1998, he was an assistant professor in Electronics Engineering Department of Ankara University in Turkey. Since July 1998, he has been with Electrical and Electronics Engineering Department of Middle East Technical University, Ankara, and currently he is a professor in the same department. Professor Tuncer's research is focused on statistical signal processing and, more specifically, on sensor arrays and multidimensional signal processing, system identification, signal processing for communications, and radar and sonar systems. He is the supervisor for the Sensor Array and Multidimensional Signal Processing Laboratory (SAM). In 2009 Dr. Tuncer became the Local Liaison Officer of The European Association for Signal Processing.

Mats Viberg (Chapter 3)

Professor Mats Viberg received the Ph.D. degree in automatic control from Linköping University in Sweden in 1989. He has held academic positions at Linköping University and visiting scholarships at Stanford University and Brigham Young University in the United States. Since 1993, Dr. Viberg has been a professor of Signal Processing at Chalmers University of Technology in Sweden. During 1999 to 2004 he served as Department Chair. His research interests are in statistical signal processing and its various applications, including antenna array signal processing, system identification, wireless communications, radar systems, and automotive signal processing.

Dr. Viberg has served in various capacities in the IEEE Signal Processing Society, including chair of the Technical Committee (TC) on Signal Processing Theory and Methods (2001–2003), vice-chair for the TC on Sensor Array and Multichannel (2009–), Associate Editor of the *Transactions on Signal Processing* (2004–2005), and member of the Awards Board (2005–2007). Professor Viberg has received two Paper Awards from the IEEE Signal Processing Society (1993 and 1999, respectively) and the Excellent Research Award from the Swedish Research Council (VR) in 2002. Dr. Viberg has been a Fellow of the IEEE since 2003, and his research group received the 2007 EURASIP European Group Technical Achievement Award. In 2008, Dr. Viberg was elected into the Royal Swedish Academy of Sciences (KVA).

Anthony J. Weiss (Chapter 10)

Anthony J. Weiss received the B.Sc. degree from the Technion–Israel Institute of Technology in 1973 and the M.Sc. and Ph.D. degrees from Tel Aviv University in Israel in 1982 and 1985, respectively—all in electrical engineering. From 1973 to 1983, he was involved in research and development of numerous projects in the fields of communications, command and control, and emitter localization. In 1985, he joined the Department of Electrical Engineering-Systems at Tel Aviv University. From 1996 to 1999 he served as the department chairman and as the chairman of IEEE Israel Section. He is currently the head of the EE School in Tel Aviv University. His research interests include detection and estimation theory, signal processing, sensor array processing, and wireless networks. Dr. Weiss also held leading scientific positions with Signal Processing Technology Ltd., Wireless on Line Ltd., and SigmaOne Ltd. Professor Weiss has published more than 100 papers in professional magazines and conferences and holds nine U.S. patents. He was a recipient of the IEEE 1983 Acoustics, Speech, and Signal Processing Society's Senior Award and the IEEE Third Millennium Medal. He has been an IEEE Fellow since 1997 and an IET Fellow since 1999.

Kung Yao (Chapter 9)

Kung Yao received the B.S.E. (Summa Cum Laude), M.A., and Ph.D. degrees in electrical engineering, all from Princeton University in New Jersey. He was a NAS-NRC Post-Doctoral Research Fellow at the University of California, Berkeley. Presently, he is a Distinguished Professor in the Electrical Engineering Department at UCLA. In 1969, he was a visiting assistant professor at Massachusetts Institute of Technology. In 1973 and 1974, he was a visiting associate professor at Eindhoven University of Technology. In 1985 to 1988, he served as Assistant Dean of the School of Engineering and Applied Science at UCLA. In 2007, he was a Royal Society Kan Tong Po visiting professor at HK Polytechnic University.

Professor Yao's research and professional interests include digital communication theory and system, digital, and array signal processing in sensor networks; statistical modeling of wireless fading channels; and systolic and VLSI algorithms, architectures, and systems. He was the coeditor of a two volume series of an IEEE reprint book, *High Performance VLSI Signal Processing*, 1997. Dr. Yao received the IEEE Signal Processing Society's 1993 Senior Award in VLSI Signal Processing and the 2008 IEEE Communications Society/Information Theory Society Joint Paper Award. He is a Life Fellow of IEEE.

T. Kaya Yasar (Chapter 4)

Temel Kaya Yasar was born in Ankara, Turkey, in 1980. During the third year of his undergraduate study in electrical and electronics engineering at Middle East Technical University, he started work at Aselsan Inc. as a member of the part-time staff. In 2005, he began to work in the Sensor Array and Multidimensional Signal Processing Laboratory (SAM) in Middle East Technical University in a project funded by The Scientific and Technological Research Council of Turkey. He received his M.Sc. degree in electrical and electronics engineering from Middle East Technical University in 2006. He is currently working for his Ph.D. in medical imaging in the Mechanical Engineering Department of the University of Illinois in Chicago.

Yimin Zhang (Chapter 6)

Yimin Zhang received his Ph.D. degree from the University of Tsukuba in Japan in 1988. Dr. Zhang joined the faculty of the Department of Radio Engineering, Southeast University, Nanjing, China, in 1988. He served as a Technical Manager at the Communication Laboratory Japan in Kawasaki from 1995 to 1997, and was a visiting researcher at ATR Adaptive Communications Research Laboratories in Kyoto, Japan, from 1997 to 1998. Since 1998, he has been with the Villanova University in Pennsylvania, where he is currently a research professor with the Center for Advanced Communications. Dr. Zhang has more than 150 publications in the area of statistical signal and array processing for communications and radar applications, including wireless communications and networking, MIMO communication/radar systems, source localization and target tracking, time-frequency analysis, blind signal processing, jammer suppression, radio frequency identification (RFID), and image processing. He is an Associate Editor for the *IEEE Signal Processing Letters*, an Associate Editor for the *Journal of the Franklin Institute*, and serves on the editorial board of the *Signal Processing* journal.

Chapter 1

Wireless Direction-Finding Fundamentals

Benjamin Friedlander

1.1 INTRODUCTION

Wireless direction finding has a long, and distinguished history going back to the very beginnings of wireless communications around the turn of the twentieth century. Many of its basic concepts had been established by that time. To put this in perspective we quote the opening sentence of the Foreword (by T. L. Eckersley, F.R.S.) to a book first published in 1922 entitled *Wireless Direction Finding* (Keen, 1938):

> Everything that the technician requires to know about the now well-established art of Radio Direction and Position Finding will be found most lucidly explained in the following chapters.

The first attempts at direction finding (DF) made use of the directional characteristics of antenna elements (dipoles, loops, etc.) (Bellini and Tosi, 1907, 1909; Marconi, 1906). However, the use of multiple antennas and phased antenna arrays did not lag far behind (Adcock, 1919; Keen, 1938).

Many advances have been made in the field of direction finding since the early 1900s, for the most part due to advances in technology rather than new concepts or principles (Travers and Hixon, 1966). The remarkable developments in electronics and in devices and components for generating and amplifying signals at higher and higher frequencies have greatly extended the capabilities and applicability of direction-finding systems. The shift from analog to digital technology has increased tremendously the flexibility of such systems and their ability to monitor and keep track of signals over a wide range of frequencies and large geographical areas. However, the basic direction-finding techniques have not changed much, with the exception of the introduction of "super-resolution" algorithms for multiple co-channel signals.

Classical and Modern Direction-of-Arrival Estimation
Copyright © 2009 by Academic Press, Inc. All rights of reproduction in any form reserved.

In this chapter we describe some of the basic direction-finding methods for antenna arrays and discuss their performance characteristics in the presence of noise, multipath, and interference. We then provide a brief introduction to co-channel multi-emitter direction finding and super-resolution, particularly the merits and drawbacks of super-resolution methods and their relationship to classical methods.

1.2 PROBLEM FORMULATION

Consider an emitter transmitting a signal $s(t)e^{jw_c t}$, where $s(t)$ is the baseband signal, and $w_c = 2\pi f_c$, where f_c is the carrier frequency. This signal is received by an array of antennas as depicted in Figure 1.1. The received signals are delayed versions of the transmitted signal

$$\mathbf{x}_{pb}(t) = \begin{bmatrix} s(t-\tau_1)e^{jw_c(t-\tau_1)} \\ s(t-\tau_2)e^{jw_c(t-\tau_2)} \\ \vdots \\ s(t-\tau_M)e^{jw_c(t-\tau_M)} \end{bmatrix} \tag{1.1}$$

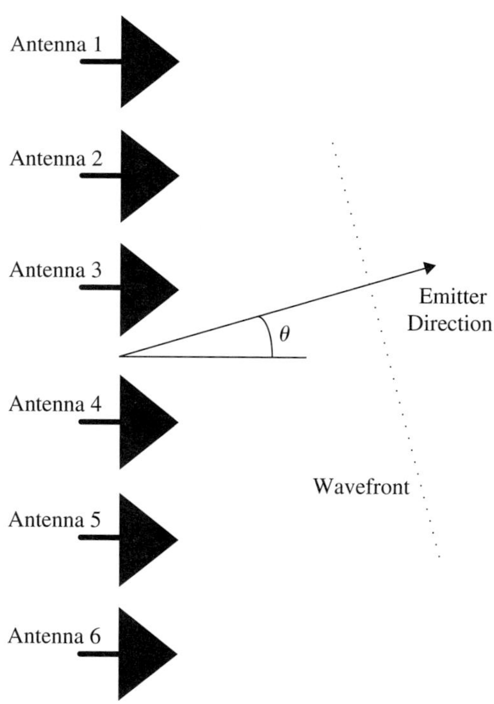

FIGURE 1.1 Six-element uniformly spaced linear array receiving signals from a single emitter at direction θ.

where $\mathbf{x}_{pb}(t)$ is the vector of the received passband signals and τ_m are the propagation delays determined by the direction of the source relative to the array. For example, in the case of a linear array

$$\tau_m = \tau_0 - (d_m/c)\sin\theta \tag{1.2}$$

where τ_0 is the propagation delay from the emitter to a reference point on the array, d_m are the distances of the array elements from that reference point, c is the speed of light, and θ is the emitter direction measured relative to the line perpendicular to the array. Without loss of generality we can assume that $\tau_0 = 0$. (This is equivalent to defining $s(t)$ as the baseband signal at the reference point of the receiver array, not as the source signal.) Thus, in the rest of this chapter we will assume that

$$\tau_m = -(d_m/c)\sin\theta \tag{1.3}$$

We note that the contribution of τ_0 to the phase $e^{-jw_c\tau_0}$ introduces an unknown random phase term that is common to the elements of the received signal vector. Because of this phase term, all direction-finding methods must be designed to be invariant to a common additive phase.

After transforming the passband signals to baseband, we have

$$\mathbf{x}(t) = \begin{bmatrix} s(t-\tau_1)e^{jw_c\tau_1} \\ s(t-\tau_2)e^{jw_c\tau_2} \\ \vdots \\ s(t-\tau_M)e^{jw_c\tau_M} \end{bmatrix} \tag{1.4}$$

where $\mathbf{x}(t)$ is the vector of received baseband signals.

Let D denote the aperture of the array in wavelengths λ. Then $D\lambda/c = D/f_c$ is the time it takes the signal to propagate over the array, so that $\tau_m < D/f_c$. It follows that, if the signal bandwidth is B and $B \ll f_c/D$, $s(t-\tau_m) \approx s(t)$. The condition $B \ll f_c/D$ or $B/f_c \ll 1/D$ implies that $s(t)$ is a narrowband signal. It follows that, under the narrowband assumption, the received baseband signal can be written as

$$\mathbf{x}(t) = s(t)\begin{bmatrix} e^{jw_c\tau_1} \\ e^{jw_c\tau_2} \\ \vdots \\ e^{jw_c\tau_M} \end{bmatrix} \tag{1.5}$$

For this reason, we define the so-called "array manifold" $\mathbf{a}(\theta)$, as the array response to a unit amplitude signal (i.e., the case where $s(t) = 1$).

$$\mathbf{a}(\theta) = \begin{bmatrix} \exp\{j2\pi d_1 \sin\theta/\lambda\} \\ \exp\{j2\pi d_2 \sin\theta/\lambda\} \\ \vdots \\ \exp\{j2\pi d_M \sin\theta/\lambda\} \end{bmatrix} \tag{1.6}$$

With no loss of generality we assume that the signal $s(t)$ has unit power, in which case the complete narrowband model for the received signal can be written as

$$\mathbf{y}(t) = \sqrt{\text{SNR}}s(t)\mathbf{a}(\theta_0) + \mathbf{v}(t) \tag{1.7}$$

where $\mathbf{v}(t)$ is a multivariate complex Gaussian noise vector with uncorrelated elements having zero mean and unit variance, θ_0 is the direction of the emitter, and SNR is the signal-to-noise ratio.

Direction-finding methods often use sampled versions of the array output, in which case

$$\mathbf{y}[k] = \sqrt{\text{SNR}}s[k]\mathbf{a}(\theta_0) + \mathbf{v}[k] \tag{1.8}$$

where k is the index of the sample, or the "snapshot." The noise $\mathbf{v}[k]$ is assumed to be independent from snapshot to snapshot. The signal $s[k]$ may or may not be independent from snapshot to snapshot, depending on its temporal correlation properties. Equation (1.8) defines the narrowband signal model used in the rest of this chapter.

1.3 DIRECTION-FINDING ALGORITHMS

We now describe three common methods for direction finding using an antenna array. The first two involve linear combinations of the received signals; the third is nonlinear.

1.3.1 Beamforming

Combining the antenna outputs so that the signals from a given direction "line up," and can thus be added coherently, is the fundamental method used in array-processing applications. Recall that the passband signal received by the array's mth element is $x_m(t) = s(t - \tau_m(\theta_0))e^{jw_c(t-\tau_m(\theta_0))}$, where the delays are functions of the emitter direction θ_0 and the array geometry.

With proper delays applied to them, the received signals line up. Assume that $x_m(t)$ is delayed by $\overline{\tau} - \tau_m(\theta_0)$ to yield $x_m(t - \overline{\tau} - \tau_m(\theta_0)) = s(t - \overline{\tau})e^{jw_c(t-\overline{\tau})}$. In other words, the properly delayed array outputs are identical and can be added up to enhance the signal-to-noise ratio at the receiver by a factor of M, the coherent array gain.

If a different set of delays is used corresponding to $\theta \neq \theta_0$, the signals will not line up and the result of the sum will be a signal with power that is smaller than in the case where $\theta = \theta_0$. Thus, the signal power at the beamformer output will be maximized in the correct direction.

The processing just described, called *delay-and-sum beamforming*, can be used for both wideband and narrowband signals. In the case where a signal is narrowband, the beamformer can be implemented using appropriate phase shifts instead of time delays. As was shown in Equation (1.5) the received signal vector is represented (up to a common factor) by the array manifold $\mathbf{a}(\theta)$. Combining the signals "in phase" is accomplished by the complex conjugate of the array manifold. Thus, we compute $|\mathbf{a}(\theta)^H\mathbf{a}(\theta_0)|$. If the assumed direction θ equals the true emitter direction, we get the largest value of the combined signals $|\mathbf{a}(\theta_0)^H\mathbf{a}(\theta_0)| = M$. If $\theta \neq \theta_0$, the signals are not added in phase and the result of the sum is smaller. This type of processing is called *frequency domain beamforming* or just *beamforming*.

In its most general form, a beamformer uses a weight vector $\mathbf{W}(\theta)$ to linearly combine the array outputs. In most applications we are interested in the magnitude or the power of the combiner output. In the latter case the output is given by

$$p(\theta) = |\mathbf{W}(\theta)^H\mathbf{y}|^2 \tag{1.9}$$

where $\mathbf{y}$ is the received signal. In the absence of noise and assuming a unit power signal, the beamformer output is

$$p(\theta) = \mathrm{SNR}|\mathbf{W}(\theta)^H\mathbf{a}(\theta_0)|^2 \tag{1.10}$$

When the beamformer is used for direction finding, its output $p(\theta)$ is computed over a range of directions θ. The direction where $p(\theta)$ reaches its largest value is the estimated emitter direction.

In this discussion we assume that $\mathbf{W}(\theta) = \mathbf{a}(\theta)$. In practice, this weight vector will be modified by a window chosen to suppress the sidelobe level of the beam pattern to a desired level. We use a normalized but nonwindowed weight vector $\mathbf{W}(\theta) = \mathbf{a}(\theta)/|\mathbf{a}(\theta)|$, which makes the noise power at the beamformer output the same as at the antenna elements.

It should be noted that the beamformer with this choice of weight vector is in fact the generalized maximum likelihood estimator of the direction θ, for the signal model

$$\mathbf{y} = \alpha_0\mathbf{a}(\theta_0) + \mathbf{v} \tag{1.11}$$

where $\mathbf{y}$ is the vector of the signals at the array output, α_0 is a complex scale factor assumed to be unknown ($\alpha_0 = \sqrt{\mathrm{SNR}}\,s$ in the previous notation), $\mathbf{a}(\theta)$ is the array manifold, θ_0 is the unknown emitter direction, and $\mathbf{v}$ is a noise vector composed of independent zero-mean unit-variance Gaussian random variables. The generalized maximum likelihood estimate is given by joint minimization of the error $|\mathbf{y} - \alpha_0\mathbf{a}(\theta_0)|^2$ over the unknowns θ_0 and α_0. Minimizing first over α_0 we have $\alpha_0 = \mathbf{a}^H(\theta_0)\mathbf{y}/\mathbf{a}^H(\theta_0)\mathbf{a}(\theta_0)$. Inserting this into the error function we have $|\mathbf{y} - \mathbf{a}(\theta_0)\mathbf{a}^H(\theta_0)/|\mathbf{a}(\theta_0)|^2\mathbf{y}|^2$. Minimizing this cost function with respect to θ_0 is equivalent to maximizing $\mathbf{y}^H\mathbf{a}(\theta_0)\mathbf{a}^H(\theta_0)\mathbf{y}/|\mathbf{a}(\theta_0)|^2$, which can be written as $|\mathbf{W}^H(\theta_0)\mathbf{y}|^2$, the output of the beamformer presented earlier. Thus, the estimate $\hat{\theta}_0$ obtained by maximizing the output power of the beamformer *is* the generalized likelihood direction estimate.

Figure 1.2 depicts the beam pattern $p(\theta)$ for 16-element circular and linear arrays in the case where $\theta_0 = 0$. As can be seen, $p(\theta)$ is maximized at the source

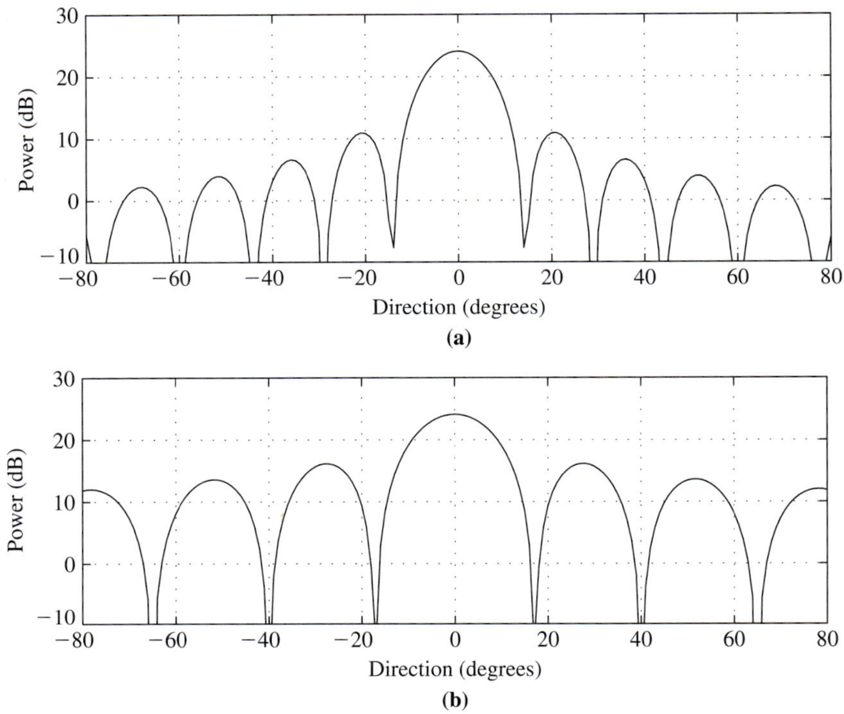

FIGURE 1.2 Beam patterns (nonwindowed) for 16-element linear (a) and circular (b) arrays with uniformly spaced elements (half-wavelength spacing).

direction θ_0. In the absence of noise (and practical imperfections), this method provides a perfect estimate of direction. In the presence of noise, the peak location is shifted, introducing a direction estimation error. The accuracy of the estimated direction will be addressed in Section 1.4.

1.3.2 Monopulse

An interesting variation of the beamformer involves a method, often referred to as *monopulse*, commonly used in radar systems for target tracking. This method involves taking the difference between the outputs of two beams pointing in slightly different directions. Let

$$b(\theta) = \frac{1}{\Delta}\left\{|\mathbf{a}^H(\theta+\Delta/2)\mathbf{y}|^2 - |\mathbf{a}^H(\theta-\Delta/2)\mathbf{y}|^2\right\} \tag{1.12}$$

denote the response of the monopulse system where $1/\Delta$ is a convenient normalization factor. In other words

$$b(\theta) = \frac{1}{\Delta}\{p(\theta+\Delta/2) - p(\theta-\Delta/2)\} \tag{1.13}$$

where $p(\theta)$ is the beampattern defined in the previous section. If Δ is small, this difference can be replaced by the derivative of $p(\theta)$:

$$b(\theta) \approx \dot{p}(\theta) \tag{1.14}$$

In fact, Δ can be on the order of one beamwidth and $b(\theta)$ will still be well approximated by $\dot{p}(\theta)$ because $b(\theta)$ is nearly linear over a significant range of angles around the zero-response point. This can be seen in Figure 1.3, which depicts the response function $b(\theta)$ for 16-element circular and linear arrays in the case where $\theta_0 = 0$. As can be seen, $b(\theta) = 0$ at the source direction θ_0. In the absence of noise (and practical imperfections), this method provides a perfect estimate of direction. In the presence of noise, the zero location is shifted, introducing a direction estimation error. The accuracy of the estimated direction will be addressed in Section 1.4.

Note that the output is positive if the emitter is to the right of the direction in which the difference beam is pointed and negative if it is to the left. This ability to determine the direction of the emitter relative to the pointing direction by the sign of the response is useful in tracking applications.

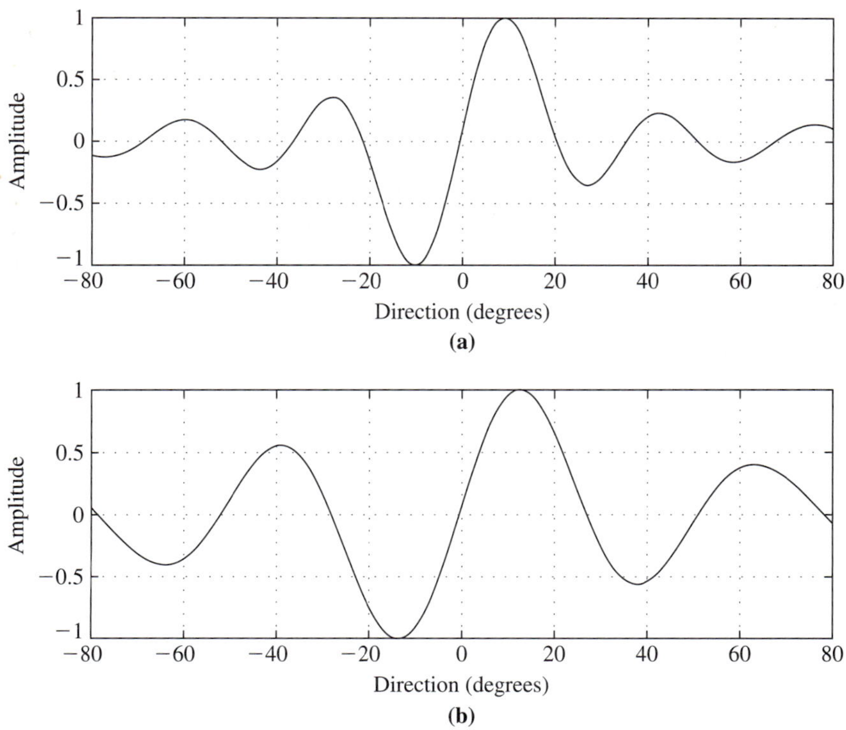

FIGURE 1.3 Response of a monopulse system for 16-element linear (a) and circular (b) arrays, with uniformly spaced elements.

1.3.3 Phase Matching

The methods just described involve a *linear* combining of the antenna outputs followed by subsequent processing. Here we consider a *nonlinear* method that uses the phase of the received signals rather than the signals themselves. In the case of a single emitter with no multipath, information about its direction resides entirely in the signals' phases (delays), not in their amplitudes. Using phase information only is therefore expected to be as accurate as using the signal itself, as well as potentially more robust to effects that cause amplitude (but not phase) impairments such as unmatched antenna and amplifier gains.

Let $\tilde{\phi}$ denote the vector of phase measurements and $\phi(\theta)$ denote the vector of phases of the array manifold $\mathbf{a}(\theta)$. The direction estimate is obtained by finding the value of θ for which the distance between the measured and assumed phase vectors is minimized. The squared norm of the difference between the phase vectors can be used as the distance measure:

$$d(\theta) = |\tilde{\phi} - \phi(\theta)|^2 \tag{1.15}$$

As discussed earlier, the phase of the received signal is the sum of the geometry/direction-related phase and a random phase component common to all antennas. This random component must be eliminated or otherwise taken care of for the direction-finding algorithm to work properly. A common way to achieve this is to measure relative rather than absolute phases, which can be done by designating one of the antennas as a phase reference and measuring phases relative to it. The phase of the reference antenna is assumed to be zero. Similarly, the array manifold $\mathbf{a}(\theta)$ is computed so that the phase of the reference antenna is zero. The performance of this method is influenced by the particular antenna chosen as the reference. An alternative approach is to eliminate dependence on a particular reference antenna by using the mean of the phase vector as the reference. In other words, we force both the measured and manifold phase vectors to have zero mean.

A more systematic approach to this problem is to include an unknown additive phase term β in the signal model and jointly estimate the unknown direction and additive phase terms. In other words,

$$\min_{\theta,\beta} |\tilde{\phi} - \phi(\theta) - \mathbf{1}\beta|^2 \tag{1.16}$$

where $\mathbf{1}$ is an $M \times 1$ vector with elements that are 1. Minimization over β yields

$$\beta = \frac{1}{M}\mathbf{1}^T(\tilde{\phi} - \phi(\theta)) \tag{1.17}$$

Inserting this into the cost function being minimized, we obtain

$$\min_{\theta} \left| \left(\mathbf{I} - \frac{1}{M}\mathbf{1}\mathbf{1}^T\right)\left(\tilde{\phi} - \phi(\theta)\right) \right|^2 \tag{1.18}$$

which is "optimal" from a statistical standpoint, but somewhat more complicated than the methods presented earlier.

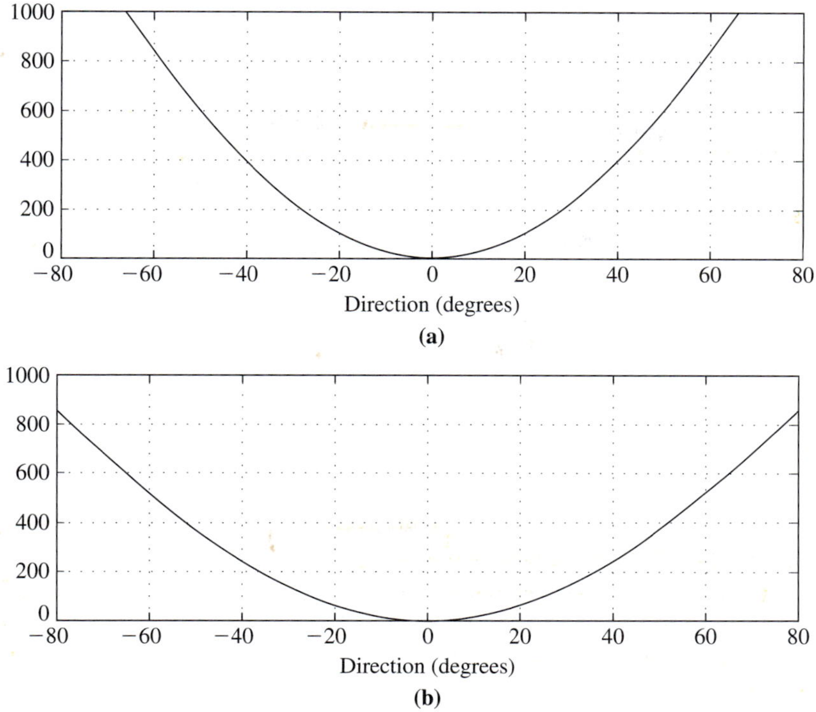

FIGURE 1.4 Phase cost function versus direction for 16-element arrays: (a) linear and (b) circular, with uniformly spaced elements.

Figure 1.4 depicts the phase cost function $d(\theta)$ in Equation (1.15) for 16-element circular and linear arrays in the case where $\theta_0 = 0$. As can be seen, $d(\theta)$ is minimized at the source direction θ_0. In the absence of noise and practical imperfections, this method provides a perfect estimate of direction. In the presence of noise, the location of the minimum is shifted, introducing a direction estimation error. The accuracy of the estimated direction will be addressed in Section 1.4.

1.4 DIRECTION-FINDING ACCURACY

The accuracy with which the direction of an emitter can be determined is an important parameter of any direction-finding system. It depends on implementation imperfections, propagation effects (multipath, variabilty of the angle of arrival), interference, signal-to-noise ratio, and so forth. In this section we consider the accuracy of an ideal system where the only effect corrupting the measurements is additive measurement noise.

Direction-finding accuracy is measured by the mean-square error (MSE) of the estimator:

$$\mathrm{MSE}\{\hat{\theta}\} = E\{|\hat{\theta} - \theta|^2\} \tag{1.19}$$

where $\hat{\theta}$ is the direction estimate and θ is the true direction. Throughout this section we consider direction-finding methods that provide *unbiased* estimates of direction. In other words,

$$E\{\hat{\theta}\} = \theta \tag{1.20}$$

In this case, the MSE equals the *variance* of the estimated direction:

$$\text{MSE}\{\hat{\theta}\} = E\{|\hat{\theta} - E\{\hat{\theta}\}|^2\} = \text{VAR}\{\hat{\theta}\} \tag{1.21}$$

In the biased case,

$$\text{MSE}\{\hat{\theta}\} = E\{|(\hat{\theta} - E\{\hat{\theta}\}) - (\theta - E\{\hat{\theta}\}|^2\} = \text{VAR}\{\hat{\theta}\} + |\theta - E\{\hat{\theta}\}|^2 \tag{1.22}$$

where $b(\theta) = \theta - E\{\hat{\theta}\}$ is the *bias* of the estimator.

1.4.1 The Cramer-Rao Bound

The Cramer-Rao bound (CRB) (Cramer, 1951) is a useful tool for assessing the accuracy of parameter estimation methods, as it provides a lower bound on the accuracy of *any* unbiased estimator. There are versions of the CRB for biased estimators as well, but we will not discuss them here.

For any unbiased estimate $\hat{\theta}$,

$$\text{MSE}\{\hat{\theta}\} = \text{VAR}\{\hat{\theta}\} \geq \text{CRB} \tag{1.23}$$

The CRB provides an algorithm-independent benchmark against which various algorithms can be compared.

In Section A.1.2 we derive the CRB for the direction-finding problem and show that

$$\text{CRB} \approx \frac{1}{2K\,\text{SNR}\,|\dot{\mathbf{a}}(\theta)|^2} \tag{1.24}$$

where K is the number of snapshots used by the estimator, SNR is the signal-to-noise ratio of the signal received at each antenna, and $\dot{\mathbf{a}}(\theta)$ is the derivative of the array manifold with respect to θ. This is a simple expression that shows the variance of the direction estimation error to be inversely proportional to the post-integration signal-to-noise ratio $K\,\text{SNR}$ and to a factor $|\dot{\mathbf{a}}(\theta)|^2$ that depends on the particular array used by the system.

Evaluating $|\dot{\mathbf{a}}(\theta)|^2$, we obtain more explicit forms of the CRB (see Section A.1.2). For an M-element linear array we get

$$\text{CRB} \approx \frac{\lambda^2}{8\pi^2 K\,\text{SNR}\cos^2\theta\,\overline{d^2}} \tag{1.25}$$

where

$$\overline{d^2} = \sum_{m=1}^{M} d_m^2 \tag{1.26}$$

and d_m is the distance of the mth antenna from the phase center of the array.

For a *uniformly* spaced linear array,

$$d_m = (m-1)\frac{L}{M-1} - \frac{L}{2} \tag{1.27}$$

where L is the array aperture. In this case, $\overline{d^2}$ can be expressed in closed form as a function of L and M.

For an M-element circular array with radius R, we have

$$\text{CRB} \approx \frac{\lambda^2}{8\pi^2 K\, \text{SNR} R^2\, \overline{c^2}(\theta)} \tag{1.28}$$

where

$$\overline{c^2}(\theta) = \sum_{m=1}^{M} \cos^2(\theta - \theta_m) \tag{1.29}$$

and θ_m are the antenna location angles on the circle.

For a *uniformly* spaced circular array it is straightforward to show that

$$\overline{c^2} = \frac{M}{2} \quad \text{for } \theta = 2(m-1)\pi/M, \quad m = 1, \ldots, M \tag{1.30}$$

1.4.2 Simulation

To evaluate the accuracy of the direction-finding algorithms described in Section 1.3, we used a Monte Carlo simulation generating 5000 independent data vectors $\mathbf{y}$, estimated the emitter direction, and computed the estimates bias and standard deviation. An 8-element uniformly spaced linear array with half-wavelength spacing was used, and the emitter was at $\theta_0 = 10°$. Estimates were based on a single snapshot. Figures 1.5 and 1.6 depict the standard deviation and bias of the estimated direction for the three methods as a function of the signal-to-noise ratio. Figure 1.5 includes the CRB for reference.

We observe that all three methods are asymptotically unbiased and statistically efficient. In other words, as the signal-to-noise ratio increases the bias tends to zero and the variance (or standard deviation) tends to the CRB. Results for a circular array are similar, and the corresponding figures are therefore omitted. In Sections A.1.4 through A.1.8 we present analysis that further validates these observations. We conclude that under ideal conditions these classical methods are as accurate as possible in the presence of measurement noise.

It should be noted that the CRB is a "tight" bound only at high signal-to-noise ratios. In other words, it is possible to design estimators with a variance that approaches the CRB asymptotically as those ratios increase. At low ratios the CRB is no longer tight and the variance of any estimator is larger—possibly much larger. In plotting the variance of the estimator as a function of the SNR, we invariably observe a threshold SNR. When the SNR exceeds this threshold, the variance closely tracks the CRB; when it falls below the threshold, the variance

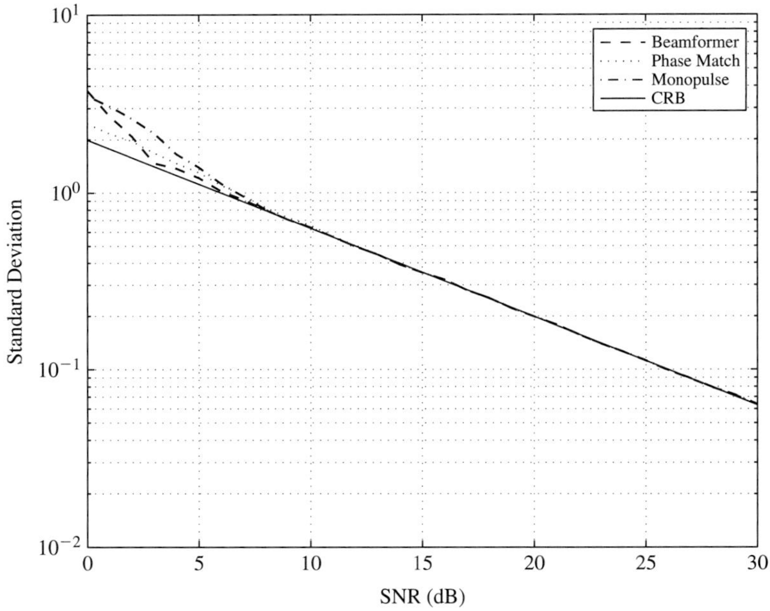

FIGURE 1.5 Standard deviation of the estimation errors for an 8-element uniformly spaced linear array using beamforming, monopulse, and phase-matching methods. The CRB is shown for reference.

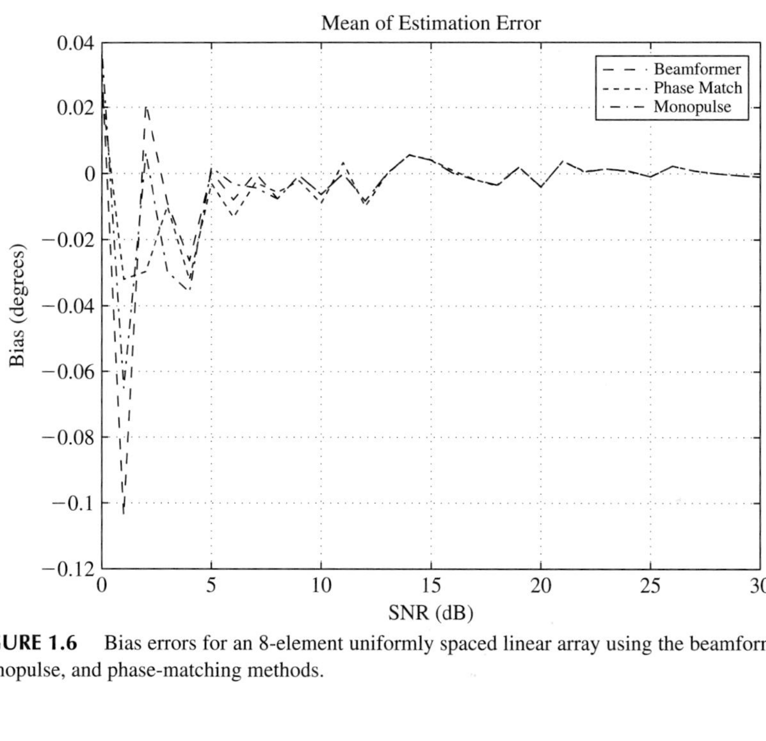

FIGURE 1.6 Bias errors for an 8-element uniformly spaced linear array using the beamforming, monopulse, and phase-matching methods.

departs sharply from the CRB. The Barankin (1949), Ziv and Zakai (1969), and other "global" bounds better indicate achievable performance at a low SNR and display this threshold phenomenon, absent in the CRB, which is purely "local." In Figures 1.5 and 1.6 the threshold SNR can be seen at around 5 dB.

1.5 MULTIPATH AND CO-CHANNEL INTERFERENCE

We considered direction finding for a single emitter with the received signal contaminated by noise only. In practice, there may be multiple co-channel signals arriving from different emitters or from a single emitter through multiple propagation paths. These spurious signals, especially if their direction is close to that of the signal of interest, degrade the accuracy of the direction-finding system. Specifically, the direction estimates provided by the methods discussed earlier become biased. In this section we study this bias as a function of some of the interfering signal parameters. For simplicity we limit our discussion to the beamformer described in Section 1.3.1.

A fundamental way to handle multiple co-channel signals is via processing techniques capable of resolving the composite signal into its individual components and estimating the direction of each component separately. We will discuss this approach in more detail in Section 1.6. Here we note only that the beamformer resolves signals with directions that are separated by more than a beamwidth, but fails to resolve those more closely spaced. In this section we focus on closely spaced received signals not resolved by the direction-finding algorithm.

Consider the case where two signals impinge on the array from directions θ_0 and θ_1. Without loss of generality, assume that the signal at θ_0 corresponds to the emitter the direction of which we want to estimate. The received signal vector is given by

$$\mathbf{y} = \sqrt{\mathrm{SNR}_0} s_0 \mathbf{a}(\theta_0) + \sqrt{\mathrm{SNR}_1} s_1 \mathbf{a}(\theta_1) + \mathbf{v} \tag{1.31}$$

where SNR_0 and SNR_1 are the signal-to-noise ratios of the emitters at θ_0 and θ_1, respectively. As before, the output of the beamformer is

$$p(\theta) = |\mathbf{W}(\theta)^H \mathbf{y}|^2 \tag{1.32}$$

where $\mathbf{W}(\theta) = \mathbf{a}(\theta)/|\mathbf{a}(\theta)|$. The expected value of $p(\theta)$ is given by

$$\begin{aligned} E\{p(\theta)\} = {} & \mathrm{SNR}_0 |\mathbf{W}(\theta)^H \mathbf{a}(\theta_0)|^2 + \mathrm{SNR}_1 |\mathbf{W}(\theta)^H \mathbf{a}(\theta_1)|^2 \\ & + 2\rho_{01}\sqrt{\mathrm{SNR}_0}\sqrt{\mathrm{SNR}_1}\,\Re\{\mathbf{W}(\theta)^H \mathbf{a}(\theta_0)(\mathbf{W}(\theta)^H \mathbf{a}(\theta_1))^*\} + 1 \end{aligned} \tag{1.33}$$

where $\rho_{01} = E\{s_0 s_1^*\}$, $E\{|s_0|^2\} = E\{|s_1|^2\} = 1$, and $\Re\{\cdot\}$ denotes the real part. This expected value represents the case where the beamformer output is averaged over multiple snapshots:

$$\hat{p}(\theta) = \frac{1}{K}\sum_{k=1}^{K} |\mathbf{W}(\theta)^H \mathbf{y}[k]|^2 \tag{1.34}$$

As K becomes large the performance of the nonaveraged beamformer nears that of the averaged beamformer (i.e., $\hat{p}(\theta) \approx E\{p(\theta)\}$).

Consider the case where the two signals are uncorrelated (i.e., $E\{s_0 s_1^*\} = 0$). Then

$$E\{p(\theta)\} = \mathrm{SNR}_0 |\mathbf{W}(\theta)^H \mathbf{a}(\theta_0)|^2 + \mathrm{SNR}_1 |\mathbf{W}(\theta)^H \mathbf{a}(\theta_1)|^2 + 1 \tag{1.35}$$

The signals are generally uncorrelated if they originate from two different emitters or if they originate from a single emitter in the presence of motion-induced Doppler.

Next consider the case where the signals originate from a single emitter in the absence of motion. In this case, they are the same expect for a complex scale factor. Thus, $s_1 = s_0 e^{-j\phi}$, where ϕ is the phase of this factor (the relative magnitudes are absorbed in SNR_0 and SNR_1. The correlation coefficient is therefore $\rho_{01} = e^{-j\phi}$, so the average beamformer output is

$$E\{p(\theta)\} = \left| \sqrt{\mathrm{SNR}_0} \mathbf{W}(\theta)^H \mathbf{a}(\theta_0) + \sqrt{\mathrm{SNR}_1} e^{-j\phi} \mathbf{W}(\theta)^H \mathbf{a}(\theta_1) \right|^2 + 1 \tag{1.36}$$

Figures 1.7, 1.8, and 1.9 illustrate the effect of the second signal on the estimated direction. They depict results for an 8-element uniformly spaced linear array with half-wavelength spacing. In Figures 1.7 and 1.9 the signal directions

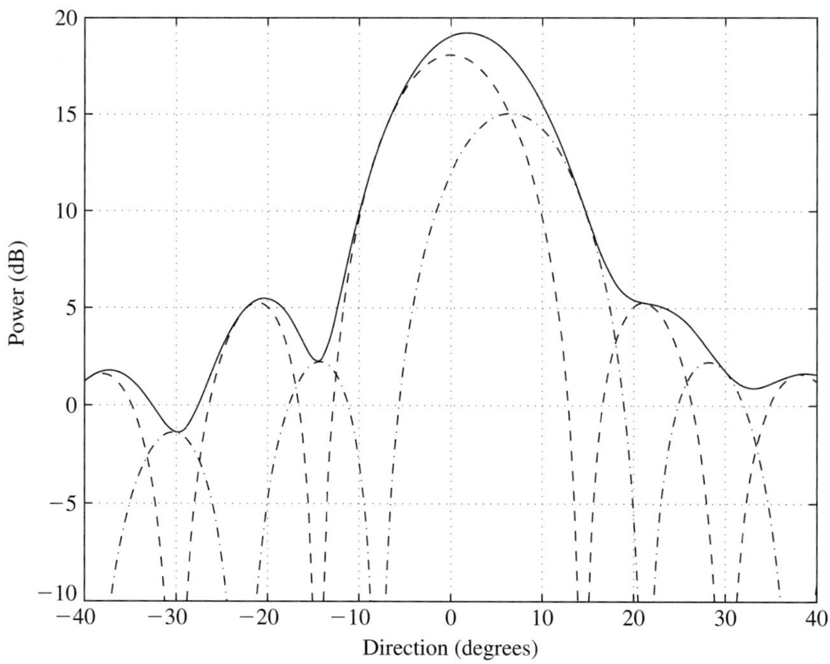

FIGURE 1.7 Composite beam response for two uncorrelated signals at $\theta_0 = 0°$ and $\theta_1 = 6.5°$. The power of the second signal is 3 dB below that of the first. The dashed and dot-dashed curves are the responses to the individual signals, and the solid curve is the composite response.

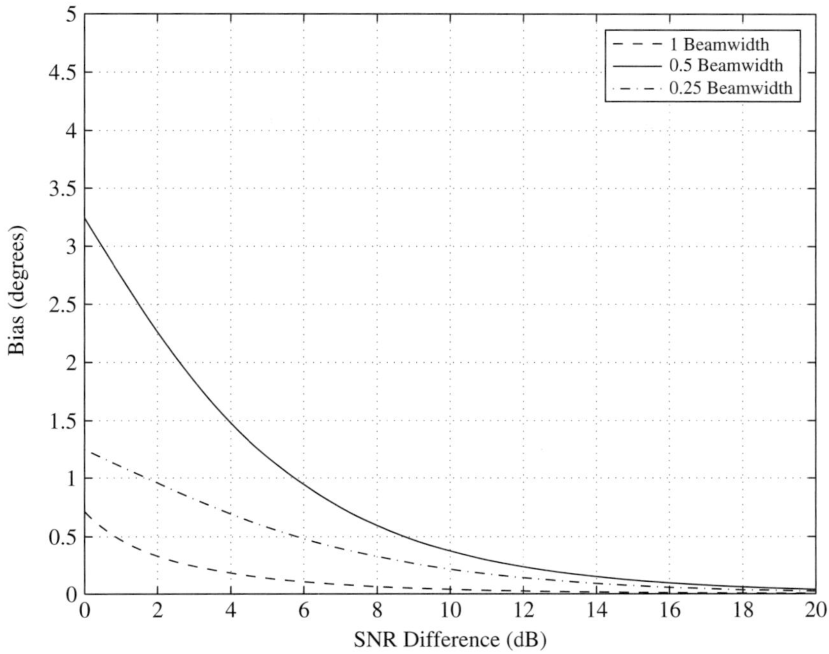

FIGURE 1.8 Bias-of-direction estimation using a beamformer for two uncorrelated signals for different angular separations: 1, 0.5, and 0.25 of a beamwidth.

are $\theta_0 = 0°$ and $\theta_1 = 6.5°$. The beamwidth of the array is 13°, so the signal directions are a half-beamwidth apart. In Figure 1.8 $\theta_0 = 0°$ and $\theta_1 = 3.250°, 6.50°$, and 13°.

Figure 1.7 depicts the beam pattern computed using Equation (1.35) and its two components corresponding to two uncorrelated signals. The power of the signal at θ_1 is half the power of the signal at θ_0—that is, the SNR difference is 3 dB. This figure illustrates two important facts: (1) The beamformer is unable to resolve the two signals, so the response has a single peak corresponding to the composite signal; (2) the peak location is between θ_0 and θ_1, providing a biased estimate of θ_0.

In Figure 1.8 the bias of the direction estimates is shown as a function of the SNR difference between the two uncorrelated signals for three values of the separation $\theta_1 - \theta_0$ between the signal directions: 1 beamwidth, 0.5 beamwidth, and 0.25 beamwidth. As expected, in all three cases the bias decreases as the SNR difference increases. Initially it increases as the separation increases, but at some point it begins to decrease. This is to be expected because the beamformer is able to resolve the two signals for separations larger than a beamwidth.

In Figure 1.9 the bias of the direction estimates is shown as a function of the SNR difference between the two correlated signals for a separation of 0.5 beamwidth. The estimates are computed by finding the peak of the beam response

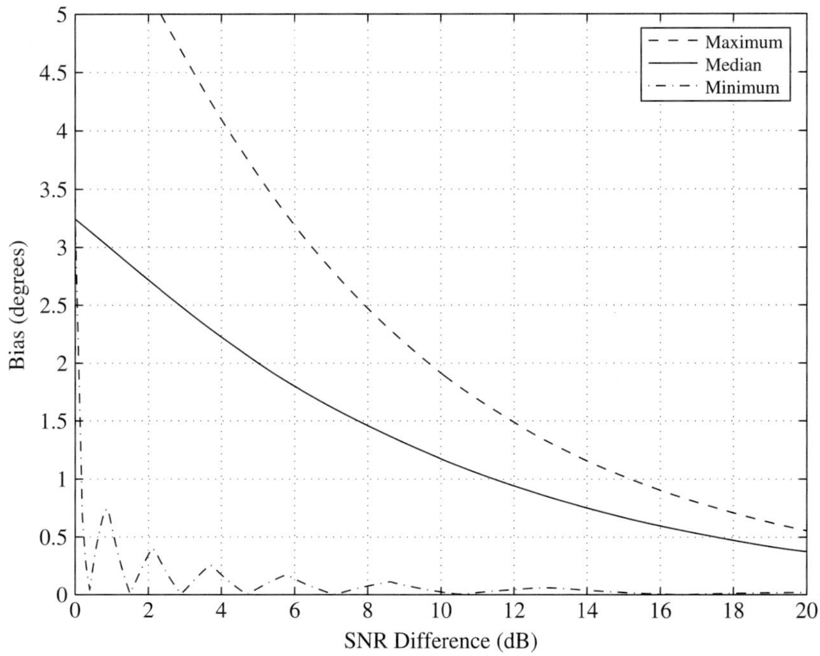

FIGURE 1.9 Bias-of-direction estimation using a beamformer for two correlated signals for a 0.5-beamwidth separation. Shown are the maximum, minimum, and median of the bias computed over all possible values of the relative phase between signals.

in Equation (1.36). Note that the beam response depends on the correlation phase ϕ. We computed the response and the corresponding direction estimates for $0 \leq \phi \leq 2\pi$, and from this we evaluated the largest, smallest, and median values of the bias $|\hat{\theta} - \theta_0|$, as shown in the figure. Note that the bias varies over a significant range of values and in some cases is larger than in the correlated case. Therefore, the bias due to correlated multipath may be larger than that due to uncorrelated multipath. More important, the magnitude of the bias is highly variable from one scenario to the next.

1.6 DIRECTION FINDING FOR MULTIPLE CO-CHANNEL EMITTERS

We considered direction-finding methods that associate a *single* direction with the received signal. When the signal is a composite of multiple components, the direction estimate of the signal of interest is biased to a smaller or larger degree. In this section we consider methods that associate *multiple* directions with the different signal components.

First we note that, when the emitters are well separated (i.e., their directions are more than a beamwidth apart), the techniques discussed earlier are in fact capable of estimating the individual component directions. Figure 1.10 depicts

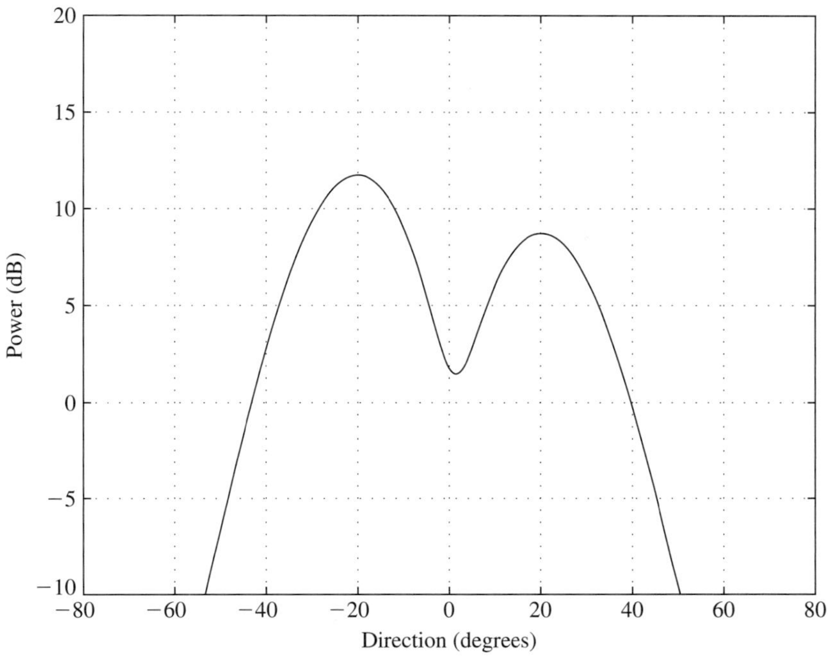

FIGURE 1.10 Beamformer response for two uncorrelated signals at $\theta_0 = -20°$ and $\theta_1 = 20°$. The power of the second signal is 3 dB below that of the first.

the response of the beamformer to the signals received from two emitters: one at direction $\theta_0 = -20$ and one at $\theta_1 = 20$. The signal from the emitter at θ_1 has half the power of the signal from the emitter at θ_0. An 8-element uniformly spaced array was used with half-wavelength element spacing. The beam response is computed using a Hamming window. It has two clearly identifiable peaks located at the emitter directions. Thus, the directions of both emitters can be reliably estimated in this case.

If the emitters are spaced less than a beamwidth apart, the beam response will have a single peak, as was shown in Figure 1.7. The beamformer will therefore fail to resolve the signals, producing a single biased direction estimate. In this case other methods are needed to obtain unbiased direction estimates of all emitters. The ability to estimate the directions of signals separated by less than a beamwidth is often called "super-resolution." Conventional methods are limited by the Rayleigh resolution limit corresponding roughly to one beamwidth.

The signal model considered in this section is

$$\mathbf{y} = \alpha_0 \mathbf{a}(\theta_0) + \alpha_1 \mathbf{a}(\theta_1) + \mathbf{v} \tag{1.37}$$

where $\mathbf{y}$ is the vector of the signals at the array output, α_0, α_1 are complex coefficients representing the magnitude and phase of the two emitters ($\alpha_p = s_p\sqrt{\mathrm{SNR}_p}, p = 0, 1$ in the earlier notation), $\mathbf{a}(\theta)$ is the array manifold, and

θ_0, θ_1 are the emitter directions. This is a straightforward extension of the model presented in Equation (1.7), and it can be extended in an obvious manner to P signals, in which case

$$\mathbf{y} = \sum_{p=0}^{P} \alpha_p \mathbf{a}(\theta_p) + \mathbf{v} \tag{1.38}$$

For simplicity, we consider here only the two-signal case.

Given the measurement $\mathbf{y}$, we want to estimate the directions θ_0 and θ_1. We assume that neither α_0 nor α_1 is known to the receiver.

1.6.1 Maximum Likelihood Direction Estimation

A conceptually straightforward solution to the problem defined previously is to compute the maximum likelihood estimator (MLE) for the unknown parameters in Equation (1.37). Assuming the noise is Gaussian, so is $\mathbf{y}$, and therefore the MLE is equivalent to the nonlinear least-squares estimator, which minimizes

$$d(\theta_0, \theta_1, \alpha_0, \alpha_1) = |\mathbf{y} - \alpha_0 \mathbf{a}(\theta_0) + \alpha_1 \mathbf{a}(\theta_1)|^2 \tag{1.39}$$

over the unknown parameters. Because $\mathbf{y}$ is a linear function of the scale factors α_0, α_1, the minimization over these parameters can be done analytically:

$$\begin{bmatrix} \alpha_0 \\ \alpha_1 \end{bmatrix} = (\mathbf{S}^H(\theta_0, \theta_1)\mathbf{S}(\theta_0, \theta_1))^{-1}\mathbf{S}^H(\theta_0, \theta_1)\mathbf{y} \tag{1.40}$$

where

$$\mathbf{S}(\theta_0, \theta_1) = [\mathbf{a}(\theta_0), \mathbf{a}(\theta_1)] \tag{1.41}$$

Inserting the result into the preceding equation, we get

$$d(\theta_0, \theta_1) = |\mathbf{y} - \mathbf{P}_{\mathbf{S}(\theta_0,\theta_1)}\mathbf{y}|^2 = |\mathbf{P}^{\perp}_{\mathbf{S}(\theta_0,\theta_1)}\mathbf{y}|^2 = \mathbf{y}^H \mathbf{P}^{\perp}_{\mathbf{S}(\theta_0,\theta_1)}\mathbf{y} \tag{1.42}$$

where $\mathbf{P}_{\mathbf{S}(\theta_0,\theta_1)}$ denotes the projection operator on the subspace $\mathbf{S}(\theta_0, \theta_1)$. Similarly, $\mathbf{P}^{\perp}_{\mathbf{S}(\theta_0,\theta_1)}$ denotes the orthogonal projection operator. The minimization of $d(\theta_0, \theta_1)$ needs to be numerical and is generally computationally intensive, requiring a two-dimensional search. In general a "brute-force" search over a selected grid of values of (θ_0, θ_1) is necessary, followed by interpolation in the neighborhood of the minimum point to compute the final estimate. (See Section A.1.9 for details regarding this interpolation step.)

Figure 1.11 depicts the maximum likelihood cost function $d(\theta_0, \theta_1)$ for the case of two emitters: one at direction $\theta_0 = -5°$ and the other at $\theta_1 = 10°$. The two signals have equal power. An 8-element uniformly spaced array was used with half-wavelength element spacing. The cost function has two minima at $-5, 10$ and $10, -5$, which is to be expected because the cost function is invariant to the ordering of the emitters. This introduces the symmetry relationship $d(\theta_0, \theta_1) = d(\theta_1, \theta_0)$.

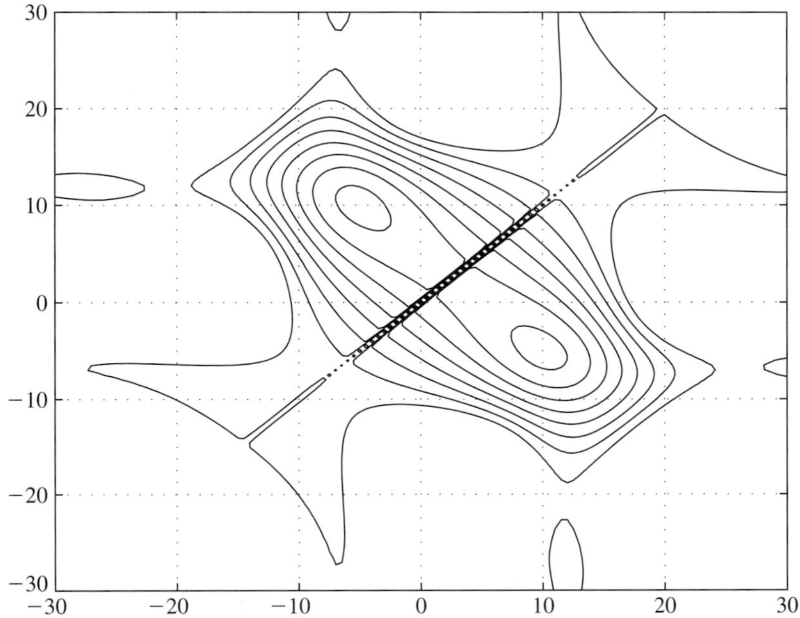

FIGURE 1.11 Maximum likelihood cost function (Equation 1.42) for an 8-element uniformly spaced array and two emitters at $\theta_0 = -5°$ and $\theta_1 = 10°$.

We considered estimating directions from a *single* snapshot of received data. If *multiple* snapshots are available, we need to distinguish between the case where the signals are correlated from snapshot to snapshot and the case where they are uncorrelated. In the uncorrelated case we have

$$\mathbf{y}[k] = \alpha_0[k]\mathbf{a}(\theta_0) + \alpha_1[k]\mathbf{a}(\theta_1) + \mathbf{v}[k], \quad k = 1, \ldots, K \tag{1.43}$$

where the scale factors $\alpha_0[k], \alpha_1[k]$ vary randomly from snapshot to snapshot. It is straightforward to show that in this case the maximum likelihood estimator requires minimization of the cost function

$$d(\theta_0, \theta_1) = \sum_{k=1}^{K} \mathbf{y}^H[k]\mathbf{P}^{\perp}_{\mathbf{S}(\theta_0,\theta_1)}\mathbf{y}[k] \tag{1.44}$$

In other words, the cost function for the case of multiple snapshots is the sum of the cost functions for the individual snapshots.

In the fully correlated case the scale factors $\alpha_0[k], \alpha_1[k]$ are fixed throughout the K snapshots so that the signal model becomes

$$\mathbf{y}[k] = \alpha_0\mathbf{a}(\theta_0) + \alpha_1\mathbf{a}(\theta_1) + \mathbf{v}[k] \tag{1.45}$$

Clearly, the maximum likelihood estimator requires minimization of the cost function in this case:

$$d(\theta_0,\theta_1)=\overline{\mathbf{y}}^H\mathbf{P}^{\perp}_{\mathbf{S}(\theta_0,\theta_1)}\overline{\mathbf{y}} \tag{1.46}$$

where

$$\overline{\mathbf{y}}=\frac{1}{K}\sum_{k=1}^{K}\mathbf{y}[k] \tag{1.47}$$

is the average of the received data vectors. In other words, the cost function for the case of multiple snapshots is the same as the cost function for a single snapshot, with the data vector replaced by the average vector.

1.6.2 Other Direction-Finding Methods

While the maximum likelihood estimator provides an optimal solution to the multi-emitter direction-finding problem, it is rarely, if ever, used in practice because of its computational complexity. Much research has been devoted to the development of alternative suboptimal but computationally feasible methods, most of which are variations on the so-called Multi-Signal Classification (MUSIC) algorithm (Bienvenu and Kopp, 1983; Schmidt, 1979, 1981; Weiss and Friedlander, 1994). It is not our intention to provide a comprehensive survey of these techniques, which would take a separate chapter. Instead, we present a brief summary of the basic MUSIC algorithm and compare it to the MLE.

Consider the case in which P signals are impinging on the array:

$$\mathbf{y}[k]=\sum_{p=1}^{P}\alpha_p[k]\mathbf{a}(\theta_p)+\mathbf{v}[k],\quad k=1,\ldots K \tag{1.48}$$

where θ_p are the signal directions, $\alpha_p[k]$ are the complex amplitudes of the signals assumed to be uncorrelated from snapshot to snapshot, and $\mathbf{v}[k]$ is a vector of zero-mean unit-variance Gaussian noise. The covariance of the received signal vector $\mathbf{R}_y=E\{\mathbf{y}[k]\mathbf{y}[k]^H\}$ is given by

$$\mathbf{R}_y=\sum_{p=1}^{P}\sigma_p^2\mathbf{a}(\theta_p)\mathbf{a}^H(\theta_p)+\mathbf{I} \tag{1.49}$$

where $\sigma_p^2=E\{|\alpha_p[k]|^2\}$ is the SNR of the pth signal. Let

$$\mathbf{R}_y=\mathbf{U}\Sigma\mathbf{U}^H \tag{1.50}$$

be the singular-value decomposition of the covariance matrix, where

$$\Sigma=\mathbf{diag}\{[\sigma_1^2+1,\ldots,\sigma_P^2+1,1,\ldots,1]\} \tag{1.51}$$

is a diagonal matrix with elements that are the singular values of $\mathbf{R}_y$. The matrix of singular vectors $\mathbf{U}$ can be partitioned as

$$\mathbf{U} = [\mathbf{U}_s, \mathbf{U}_n] \tag{1.52}$$

where the $M \times P$ matrix $\mathbf{U}_s$ contains the singular vectors corresponding to P largest singular values, and the matrix $\mathbf{U}_n$ contains the singular vectors corresponding to $M - P$ smallest singular values. The range space of $\mathbf{U}_s$ equals the space spanned by the signal

$$\mathcal{R}\{\mathbf{U}_s\} = [\mathbf{a}(\theta_1), \ldots, \mathbf{a}(\theta_P)] \tag{1.53}$$

so we refer to $\mathbf{U}_s$ as the "signal subspace." Similarly, $\mathbf{U}_n$ is the "noise subspace." Because $\mathbf{U}$ is a unitary matrix the signal and noise subspaces are orthogonal, so that

$$\mathbf{U}_s^H \mathbf{U}_n = 0 \tag{1.54}$$

It follows that

$$\mathbf{a}^H(\theta)\mathbf{U}_n = 0 \quad \text{for } \theta = \theta_1, \ldots, \theta_P \tag{1.55}$$

For this reason we compute

$$S(\theta) = \frac{1}{|\mathbf{a}^H(\theta)\mathbf{U}_n|^2} \tag{1.56}$$

which will have peaks (infinitely high in the noise-free case) at the signal directions.

In practice, the covariance matrix $\mathbf{R}_y$ is not known. However, it can be estimated from the available data by forming the sample covariance matrix,

$$\hat{\mathbf{R}}_y = \sum_{k=1}^{K} \mathbf{y}[k]\mathbf{y}[k]^H \tag{1.57}$$

The singular-value decomposition of the estimated covariance matrix $\hat{\mathbf{R}}_y = \hat{\mathbf{U}}\hat{\Sigma}\hat{\mathbf{U}}^H$ and the estimated noise subspace are obtained by partitioning $\hat{\mathbf{U}}$ as $\hat{\mathbf{U}} = [\hat{\mathbf{U}}_s, \hat{\mathbf{U}}_n]$. The estimated noise subspace is then used to compute the so-called "MUSIC spectrum":

$$\hat{S}(\theta) = \frac{1}{|\mathbf{a}^H(\theta)\hat{\mathbf{U}}_n|^2} \tag{1.58}$$

Figure 1.12 depicts the MUSIC spectrum for an 8-element uniformly spaced linear array with half-wavelength spacing, signal directions $\theta_0 = 5°, \theta_1 = 10°$, and $\text{SNR}_0 = \text{SNR}_1 = 10\,\text{dB}$, $K = 100$. Note the sharp peaks at the emitter directions.

The main advantage of MUSIC is computational: It requires only a one-dimensional search, whereas the MLE requires a P-dimensional search. Certain

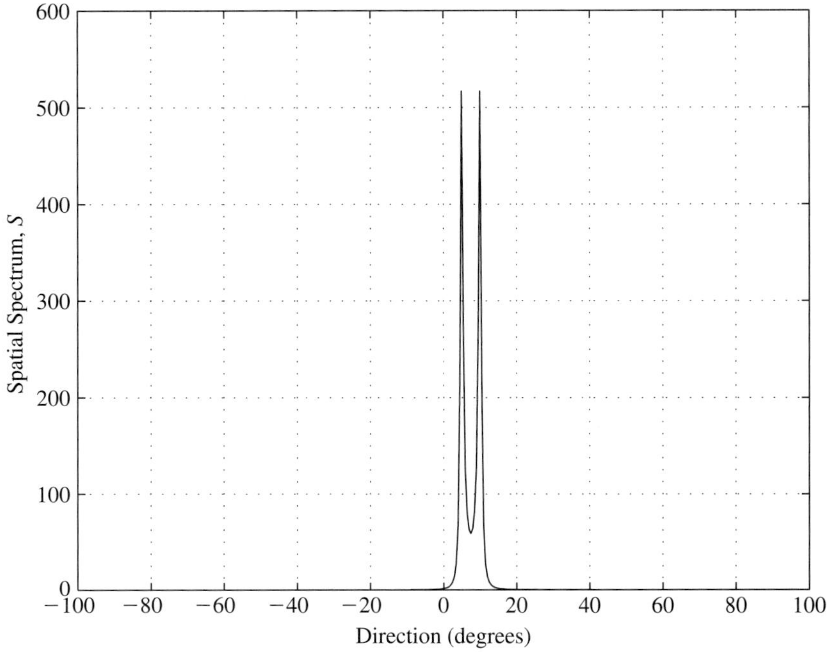

FIGURE 1.12 MUSIC spectrum for an 8-element uniformly spaced linear array with emitters at directions $\theta_0 = 5°$ and $\theta_1 = 10°$, with $\text{SNR}_0 = \text{SNR}_1 = 10\,\text{dB}$, using $K = 100$ snapshots.

variations of MUSIC, such as Root-MUSIC, eliminate the search altogether, replacing it with root solving (Friedlander, 1993; Weiss and Friedlander, 1993).

MUSIC is a suboptimal algorithm and generally is not statistically efficient. However, as the SNR of all the signals approaches infinity, the error variance of MUSIC approaches the CRB (Friedlander, 1990; Porat and Friedlander, 1988). Furthermore, in most scenarios of practical interest its accuracy is close to optimal.

The MUSIC algorithm and all of its many variations are based on the estimated covariance of the received data. To form a reliable estimate of $\mathbf{R}_y$, it is necessary to collect a sufficiently large number of snapshots—a simple rule of thumb is, at the very least, $K > 3M$. However, to separate closely spaced signals a much larger number of snapshots may be required—typically tens or hundreds. This is in contrast to the MLE, which can operate, if necessary, with a single snapshot.

The requirement of a large number of snapshots is generally disadvantageous. It slows down the response time of the system and limits the number of emitters that can be monitored given finite system resources. More important, the system may fail to estimate the directions of short-lived signals because collecting a sufficient number of snapshot is not possible.

1.6.3 Accuracy

The maximum likelihood estimator is known to be asymptotically efficient (Cramer, 1951; Fisher, 1922; Rao, 1965). Therefore, at a high SNR its performance approaches the CRB. The CRB for the multiple emitter case is summarized in Section A.1.3.

Figure 1.13 depicts the standard deviation calculated using the CRB for the case of two emitters at directions $\theta_0 = \Delta/2$ and $\theta_1 = -\Delta/2$, where Δ is the angular separation. An 8-element uniformly spaced array was used with half-wavelength element spacing and signal-to-noise ratios of $\mathrm{SNR}_0 = \mathrm{SNR}_1 = 30$ dB. The figure demonstrates the fact that the error variance increases sharply when emitter separation is smaller than a beamwidth (the 3-dB beamwidth is 13° in this case). When the separation exceeds the beamwidth, the error variance is essentially independent of it and is equal to the variance of the single-emitter case.

Figure 1.14 compares the standard deviation of the maximum likelihood estimator, obtained via Monte Carlo simulation, to the CRB for two equal-power emitters at directions $\theta_0 = 3°$ and $\theta_1 = 10°$. As expected, the standard deviation of the MLE approaches the bound at a high SNR (Friedlander, 1989) but departs from it below a threshold SNR with value that is around 15 dB in this case. We note that the threshold SNR increases as the emitter separation decreases (not shown here). Thus, reliable direction finding for closely spaced signals is only

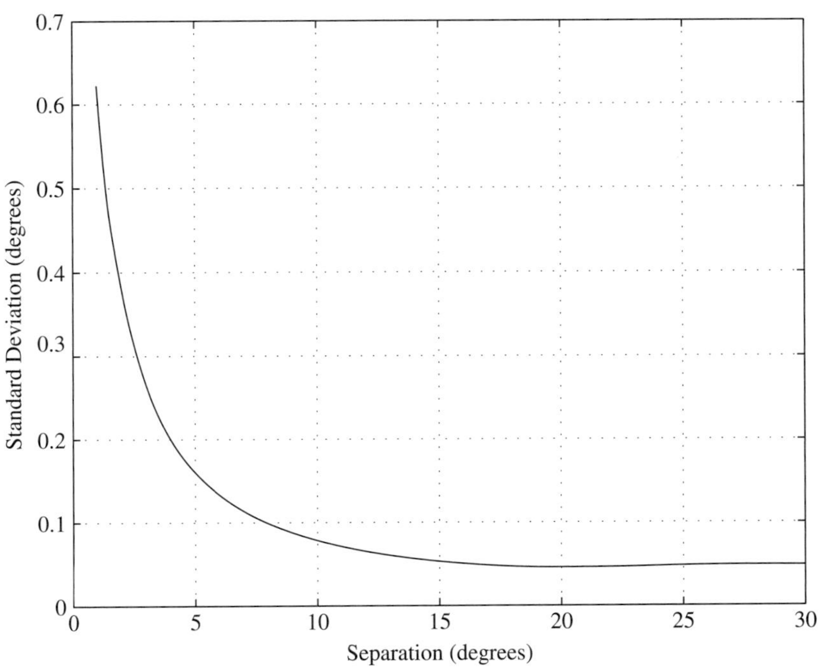

FIGURE 1.13 CRB as a function of emitter separation for two equal-power signals that use an 8-element uniformly spaced array.

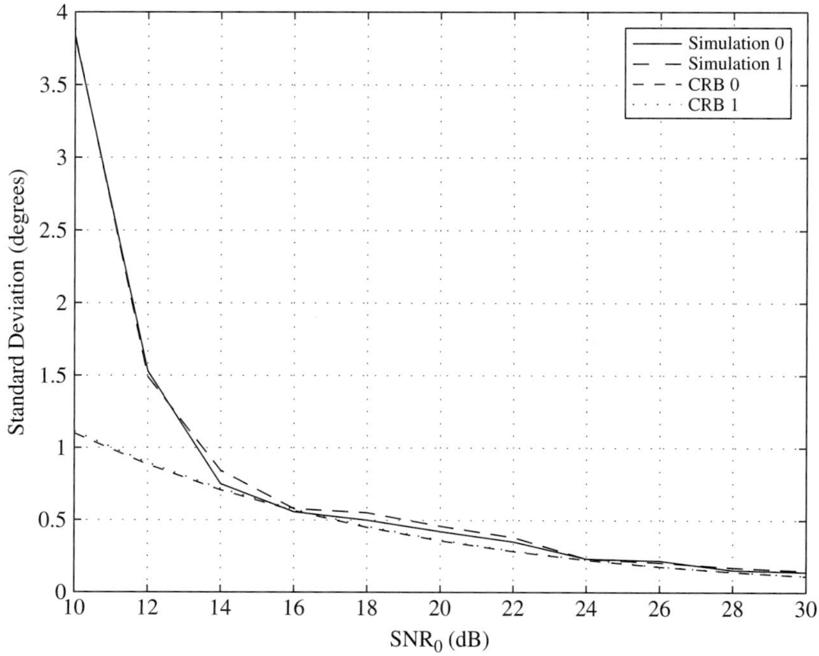

FIGURE 1.14 Experimental and theoretical standard deviation of direction estimates for two equal-power emitters at $\theta_0 = 3°$ and $\theta_1 = 10°$, using an 8-element uniformly spaced array.

feasible when the SNR is sufficiently high (Weinstein and Weiss, 1988; Weiss and Weinstein, 1985). This important point will be discussed in more detail in the following section.

1.6.4 Resolution

Estimation accuracy is not sufficient to characterize the performance of the direction-finding system when multiple co-channel signals are present. We must also consider the issue of resolution—the ability of the system to distinguish between tightly spaced emitters (Helstrom, 1955).

Resolution is closely related to determination of the number of distinct components in the composite signal received by the array. Whereas accuracy is related to estimation, resolution is related to detection or decision (Kay, 1998; Van Trees, 1968). To estimate emitter directions it is necessary to determine the number of directions to be estimated. Applying an MLE designed for P signals to the received data will produce P direction estimates regardless of how many signals are present. Thus, a separate detection/decision step is needed to determine how many directions can be estimated meaningfully.

Knowing the number of emitters present is not sufficient. Assume, for example, that we know that two signals are present: one strong (high SNR) and the other weak (low SNR). Given this knowledge we use the MLE introduced earlier to estimate two directions. The MLE will produce two estimates regardless of how low the SNR of the weak signal may be. It is evident that one of them will not be meaningful if the SNR of the weak signal is too low. Thus, rather than how many signals are present, a better question would be how many signal directions can be reliably estimated *given the available data*. This is a subtle but important distinction: The first question presumes a unique "true" answer independent of system parameters; the second has different answers depending on the SNR, the separation between signals, and possibly some other factors.

This question can be formulated more precisely as a hypothesis-testing or detection problem. For simplicity we consider only the case of two signals; the same idea can be extended to more signals, although the details become more complicated. Consider the following binary hypothesis-testing problem:

$$H_0: \quad \mathbf{y} = \alpha_0 \mathbf{a}(\theta_0) + \mathbf{v} \tag{1.59}$$

$$H_1: \quad \mathbf{y} = \alpha_1 \mathbf{a}(\theta_1) + \alpha_2 \mathbf{a}(\theta_2) + \mathbf{v} \tag{1.60}$$

Given the data $\mathbf{y}$ we decide whether it was generated by a *single* signal with unknown direction θ_0 and unknown amplitude α_0, or by *two* signals with unknown directions θ_1, θ_2 and unknown amplitudes α_1, α_2.

Different detection techniques can be used to solve this problem. The generalized likelihood ratio (GLR) detector provides one solution (see Section A.1.10 and Friedlander, 2009). Recently a Bayesian detector was used to obtain practical closed-form results (Amar and Weiss, 2007, 2008). Given a particular scenario (directions, amplitudes, or signal-to-noise ratios), the detector can correctly determine which of the two hypotheses is true with some probability P_c. We will say that the two signals are resolved if the probability of correct detection equals or exceeds some nominal value, say $P_c > 0.8$. The *resolution limit* can now be defined as the angular separation Δ_r between the signals for which P_c equals its nominal value.

A number of other attempts have been made to define the resolution limit. One is based on estimation accuracy. Assume two emitters at directions θ_0 and θ_1, where the directions are estimated with standard deviations σ_0 and σ_1, respectively. These standard deviations are sometimes approximated by the corresponding values of the CRB. According to Lee and Wengrovitz (1990) the signals are resolvable if half of the separation is smaller than the largest of σ_0, σ_1. Although this approach provides a simple expression for the resolution limit, it ignores the correlation between the direction estimates, which increases as the separation decreases. To overcome this difficulty, Smith (2005) proposed estimating the angular separation (rather than the individual directions) and evaluating the standard deviation of this estimate using the CRB. The signals are said to be resolvable if this standard deviation is smaller than the separation.

A different approach to determining resolution is to consider a cost function that exhibits two peaks at approximately the true directions (Kaveh and Barabell, 1986; Marple, 1977). When the separation decreases the two peaks merge into one. The resolution limit is defined as the separation at which the expected value of the cost function at the midpoint between the two peaks equals the expected values of the peaks.

Yet another approach is based on the performance of the GLR test for determining the number of signals. This method has been used to determine the resolution of two radar signals (Root, 1962).

A review of this literature shows that the resolution limit of the MUSIC algorithm normalized by the beamwidth is inversely proportional to the fourth root of the signal-to-noise ratio (for the case of two equal-power signals):

$$\frac{\Delta_r}{BW} \sim \frac{1}{\mathrm{SNR}^{\frac{1}{4}}} \tag{1.61}$$

This relationship indicates that super-resolution can only be achieved at high signal-to-noise ratios. For instance, resolving signals half a beamwidth apart requires a 6-dB higher SNR than is needed to resolve signals a full beamwidth apart. Similarly, resolution one-tenth of a beamwidth requires a 40-dB higher SNR than is needed for a one-beamwidth resolution.

Amar and Weiss (2007) showed that a lower bound on the resolution limit normalized by the beamwidth is inversely proportional to the square root (rather than the fourth root) of the signal-to-noise ratio. A similar relationship of resolution and SNR is observed when using the model order selection criteria known as the Akaike information criterion (AIC) and the minimum description length (MDL) to determine the number of signals (Wax and Kailath, 1985). However, it appears that under most circumstances the practically achievable resolution limit is inversely proportional to the fourth root of the SNR.

1.7 DISCUSSION

Classical methods of direction finding using antenna arrays are optimal when the received signal has a single component—that is, a single emitter and no multipath. In fact, the beamformer can be interpreted as the generalized maximum likelihood direction estimator, as was shown earlier. In the case where the received signal has multiple components of which the directions are well separated (more than a beamwidth apart) the classical methods are capable of resolving the signal components and optimally estimating their directions.

When using arrays with relatively large numbers of antennas and large apertures, the correspondingly small beamwidth generally provides sufficient resolving power to meet system requirements. Also, the probability of multiple co-channel emitters falling within a given beam decreases as the size of the beamwidth decreases, making it unlikely that more than one signal will be present at any given time. In such situations only classical methods are needed.

The more challenging situation is when arrays with a small number of elements and a small aperture must be used either because of cost considerations or because of physical constraints such as limited space to place the antennas in mobile and airborne applications. In this case, the beamwidth and hence the Rayleigh resolution are relatively large. Also, the larger the beamwidth, the higher the probability of having multiple co-channel emitters fall within a beam.

As discussed earlier, the presence of multiple unresolved signal components introduces bias errors that limit the accuracy of the direction-finding system to some significant fraction of a beamwidth. These errors are a function of the angular distribution and relative powers of the co-channel signal components and are present even when the SNR is very high. It is desirable to reduce or eliminate them.[1]

High accuracy and high resolution with small aperture arrays and a small number of antennas are the "holy grails" of direction finding. Achieving them has motivated extensive research on direction-finding methods that produce fraction-of-a-beamwidth resolution, or super-resolution. Since the 1980s many such algorithms have been developed and studied in the literature (Bienvenu and Kopp, 1983; Schmidt, 1979, 1981). Their performance characteristics are now fairly well understood. In addition to analysis and simulations work, substantial experimental work has been done in this area.

In spite of extensive research and development over a substantial length of time and the demonstrated superiority of advanced super-resolution direction-finding methods, it appears that their practical use is limited. In the rest of this section we attempt to identify the possible reasons for this situation.

Consider the following characteristics of super-resolution techniques:

Relatively high computational complexity. Covariance-based methods such as MUSIC require significantly more computations than classical methods. The computational requirement of MLE and its variations increase exponentially in the number of simultaneously estimated directions.

Accurate calibration. Required accuracy increases with desired resolution (Porat and Friedlander, 1997; Weiss and Friedlander, 1989a,b). Classical methods also require calibration but are more robust to errors.

Relatively high costs. These are in part due to the two issues justly mentioned.

Multiple snapshots. Covariance-based techniques such as MUSIC require a relatively large number of snapshots to produce direction estimates. Classical methods are able to operate with a single snapshot but can fully benefit from multiple snapshots when these are available.

High SNR. The SNR required by MUSIC and its variations is inversely proportional to the *fourth power* of the desired resolution limit.

The first three listed items are not fundamental limitations, and they are amenable to technological solutions. The ever increasing power and decreasing cost of processors makes it possible to implement increasingly more sophisticated and computationally demanding algorithms over time. The requirement of multiple snapshots is more problematic but can, in principle, be eliminated

1. The inability to resolve signals is, of course, a serious problem when attempting to classify or copy them. However, here we focus only on the direction-finding aspect of the problem.

with use of the MLE, at least for a (very) small number of signal components. However, the requirement for a high SNR is fundamental and cannot be "solved."

To better illustrate the SNR issue, we consider the following example. Figure 1.15 depicts the mean-square error of the beamformer for two emitters: one at direction $\theta_0 = -1°$ and the other at $\theta_1 = 1°$. The two signals have equal power. An 8-element uniformly spaced array was used with half-wavelength element spacing, and $K = 100$ snapshots were used to produce each estimate. The beamwidth of this array is approximately 13°, so the emitters are about 15% of a beamwidth apart. The beamformer cannot resolve the two emitters and instead estimates the midpoint between the two directions. The threshold SNR of the beamformer is seen to be approximately $\mathrm{SNR}_{bf} = -15$ dB. This is consistent with the earlier example in Figures 1.5 and 1.6, where the threshold was observed around 5 dB using a single snapshot. $K = 100$ snapshots provides an additional 20-dB processing gain, which reduces the threshold to -15 dB.

Next we apply the MUSIC algorithm to the same data set. MUSIC resolves the two signals with some probability P_r. Figure 1.16 depicts resolution probability as a function of SNR. We define (somewhat arbitrarily) the threshold SNR of the MUSIC algorithm to be that corresponding to $P_r = 0.8$, which is seen to be at $\mathrm{SNR}_{music} = 20$ dB. In this case we have a large difference between the two

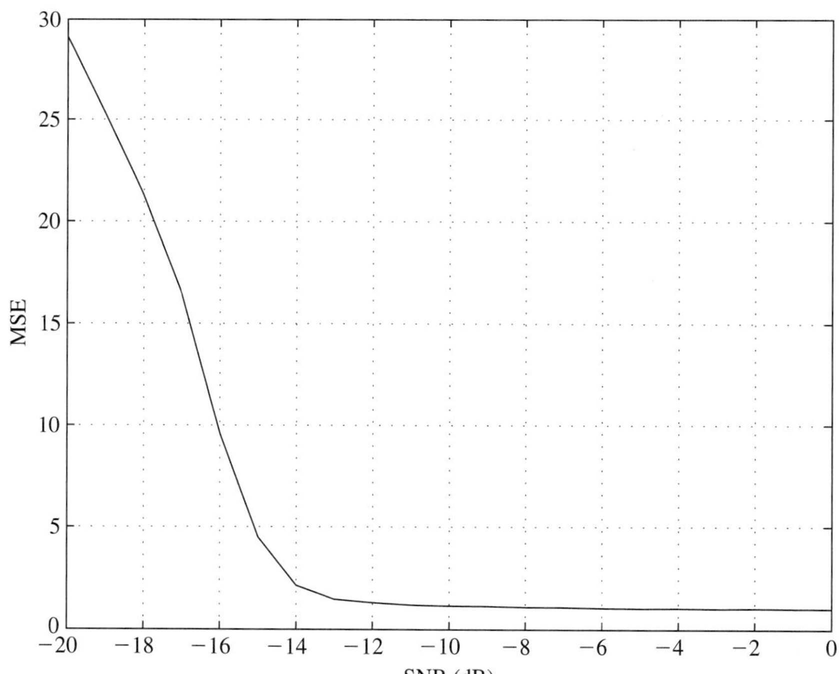

FIGURE 1.15 MSE of the direction estimate produced by a beamformer for two equal-power emitters at $\theta_0 = -1°$ and $\theta_1 = 1°$, $K = 100$ snapshots, and an 8-element uniformly spaced linear array. The SNR threshold is around -15 dB.

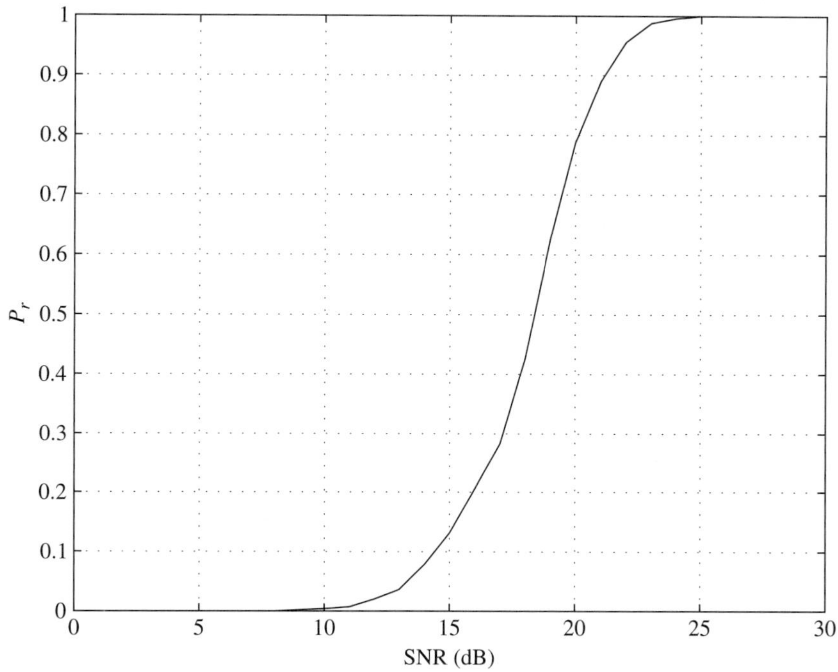

FIGURE 1.16 Resolution probability for two signals using the MUSIC algorithm for two equal-power emitters at $\theta_0 = -1°$ and $\theta_1 = 1°$, $K = 100$ snapshots, and an 8-element uniformly spaced linear array.

thresholds: $\text{SNR}_{music} - \text{SNR}_{bf} = 35$ dB. Various techniques have been proposed to reduce the SNR threshold of MUSIC (Lee and Wengrovitz, 1990) so that the actual difference may be smaller than just indicated. However, the point here is that there is a very substantial difference between the two threshold SNRs in general.

Thus, we have the following situation. If the SNR is lower than SNR_{bf}, the emitter directions cannot be reliably estimated by any method. If the SNR is higher than SNR_{bf} but lower than SNR_{music}, the emitters cannot be resolved but their general direction (in this case the direction of the midpoint) can be reliably estimated. If the SNR is higher than SNR_{music}, the emitters are reliably resolved by the MUSIC algorithm. It is only in the latter case that MUSIC provides a significant advantage over the beamformer. More generally, *any* super-resolution technique will have some threshold SNR below which it cannot resolve the two emitters.

We refer to emitters with an SNR higher than SNR_{bf} as *estimable* and emitters with an SNR higher than that required by whatever super-resolution technique is being used as *resolvable*. In any real-world scenario the direction-finding system will receive signals of interest from some emitters that are resolvable, some that are estimable but not resolvable, and some that are not estimable. (We ignore the latter class of emitters for now.) An important question is what fraction of the observed emitters are resolvable. This is the fraction of cases for which the super-resolution methods are effective. The answer will depend, of course, on the parameters of the direction-finding system and the specific surveillance scenario, so it is not possible to answer this question in general. However, it is likely that in many direction-finding applications this fraction will not be large. The SNR of signals of interest received by the array will have some distribution, which will often have more low and moderate SNR signals (corresponding to longer ranges, weaker transmitters, emitters obscured by hills or buildings, and the like) than high-SNR emitters (relatively nearby, high-power transmitters, line-of-sight propagation). Generally it is desirable to estimate the directions of all signals present, not just those with a high SNR. Thus, only a fraction of the signals will be resolvable.

To better illustrate this point we simulated the following "toy" scenario: Five emitters are randomly distributed in angle and their SNRs are randomly distributed between -15 dB and 30 dB. An 8-element uniformly spaced array is used to collect $K = 100$ snapshots. The MUSIC algorithm and the beamformer estimate

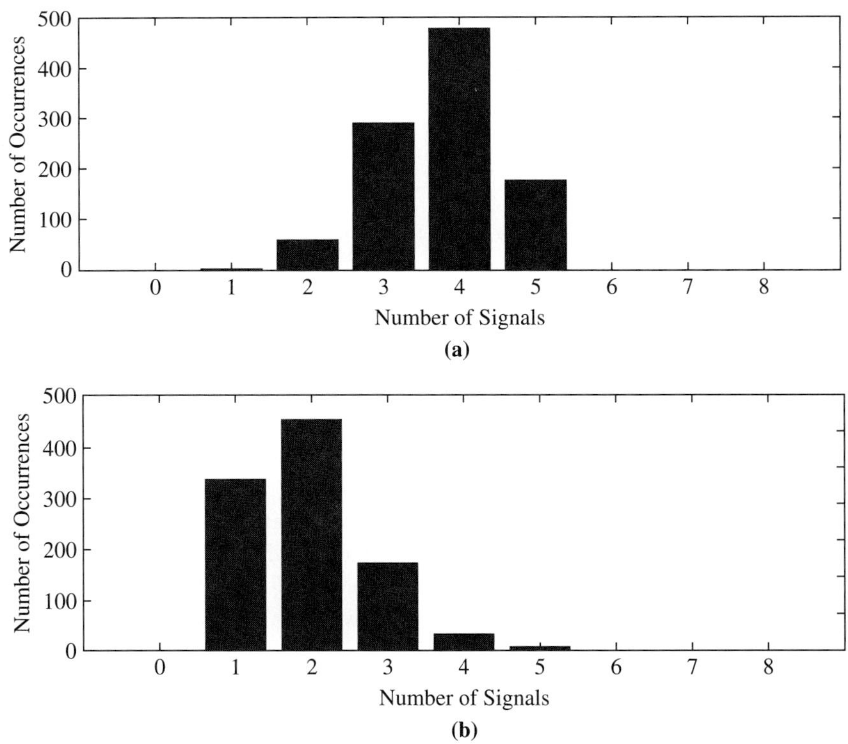

FIGURE 1.17 Histogram of the number of signals resolved by MUSIC (a) and by a beamformer (b) for a random scenario with five emitters using an 8-element uniformly spaced array with $K = 100$ snapshots, based on 1000 random experiments, $-15 \leq \text{SNR} \leq 30$.

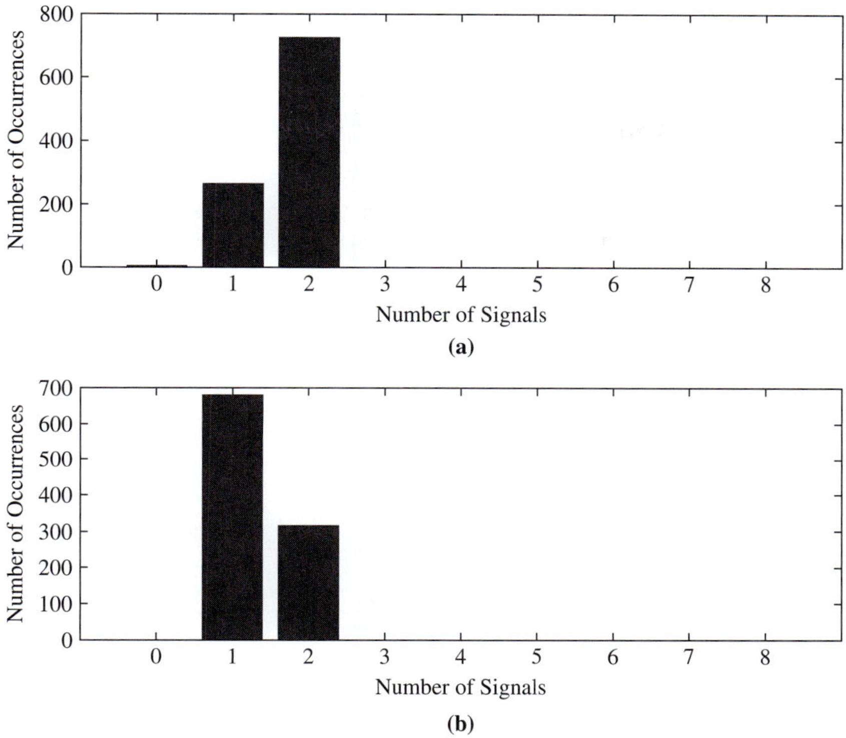

FIGURE 1.18 Histogram of the number of signals resolved by MUSIC (a) and by a beamformer (b) for a random scenario with two emitters using an 8-element uniformly spaced array with $K = 100$ snapshots, based on 1000 random experiments.

directions, and the *number* of signals resolved by each algorithm is recorded. This experiment is repeated for $N = 1000$ randomly chosen scenarios (angles and SNRs). The histogram of the number of resolved signals is depicted in Figure 1.17. Figure 1.18 depicts the results of the same experiment except that in this case only two emitters are present. Figure 1.19 shows the results of the experiment in Figure 1.17 except that the SNRs are distributed between −15 dB and 15 dB.

As is to be expected, MUSIC resolves more signals than the beamformer does. The beamformer will only resolve signals separated by more than a beamwidth, while MUSIC will resolve signals with much smaller separation. However, the difference is not as decisive as one may expect and it depends on the parameters of the scenario—for example, the number of emitters and the SNR distribution. Whether or not the difference is sufficient to justify using a super-resolution method with its associated costs is a subject to be determined by the system designer.

We rarely, if ever, find in the literature an evaluation of the relative performance of classical and advanced direction-finding methods in the context of

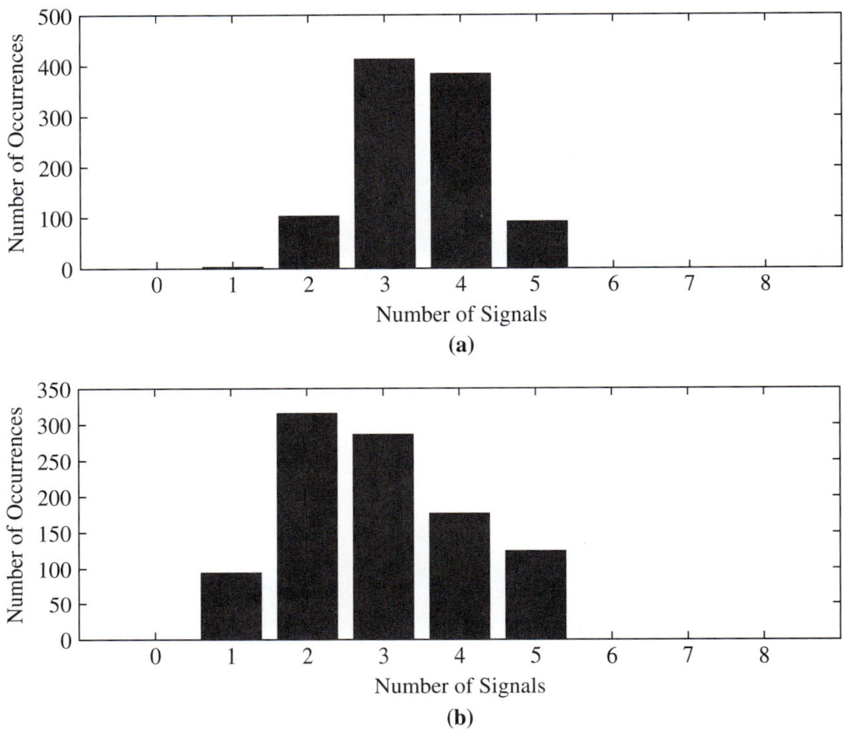

FIGURE 1.19 Histogram of the number of signals resolved by MUSIC (a) and by a beamformer (b) for a random scenario with five emitters using an 8-element uniformly spaced array with $K = 100$ snapshots, based on 1000 random experiments, $-15 \leq \text{SNR} \leq 10$.

a direction-finding scenario with realistic mixtures of emitters with different directions and powers. Results are usually presented for one or a few selected scenarios—most frequently for two equal-power emitters. These results are certainly important and instructive, but they may create unrealistic expectations about the performance of super-resolution methods in real-world deployments.

APPENDIX

A.1.1 The Array Manifold and Its Derivatives

The array manifold $\mathbf{a}(\theta)$ plays a key role in the performance of direction-finding systems. Here we present the form of the manifolds for linear and circular arrays, and we calculate their first- and second-order derivatives, which are needed to compute the accuracy of direction estimation achieved by different algorithms.

For simplicity we assume that the antennas and all the emitters are in the same plane, so that the array manifold can be characterized by azimuth only.

This can be extended in a straightforward manner to the three-dimensional case, where the manifold is a function of both azimuth and elevation. We consider the far-field narrowband model discussed earlier.

Linear Array

Consider a linear array with M omnidirectional antenna elements the distance of which is d_m from a common reference point in the array. Let θ be the direction measured from the line perpendicular to the array. The array manifold is given by

$$\mathbf{a}(\theta) = \begin{bmatrix} \exp\{j2\pi d_1 \sin\theta/\lambda\} \\ \exp\{j2\pi d_2 \sin\theta/\lambda\} \\ \vdots \\ \exp\{j2\pi d_M \sin\theta/\lambda\} \end{bmatrix} \tag{1.62}$$

Note that $|\mathbf{a}(\theta)|^2 = M$. If the antenna elements are directional and have identical patterns $g(\theta)$, the corresponding array manifold is obtained by multiplying the elements of $\mathbf{a}(\theta)$ in Equation (1.62) by $g(\theta)$.

The first derivative of the array manifold is given by

$$\dot{\mathbf{a}}(\theta) = \frac{d}{d\theta}\mathbf{a}(\theta) = j(2\pi/\lambda)\cos\theta\mathbf{D}\mathbf{a}(\theta) \tag{1.63}$$

where

$$\mathbf{D} = \mathbf{diag}\{([d_1, d_2, \ldots, d_M])\} \tag{1.64}$$

is a diagonal matrix containing the element locations on the diagonal. Note that

$$\mathbf{a}(\theta)^H \dot{\mathbf{a}}(\theta) = j(2\pi/\lambda)\cos\theta \sum_{m=1}^{M} d_m \tag{1.65}$$

Without loss of generality the distances d_m can be defined relative to the phase center of the array, in which case $\sum_{m=1}^{M} d_m = 0$, and therefore

$$\mathbf{a}(\theta)^H \dot{\mathbf{a}}(\theta) = 0 \tag{1.66}$$

Note also that

$$\text{real}\{\mathbf{a}(\theta)^H \dot{\mathbf{a}}(\theta)\} = 0 \tag{1.67}$$

even if the distances d_m are not defined relative to the phase center of the array.

The norm of $\dot{\mathbf{a}}(\theta)$ is given by

$$|\dot{\mathbf{a}}(\theta)|^2 = \dot{\mathbf{a}}(\theta)^H \dot{\mathbf{a}}(\theta) = (2\pi/\lambda)^2 \cos^2\theta\,\overline{d^2} \tag{1.68}$$

where

$$\overline{d^2} = \sum_{m=1}^{M} d_m^2 \tag{1.69}$$

is the array's "moment of inertia." This variable plays a key role in the performance of the direction-finding system.

The second derivative of the array manifold is given by

$$\ddot{\mathbf{a}}(\theta) = -(2\pi/\lambda)^2 \cos^2\theta \mathbf{D}^2\mathbf{a}(\theta) - j(2\pi/\lambda)\cos\theta\sin\theta\mathbf{D}\mathbf{a}(\theta) \tag{1.70}$$

It follows that

$$\mathbf{a}(\theta)^H\ddot{\mathbf{a}}(\theta) = -(2\pi/\lambda)^2\cos^2\theta\,\overline{d^2} - j(2\pi/\lambda)\cos\theta\sin\theta\sum_{m=1}^{M} d_m \tag{1.71}$$

The second term is zero, assuming that $\sum_{m=1}^{M} d_m = 0$, so that

$$\mathbf{a}(\theta)^H\ddot{\mathbf{a}}(\theta) = -(2\pi/\lambda)^2\cos^2\theta\,\overline{d^2} \tag{1.72}$$

Note that

$$\text{real}\{\mathbf{a}(\theta)^H\ddot{\mathbf{a}}(\theta)\} = -(2\pi/\lambda)^2\cos^2\theta\,\overline{d^2} \tag{1.73}$$

even if we do not assume that $\sum_{m=1}^{M} d_m = 0$.

Circular Array

Consider a circular array with radius R and M antenna elements. For omnidirectional elements the array manifold is given by

$$\mathbf{a}(\theta) = \begin{bmatrix} \exp\{j2\pi R\sin(\theta-\theta_1)/\lambda\} \\ \exp\{j2\pi R\sin(\theta-\theta_2)/\lambda\} \\ \vdots \\ \exp\{j2\pi R\sin(\theta-\theta_M)/\lambda\} \end{bmatrix} \tag{1.74}$$

where θ_m are the element angles relative to the center of the circle. Note that $|\mathbf{a}(\theta)|^2 = M$.

For elements with identical directional patterns $g(\theta)$ oriented so that the antenna boresight is along the radial line from the circle center to the antenna location, we have

$$\mathbf{a}(\theta) = \begin{bmatrix} g(\theta-\theta_1)\exp\{j2\pi R\sin(\theta-\theta_1)/\lambda\} \\ g(\theta-\theta_2)\exp\{j2\pi R\sin(\theta-\theta_2)/\lambda\} \\ \vdots \\ g(\theta-\theta_M)\exp\{j2\pi R\sin(\theta-\theta_M)/\lambda\} \end{bmatrix} \tag{1.75}$$

In the following we consider only arrays with omnidirectional elements.

The first derivative of the array manifold is given by

$$\dot{\mathbf{a}}(\theta) = \frac{d}{d\theta}\mathbf{a}(\theta) = j(2\pi R/\lambda)\mathbf{C}(\theta)\mathbf{a}(\theta) \tag{1.76}$$

where

$$\mathbf{C}(\theta) = \mathbf{diag}\{([\cos(\theta-\theta_1), \cos(\theta-\theta_2), \ldots, \cos(\theta-\theta_M)])\} \tag{1.77}$$

is a diagonal matrix. Note that

$$\mathbf{a}(\theta)^H \dot{\mathbf{a}}(\theta) = j(2\pi R/\lambda)\sum_{m=1}^{M}\cos(\theta-\theta_m) \tag{1.78}$$

A circular array with uniformly spaced elements has its phase center at the center of the circle. In other words, if $\theta_m = 2\pi m/M + \theta_0$, then $\sum_{m=1}^{M}\cos(\theta-\theta_m) = 0$. Therefore, $\mathbf{a}(\theta)^H \dot{\mathbf{a}}(\theta) = 0$. Also note that

$$\text{real}\{\mathbf{a}(\theta)^H \dot{\mathbf{a}}(\theta)\} = 0 \tag{1.79}$$

for any circular array.

The norm of $\dot{\mathbf{a}}(\theta)$ is

$$|\dot{\mathbf{a}}(\theta)|^2 = \dot{\mathbf{a}}(\theta)^H \dot{\mathbf{a}}(\theta) = (2\pi R/\lambda)^2 \overline{c^2}(\theta) \tag{1.80}$$

where

$$\overline{c^2}(\theta) = \sum_{m=1}^{M}\cos^2(\theta-\theta_m) \tag{1.81}$$

This variable plays a key role in the performance of the direction-finding system.

The second derivative of the array manifold is given by

$$\begin{aligned}\ddot{\mathbf{a}}(\theta) = &-(2\pi R/\lambda)^2 \mathbf{C}^2(\theta)\mathbf{a}(\theta) \\ &-j(2\pi R/\lambda)\mathbf{diag}\{([\cos(\theta-\theta_1)\sin(\theta-\theta_1),\ldots, \\ &\cos(\theta-\theta_M)\sin(\theta-\theta_M)])\}\mathbf{a}(\theta)\end{aligned} \tag{1.82}$$

and

$$\mathbf{a}(\theta)^H \ddot{\mathbf{a}}(\theta) = -(2\pi R/\lambda)^2 \overline{c^2}(\theta) - j(2\pi R/\lambda)\sum_{m=1}^{M}\cos(\theta-\theta_m)\sin(\theta-\theta_m) \tag{1.83}$$

For a uniformly spaced circular array, $\sum_{m=1}^{M}\cos(\theta-\theta_m)\sin(\theta-\theta_m) = 0$, so that

$$\mathbf{a}(\theta)^H \ddot{\mathbf{a}}(\theta) = -(2\pi R/\lambda)^2 \overline{c^2}(\theta) \tag{1.84}$$

Note that

$$\text{real}\{\mathbf{a}(\theta)^H \ddot{\mathbf{a}}(\theta)\} = -(2\pi R/\lambda)^2 \overline{c^2}(\theta) \tag{1.85}$$

for any circular array, whether or not it is uniformly spaced.

A.1.2 The Cramer-Rao Bound

The Cramer-Rao bound (CRB) on the variance of direction estimation errors provides a useful characterization of the achievable accuracy of DF systems. In this section we derive the CRB for linear and circular arrays.

Let $\mathbf{y}$ be a multivariate complex Gaussian vector with zero mean and covariance matrix $\mathbf{R}_y$. Assume that $\mathbf{R}_y$ depends on a single unknown parameter θ. Then the Fisher information matrix (Bangs, 1971) (a scalar in this case) is given by

$$\mathbf{F} = \text{trace}\left\{\mathbf{R}_y^{-1}\frac{\partial \mathbf{R}_y}{\partial\theta}\mathbf{R}_y^{-1}\frac{\partial \mathbf{R}_y}{\partial\theta}\right\} \tag{1.86}$$

where

$$\mathbf{R}_y = \text{SNR}\,\mathbf{a}(\theta)\mathbf{a}(\theta)^H + \mathbf{I} \tag{1.87}$$

Given K independent measurements $\mathbf{y}[1],\ldots,\mathbf{y}[K]$, the CRB is given by

$$\text{CRB} = \frac{1}{K}\mathbf{F}^{-1} \tag{1.88}$$

To calculate the CRB note that

$$\frac{\partial \mathbf{R}_y}{\partial\theta} = \text{SNR}\,\dot{\mathbf{a}}(\theta)\mathbf{a}(\theta)^H + \text{SNR}\,\mathbf{a}(\theta)\dot{\mathbf{a}}(\theta)^H \tag{1.89}$$

and therefore

$$\begin{aligned}\mathbf{F} = \text{SNR}^2\,\text{trace}\{&(\mathbf{R}_y^{-1}\dot{\mathbf{a}}(\theta)\mathbf{a}(\theta)^H + \mathbf{R}_y^{-1}\mathbf{a}(\theta)\dot{\mathbf{a}}(\theta)^H)(\mathbf{R}_y^{-1}\dot{\mathbf{a}}(\theta)\mathbf{a}(\theta)^H\\ &+\mathbf{R}_y^{-1}\mathbf{a}(\theta)\dot{\mathbf{a}}(\theta)^H)\}\end{aligned} \tag{1.90}$$

or

$$\begin{aligned}\mathbf{F} = \text{SNR}^2\,[&(\mathbf{a}(\theta)^H\mathbf{R}_y^{-1}\dot{\mathbf{a}}(\theta))^2 + 2(\mathbf{a}(\theta)^H\mathbf{R}_y^{-1}\mathbf{a}(\theta))(\dot{\mathbf{a}}(\theta)^H\mathbf{R}_y^{-1}\dot{\mathbf{a}}(\theta))\\ &+(\dot{\mathbf{a}}(\theta)\mathbf{R}_y^{-1}\mathbf{a}(\theta))^2]\end{aligned} \tag{1.91}$$

The first and third terms are zero because

$$\begin{aligned}\mathbf{a}(\theta)^H\mathbf{R}_y^{-1}\dot{\mathbf{a}}(\theta) &= \mathbf{a}(\theta)^H\dot{\mathbf{a}}(\theta)\\ &\quad -\frac{1}{1/\text{SNR} + \mathbf{a}(\theta)^H\mathbf{a}(\theta)}(\mathbf{a}(\theta)^H\mathbf{a}(\theta))(\mathbf{a}(\theta)^H\dot{\mathbf{a}}(\theta)) = 0\end{aligned} \tag{1.92}$$

since $\mathbf{a}(\theta)^H\dot{\mathbf{a}}(\theta) = 0$, as was shown earlier. Similarly,

$$\dot{\mathbf{a}}(\theta)^H\mathbf{R}_y^{-1}\mathbf{a}(\theta) = (\mathbf{a}(\theta)^H\mathbf{R}_y^{-1}\dot{\mathbf{a}}(\theta))^H = 0 \tag{1.93}$$

Equation (1.92) follows from Equation (1.91) using the well-known matrix identity

$$(A+BCD)^{-1} = A^{-1} - A^{-1}B(C^{-1}+DA^{-1}B)^{-1}DA^{-1} \tag{1.94}$$

from which it follows that

$$\mathbf{R}_y^{-1} = \mathbf{I} - \frac{1}{1/\text{SNR} + \mathbf{a}(\theta)^H\mathbf{a}(\theta)}\mathbf{a}(\theta)\mathbf{a}(\theta)^H \tag{1.95}$$

We conclude that

$$\mathbf{F} = 2\,\text{SNR}^2\,(\mathbf{a}(\theta)^H\mathbf{R}_y^{-1}\mathbf{a}(\theta))(\dot{\mathbf{a}}(\theta)^H\mathbf{R}_y^{-1}\dot{\mathbf{a}}(\theta)) \tag{1.96}$$

Using Equation (1.95),

$$\mathbf{a}(\theta)^H \mathbf{R}_y^{-1} \mathbf{a}(\theta) = \mathbf{a}(\theta)^H \mathbf{a}(\theta) - \frac{1}{1/\mathrm{SNR} + \mathbf{a}(\theta)^H \mathbf{a}(\theta)} (\mathbf{a}(\theta)^H \mathbf{a}(\theta))(\mathbf{a}(\theta)^H \mathbf{a}(\theta)) \tag{1.97}$$

or

$$\mathbf{a}(\theta)^H \mathbf{R}_y^{-1} \mathbf{a}(\theta) = |\mathbf{a}(\theta)|^2 - \frac{|\mathbf{a}(\theta)|^4}{1/\mathrm{SNR} + |\mathbf{a}(\theta)|^2} = \frac{1}{\mathrm{SNR}} \frac{|\mathbf{a}(\theta)|^2}{1/\mathrm{SNR} + |\mathbf{a}(\theta)|^2} \tag{1.98}$$

and

$$\dot{\mathbf{a}}(\theta)^H \mathbf{R}_y^{-1} \dot{\mathbf{a}}(\theta) = \dot{\mathbf{a}}(\theta)^H \dot{\mathbf{a}}(\theta) - \frac{1}{1/\mathrm{SNR} + \mathbf{a}(\theta)^H \mathbf{a}(\theta)} (\dot{\mathbf{a}}(\theta)^H \mathbf{a}(\theta))(\mathbf{a}(\theta)^H \dot{\mathbf{a}}(\theta)) \tag{1.99}$$

The second term is zero, so

$$\dot{\mathbf{a}}(\theta)^H \mathbf{R}_y^{-1} \dot{\mathbf{a}}(\theta) = |\dot{\mathbf{a}}(\theta)|^2 \tag{1.100}$$

Finally,

$$\mathbf{F} = 2\,\mathrm{SNR}|\dot{\mathbf{a}}(\theta)|^2 \frac{|\mathbf{a}(\theta)|^2}{1/\mathrm{SNR} + |\mathbf{a}(\theta)|^2} \tag{1.101}$$

Note that for arrays with omnidirectional antennas, we have $|\mathbf{a}(\theta)|^2 = M$, and therefore

$$\frac{|\mathbf{a}(\theta)|^2}{1/\mathrm{SNR} + |\mathbf{a}(\theta)|^2} = \frac{M\,\mathrm{SNR}}{M\,\mathrm{SNR} + 1} \tag{1.102}$$

If $M\,\mathrm{SNR} \gg 1$, then $\frac{M\,\mathrm{SNR}}{M\,\mathrm{SNR}+1} \approx 1$, and therefore

$$\mathbf{F} \approx 2\,\mathrm{SNR}|\dot{\mathbf{a}}(\theta)|^2 \tag{1.103}$$

The CRB is then given by

$$\mathrm{CRB} \approx \frac{1}{2K\mathrm{SNR}|\dot{\mathbf{a}}(\theta)|^2} \tag{1.104}$$

In other words, the CRB is inversely proportional to the number of snapshots K, the SNR, and the squared norm of $\dot{\mathbf{a}}(\theta)$. This norm was evaluated in the previous section for linear and circular arrays. Thus, for a linear array we have (1.68),

$$\mathrm{CRB} \approx \frac{1}{2K\mathrm{SNR}(2\pi/\lambda)^2 \cos^2\theta\, \overline{d^2}} \tag{1.105}$$

and for a circular array (1.80),

$$\mathrm{CRB} \approx \frac{1}{2K\mathrm{SNR}(2\pi R/\lambda)^2 \overline{c^2}(\theta)} \tag{1.106}$$

A.1.3 The Cramer-Rao Bound for Multiple Signals

Let $\mathbf{y}$ be a multivariate complex Gaussian vector with zero mean and covariance matrix $\mathbf{R}_y$. Assume that $\mathbf{R}_y$ depends on a vector of unknown parameters $\theta_1, \ldots, \theta_P$. Then the Fisher information matrix (Fisher, 1922) is given by

$$\mathbf{F}_{m,n} = \text{trace}\left\{ \mathbf{R}_y^{-1} \frac{\partial \mathbf{R}_y}{\partial \theta_m} \mathbf{R}_y^{-1} \frac{\partial \mathbf{R}_y}{\partial \theta_n} \right\} \tag{1.107}$$

where

$$\mathbf{R}_y = \sum_{p=1}^{P} \text{SNR}_p \, \mathbf{a}(\theta_p)\mathbf{a}(\theta_p)^H + \mathbf{I} \tag{1.108}$$

Given K measurements $\mathbf{y}[1], \ldots, \mathbf{y}[K]$, the CRB is given by

$$\text{CRB} = \frac{1}{K}\mathbf{F}^{-1} \tag{1.109}$$

where $\mathbf{F}$ is a $P \times P$ matrix with (m,n)-th element that is $F_{m,n}$. To calculate the CRB note that

$$\frac{\partial \mathbf{R}_y}{\partial \theta_m} = \text{SNR}_m \, \dot{\mathbf{a}}(\theta_m)\mathbf{a}(\theta_m)^H + \text{SNR}_m \, \mathbf{a}(\theta_m)\dot{\mathbf{a}}(\theta_m)^H \tag{1.110}$$

The CRB just presented assumes that the SNRs of the different signals are known. It is of interest to consider the case where these SNRs are unknown and need to be jointly estimated with the directions θ_p. In this case the Fisher information matrix will consist of four $P \times P$ blocks:

$$\mathbf{F} = \begin{bmatrix} \mathbf{F}_{\theta,\theta} & \mathbf{F}_{\theta,\text{SNR}} \\ \mathbf{F}_{\text{SNR},\theta} & \mathbf{F}_{\text{SNR},\text{SNR}} \end{bmatrix} \tag{1.111}$$

where the elements of $\mathbf{F}_{\theta,\theta}$ are (1.107)

$$[\mathbf{F}_{\theta,\theta}]_{m,n} = \text{trace}\left\{ \mathbf{R}_y^{-1} \frac{\partial \mathbf{R}_y}{\partial \theta_m} \mathbf{R}_y^{-1} \frac{\partial \mathbf{R}_y}{\partial \theta_n} \right\} \tag{1.112}$$

The elements of $\mathbf{F}_{\text{SNR},\text{SNR}}$ are

$$[\mathbf{F}_{\text{SNR},\text{SNR}}]_{m,n} = \text{trace}\left\{ \mathbf{R}_y^{-1} \frac{\partial \mathbf{R}_y}{\partial \text{SNR}_m} \mathbf{R}_y^{-1} \frac{\partial \mathbf{R}_y}{\partial \text{SNR}_n} \right\} \tag{1.113}$$

The elements of $\mathbf{F}_{\theta,\text{SNR}}$ are

$$[\mathbf{F}_{\theta,\text{SNR}}]_{m,n} = \text{trace}\left\{ \mathbf{R}_y^{-1} \frac{\partial \mathbf{R}_y}{\partial \theta_m} \mathbf{R}_y^{-1} \frac{\partial \mathbf{R}_y}{\partial \text{SNR}_n} \right\} \tag{1.114}$$

and the elements of $\mathbf{F}_{\text{SNR},\theta}$ are

$$[\mathbf{F}_{\text{SNR},\theta}]_{m,n} = \text{trace}\left\{ \mathbf{R}_y^{-1} \frac{\partial \mathbf{R}_y}{\partial \text{SNR}_m} \mathbf{R}_y^{-1} \frac{\partial \mathbf{R}_y}{\partial \theta_n} \right\} \tag{1.115}$$

To calculate these entries note that

$$\frac{\partial \mathbf{R}_y}{\partial \mathrm{SNR}_m} = \mathbf{a}(\theta_m)\mathbf{a}(\theta_m)^H \tag{1.116}$$

The derivative $\frac{\partial \mathbf{R}_y}{\partial \theta_m}$ is given in (1.110).

A.1.4 Sensitivity Analysis

Various methods have been developed for estimating the direction of an emitter from measurements of the array outputs. It is of interest to study the sensitivity of these estimates to measurement errors due to noise. In this section we derive some basic expressions for this sensitivity. In the following sections these expressions will be used to evaluate the estimation error variance for three specific direction-finding methods.

Assume that a parameter θ is estimated by maximizing (or minimizing) a known function $f(\theta)$. In the absence of noise, $\theta = \theta_0$ is the value that maximizes (or minimizes) that function. Therefore, the derivative of $f(\theta)$ is zero at θ_0 (i.e., $\dot{f}(\theta_0) = 0$). Expanding $\dot{f}(\theta)$ in a first-order Taylor series, we obtain

$$\dot{f}(\theta) = \dot{f}(\theta_0) + \ddot{f}(\theta_0)(\theta - \theta_0) \tag{1.117}$$

Because $\dot{f}(\theta_0) = 0$,

$$\dot{f}(\theta) = \ddot{f}(\theta_0)(\theta - \theta_0) \tag{1.118}$$

In the absence of noise the value of θ for which $\dot{f}(\theta) = 0$ is θ_0. In the presence of noise we have $\dot{\tilde{f}}(\theta)$ instead of $\dot{f}(\theta)$, so that

$$\dot{\tilde{f}}(\theta) = \dot{f}(\theta) + e = \ddot{f}(\theta_0)(\theta - \theta_0) + e = 0 \tag{1.119}$$

where e is the noise (error) term. The value of θ optimizing $\tilde{f}(\theta)$ is then

$$\hat{\theta} = \theta_0 + \frac{e}{\ddot{f}(\theta_0)} \tag{1.120}$$

The estimation error variance is therefore given by

$$E\{|\hat{\theta} - \theta_0|^2\} = \frac{1}{|\ddot{f}(\theta_0)|^2} E\{|e|^2\} \tag{1.121}$$

This analysis is valid only if the errors are sufficiently small, because the first-order Taylor series expansion is a good approximation only for small deviations from θ_0.

Next consider the case where θ is estimated by the value for which some function $f(\theta) = 0$. Then

$$f(\theta) = f(\theta_0) + \dot{f}(\theta_0)(\theta - \theta_0) \tag{1.122}$$

Because $f(\theta_0) = 0$, we have

$$f(\theta) = \dot{f}(\theta_0)(\theta - \theta_0) \tag{1.123}$$

In the presence of noise we have $\tilde{f}(\theta)$ instead of $f(\theta)$, so that

$$\tilde{f}(\theta) = f(\theta) + \epsilon = \dot{f}(\theta_0)(\theta - \theta_0) + \epsilon \tag{1.124}$$

Solving for the value of θ, which makes $\tilde{f}(\theta) = 0$, we get

$$\hat{\theta} - \theta_0 = \frac{\epsilon}{\dot{f}(\theta_0)} \tag{1.125}$$

The estimation error variance is given by

$$E\{|\theta - \theta_0|^2\} = \frac{1}{|\dot{f}(\theta_0)|^2} E\{|\epsilon|^2\} \tag{1.126}$$

As before, this analysis is only valid for small errors (i.e., high signal-to-noise ratio).

Equations (1.121) and (1.126) show that the estimation error variance is a simple function of the derivative of the cost function $f(\theta)$ and the variance of the measurement error.

A.1.5 Calculation of $E\{|e|^2\}$ for a Linear Combiner

Direction-finding systems such as the beamformer use a linear combination of the array output to obtain information about emitter direction. In this section we analyze the variance of the measurement error involved in the direction estimation for a general linear combiner characterized by a weight vector $\mathbf{W}(\theta)$. For convenience we assume that the weight vector is normalized to have unit norm $|\mathbf{W}(\theta)|^2 = 1$.

These weights are applied to the measurements $\mathbf{y}$ at the array output (Equation 1.7) to yield

$$\underbrace{\mathbf{W}^H(\theta)\mathbf{y}}_{\tilde{g}(\theta)} = \underbrace{\sqrt{\text{SNR}}\,\mathbf{W}^H(\theta)\mathbf{a}(\theta_0)}_{g(\theta)} + \underbrace{\mathbf{W}^H(\theta)\mathbf{v}}_{u(\theta)} \tag{1.127}$$

Because the weight vector is normalized, we have

$$E\{|u(\theta)|^2\} = \mathbf{W}^H(\theta)\underbrace{E\{\mathbf{v}\mathbf{v}^H\}}_{\mathbf{I}}\mathbf{W}(\theta) = 1 \tag{1.128}$$

Consider the case where θ is estimated by evaluating the magnitude of the linear combiner output, which we denote $f(\theta)$ in the noise-free case and $\tilde{f}(\theta)$ in the noisy case. Thus,

$$f(\theta) = |g(\theta)|^2 \tag{1.129}$$

and

$$\begin{aligned}\tilde{f}(\theta) &= [g(\theta)+u(\theta)][g(\theta)+u(\theta)]^* \\ &= \underbrace{g(\theta)g^*(\theta)}_{f(\theta)} + g(\theta)u^*(\theta) + g^*(\theta)u(\theta) + u(\theta)u^*(\theta)\end{aligned} \tag{1.130}$$

$$\dot{\tilde{f}}(\theta) = \dot{f}(\theta) + \underbrace{\dot{g}(\theta)u^*(\theta) + \dot{g}^*(\theta)u(\theta) + \dot{u}(\theta)u^*(\theta) + g(\theta)\dot{u}^*(\theta) + g^*(\theta)\dot{u}(\theta) + u(\theta)\dot{u}^*(\theta)}_{e} \tag{1.131}$$

Under the assumption that $u(\theta), u^*(\theta), \dot{u}(\theta), \dot{u}^*(\theta)$ are uncorrelated random variables, it is straightforward to check that the expected value of the cross-product of any two of the six terms that e comprises is zero. Therefore,

$$\begin{aligned}E\{|e|^2\} = {} & 2|\dot{g}(\theta)|^2 E\{|u(\theta)|^2\} + 2|g(\theta)|^2 E\{|\dot{u}(\theta)|^2\} \\ & + 2E\{|u(\theta)|^2\}E\{|\dot{u}(\theta)|^2\}\end{aligned} \tag{1.132}$$

or

$$E\{|e|^2\} = 2|\dot{g}(\theta)|^2 + 2|g(\theta)|^2 E\{|\dot{\mathbf{W}}(\theta)|^2\} + 2E\{|\dot{\mathbf{W}}(\theta)|^2\} \tag{1.133}$$

Inserting the expressions for $g(\theta)$ and $\dot{g}(\theta)$, we get

$$\begin{aligned}E\{|e|^2\} = {} & 2\,\mathrm{SNR}\,|\dot{\mathbf{W}}^H(\theta)\mathbf{a}(\theta_0)|^2 \\ & + 2\,(\mathrm{SNR}\,|\mathbf{W}^H(\theta)\mathbf{a}(\theta_0)|^2 + 1)E\{|\dot{\mathbf{W}}(\theta)|^2\}\end{aligned} \tag{1.134}$$

For a high SNR this equation can be simplified to

$$E\{|e|^2\} \approx 2\,\mathrm{SNR}\,|\dot{\mathbf{W}}^H(\theta)\mathbf{a}(\theta_0)|^2 + 2\,\mathrm{SNR}\,|\mathbf{W}^H(\theta)\mathbf{a}(\theta_0)|^2 E\{|\dot{\mathbf{W}}(\theta)|^2\} \tag{1.135}$$

Note that $u(\theta), u^*(\theta)$ are uncorrelated because $\mathbf{v}$ and consequently $u(\theta)$ are circularly symmetric complex Gaussian random variables. For the same reason $\dot{u}(\theta), \dot{u}^*(\theta)$ are uncorrelated, as are $u(\theta), \dot{u}^*(\theta)$ and $\dot{u}(\theta), u^*(\theta)$. The correlation of $u(\theta)$ and $\dot{u}(\theta)$ is given by

$$E\{u(\theta)\dot{u}^*(\theta)\} = \mathbf{W}^H(\theta)\dot{\mathbf{W}}(\theta) \tag{1.136}$$

These will be uncorrelated if $\mathbf{W}^H(\theta)\dot{\mathbf{W}}(\theta) = 0$. The cross-term involving $u(\theta)\dot{u}^*(\theta)$ appears in conjunction with a term involving $u^*(\theta)\dot{u}(\theta)$. Therefore, the condition required for $u(\theta)$ and $\dot{u}(\theta)$ is that $\mathrm{real}\{\mathbf{W}^H(\theta)\dot{\mathbf{W}}(\theta)\} = 0$. This condition will always be satisfied in the cases discussed here.

Next consider the case where θ is estimated by evaluating the real part of the linear combiner output rather than its magnitude. In this case $f(\theta) = \mathrm{real}\{\mathbf{W}^H(\theta)\mathbf{y}\}$ and

$$\tilde{f}(\theta) = \mathrm{real}\{\mathbf{W}^H(\theta)\mathbf{y}\} = \underbrace{\sqrt{\mathrm{SNR}}\,\mathrm{real}\{\mathbf{W}^H(\theta)\mathbf{a}(\theta_0)\}}_{f(\theta)} + \underbrace{\mathrm{real}\{\mathbf{W}^H(\theta)\mathbf{v}\}}_{\epsilon} \tag{1.137}$$

The variance of the error is one-half of the variance of the complex random variable $\mathbf{W}^H(\theta)\mathbf{v}$, which has unit variance. Therefore,

$$E\{|\epsilon|^2\} = \frac{1}{2} \tag{1.138}$$

A.1.6 Estimation Error Variance for Beamforming

In this section we derive the estimation error variance for a direction-finding system that uses a beamformer and estimates direction by the angle at which maximum power is observed. We consider the case where the beamformer weight vector is the nonwindowed normalized array manifold:

$$\mathbf{W}(\theta) = \frac{\mathbf{a}(\theta)}{|\mathbf{a}(\theta)|} \tag{1.139}$$

To evaluate the estimation error variance we use Equation (1.121), which requires calculation of $\ddot{f}(\theta)$ and $E\{|e|^2\}$. First we use Equation (1.135) to evaluate $E\{|e|^2\}$. Note that

$$\dot{\mathbf{W}}^H(\theta)\mathbf{a}(\theta_0) = \frac{\dot{\mathbf{a}}^H(\theta)\mathbf{a}(\theta_0)}{|\mathbf{a}(\theta)|} = 0 \tag{1.140}$$

Therefore,

$$E\{|e|^2\} \approx 2\text{SNR}\left|\frac{\mathbf{a}^H(\theta)\mathbf{a}(\theta_0)}{|\mathbf{a}(\theta)|}\right|^2 E\left\{\left|\frac{\dot{\mathbf{a}}(\theta)}{|\mathbf{a}(\theta)|}\right|^2\right\} = 2\text{SNR}\,|\dot{\mathbf{a}}(\theta)|^2 \tag{1.141}$$

In this case,

$$f(\theta) = |g(\theta)|^2 = \text{SNR}(\mathbf{W}^H(\theta)\mathbf{a}(\theta_0))(\mathbf{W}^H(\theta)\mathbf{a}(\theta_0))^* \tag{1.142}$$

The first derivative of $f(\theta)$ is given by

$$\begin{aligned}\dot{f}(\theta) = {} & \text{SNR}(\dot{\mathbf{W}}^H(\theta)\mathbf{a}(\theta_0))(\mathbf{W}^H(\theta)\mathbf{a}(\theta_0))^* \\ & + \text{SNR}\,\mathbf{W}^H(\theta)\mathbf{a}(\theta_0)(\dot{\mathbf{W}}^H(\theta)\mathbf{a}(\theta_0))^*\end{aligned} \tag{1.143}$$

The second derivative of $f(\theta)$ is then given by

$$\begin{aligned}\ddot{f}(\theta) = {} & 2\,\text{SNR}\,\text{real}\left\{(\ddot{\mathbf{W}}^H(\theta)\mathbf{a}(\theta_0))(\mathbf{W}^H(\theta)\mathbf{a}(\theta_0))^*\right\} \\ & + 2\,\text{SNR}(\dot{\mathbf{W}}^H(\theta)\mathbf{a}(\theta_0))(\dot{\mathbf{W}}^H(\theta)\mathbf{a}(\theta_0))^*\end{aligned} \tag{1.144}$$

Since $\dot{\mathbf{W}}^H(\theta)\mathbf{a}(\theta_0) = 0$, the second term is zero and we are left with

$$\ddot{f}(\theta) = 2\,\text{SNR}\,\text{real}\left\{(\ddot{\mathbf{a}}^H(\theta)\mathbf{a}(\theta_0))\right\} \tag{1.145}$$

The direction estimation error variance is then given by

$$E\{|\theta - \theta_0|^2\} = \frac{1}{|\ddot{f}(\theta_0)|^2} E\{|e|^2\} = \frac{|\dot{\mathbf{a}}(\theta)|^2}{2\,\text{SNR}(\text{real}\{\ddot{\mathbf{a}}^H(\theta)\mathbf{a}(\theta_0)\})^2} \tag{1.146}$$

This result is for an estimate based on a single snapshot. If K snapshots are used, the estimated directions can be averaged, reducing the variance by a factor of K. Thus,

$$E\{|\theta-\theta_0|^2\}=\frac{|\dot{\mathbf{a}}(\theta)|^2}{2K\,\mathrm{SNR}(\mathrm{real}\{\ddot{\mathbf{a}}^H(\theta)\mathbf{a}(\theta_0)\})^2} \tag{1.147}$$

Using Equations (1.68) and (1.73) we conclude that for a linear array

$$E\{|\theta-\theta_0|^2\}=\frac{1}{2K\,\mathrm{SNR}(2\pi/\lambda)^2\cos^2\theta\,\overline{d^2}} \tag{1.148}$$

Using Equations (1.80) and (1.85) we conclude that for a circular array

$$E\{|\theta-\theta_0|^2\}=\frac{1}{2K\,\mathrm{SNR}(2\pi R/\lambda)^2\overline{c^2}(\theta)} \tag{1.149}$$

These results are identical to the corresponding CRB Equations (1.105) and (1.106).

A.1.7 Estimation Error Variance for Phase Matching

A common direction-finding method involves measuring the phases of the signals arriving at the array and matching them to the "phase manifold"—that is, the phases of the array manifold $\mathbf{a}(\theta)$. In this section we evaluate the estimation error of this method for a linear and circular array.

Linear Array

The model of the measured phases ϕ_m is given by

$$\begin{bmatrix}\phi_1\\ \phi_2\\ \vdots\\ \phi_m\end{bmatrix}=(2\pi/\lambda)\begin{bmatrix}d_1\\ d_2\\ \vdots\\ d_m\end{bmatrix}\sin\theta+\begin{bmatrix}u_1\\ u_2\\ \vdots\\ u_m\end{bmatrix} \tag{1.150}$$

where $(2\pi/\lambda)d_m\sin\theta$ are the expected phases for an emitter at direction θ, and u_m are the phase measurement errors assumed to be complex Gaussian with zero mean and variance $E\{u_m^2\}=\sigma^2$.

The estimated direction is obtained by minimizing the distance between the measured phase vector and the expected phase vector. This is accomplished by the following least-squares solution of Equation (1.150):

$$\sum_{m=1}^{M}\phi_m d_m=(2\pi/\lambda)\sin\theta\overline{d^2}+u \tag{1.151}$$

where $\overline{d^2} = \sum_{m=1}^{M} d_m^2$, as defined in Equation (1.69), and the error u,

$$u = \sum_{m=1}^{M} d_m u_m \tag{1.152}$$

is a zero-mean complex Gaussian random variable with variance

$$E\{u^2\} = \sum_{m=1}^{M} d_m^2 \sigma^2 = \overline{d^2}\sigma^2 \tag{1.153}$$

Solving for the sine of the direction we have

$$\sin\hat{\theta} = \underbrace{\frac{\sum_{m=1}^{M} \phi_m d_m}{(2\pi/\lambda)\overline{d^2}}}_{\sin\theta_0} + e \tag{1.154}$$

where the error term is

$$e = \frac{u}{(2\pi/\lambda)\overline{d^2}} \tag{1.155}$$

The variance of this error is given by

$$E\{e^2\} = \frac{E\{u^2\}}{(2\pi/\lambda)^2(\overline{d^2})^2} = \frac{\sigma^2}{(2\pi/\lambda)^2\overline{d^2}} \tag{1.156}$$

Note that this error induces a direction error that can be identified from the Taylor expansion

$$\sin\hat{\theta} = \sin\theta_0 + \underbrace{\cos\theta_0(\hat{\theta} - \theta_0)}_{e} \tag{1.157}$$

Therefore,

$$E\{(\hat{\theta} - \theta_0)^2\} = \frac{E\{e^2\}}{\cos^2\theta_0} \tag{1.158}$$

and

$$E\{(\hat{\theta} - \theta_0)^2\} = \frac{\sigma^2}{(2\pi/\lambda)^2 \cos^2\theta \overline{d^2}} \tag{1.159}$$

It remains to relate the phase measurement errors u_m to the signal measurement errors v_m (the elements of $\mathbf{v}$ in Equation 1.7). The two errors are related by

$$\sqrt{\text{SNR}}e^{j(\overline{\phi}_m + u_m)} = \sqrt{\text{SNR}}e^{j\overline{\phi}_m} + v_m \tag{1.160}$$

where $\overline{\phi}_m = (2\pi/\lambda)d_m \sin\theta$. This nonlinear relationship can be approximated at a high signal-to-noise ratio by the following linear relationship:

$$\begin{aligned}\sqrt{\text{SNR}}e^{j(\overline{\phi}_m + u_m)} &\approx \sqrt{\text{SNR}}e^{j\overline{\phi}_m}(1 + ju_m)\\ &= \sqrt{\text{SNR}}(e^{j\overline{\phi}_m} + e^{j(\overline{\phi}_m + \pi/2)}u_m)\end{aligned} \tag{1.161}$$

Comparing this to the signal model, we see that

$$v_m = \sqrt{\mathrm{SNR}}\, e^{j(\overline{\phi}_m + \pi/2)} u_m \tag{1.162}$$

and therefore

$$\sigma^2 = E\{u_m^2\} = \frac{1}{2\,\mathrm{SNR}} \tag{1.163}$$

where the factor of 2 is due to the fact that u_m is real while v_m is complex. We conclude that

$$E\{(\hat{\theta} - \theta_0)^2\} = \frac{1}{2\,\mathrm{SNR}(2\pi/\lambda)^2 \cos^2\theta \overline{d^2}} \tag{1.164}$$

This result is for an estimate based on a single snapshot. If K snapshots are available, the estimates will be averaged, yielding a K-fold variance reduction, which result is identical to the CRB for the linear array.

Circular Array

The model of the measured phases ϕ_m is given by

$$\begin{bmatrix} \phi_1 \\ \phi_2 \\ \vdots \\ \phi_m \end{bmatrix} = (2\pi R/\lambda) \begin{bmatrix} \sin(\theta - \theta_1) \\ \sin(\theta - \theta_2) \\ \vdots \\ \sin(\theta - \theta_m) \end{bmatrix} + \begin{bmatrix} u_1 \\ u_2 \\ \vdots \\ u_m \end{bmatrix} \tag{1.165}$$

where $(2\pi R/\lambda)\sin(\theta - \theta_m)$ are the expected phases for an emitter at direction θ, and u_m are the phase measurement errors assumed to be complex Gaussian with zero mean and variance $E\{u_m^2\} = \sigma^2$.

The estimated direction is obtained by minimizing the distance between the measured phase vector and the expected phase vector. This involves solving a nonlinear least-squares problem. To analyze the error variance we linearize these equations by expanding the phase in a Taylor series around the emitter direction θ_0:

$$\begin{bmatrix} \phi_1 \\ \phi_2 \\ \vdots \\ \phi_M \end{bmatrix} = (2\pi R/\lambda) \begin{bmatrix} \sin(\theta_0 - \theta_1) \\ \sin(\theta_0 - \theta_2) \\ \vdots \\ \sin(\theta_0 - \theta_M) \end{bmatrix} + (2\pi R/\lambda) \begin{bmatrix} \cos(\theta_0 - \theta_1) \\ \cos(\theta_0 - \theta_2) \\ \vdots \\ \cos(\theta_0 - \theta_M) \end{bmatrix} (\theta - \theta_0) \tag{1.166}$$

Comparing Equations (1.165) and (1.166) we see that

$$\begin{bmatrix} u_1 \\ u_2 \\ \vdots \\ u_M \end{bmatrix} = (2\pi R/\lambda) \begin{bmatrix} \cos(\theta_0 - \theta_1) \\ \cos(\theta_0 - \theta_2) \\ \vdots \\ \cos(\theta_0 - \theta_M) \end{bmatrix} (\theta - \theta_0) \tag{1.167}$$

We can now solve a linear least-squares problem for the estimated direction θ to obtain

$$u = \sum_{m=1}^{M} \cos(\theta_0 - \theta_m) u_m = (\theta - \theta_0)(2\pi R/\lambda) \sum_{m=1}^{M} \cos^2(\theta_0 - \theta_m) \tag{1.168}$$

or

$$(\theta - \theta_0) = \frac{u}{(2\pi R/\lambda)\overline{c^2}(\theta_0)} \tag{1.169}$$

Note that u has zero mean and variance

$$E|u|^2 = \sum_{m=1}^{M} \cos^2(\theta_0 - \theta_m)\sigma^2 = \overline{c^2}(\theta_0)\sigma^2 \tag{1.170}$$

Therefore,

$$E\{(\hat{\theta} - \theta_0)^2\} = \frac{E|u|^2}{(2\pi R/\lambda)^2(\overline{c^2}(\theta_0))^2} = \frac{\sigma^2}{(2\pi R/\lambda)^2\overline{c^2}(\theta_0)} \tag{1.171}$$

To relate the phase measurement errors to the signal measurement errors we proceed exactly as for the case of the linear array to get

$$\sigma^2 = \frac{1}{2\text{SNR}} \tag{1.172}$$

and therefore

$$E\{(\hat{\theta} - \theta_0)^2\} = \frac{1}{2\text{SNR}(2\pi R/\lambda)^2\overline{c^2}(\theta_0)} \tag{1.173}$$

As before, this result is for an estimate based on a single snapshot. If K snapshots are available, the estimates will be averaged, yielding a K-fold variance reduction, which result is identical to the CRB for the circular array.

A.1.8 Estimation Error Variance for Monopulse

It was shown in Equations (1.12), (1.13), and (1.14), that the response $b(\theta)$ of the monopulse system can be approximated by the first derivative $\dot{p}(\theta)$ of the beamformer response $p(\theta)$. This approximation holds for a range of angles around the point where $\dot{p}(\theta) = 0$; it can therefore be used for computing the error variance provided that the errors are sufficiently small (i.e., in the high SNR region).

The direction estimate of the monopulse system is the direction θ for which $b(\theta) = \dot{p}(\theta) = 0$. This is mathematically identical to the value of θ, for which $p(\theta)$ is maximized. In other words, it is identical to the direction estimate generated by the beamformer.

To be more specific, following Equation (1.126), the direction error variance is given by

$$E\{|\theta-\theta_0|^2\}=\frac{1}{|\ddot{p}(\theta_0)|^2}E\{|\epsilon|^2\} \tag{1.174}$$

where we substituted $f(\theta)=\dot{p}(\theta)$. This equation is identical to Equation (1.121).

The calculation of the error variance follows from Equations (1.130) and (1.131), where in this case we substitute $\dot{f}(\theta)=b(\theta), f(\theta)=p(\theta)$, and their noisy versions $\tilde{\dot{f}}(\theta)=\tilde{b}(\theta)$, $\tilde{f}(\theta)=\tilde{p}(\theta)$. The rest of the calculations are identical to those in Section A.1.6 to finally yield

$$E\{|\theta-\theta_0|^2\}=\frac{|\dot{\mathbf{a}}(\theta_0)|^2}{2\,\mathrm{SNR}\,\mathrm{real}\{\ddot{\mathbf{a}}^H(\theta_0)\mathbf{a}(\theta_0)\}^2} \tag{1.175}$$

Inserting the appropriate expression for the derivatives of the array manifolds of linear and circular arrays, we get the results presented in Equations (1.148) and (1.149). These results are identical to the corresponding CRB expressions (1.105 and 1.106).

A.1.9 Quadratic Interpolation for the MLE

Assume that we have computed values of a two-dimensional function $d(\cdot,\cdot)$ in the neighborhood of a given point (θ_0,θ_1). Specifically, we have five values of the function: $d(\theta_0,\theta_1)$, $d(\theta_0\pm\Delta,\theta_1)$, and $d(\theta_0,\theta_1\pm\Delta)$ for some step size Δ. We further assume that the peak of this function occurs in this neighborhood and we want to calculate the exact peak location.

Let

$$\hat{\theta}_0=\theta_0+\tilde{\theta}_0,\quad \hat{\theta}_1=\theta_1+\tilde{\theta}_1 \tag{1.176}$$

be a point in this neighborhood. Expanding the function in a first-order Taylor series around (θ_0,θ_1), we get

$$d(\hat{\theta}_0,\hat{\theta}_1)=d(\theta_0,\theta_1)+b_1\tilde{\theta}_0+b_2\tilde{\theta}_1+b_3\tilde{\theta}_0^2+b_4\tilde{\theta}_1^2+b_5\tilde{\theta}_0\tilde{\theta}_1 \tag{1.177}$$

Inserting the five values of the function and the corresponding values of $(\tilde{\theta}_0,\tilde{\theta}_1)$, we obtain a set of linear equations that can be solved for the Taylor coefficients $b_1,\ldots,b_5$.

The optimum point must satisfy

$$\frac{\partial d(\hat{\theta}_0,\hat{\theta}_1)}{\partial\tilde{\theta}_0}=b_1+2b_3\tilde{\theta}_0+b_5\tilde{\theta}_1=0 \tag{1.178}$$

and

$$\frac{\partial d(\hat{\theta}_0,\hat{\theta}_1)}{\partial\tilde{\theta}_1}=b_2+2b_4\tilde{\theta}_1+b_5\tilde{\theta}_0=0 \tag{1.179}$$

This can be written as

$$\begin{bmatrix} 2b_3 & b_5 \\ b_5 & 2b_4 \end{bmatrix} \begin{bmatrix} \tilde{\theta}_0 \\ \tilde{\theta}_1 \end{bmatrix} = - \begin{bmatrix} b_1 \\ b_2 \end{bmatrix} \tag{1.180}$$

which is solved to yield $(\tilde{\theta}_0, \tilde{\theta}_1)$ and hence the peak location.

A.1.10 Generalized Likelihood Ratio Detector

Consider the binary hypothesis-testing problem defined in Equations (1.59) and (1.60). Let $\mathbf{Y} = [\mathbf{y}[1], \ldots, \mathbf{y}[K]]$ be the $M \times K$ received data matrix collected over K snapshots. The generalized likelihood ratio (GLR) detector for a binary hypothesis-testing problem with unknown parameters involves the calculation of the likelihood ratio

$$LR = \frac{\max_{p_1} f(\mathbf{Y}, H_1, p_1)}{\max_{p_1} f(\mathbf{Y}, H_0, p_0)} \tag{1.181}$$

where $f(\mathbf{Y}, p_0|H_0)$ is the probability density function (p.d.f.) of the measurements under hypothesis H_0, with p_0 being the "nuisance parameters" affecting the p.d.f. Similarly, $f(\mathbf{Y}, p_1|H_1)$ is the p.d.f. under H_1, with p_1 being the corresponding "nuisance parameters."

In the case where $f(\cdot)$ is a multivariate Gaussian distribution function, the log likelihood function has the form

$$L = \text{trace}\left\{\mathbf{Y}^H (\mathbf{R}_0^{-1}(\hat{p}_0) - \mathbf{R}_1^{-1}(\hat{p}_1))\mathbf{Y}\right\} + K \log \frac{|\mathbf{R}_0(\hat{p}_0)|}{|\mathbf{R}_1(\hat{p}_1)|} \tag{1.182}$$

where $\mathbf{R}_0$ and $\mathbf{R}_1$ are the covariance matrices of the data vector under H_0 and H_1, respectively. The data is assumed to have zero mean under both hypotheses. The variables $\hat{p}_0, \hat{p}_1$ are the maximum likelihood estimates of p_0 and p_1 under H_0 and H_1, respectively.

In the resolution problem considered in this chapter, we have $p_0 = [\alpha_0, \theta_0]$ and $p_1 = [\alpha_1, \theta_1, \alpha_2, \theta_2]$. The covariance matrices are

$$\mathbf{R}_0 = \alpha_0 \mathbf{a}(\theta_0)\mathbf{a}^H(\theta_0) + \sigma^2 \mathbf{I} \tag{1.183}$$

and

$$\mathbf{R}_1 = \alpha_1 \mathbf{a}(\theta_1)\mathbf{a}^H(\theta_1) + \alpha_2 \mathbf{a}(\theta_2)\mathbf{a}^H(\theta_2) + \sigma^2 \mathbf{I} \tag{1.184}$$

To calculate the log likelihood ratio, we need the maximum likelihood estimates of α_0, θ_0 under H_0 and of $\theta_1, \theta_2, \alpha_1, \alpha_2$ under H_1. The maximum likelihood estimate of θ_0 can be obtained using the beamformer and searching for the direction of the maximum output power. The estimate of α_0 is then given by

$$\hat{\alpha}_0 = \mathbf{a}^H(\hat{\theta}_0) \sum_{k=1}^{K} \mathbf{y}[k] / |\mathbf{a}^H(\hat{\theta}_0)|^2 \tag{1.185}$$

The maximum likelihood estimate θ_1, θ_2 is given by numerical minimization of the cost function

$$d(\theta_1, \theta_2) = \sum_{k=1}^{K} \mathbf{y}^H[k]\mathbf{P}^{\perp}_{\mathbf{S}(\theta_1,\theta_2)}\mathbf{y}[k] \tag{1.186}$$

where

$$\mathbf{S}(\theta_1, \theta_2) = [\mathbf{a}(\theta_1), \mathbf{a}(\theta_2)] \tag{1.187}$$

After evaluating $\hat{\theta}_1, \hat{\theta}_2$, the estimates of α_1, α_2 are given by

$$\begin{bmatrix} \hat{\alpha}_1 \\ \hat{\alpha}_2 \end{bmatrix} = (\mathbf{S}^H(\hat{\theta}_1, \hat{\theta}_2)\mathbf{S}(\hat{\theta}_1, \hat{\theta}_2))^{-1}\mathbf{S}^H(\hat{\theta}_1, \hat{\theta}_2)\frac{1}{K}\sum_{k=1}^{K}\mathbf{y}[k] \tag{1.188}$$

Inserting the estimated parameters into Equations (1.183) and (1.184), we can now calculate $\mathbf{R}_0$ and $\mathbf{R}_1$ and the log likelihood ratio L.

The GLR detector operates by comparing the log likelihood ratio to a threshold value Th. If $L > Th$, we choose H_1 (i.e., we decide that two signals are present). If $L < Th$, we choose H_0 (i.e., we decide that only one signal is present).

The detector can make two types of error: deciding H_1 when H_0 is true, or deciding H_0 when H_1 is true. The probabilities of these events are $Pr\{L > Th|H_0\}$ and $Pr\{L < Th|H_1\}$, respectively. The overall error probability P_e is then given by

$$P_e = Pr\{L > Th|H_0\}Pr\{H_0\} + Pr\{L < Th|H_1\}Pr\{H_1\} \tag{1.189}$$

where $Pr\{H_0\}$ and $Pr\{H_1\}$ are the a priori probabilities of having one or two signals.

Assuming equal a priori probabilities, we have

$$P_e = 0.5(Pr\{L > Th|H_0\} + Pr\{L < Th|H_1\}) \tag{1.190}$$

Note that this probability is a function of the selected threshold. It is always possible to select this threshold so as to minimize error probability. We will refer to this value of P_e as the minimum probability of error (MPE).

For a given signal-to-noise ratio the *resolution limit* can be defined as the separation between two signals for which the MPE equals a desired value—say MPE $= 0.1$. Evaluating the resolution limit for different SNRs shows that in general the resolution limit normalized by the beamwidth is inversely proportional to the fourth root of the SNR (Friedlander, 2009).

REFERENCES

Adcock, F., 1919. British Patent Specification No. 130490.

Amar, A., Weiss, A.J., 2007. Resolution of closely spaced deterministic signals with given success rate. IEEE Int. Symp. Inform. Theory 2007, Nice.

Amar, A., Weiss, A.J., 2008. Fundamental resolution limits of closely spaced random signals. IET Radar, Sonar and Navigation 2 (3), 170–179.

Bangs, W., 1971. Array processing with generalized beamformers. Ph.D. dissertation, Yale University.

Barankin, E.W., 1949. Locally best unbiased estimates. Ann. Math. Stat. 20, 477–501.

Bellini, E., Tosi, A., 1907. A directive system of wireless telegraphy. Elect. Eng. 2 (11), 771–775.

Bellini, E., Tosi, A., 1909. The radio direction finder. Atti dell' Assoc. Elettr. Ital. 13, 5–16.

Bienvenu, G., Kopp, L., 1983. Optimality of high resolution array processing using the eigensystem approach. IEEE Trans. Acoust. Speech Signal Process. ASSP-31, 1235–1248.

Cramer, H., 1951. Mathematical methods of statistics. Princeton University Press.

Fisher, R.A., 1922. On the mathematical foundations of theoretical statistics. Philosophical Transactions of the Royal Society of London. (A) 222, 309–368.

Friedlander, B., 1989. Sensitivity analysis of the maximum likelihood direction-finding algorithm. IEEE Trans. Aerosp. Electron. Syst. 26 (6), 953–968.

Friedlander, B., 1990. A sensitivity analysis of the MUSIC algorithm. IEEE Trans. Acoust. Speech Signal Process. 38 (10), 1740–1751.

Friedlander, B., 1993. The root-MUSIC algorithm for direction finding with interpolated arrays. Eur. J. Signal Process. 30, 15–29.

Friedlander, B., 2009. On the resolution limits of direction finding systems. In press.

Helstrom, C.W., 1955. The resolution of signals in white Gaussian noise. Proc. IRE 43 (9), 1111–1118.

Kaveh, M., Barabell, A.J., 1986. The statistical performance of the MUSIC and the minimum-norm algorithms in resolving plane waves in noise. IEEE Trans. Acoust. Speech Signal Process. 34 (2), 331–341.

Kay, S., 1998. Fundamentals of Statistical Signal Processing, Volume II: Detection Theory. Prentice-Hall.

Keen, R., 1938. Wireless Direction Finding. Iliffe & Sons, Dorset House.

Lee, H.B., Wengrovitz, M.S., 1990. Resolution threshold of beamspace MUSIC for two closely spaced emitters. IEEE Trans. Acoust. Speech Signal Process. 38 (9), 1545–1559.

Marconi, G., 1906. On methods whereby the radiation of electric waves may be mainly confined to certain directions and whereby the receptivity of a receiver may be restricted to electric waves emanating from certain directions. Proc. Roy. Soc. 77, 413–421.

Marple, L., 1977. Resolution of conventional Fourier, autoregressive, and special ARMA methods of spectrum analysis. IEEE Int. Conf. Acoust. Speech Signal Process. 2, 74–77.

Porat, B., Friedlander, B., 1988. Analysis of the asymptotic relative efficiency of the MUSIC algorithm. IEEE Trans. Acoust. Speech Signal Process. 36 (4), 532–544.

Porat, B., Friedlander, B., 1997. Accuracy requirements in off-line array calibration. IEEE Trans. Aerosp. Electron. Syst. 33 (2), 545–556.

Rao, C.R., 1965. Linear Statistical Inference and Its Application. Wiley (Chapter 5).

Root, W.L., 1962. Radar resolution of closely spaced targets. IRE Trans. Mil. Electron. MIL-6, 197–204.

Schmidt, R.O., 1979. Multiple emitter location and signal parameter estimation. In: Proc. RADC Spectral Estimation Workshop.

Schmidt, R.O., 1981. A signal subspace approach to multiple emitter location and spectral estimation. Ph.D. dissertation, Stanford University.

Smith, S.T., 2005. Statistical resolution limits and the complexified Cramer-Rao bound. IEEE Trans. Signal Process. 53 (5), 1597–1609.

Travers, D.N., Hixon, S.M., 1966. Abstracts of the available literature on radio direction finding 1899–1965. Southwest Research Institute, San Antonio.

Van Trees, H.L., 1968. Detection, Estimation, and Modulation Theory—Part I. Wiley.

Wax, M., Kailath, T., 1985. Detection of signals by information theoretic criteria. IEEE Trans. Acoust. Speech Signal Process. ASSP-33 (2), 387–392.

Weinstein, E., Weiss, A.J., 1988. A general class of lower bounds in parameter estimation. IEEE Trans. Inform. Theory. 34 (2), 338–342.

Weiss, A.J., Friedlander, B., 1989a. Array shape calibration using sources in unknown locations: a maximum likelihood approach. IEEE Trans. Acoust. Speech Signal Process. 37 (12), 1958–1966.

Weiss, A.J., Friedlander, B., 1989b. Eigenstructure methods for direction finding with sensor gain and phase uncertainties. J. Circuits, Syst. Signal Process. 9 (3), 271–300.

Weiss, A.J., Friedlander, B., 1993. On the Cramer-Rao bound for direction finding of correlated signals. IEEE Trans. Signal Process. 41 (1), 495–499.

Weiss, A.J., Friedlander, B., 1994. Effects of modeling errors on the resolution threshold of the MUSIC algorithm. IEEE Trans. Signal Process. 42 (6), 1519–1526.

Weiss, A.J., Weinstein, E., 1985. Lower bounds on the mean square error in random parameter estimation. IEEE Trans. Inform. Theory IT-31 (5), 680–682.

Ziv, J., Zakai, M., 1969. Some lower bounds on signal parameter estimation. IEEE Trans. Inform. Theory. 15 (3), 386–391.

Chapter 2

Practical Aspects of Design and Application of Direction-Finding Systems

Franz Demmel

2.1 INTRODUCTION

The rapid development of radio communications has resulted in the increasing importance of direction finding, at the same time significantly boosting the associated requirements and complexity of direction-finding (DF) systems. Grasping this level of complexity requires a sufficiently broad theoretical foundation and (more than ever) the ability to apply it to practical situations. Ultimately, direction finding is an interdisciplinary field, and its superstructure in signal theory falls in the field of *array signal processing*. However, antennas and wave propagation, radio frequency (RF) circuit technology, data communications, software engineering, and the like, are also essential in this context. Practical implementation of ideas and concepts always involves consideration of applicable economic constraints as well as management of issues such as size, weight, and operation. The following sections will discuss how these different aspects interact as part of any practical implementation of direction-finding systems.

2.2 APPLICATION OF DIRECTION-FINDING SYSTEMS

Direction finding plays a key role in many areas of radio communications. The great majority of these involve *estimating the direction of an emitter* relative to a specified reference direction (true north, magnetic north, heading of a vehicle, airplane, or ship).

Copyright © 2009 by Academic Press, Inc. All rights of reproduction in any form reserved.

In direction finding, the direction of the emitter is specified as in navigation as follows:

- Azimuth angle α computed in the clockwise direction from the reference direction
- Elevation angle ε from the horizontal plane in the direction of the zenith

For characterizing antenna properties and for theoretical considerations, angular notation is used for polar coordinates:

- Azimuth angle ϕ computed in the counterclockwise direction from the x-axis
- Zenith or polar angle θ computed from the positive z-axis

We thus have

$$\alpha = \pi/2 - \phi$$
$$\varepsilon = \pi/2 - \theta$$

In any case, the *wave incidence angles* represent merely an intermediate result for the user because the estimated directions of the incoming waves are in general different from the direction of the emitter of interest. Any practical implementation of a direction-finding system must take this fact into account, primarily by minimizing the following influences:

- Multipath propagation
- Co-channel interfering signals
- Interference generated internally in the system

One of the most important uses of direction finders involves fixing the position of emitters. This can include rescuing shipwrecked persons and locating unknown, illegitimate signals such as undesired emissions from an industrial facility or unlicensed base stations, or signals to be used to remotely detonate improvised explosive devices (IEDs). In the area of military radio reconnaissance, radio position fixing plays a key role in gaining tactical structure pictures and thus in assessing the potential threat of an enemy.

A popular method for obtaining exact position results involves the use of multiple direction finders. The bearings they provide are used in a center station to determine location through triangulation.

Single-station locators (SSLs) use only a single direction finder, although in actual practice two techniques are used. Position fixing with shortwave signals propagated via the ionosphere involves the use of direction finders that measure the elevation of the incident wave in addition to the azimuth. Based on an estimate of the virtual height H of the ionosphere's reflecting layer, the distance to the emitter can be determined; the intersection with the line of bearing (LOB) then yields the emitter location.

Single-station location is also possible with mobile direction finders (called "running fix"). Here, too, position is determined through triangulation, with the LOB being tracked to the intersection as determined from different positions

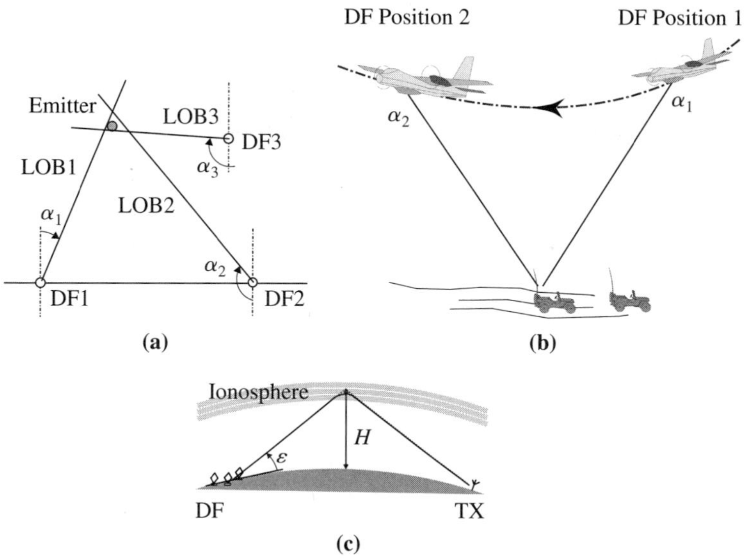

FIGURE 2.1 Locating emitters using direction finders: (a) triangulation; (b) single-station location by running fix; (c) single-station location by estimation of elevation angle and virtual height of the reflecting ionosphere layer in the high-frequency range.

of the mobile direction finder. Figure 2.1 provides an overview of the different techniques.

Position fixing for emitters is often a multistep process. Direction finders distributed countrywide make it possible to fix the position of an emitter through triangulation with an accuracy of several kilometers (typically 1–3% of the distance between direction finders). The next step in localizing the emitter is the use of direction finders installed in vehicles. Finally, portable direction finders are used to focus the search on the last 100 meters (e.g., in buildings).

However, direction finders have other important applications besides fixing the position of emitters. Information about the angle of incidence of radio signals can be exploited very well for the following purposes:

- Spatially selective searches for signals
- Separation of signals (e.g., for resolution of complex radar scenarios in EW)
- Segmentation of spectra with the aid of wideband direction finders, allowing improved estimation of the center frequency and the bandwidth of unknown signals
- Control of directional antennas in communications systems (e.g., for spatial division multiple access (SDMA))

Many of the latest applications require broad or multichannel direction finding. One important example is searching for unknown signals and estimating their parameters, since in most cases the spatial spectrum of an emitter exhibits much higher contrast than does, say, the power spectrum. Figure 2.2 shows by way of example the power and spatial spectra of a conventional narrowband signal in comparison to the spectra of a signal with spectrum spreading.

Because spectral components with the same bearing have a high probability of arriving from the same emitter, direction finders are very efficient for segmenting signals. The example shown in Figure 2.3 illustrates how this works even with

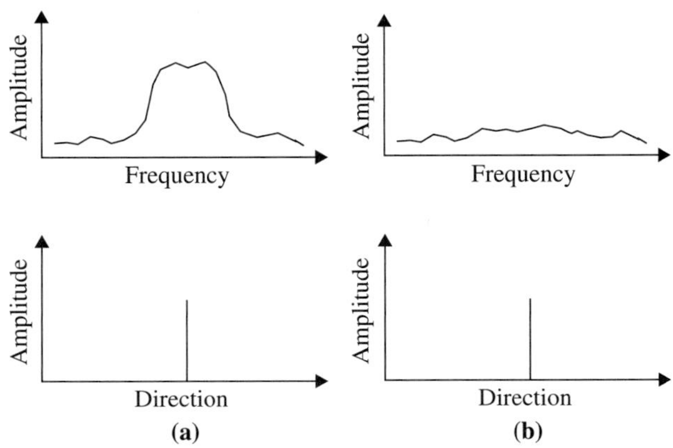

FIGURE 2.2 Power spectrum and spatial spectrum of a conventional narrowband signal (a) and a wideband signal (b). The contrast remains unchanged in the spatial spectrum.

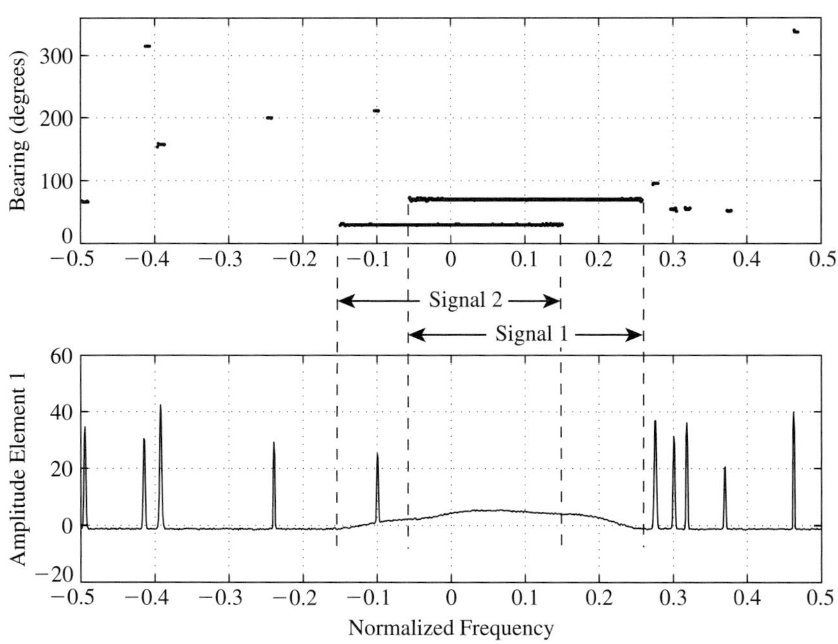

FIGURE 2.3 Signal segmentation of dense scenarios through evaluation of the spatial spectrum.

very dense frequencies: While the power spectrum provides no indication of the different emitters, the spatial spectrum allows them to be clearly distinguished.

2.3 TYPICAL SYSTEM DESIGN—OVERVIEW

All of the latest direction-finding systems use the structure shown in Figure 2.4. As shown, the N output signals from spatially distributed antenna elements are mapped in an RF network among H coherent measurement channels, and digitized and fed to a signal processing unit where they are used to estimate parameters of interest such as

- Number of incident waves (detection)
- Azimuth and elevation of detected waves
- Polarization of detected waves
- Power levels, modulation signals, and the like

In the simplest case, a measurement channel is assigned to each antenna element: $H = N$. For economic reasons, it is often necessary to minimize the number of measurement channels. In this case, the RF network can be implemented as a multiplexer or beamformer, for example. If a multiplexer is used, the antenna

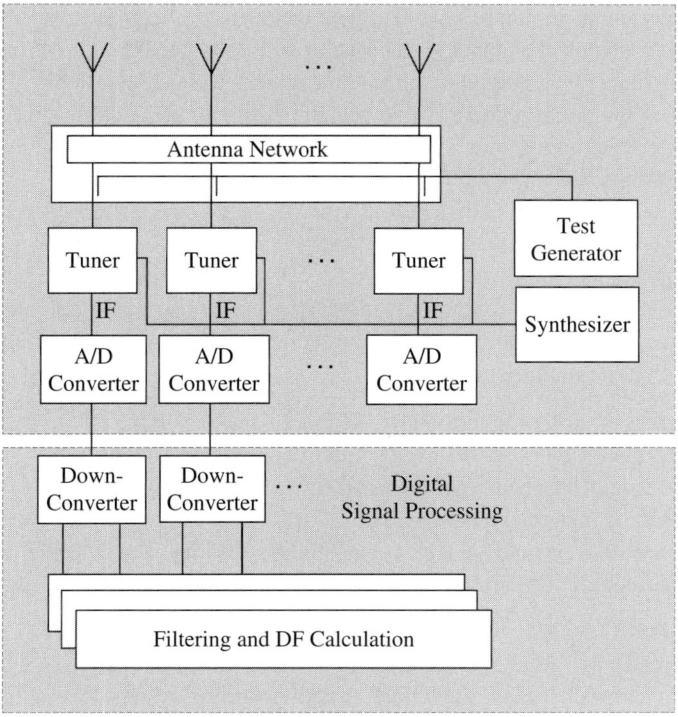

FIGURE 2.4 Typical block diagram of a DF system.

array is divided into subarrays, with H elements each, and connected sequentially to the measurement channels. This results in an increase in measurement time by the number of sequential measurement steps so that the necessary minimum signal duration also increases, meaning that a compromise between expense and measurement time is required.

Prior to digitization of the signals, filtering and—if the analog/digital converters (ADCs) do not allow processing at the frequency of the antenna signals—conversion to a suitable intermediate frequency (IF) are required (tuner). The input signal is mixed in one or more stages with an unmodulated signal, and the result is filtered at the sum or difference frequency (corresponding to the intermediate frequency). For subsequent evaluation of the phase differences, coherent behavior is required from the measurement channels. Accordingly, a common local oscillator (synthesizer) is used in most cases for the frequency conversion along with common clock sources for analog-to-digital (A/D) conversion.

To reduce slowly fluctuating amplitude and phase deviations (e.g., due to temperature and aging), each measurement channel can be supplied with defined calibration signals. With these, the transmission characteristics of each measurement path can be assessed at regular intervals and stored for subsequent compensation of the antenna measured values.

At the digital signal processing level, the signals are first down-converted to the complex baseband. This is followed by the main filtering, which, depending on the particular application, is handled either in the form of a filter bank or for a selected central channel in the case of simple systems for direction finding involving narrowband signals. Filter banks are used in wideband direction finders. During the next processing step, a quantity of samples determined based on the selected averaging time is collected and fed to the algorithms for estimation of the bearings and any other parameters of interest.

This part of the processing typically involves the use of field programmable gate arrays (FPGAs) because of the necessary high processing speed and the tight coupling with hardware. Standard processors such as those found in today's PCs are typically used for further processing of the estimated parameters. All of the computations not subject to any challenging real-time requirements are performed here, including

- Computation of position and compass data
- Computation of platform-dependent compensation data
- Setting of hardware parameters
- Management of system hardware
- Execution of internal test routines in the system
- Data output (typically via an Ethernet interface)

Standard PCs are used for displaying results and to output and/or record the waveforms at the baseband level, and for the demodulated signals. Figure 2.5 shows the components in a practical direction-finding system for the frequency range from 20 MHz to 3 GHz.

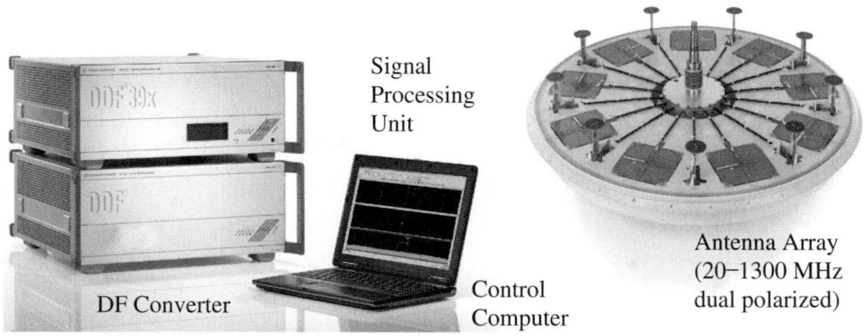

FIGURE 2.5 Practical implementation of the components in a direction-finding system for the VHF/UHF range. (Photo courtesy of Rohde & Schwarz.)

2.4 PERFORMANCE PARAMETERS

The main requirements for a direction-finding system are as follows:

- High accuracy
- High sensitivity
- Sufficient large-signal immunity
- Immunity to distortion of the received wavefield as a result of multipath propagation
- Stable performance in case of co-channel interfering signals
- Short minimum signal duration
- High search speed and probability of intercept in scanning direction finders

2.4.1 Accuracy

The accuracy of a direction finder is characterized by the bearing errors that result from variations in emitter position and frequency. The signal-to-noise ratio (SNR) of the signal is selected to be high enough that the possibility of errors caused by noise can be neglected. Conventionally, the root mean-square (RMS) value is used to characterize the errors. In the frequency range from 1 MHz to 3 GHz, errors of between 1° and 2° RMS can be expected in commercial direction-finding systems.

2.4.2 Sensitivity

The sensitivity of a direction finder means its ability to produce accurate direction-finding results even when faced with low-amplitude signals. Conventionally, the RMS value of the electrical field strength of a narrowband signal is specified for which a certain bearing error is not exceeded (e.g., 3° RMS). The best possible sensitivity that can be attained is dependent primarily on the size of the antenna array, the gain and noise factor of the antenna elements, the noise factor of the DF converter, and measurement time. What are known as the Cramer-Rao limit bounds (CRLB) are of great practical interest when estimating this limit (Van Trees, 2002). If we consider a case important in practice—a uniform circular array (UCA) with N omnidirectional elements and a diameter D at wavelength λ—we obtain the following for the lower limit of the variance σ_ϕ^2 of the azimuth bearing (Van Trees, 2002):

$$\sigma_\phi^2 \geq \frac{\lambda^2}{\pi^2 D^2 N K}\left(\frac{1}{\text{SNR}}+\frac{1}{N\,\text{SNR}^2}\right) \tag{2.1}$$

We assume a wave at an angle of elevation of 0. Equation (2.1) does not depend on the azimuth angle ϕ because of the balanced symmetric nature of UCAs (Manikas, 2004). K is the number of snapshots used to compute the bearing, and SNR is the signal-to-noise ratio at the output of one of the N receiving paths, which are assumed to be identical:

$$\text{SNR} = \frac{P}{P_n} = \frac{E^2}{E_n^2} \tag{2.2}$$

E is the signal field strength, and E_n is the equivalent noise field strength of the instrument noise—that is, the field strength that produces the same output noise power P_n assuming a noise-free system.

For high SNR,

$$\frac{1}{\text{SNR}} \gg \frac{1}{N\,\text{SNR}^2}$$

we obtain

$$E \geq E_n \frac{\lambda}{\sigma_\phi \pi D \sqrt{N K}} \tag{2.3}$$

For example, a system with a 5-element antenna with $D/\lambda = 0.5$, a bearing variance of $\sigma_\phi = 2° \triangleq 0.0349$ rad, and one snapshot would require a field strength of $E \geq 4.08\,E_n$.

2.4.3 Large-Signal Immunity

The large-signal immunity of a direction finder characterizes its behavior in the presence of high-amplitude signals outside the frequency channel containing the signal of interest. Because of the increase in transmitter density, this aspect plays an increasingly critical role in the implementation of practical direction-finding systems.

Saturation effects, inter-modulation, and reciprocal mixing are the main factors influencing large-signal immunity. *Saturation effects* are relevant primarily

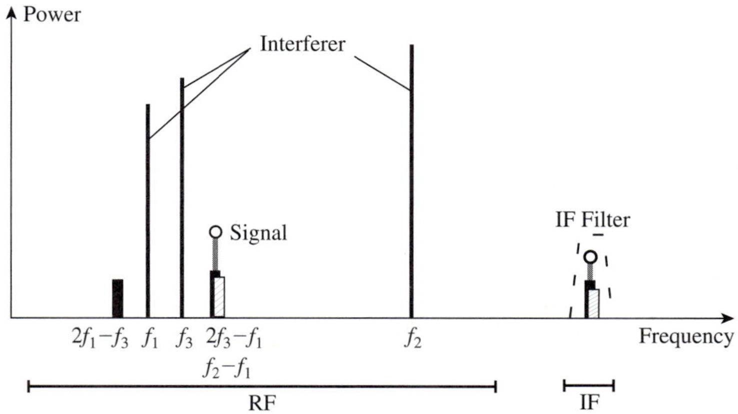

FIGURE 2.6 Inter-modulation products of the second order (frequency $f_2 - f_1$) and third order (frequency $2f_3 - f_1$) lying within the receiving band.

with A/D conversion. *Inter-modulation* is an effect where two or more large signals are mixed because of the nonlinear transfer characteristics of electronic devices such as amplifiers and mixers (Bussgang et al., 1974). The mixing products fall within the tuned frequency band (Figure 2.6).

Inter-modulation is typically characterized using what is known as the "intercept points" IP_k of order k. In real-world systems, inter-modulation products of the second and third orders predominate.

Using the intercept point IP_2 we obtain the following relationship between the power of the second-order mixing products P_{m2} at frequency f_{m2} and the interfering signals P_{I1} and P_{I2} at frequencies f_1 and f_2:

$$P_{m2} = \frac{P_{I1} P_{I2}}{\mathrm{IP}_2} \tag{2.4}$$

assuming small-signal conditions and the frequency relationship

$$f_{m2} = |f_1 \pm f_2| \tag{2.5}$$

For the power of the third-order mixing products, we obtain

$$P_{m3} = \frac{P_{I1}^p P_{I2}^q P_{I3}^r}{\mathrm{IP}_3^2} \quad p + q + r = 3 \tag{2.6}$$

The frequency relation between interferers and mixing product satisfies

$$f_{m3} = |pf_1 \pm qf_2 \pm rf_3| \tag{2.7}$$

Conventionally, the intercept point is specified as a power level refered to 1 mW in dBm, $\mathrm{IP}_k(\mathrm{dBm}) = 10\lg(\mathrm{IP}_k/1mW)$:

$$\begin{aligned} \mathrm{IP}_2 &= P_{I1} + P_{I2} - P_{m2} \quad (\mathrm{dBm}) \\ \mathrm{IP}_3 &= \frac{pP_{I1} + qP_{I2} + rP_{I3} - P_{m3}}{2} \quad (\mathrm{dBm}) \end{aligned} \tag{2.8}$$

Typical values of the antenna output/DF converter interface are around 50 dBm for the second-order intercept point IP_2 and 10 dBm for the third-order intercept point IP_3.

Reciprocal mixing arises in sampling systems such as ADCs and mixers. It is due to phase fluctuations in the sample signal (e.g., clock generator or local oscillator). These fluctuations cause a widening of the signal in the frequency domain, which is known as sideband noise. If a strong interfering signal now reaches the sampling device, the sideband noise will be mixed into its sidebands. If the signal of interest is located in these sidebands, there will be a reduction in the SNR (Figure 2.7).

The noise power P_{np} caused by reciprocal mixing within a narrow receiving band of width B is determined by

$$P_{np} = P_I \, B \, 10^{a_{pn}/10} \tag{2.9}$$

where a_{pn} is the spectral power density of the single-sideband phase noise of the local oscillator or other sampling device related to the carrier (usually stated in dBc/Hz). P_I is the power of the interfering signal.

Consider the phase noise of a high-quality local oscillator specified as −120 dBc/Hz at a frequency offset of 100 kHz. The receiving bandwidth is 10 kHz that occurs as an interfering signal with a power level of −30 dBm at a distance of 100 kHz from the tuned frequency. We get a noise power of $P_{np}(\mathrm{dBm}) = P_I(\mathrm{dBm}) + 10\lg B + a_{pn} = -110$ dBm due to reciprocal mixing.

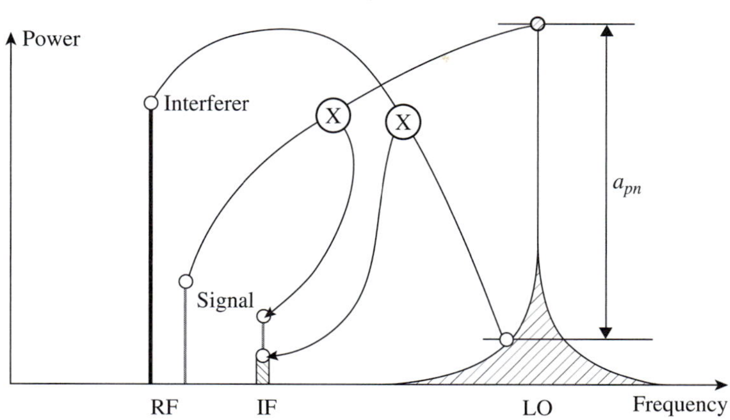

FIGURE 2.7 Reciprocal mixing.

Compared with a device having a noise figure of 10 dB we obtain a noise power of just −124 dBm.

2.4.4 Immunity to Distortion of the Received Wavefield Due to Multipath Propagation

Under normal circumstances, wave propagation between the emitter and the direction finder is disrupted by buildings, mountains, hills, and the like. Even if we have direct line of sight, secondary waves still arise, because of reflection, refraction, and diffraction, which are then superimposed on the direct wave at the receiving location as interfering fields. If the interfering wave component has a lower power level than the desired wave component, the bearing error can be minimized by choosing suitable dimensions in the direction finder. In the simplest case involving single-wave algorithms, we just need to select a sufficiently large antenna aperture. With high-resolution direction-finding algorithms we can separate the dominant wave angle of incidence so that it becomes possible to estimate the direction of the emitter (e.g., by employing propagation models). The disadvantage here is the large amount of information required in advance (besides the significant computational effort).

Special attention must be paid to the possibility of errors known as "wild bearings." These are caused by violation of the Spatial Sampling theorem above a critical frequency and starting at a certain amplitude ratio for the secondary to primary waves, usually because too few antenna elements are available within the antenna aperture or because coupling effects within the antenna array lead to mutual dependencies between the element outputs.

2.4.5 Immunity to Co-Channel Interfering Signals

Unlike multipath interference, co-channel interfering signals usually occur in the form of uncorrelated signals within the frequency band that contains the signal of interest. This can be caused, for example, by electrical equipment and systems (sparks on switches and power sinks), overshoots in transmitters, or defects in transmitting systems. Correlated co-channel interfering signals occur when fixing the position of common-wave emitters.

2.4.6 Short Minimum Signal Duration

Advanced transmission techniques sometimes use time division multiple access (TDMA), frequency hopping, or burst transmission. Direction finding is possible with such signals only if we can process the antenna signals with sufficient speed and with the required signal-to-noise ratio and make them available for use in computing the bearings. Short signal durations also imply that few snapshots are available for averaging so that larger antenna systems are needed to approach the same sensitivity available with continuous signals.

2.4.7 High Search Speed and Probability of Intercept in Scanning Direction Finders

For dependable intercept and direction finding with signals having an unknown frequency, the possible frequency range must be processed within an interval of time that is as short as possible compared to the transmit duration of the signal of interest.

2.5 ANTENNA ARRAY DESIGN

The characteristics we can obtain with direction finders are dependent exclusively on the characteristics of the antenna array in addition to the system's intrinsic noise:

- Sensitivity (field strength value for a certain bearing fluctuation)
- Probability of wild bearings (bearing jumps, outliers)

The standard model of *array signal processing* expresses this fact in the narrowband approximation for N sensors and M waves as the relationship between the measurement vector $\mathbf{x}(t)$ and the array manifold vectors $\mathbf{a}(\phi_j, \theta_j)$, signals $s_j(t)$, and the noise vector $\mathbf{n}(t)$ (Van Trees, 2002; Wax, 1995):

$$\mathbf{x}(t) = \sum_{j=1}^{M} s_j(t)\mathbf{a}(\phi_j, \theta_j) + \mathbf{n}(t) \tag{2.10}$$

The summation occurs over the M waves superimposed on the frequency channel of interest. The components $a_i(\phi, \theta)$ of the array manifold vectors (or steering vectors) represent the complex relative directional characteristics of the antenna outputs

$$\mathbf{a}(\phi, \theta) = [a_1(\phi, \theta) \ldots a_i(\phi, \theta) \ldots a_N(\phi, \theta)]^T \tag{2.11}$$

where ϕ represents the azimuth angle and θ the zenith angle. The continuum spanned by the vectors $\mathbf{a}(\phi, \theta)$ in the N-dimensional complex vector space $\mathbb{C}^N$ is the *array manifold* and shows, so to speak, the directional characteristics of the N array outputs at a glance.

The model (2.10) illustrates the key importance of the array manifold vectors:

- They transform the change in wave angle of incidence into a change in measurement vector, which we can take advantage of depending on the signal and noise power. This is expressed by the Cramer-Rao limit bound for attainable direction-finding sensitivity (Manikas, 2004; Van Trees, 2002).
- If, for *different* wave angles of incidence, array manifold vectors assume similar directions in $\mathbb{C}^N$ (potentially first-order ambiguities), even insignificant interference (e.g., due to noise) will cause wild bearings.
- If linear combinations of two array manifold vectors with a different wave angle of incidence assume similar directions in $\mathbb{C}^N$-like array manifold vectors for other wave angles of incidence, then we have potentially second-order

ambiguities, which can result in wild bearings depending on the signal scenario. This also applies to higher orders of ambiguity.

These aspects determine the main design criteria for a DF antenna array:

Largest possible antenna aperture. Large changes in the array manifold vectors for changes in the wave angle of incidence are achieved using antenna dimensions that are large with respect to wavelength. From this, the *array geometry* can be derived: Circular arrays are useful for panoramic vision in the azimuth, while linear arrays are useful for intercept angles below 90°. For intercept angles greater than 90°, it makes sense to use partial circular arrays.

Number of antenna elements. For a given aperture size, ambiguities can be avoided by including a sufficient number of antenna elements ($\triangleq$-dimension number of the array manifold). The Spatial Sampling theorem provides a clear justification for this requirement. In any practical implementation, we must naturally take into account the fact that the coupling between the elements will increase with a larger number of elements. This can lead to a loss of sensitivity. More elements also entail a larger number of measurement paths if we do not wish to increase the measurement time.

We must also recall that in practice the parameter space generally extends well beyond the wave angles of incidence ϕ and θ. In particular, the polarization dependency of the array manifold vectors and the influence of different support platforms (e.g., mast, vehicle) on the array manifold vectors result in bearing errors if these issues are not taken into account when designing the antenna or characterizing the array manifold.

A proven design approach involves configuring the antenna array so that the sensitivity of the array manifold vectors with respect to the parameters to be assessed is maximized and the sensitivity with respect to parameters not of interest is minimized.

Example 2.1

We want to design an antenna array for use in estimating the azimuth of an emitter. Since we are not interested in elevation, we can use an antenna geometry that extends horizontally. Also, we are not interested in polarization, so we should select our antenna elements based on the preferably expected polarization and (through careful decoupling of the feed lines and support structures) ensure that the array manifold vectors are largely independent of any possible fluctuations in the polarization parameters. Because we will use the antenna on masts as well as on vehicles, we must also strive for good decoupling between the antenna elements and the support structure. This is achieved primarily through the use of symmetrical antenna elements with very high common-mode rejection.

A second workable approach involves taking into account the parameter space as fully as possible when characterizing the antenna vectors. The disadvantage here is the increased number of unknowns, which when reestimated will require a larger number of antenna elements and thus a higher implementation cost. In addition, the cost of measuring, storing, and managing the parameters increases as the number of parameters to be taken into account increases.

2.5.1 Design Example

Figure 2.8 shows an antenna array for direction finding predominantly with vertically polarized signals in the frequency range from 20 MHz to 3 GHz. The outer circular array antenna has a diameter of approximately 1 m and covers the frequency range from 20 MHz to 1.3 GHz. It consists of nine dipole elements equipped with a high-impedance differential amplifier at the dipole feedpoint to optimize gain and sensitivity. The top capacitances of the dipole elements also help to boost the gain at low frequencies. At high frequencies, the radiators can be electrically shortened using integrated switches to reduce mutual coupling and minimize backscatter within the operating frequency range of the inner antenna array. The elements exhibit an approximately omnidirectional characteristic in the horizontal plane, while the directional characteristic of the phases corresponds to the expected cosinusoidal shape:

$$a_i(\phi,\theta) \approx \sin\theta e^{jk_0 R \sin\theta \cos\left(\phi - \frac{2\pi(i-1)}{N}\right)} \tag{2.12}$$

where R is the array radius, $k_0 = \frac{2\pi}{\lambda}$, and the element number $N = 9$.

FIGURE 2.8 Antenna array used for direction finding in the frequency range from 20 MHz to 3 GHz. The upper half of the radome and the lightning rod in the center of the antenna have been removed. (Photo courtesy of Rohde & Schwarz.)

The inner circular array consists of eight dipole elements arranged in front of a cylindrical reflector. The omnidirectional reference element is implemented through the dipole elements' summation. The pin located in the center of the antenna is used to accommodate a lightning rod.

2.6 NUMBER OF ANTENNA ELEMENTS AND PROCESSING CHANNELS

To obtain unambiguous DF results, the following parameters must satisfy certain conditions:

- Number N of antenna elements
- Distance between antenna elements
- Number M of waves to be resolved
- Rank η of the signal's correlation matrix
- Array manifold

Wax and Ziskind (1989) make the basic assumption that any arbitrary set of N array manifold vectors $\mathbf{a}_i(\mathbf{\Theta}), i=1,2,\ldots,N$ with disjunct parameter vectors $\mathbf{\Theta}$ are linearly independent. Based on this assumption and where the rank of the noise-free correlation matrix $\mathbf{S}$ of the signals is

$$\eta = rank\left(\mathbf{SS}^H\right) \leq M \tag{2.13}$$

it is shown that

$$M < \frac{N+\eta}{2} \tag{2.14}$$

waves can be subjected to unambiguous direction finding (Wax and Ziskind, 1989). For the limiting case of uncorrelated signals ($\eta = M$) we obtain the known criterion $M < N$ or $M \leq N-1$.

The starting assumption is critical for antenna array implementation since the linear independence of the array manifold vectors must be maintained with the sparsest possible configuration of the antenna array, taking into account error influences. From a practical point of view, this aspect plays a predominant role in determining the minimum number of elements for a given array size.

Maximum sensitivity and minimum required signal duration can be ensured if each antenna element is processed using a separate processing channel. If we wish to simultaneously obtain good large-signal characteristics, such solutions are too costly in many cases. Thus, we have to deal with a reduced number of processing channels and divide the antenna array into subarrays. Coherent measurements are made within the subarrays, which are processed sequentially. Figure 2.9 shows the basic block diagram of this kind of array data acquisition. Direction-of-arrival (DOA) estimation methods are treated in Sheinvald and Wax (1999) and Worms (2002).

Use of only a *single* measurement path represents a special case. Its practical significance is that existing receivers can be extended with relative ease to

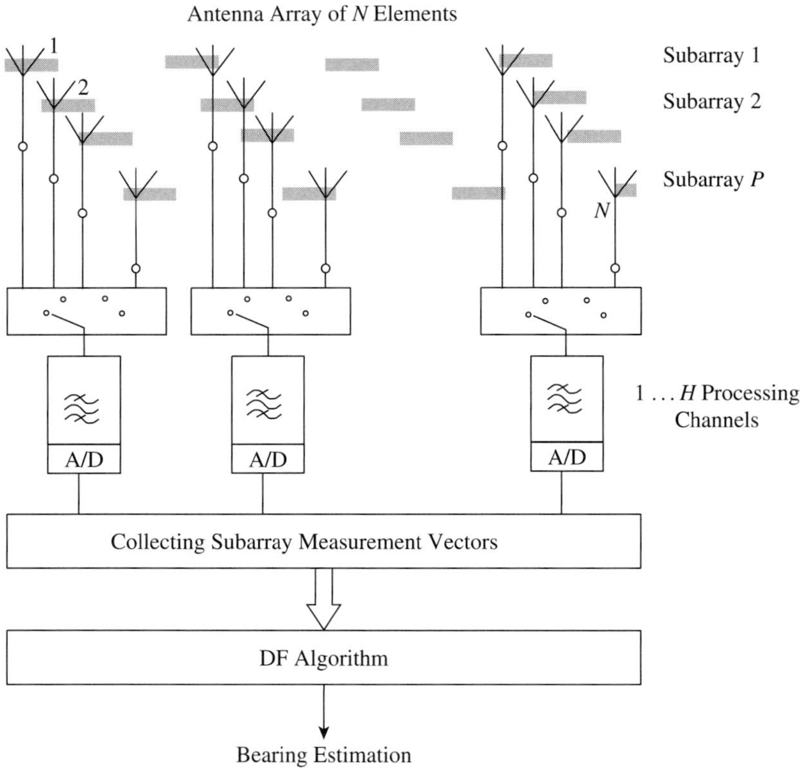

FIGURE 2.9 Sequential measurement of N antenna elements with H processing channels.

include a direction-finding function (Demmel and Unselt, 2002). Since the phase reference from a second measurement path is lacking, the phase information must be obtained indirectly through amplitude measurements in this case. The basic principle involves performing this amplitude measurement in several steps with different antenna states and then computing the wave angles of incidence based on changes in the measured signal power and known changes in state.

Figure 2.10 shows an example of this. The phase difference β_{1k} between reference element 1 and element $k \in \{2 \ldots N\}$ is determined by adding up the element outputs, which are phase-modulated in four states (0°, 90°, 180°, and 270°) with reference element 1. The $4(N-1)$ measurement of the square magnitudes of the antenna output are required.

For one measurement cycle, we have

- Measurement 1: $y_{k1} = |x_1 + x_k|^2 = |x_1|^2 + |x_k|^2 + x_1 x_k^* + x_k x_1^*$
- Measurement 2: $y_{k2} = |x_1 - x_k|^2 = |x_1|^2 + |x_k|^2 - x_1 x_k^* - x_k x_1^*$
- Measurement 3: $y_{k3} = |x_1 - jx_k|^2 = |x_1|^2 + |x_k|^2 + jx_1 x_k^* - jx_k x_1^*$
- Measurement 4: $y_{k4} = |x_1 + jx_k|^2 = |x_1|^2 + |x_k|^2 - jx_1 x_k^* + jx_k x_1^*$

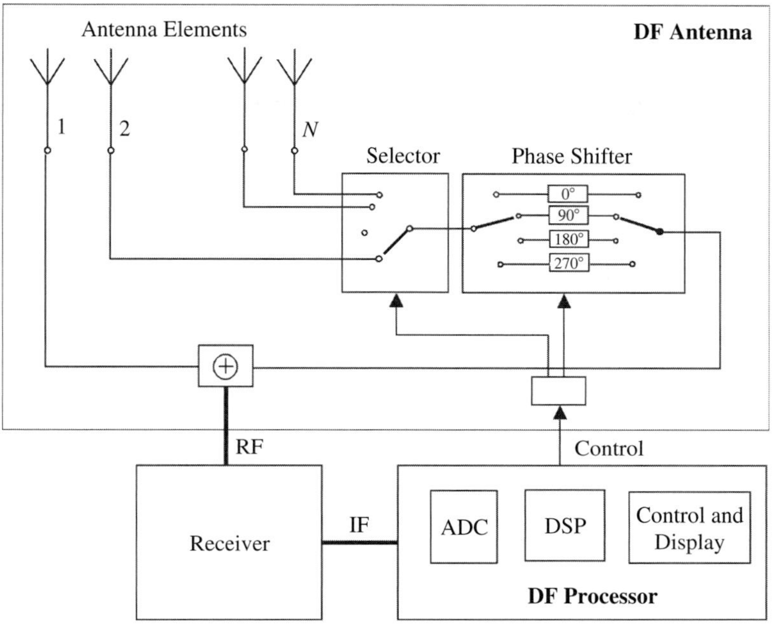

FIGURE 2.10 Example of DF with N antenna elements and a single processing channel.

By taking the difference, we obtain the real and imaginary parts of the conjugate complex products $x_1 x_k^*$:

$$\begin{aligned} y_{k1} - y_{k2} &= 4\,\mathrm{Re}\left(x_1 x_k^*\right) \\ y_{k3} - y_{k4} &= -4\,\mathrm{Im}\left(x_1 x_k^*\right) \end{aligned} \quad (2.15)$$

For the phase difference:

$$\beta_{1k} = \arctan\left(-\frac{y_{k3} - y_{k4}}{y_{k1} - y_{k2}}\right) \quad (2.16)$$

Another possibility for direction finding using only a single measurement path involves the use of antenna arrays consisting of parasitic radiators. This published technique is also referred to as an electronically steerable parasitic array radiator (ESPAR) antenna (Cheng and Ohira, 2006) and is based on the idea of intentionally influencing the directional characteristic of the antenna element connected to the measurement channel by changing the impedances on the parasitic radiators. Based on the measured amplitude changes at the output, we can then compute the wave angles of incidence.

2.7 MULTICHANNEL RECEIVERS

To achieve sufficient phase and amplitude balance, identical processing channels are needed to implement the measurement paths. These channels convert the bandpass signals from the antenna outputs into signal samples in the complex baseband. Basically, two methods can be used here.

Direct receiver. The antenna signals are band-limited to a bandwidth B (reducing the input power and avoiding problems with aliasing) and are then fed directly to the A/D converters and digitized using a suitable sampling rate $f_S \geq 2B$ (Figure 2.11). If possible, the sampling clock should be taken from a common source as shown in the figure to ensure the required sampling synchronization. If clock synchronization cannot be managed via a direct connection, the sampling clock must be derived from precise timing standards.

Down-conversion receiver. If direct A/D conversion is not possible for technical reasons, the frequency band of interest must first be shifted to a lower frequency (LF). In that case, the antenna signals are mixed with a signal generated in the local oscillator (LO). Based on the mixing products that arise, a frequency band that is suitable for further processing is then filtered out with center frequency f_{IF} and fed to the A/D converter. If the frequency is shifted to the center frequency $f_{IF} = 0$, then we speak of a "direct conversion receiver" or "homodyne receiver." However, a "heterodyne receiver," or "superhet," has fewer problems with undesired mixing products. It uses an IF with a filter to provide good rejection of undesired mixing products as

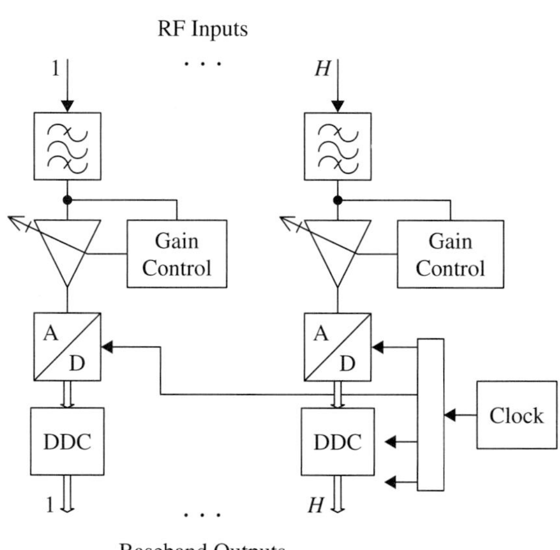

FIGURE 2.11 Direct receiver with multiple channels.

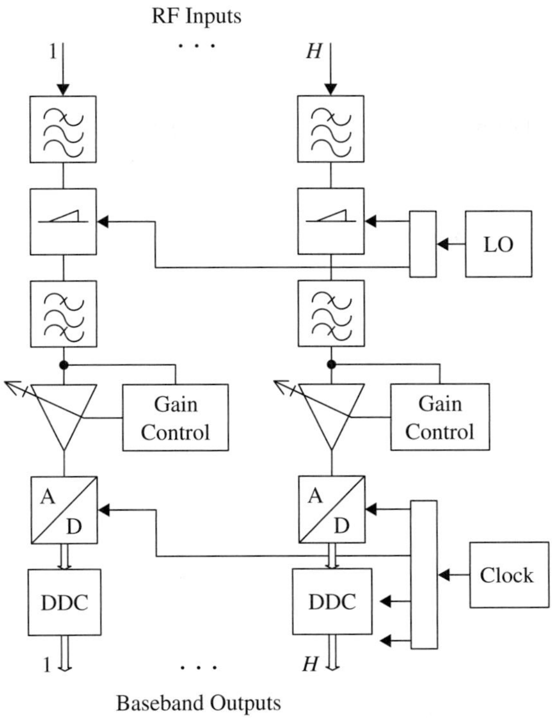

FIGURE 2.12 Multichannel superhet.

well as a favorable frequency for the A/D conversion process (Pace, 2000). Figure 2.12 is a schematic diagram of a multichannel superhet receiver.

Besides providing for sampling synchronization during A/D conversion, we must also ensure that the mixers in each processing channel are driven with the same phase. In both concepts, adaptation of the receiver's dynamic range to the current signal scenario plays an important role. A separate detector is provided in each processing channel to deliver the signal needed for fast automatic gain control (AGC) to avoid overdriving the A/D converter.

The following issues are critical in the receiver design:

- No undesired mixing products should appear in the IF band.
- The number of mixing stages should be minimized.
- Each mixing stage should have as little power applied to it as possible.

Since these issues entail relatively stringent limits when selecting the LF frequencies, the A/D converter is often implemented in bandpass sampling mode—an IF band from 47 MHz to 67 MHz and a sampling frequency of 76 MHz in the A/D converter, for example.

2.8 WIDEBAND DIRECTION FINDING

A wideband signal is any signal with energy distributed over a bandwidth that is not small in comparison to the signal's center frequency. Direction finding of that kind of signal requires methods that are not addressed by the procedures traditionally used for estimating a signal's direction of arrival.

2.8.1 General

Spectrum spreading is used increasingly in radio communications nowadays. It includes

- Frequency hopping (FH)
- Direct-sequence spread spectrum (DSSS)

In many applications time-compressed (i.e., wideband) burst signals are also important. Figure 2.13 illustrates the basic functioning of such low-probability-of-intercept (LPI) techniques in the frequency time domain. Conventional continuous signals are also shown by way of reference.

Using frequency hopping, the message to be transmitted is divided into small sections (typically lasting a few milliseconds depending on the frequency range) and transmitted using carrier frequencies that are varied in a pseudo-random manner. The receiver tracks the agreed pseudo-random sequence in order to recover the message content.

In DSSS radio systems, the message signal undergoes an additional high-speed pseudo-random phase modulation, which causes the desired signal to be distributed over a wide-frequency band. Direction finding with signals of this sort necessitates bandwidths that are generally no longer small in comparison to the center frequency. Accordingly, our narrowband approximation from the signal model (2.7) no longer holds, and we must now take into account the frequency response of the antenna characteristics.

Frequency response compensation is commonly referred to as *focusing* in technical publications. Two approaches are possible (Krolik, 1991). In *coherent focusing*, the measured antenna signals are multiplied by a compensation

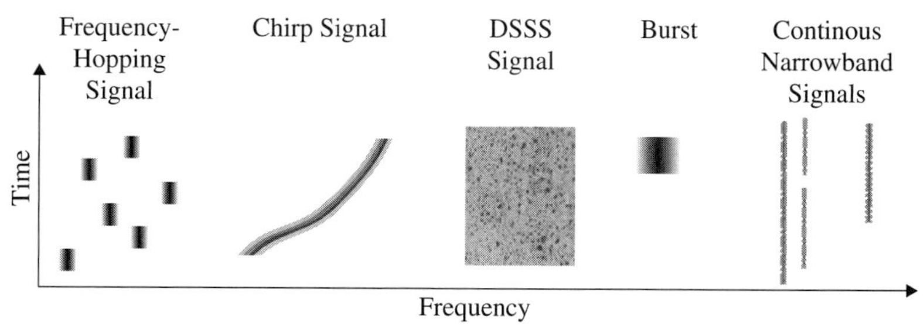

FIGURE 2.13 Typical LPI signals and classic continuous signals in the frequency time domain.

term prior to computation of the bearings. This term compensates for the delay differences between antenna elements. Accordingly, information is needed in advance about the approximate wave angles of incidence. In *noncoherent focusing*, the frequency range we are monitoring is divided into subranges using a filter bank. Direction finding is then performed based on the narrowband model. Related parts of the bearing spectrum obtained in this manner are combined and allocated to an emitter.

In real-world applications, noncoherent focusing is most commonly used for the following reasons:

- No a priori information is required.
- Almost all of the signals are already separated in spectral terms, and even with densely occupied scenarios, the signals can be separated in most cases with simple single-wave algorithms.
- Besides the wave angles of incidence, we can obtain estimates of the frequency ranges occupied by the emitters.

Figure 2.14 illustrates the functional principle of a wideband direction finder that uses noncoherent focusing.

Following A/D conversion and digital down-conversion to the baseband, the real and imaginary parts of the wideband signal in each measurement path are fed to a digital filter bank, which is conventionally implemented as a fast Fourier transform (FFT) or polyphase filter bank. The filter bank breaks down the wideband signal with bandwidth B_{RT} into J_F narrowband channels with bandwidth B_C, where $B_{RT} = J_F B_C$.

In the next step in the processing chain, the filter bank outputs from each measurement path are combined to compute the correlation matrices. The number of snapshots used to estimate the correlation matrices is typically determined by the operator in the form of an averaging time. The minimum size of the correlation matrices depends on the particular DF algorithm: Techniques based on single-wave models need an $N \times 1$ vector for each frequency channel, while multichannel algorithms require up to $(N^2+N)/2$ correlation terms per frequency channel.

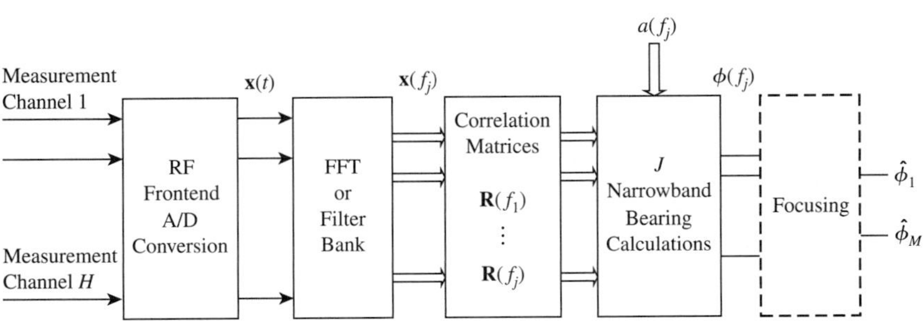

FIGURE 2.14 Functional principle of a wideband direction finder.

The correlation matrices and the frequency-dependent array manifold vectors form the input data for the bearing computation. The bearing spectrum obtained in this manner is then displayed with a graphical user interface and/or combined for emitters using cluster techniques (focusing). Figure 2.15 is an example of this type of results presentation.

Using today's technology in combination with the necessary dynamic range, the real-time bandwidth is limited to several times 10 MHz. If the center frequency of the broadband signal is unknown, the frequency ranges we must analyze typically become much larger—several times 100 MHz in the V/UHF range and several GHz in the SHF range. We can implement a system for interception and direction finding over this larger frequency range through

- Multiple filter bank receivers with shifted center frequencies
- Sequential shifting of the filter bank frequencies (frequency scan)

The first method ensures the maximum probability of intercept, but is costly to implement. The second method is significantly more economical to implement and is thus commonly used even though the probability of intercept decreases as the search range widens. The shifting of the filter bank frequency is typically carried out by changing the receiving center frequency of the RF converter (shifting of the LO frequency), whereas the A/D converter and the digital filter bank

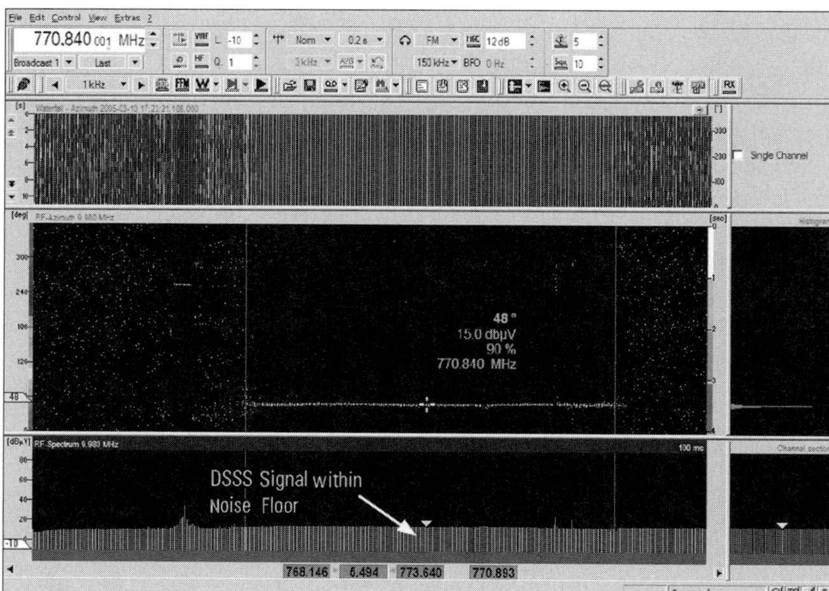

FIGURE 2.15 Graphical results presentation for a wideband direction finder. The spectrum of bearings allows detection, DF processing, and frequency estimation of the spread-spectrum signal buried in the noise of a single receiving channel (lower part of the display). (Screenshot by Rohde & Schwarz.)

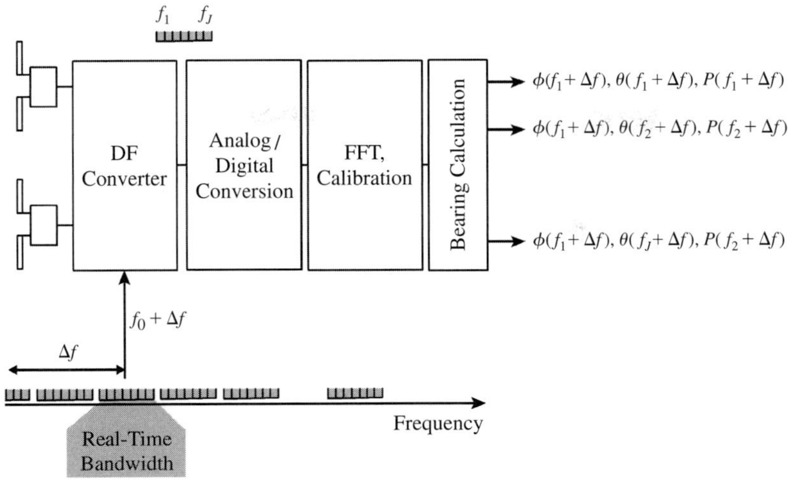

FIGURE 2.16 Basic operation of a wideband direction finder during scanning.

operate at fixed IF center frequencies. Figure 2.16 is a schematic representation of how this works.

2.8.2 Interception of Frequency-Hopping Signals

Because of its great significance in actual practice, we now examine the probability of intercept of a frequency-hopping signal using a scanning wideband direction finder.

We assume that the frequency-hopping transmitter has J_H channels with channel hopping in a random sequence. The direction finder's filter bank covers J_F channels. During scanning, the filter bank is shifted in L steps so that during one search cycle $J_{SC} = LJ_F$ channels are covered. We also assume that the frequency channels of the frequency-hopping transmitter and the search range of the direction finder overlap by J_J channels (Figure 2.17).

The length of a hop is equal to t_H. We assume that the direction finder's measurement time for a snapshot is equal to t_M, where $t_M > t_H$. Changing the direction finder's center frequency (shifting the filter bank) requires a duration t_S.

The probability P_1 of detecting a single FH burst with the direction finder is

$$P_1 = \frac{J_C J_F}{J_H J_{SC}} \tag{2.17}$$

The average number n_{av} of valid detection attempts (i.e., the measurement takes place within one FH burst) is obtained as follows (Höring, 1998):

$$n_{av} = \frac{t_H - t_M}{t_S + t_M} \tag{2.18}$$

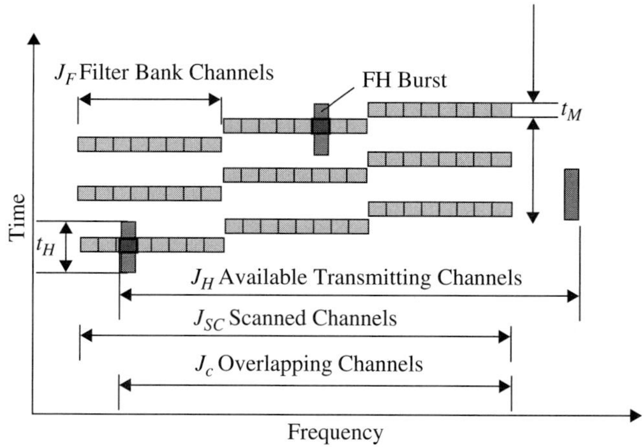

FIGURE 2.17 Interception of a frequency-hopping signal using a scanning filter bank receiver.

The detection of a burst for n_{av} valid detection attempts is then

$$\begin{aligned} &P_{1h} = P_1 n_{av} = \frac{J_C J_F}{J_H J_{SC}} \left(\frac{t_H - t_M}{t_S + t_M}\right) && t_M < t_H < \left(\frac{J_{SC}}{J_F}(t_M + t_S) + t_M\right) \\ &P_{1h} = \frac{J_C}{J_H} && t_H \geq \left(\frac{J_{SC}}{J_F}(t_M + t_S) + t_M\right) \\ &P_{1h} = 1 && J_{SC} \geq J_H, \quad t_H \geq \left(\frac{J_{SC}}{J_F}(t_M + t_S) + t_M\right) \end{aligned} \tag{2.19}$$

The probability of detection of k bursts with multiple measurements and transmission of a total of N bursts is computed based on the binomial distribution

$$P(z \geq k) = 1 - \sum_{i=0}^{k-1} \binom{N}{k} P_{1h} (1 - P_{1h})^{N-i} \tag{2.20}$$

where N is obtained using the duration t_{TX} of the FH transmission as

$$N = \frac{t_{TX}}{t_H} \tag{2.21}$$

Example 2.2

An FH transmitter is operating in the frequency range from 50 MHz to 80 MHz with 500 hops/s ($t_H = 2$ ms) with a frequency grid of 25 kHz ($J_H = 1200$). The direction finder has a real-time bandwidth of 10 MHz and scans the range from 20 MHz to 90 MHz using 25-kHz-wide filter bank channels (i.e., $J_F = 400$, $J_{SC} = 2800$). The measurement time for the direction finder is equal to $t_M = 0.32$ ms; the settling time t_S is equal to 1 ms.

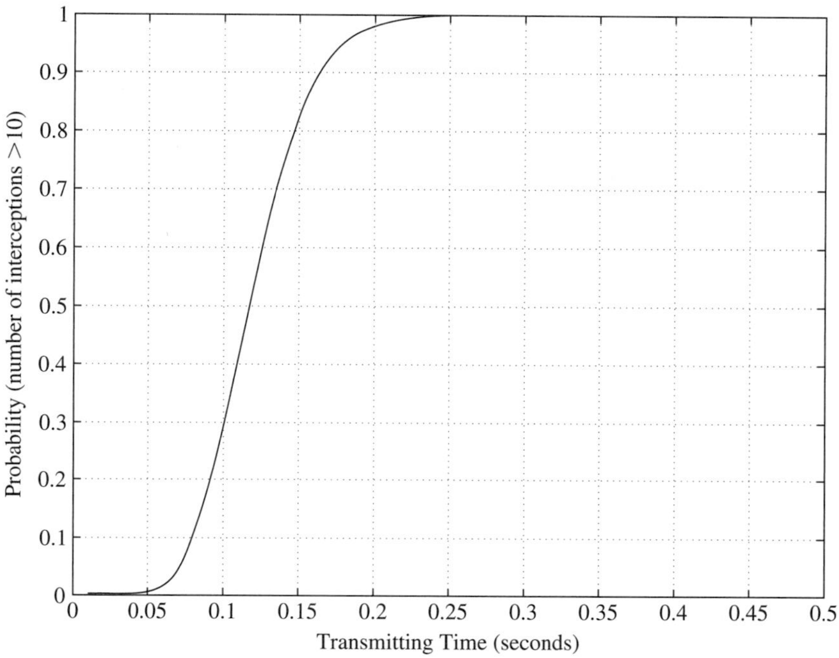

FIGURE 2.18 Probability of detection of and direction finding for at least 10 hops of a frequency-hopping transmitter as a function of transmit duration.

We want to know the probability of detecting and determining the direction of at least $k = 10$ FH bursts during a specified time interval. Figure 2.18 shows the resulting probability as a function of the transmit duration t_{TX}.

2.9 IMPLEMENTATION ASPECTS OF HIGH-RESOLUTION DIRECTION FINDING

In spite of the attractive features of high-resolution methods and the large amount of literature on the subject, most of the currently used direction-finding techniques are based on classical single-wave algorithms. Possible reasons will be discussed in the following section.

2.9.1 Conventional versus High-Resolution Direction Finding

If unwanted interfering waves are present in the frequency channel of interest, in addition to the desired wave, bearing errors will occur using conventional DF techniques based on single-wave models depending on the antenna configuration (geometry) used. There are two approaches for solving this problem:

- If the interfering wave component has less power than the desired wave component, we can minimize the bearing error through the use of suitable dimensioning of the direction finder (especially by selecting a sufficiently large antenna base).

- If the interfering wave component is larger than or equal to the desired wave component, we also have to determine the interfering waves in order to eliminate them in turn. When using conventional beamforming algorithms, this means that we must also evaluate the secondary maxima in the DF function. We reach the relevant limits when the ratio between the primary maximum and the secondary maxima of the directional characteristic becomes too small or when the angular difference between the wanted and interfering waves is less than the width of the main lobe. By optimizing the weighting factors, we can reduce the level of the secondary maxima, but at the expense of simultaneous widening of the primary maximum.

The objective of high-resolution (HR) direction-finding techniques is to circumvent this disadvantage. Of course, in practical applications we must keep in mind that even if we know the quantity and direction of the waves, we still do not have definitive information about the direction of the emitter of interest.

If the interfering waves are correlated with the desired wave (i.e., if there is multipath propagation), emitter direction must be estimated based on the distribution of the directions and amplitudes of the waves we are analyzing. In many cases, this can be handled more efficiently using beamforming or correlation techniques since the estimation of the emitter direction from the coherent secondary waves is implicitly contained in the formation of the bearings here (assuming we have a sufficiently large antenna aperture).

If we are dealing with uncorrelated signals from different sources, unambiguous assignment is possible only if all of the signals except the one of interest can be suppressed through suitable shaping of the antenna's directional characteristic (signal copy filter, null steering, adaptive antenna), so that we can, for example, use modulation parameters to make the assignment.

With the current state of the art, it is obvious that HR techniques are used as an *extension* instead of a replacement for single-wave techniques. This is due on the one hand to the fact that when exploiting the spectral and timing structure of the signals, separation and direction finding with single-wave algorithms are successful in the majority of cases. On the other hand, radio scenarios are highly transient so that general usage of a certain HR technique fails because of inconsistent estimation of the number of involved waves. Accordingly, HR techniques are used more for targeted analysis of overlapping signals or for investigation of narrow sections of the receive spectrum and not so much with a direction finder's search mode. Figure 2.19 shows an example.

2.9.2 Practical Limitations of High-Resolution Methods

When implementing HR techniques, it is essential to minimize

- The deviation of the actual array manifold vectors from those used to estimate the direction (modeling errors)
- The measurement errors for the array signals

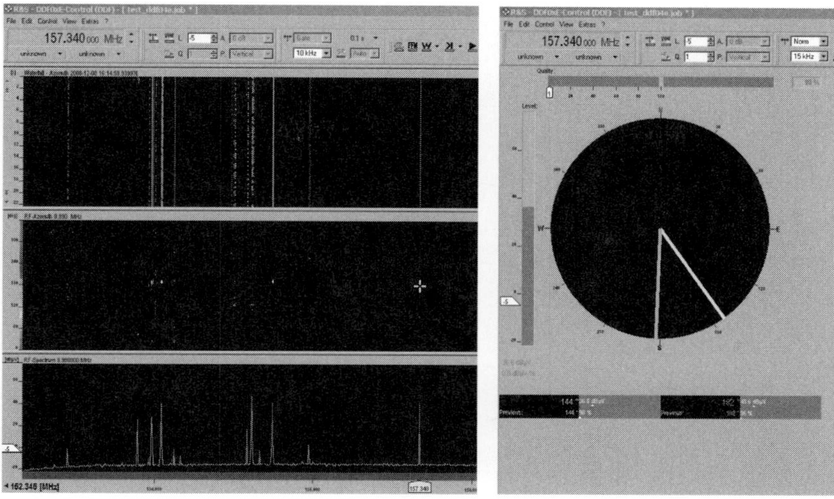

(a) (b)

FIGURE 2.19 Bearing spectrum and evaluation of a spectral section with high-resolution direction finding (MUSIC applied to a 9-element UCA 3 m in diameter). (a) Display obtained by a single-wave beamforming algorithm showing a result of 179°. (b) HR evaluation showing the hidden signal impinging at 144° and the correct value of 182° of the stronger signal. (Screenshot by Rohde & Schwarz.)

These errors have similar consequences, yet they have entirely different causes. Measurement errors for the array signals can be kept relatively low by carefully designing the measurement paths and ensuring good performance for internal calibration. Typical values we can attain are around 0.3-dB RMS amplitude error and 0.5° RMS phase error for frequencies up to about 3 GHz. At higher frequencies, we can expect to see an approximately linear increase in these errors, which in comparison with modeling errors, play a secondary role.

Modeling errors (see next section) represent the greatest challenge in implementing HR techniques (Brandwood, 1994; Friedlander, 1990). For example, if we consider the limit of MUSIC's resolution capability with a circular array antenna having diameter D, the result, in accordance with Brandwood (1994), is approximately as follows:

$$\Delta\phi \approx \frac{4.757\lambda\sqrt{\sigma_P}}{\pi D} \tag{2.22}$$

$\Delta\phi$ is the angular difference between two signals for the case in which the two maxima are no longer formed in the DF function; σ_P is the standard deviation of the phase errors in radians. If we compare the angular difference $\Delta\phi$ with the null-width BW_{UCA} of a conventionally shaped directional characteristic, we obtain the following, also in accordance with Brandwood (1994):

$$\frac{\Delta\phi}{BW_{UCA}} \approx 1.98\sqrt{\sigma_P} \tag{2.23}$$

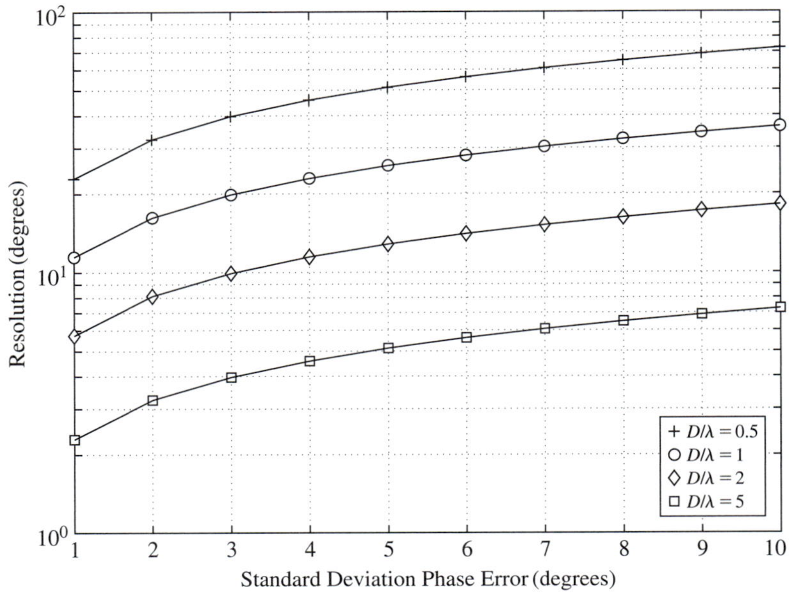

FIGURE 2.20 Limit of MUSIC's resolution capability due to phase errors for different diameters of a UCA.

A standard deviation of the modeling error of, say, 5° (0.0873 radians) and an antenna size $D = \lambda$ limits the ability to distinguish two signals to $\Delta\phi = 25.6°$. This corresponds to about 60% of the null width of the lobe of a conventionally shaped antenna pattern.

Smaller phase errors and the associated increase in possible resolution are difficult to obtain with certainty, particularly over longer time intervals. Figure 2.20 shows the influence of phase errors and antenna size on the resolution capability according to Equation (2.22).

2.10 ERROR SOURCES

Direction finding accuracy is influenced by a number of factors:

- Wave propagation (typically disrupted by obstacles such as mountains or buildings).
- Signals radiated by emitters are modulated or limited in time and their carrier frequency is often unknown. Also, their polarization is not clearly defined.
- The received field has additional superimposed noise and co-channel interfering signals.
- Noise caused by components of the DF system.
- Measurement errors.
- Modeling errors.

2.10.1 Multipath Wave Propagation

As already mentioned, the simple case of a plane wave seldom occurs in practice. In a real-world environment, additional waves typically must be taken into account. These are generated by other emitters in the same frequency channel (incoherent interference) or arise as secondary waves due to reflection, refraction, or diffraction (coherent in-channel interference). In general, a large number of waves are involved. Figure 2.21 shows, for example, the azimuth distribution of the waves generated by a mobile transmitter in a built-up area. The direct wave component with amplitude 1 arrives from an angle of 90°.

Figure 2.22 shows the resultant wave front in the form of a contour plot for phase and amplitude. If most of the waves arrive from the direction of the emitter, the bearing error can be sufficiently reduced by increasing the aperture of the antenna system. This effect is illustrated in Figure 2.23 for an interferometer direction finder.

2.10.2 Polarization Errors

The antennas used by the emitters we are tracking will usually have a distinct polarization (often vertical), but it is influenced by the antenna support itself. Examples include antennas mounted on vehicles, aircraft, and even man-pack radios. In the case of direction finding for interference sources, there is no defined polarization since it depends on the spatial position of the conductors producing the emission. During transmission over the propagation path, the waves are

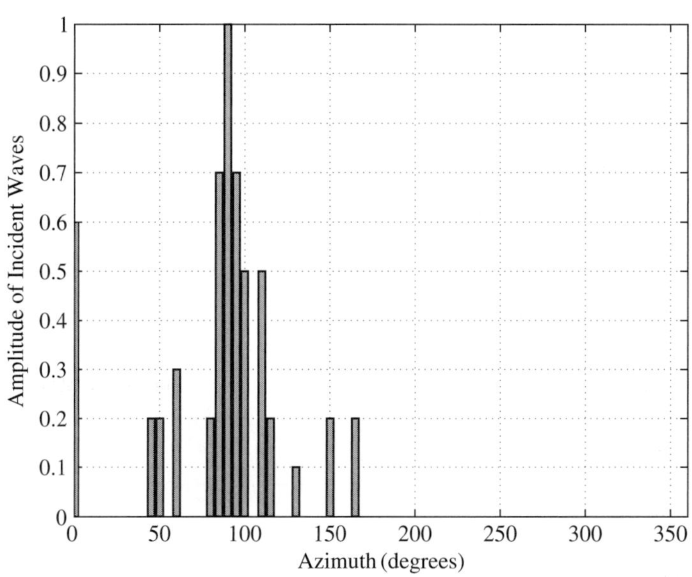

FIGURE 2.21 Distribution of wave amplitudes versus incident angle as produced by an emitter located in a suburban area.

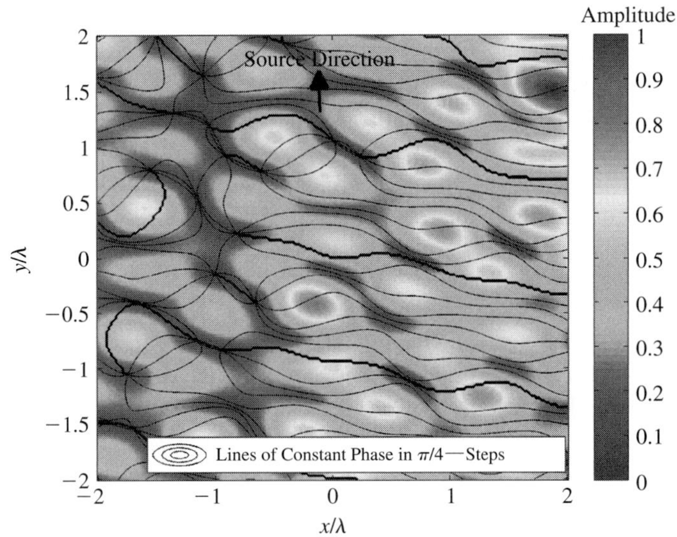

FIGURE 2.22 Contour plot of electric field strength resulting from superposition of the multicomponent wavefield from Figure 2.21.

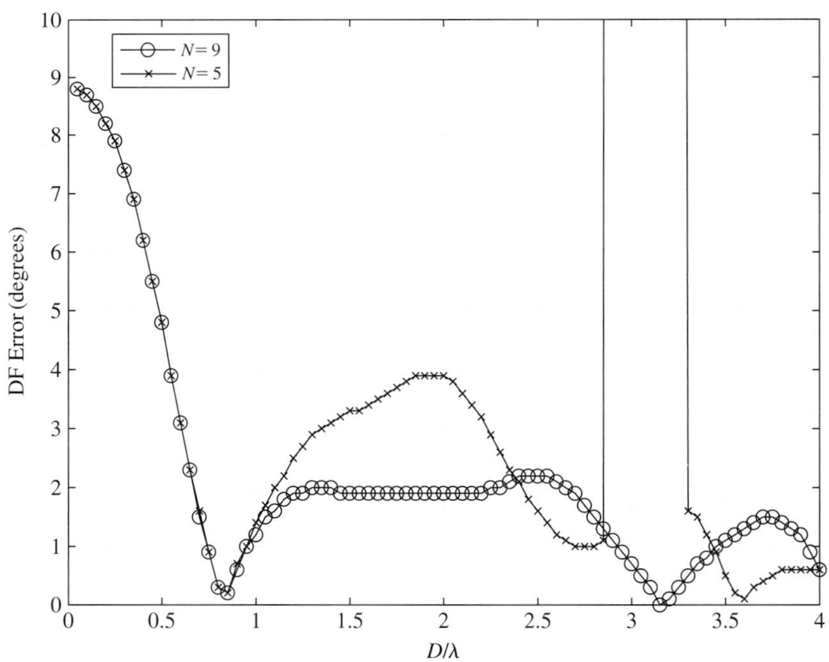

FIGURE 2.23 Bearing error of a 9-element and a 5-element interferometer in a multiwavefield versus antenna aperture. In the small-aperture region the bearing errors are independent of the number of elements, but the 5-element array fails in the large-aperture region ($D/\lambda > 1$).

subject to diffraction and reflection, which results in additional changes in polarization.

In the DF antenna, the antenna elements as well as the support structure are excited with field components having nominal polarization direction (e.g., vertical or horizontal) as well as those that are cross-polarized. If these cross-polarized field components are not sufficiently rejected, bearing errors will arise depending on the components' relative magnitude and phase.

2.10.3 Internal Noise

We discussed the influence of internal (instrument) noise in the preceding sections. This noise is caused by random-charge carrier movements in electronic modules, such as amplifiers, mixers, and A/D converters, and in resistors and other lossy components. From a systems perspective, it is useful to imagine all of the noise components in a processing channel arising in the antenna element.

This equivalent noise source has the available power

$$P_{n0} = F_S k T_0 B \tag{2.24}$$

and produces at the output of the processing channel the same signal-to-noise ratio with measurement bandwidth B as the noise sources' actual distribution. F_S is the system noise factor for a processing channel, k is the Boltzmann constant ($1.38 \times 10^{-23}\, Ws/K$), and $T_0 = 290$ Kelvin.

Since we can assume that the noise sources in the processing channels are uncorrelated, for the same transfer power gain G_T and the same noise in the processing channels, we obtain the following for the covariance matrix of the internal noise:

$$\mathbf{R}_n = \sigma_n^2 \mathbf{I} = F_S k T_0 B G_T \,\mathbf{I} \tag{2.25}$$

where **I** is the identity matrix.

2.10.4 External Noise and Interference

Particularly at frequencies below about 100 MHz, external noise plays a role that cannot be ignored. It is caused by thunderstorms (atmospheric noise), machinery (human-made noise), and processes in the universe (galactic noise). Figure 2.24 illustrates the variation in the noise factor $F_{ex}(dB) = 10\lg(F_{ex})$ for external noise versus frequency in accordance with ITU (2007). The noise factor F_{ex} corresponds to the ratio of the available noise power P_{n0ex} at the output of a short antenna to thermal noise power at the temperature $T_0 = 290$ Kelvin, where both power levels are measured with bandwidth B:

$$F_{ex} = \frac{P_{n0ex}}{k T_0 B} \tag{2.26}$$

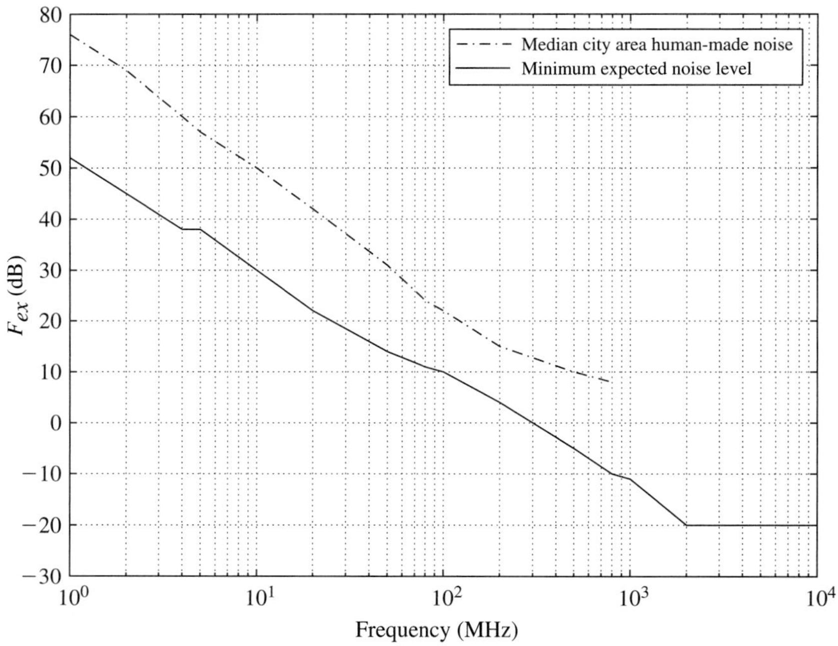

FIGURE 2.24 External noise in accordance with Recommendation ITU-R P.372-9. (*Source:* ITU, 2007, with permission.)

The antenna array's elements can maximally extract the available external noise power $P_{n0ex} = F_{ex}kT_0B$ from the received wavefield. The actually measured external noise power P_{nRex} is a function of the transfer power gain G_T of the processing channel:

$$P_{nex} = F_{ex}G_TkT_0B \tag{2.27}$$

While we can assume that the noise generated in the processing channels is uncorrelated, the external noise, which is coupled in, is partially correlated. The correlation between antenna elements is a function of the distance related to the wavelength and the spatial intensity distribution of the external noise (Hudson, 1981). For example, for the covariance between elements i and j for isotropic distribution of the noise sources, we have

$$R_{nex\,i,j} = P_{nex}\frac{\sin\left(k_0 d_{i,j}\right)}{k_0 d_{i,j}} \tag{2.28}$$

Discrete external interference sources occur in the form of in-band and out-of-band interfering signals and are represented as correlated sources in the processing channels. In the case of strong out-of-band signals, interference is often caused by inter-modulation and reciprocal mixing.

2.10.5 Measurement Errors

Different gain and phase in the receive sections cause bearing errors that become greater as the antenna aperture referred to the wavelength becomes smaller. As already mentioned, the receive sections of most multichannel direction finders are calibrated for amplitude and phase balance with the aid of a test generator prior to DF operation. The transmission parameters in each section are measured in terms of their magnitude and phase and the level and phase differences are stored. During the DF process, the measured values are corrected using the stored values before the bearing is calculated. The statistical measurement uncertainty that arises as a result of the A/D converter's limited resolution is normally added to the internal noise sources.

2.10.6 Modeling Errors

Modeling errors arise primarily because of interactions in the near field of the antenna array, that is,

- Mutual coupling between antenna elements
- Coupling between antenna elements and mechanical support elements, including cabling
- Reflection and diffraction caused by antenna support structures (e.g., mast, housing)
- Reflection and diffraction caused by objects in the vicinity of the antenna array (additional antennas, lamps, power lines, cars, etc.)

It is relatively easy to take into account the coupling between the antenna elements either by measurement or theoretical modeling (Friedlander and Weiss, 1991). However, it is significantly harder to take into account the influences of the support construction and the objects situated in the immediate vicinity of the antenna since they are highly dependent on the wave angle of incidence and frequency. Moreover, they are themselves subject to influences that are not constant over time and cannot be measured adequately because of their complexity (temperature, humidity, corrosion of contacts, etc.).

2.11 TEST AND MEASUREMENT PROCEDURES

By examining the test procedures used to measure the relevant performance parameters in direction finders, we can build the foundation needed to measure and understand the significance of these parameters in real-world applications. Based on the divergent requirements for the test environment, we categorize our test techniques as outdoor and indoor.

2.11.1 Outdoor Tests

With outdoor tests, the focus is on measuring the parameters that are closely related to the properties of the antenna system.

Test Range

The test range basically consists of a rotary stand where we can mount the DF antenna along with the required number of test transmitters with the transmit antenna (Figure 2.25).

The size of the test range is oriented primarily toward the required distance d between the DF antenna and the test transmitter. This is based on the requirement that the electromagnetic field in the area of the DF antenna must approximately contain only components present in the far field of the transmit antenna ($d \rightarrow \infty$). As we know, the surfaces with equal phase are curved in the near field of a transmit antenna, and additional field components exist as well. Accordingly, we must take both of these circumstances into account when selecting this distance.

The curvature of the phase planes is independent of the design of the transmit antenna if it is small in comparison to the distance being considered. In this case, we can consider the transmit antenna as a point radiator approximation. The influence of the curvature is then dependent only on the extension A of the DF antenna and the wavelength λ, since the phase deviation δ with respect to a planar wave increases toward the edge of the DF antenna. Figure 2.26 illustrates the underlying geometry.

For $b \ll A$ and $b \ll d$ we have

$$b \cong \frac{A^2}{8d} \tag{2.29}$$

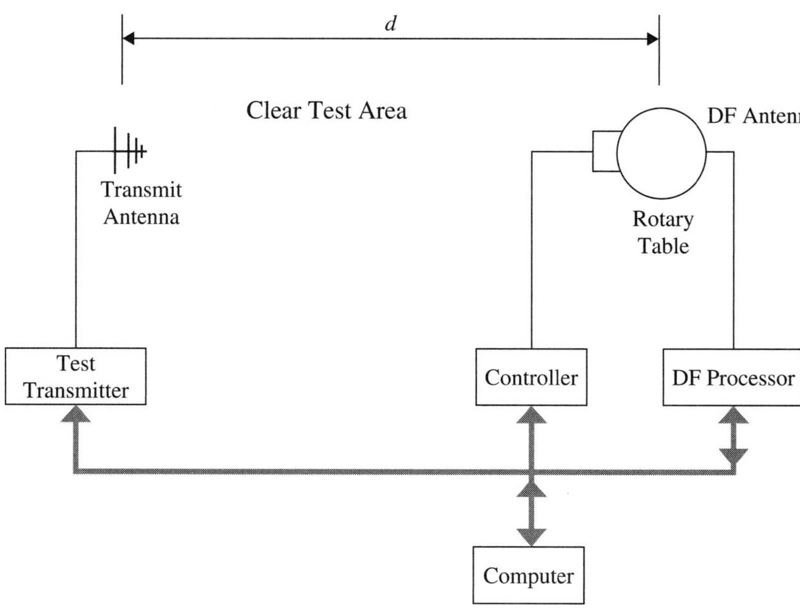

FIGURE 2.25 Test range for direction-finder measurements.

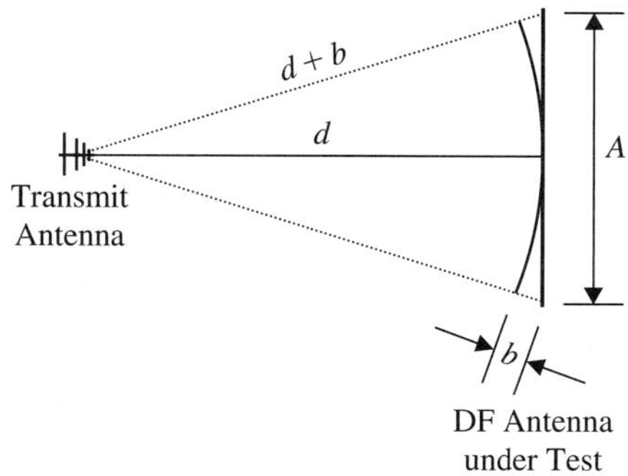

FIGURE 2.26 Geometry underlying the distance requirement.

The phase deviation at the edge of the DF antenna aperture (width A) is thus equal to

$$\delta \cong \frac{\pi A^2}{4d\lambda} \tag{2.30}$$

and the minimum distance for a specified phase deviation is

$$d \geq \frac{\pi A^2}{4\delta_{tol}\lambda} \tag{2.31}$$

A point of reference for the tolerable phase deviation δ_{tol} is provided by the tolerances of the antenna array's edge elements at the shortest operating wavelength. For example, if the tolerances are specified as 10° at a frequency of 1 GHz, 2° of additional phase deviation due to the test distance have practically no influence on the measurement result. For a 1-m DF antenna, a distance of $d \geq 75$ m is thus sufficient.

For wavelengths that are large in comparison to the aperture, the curvature of the wavefronts plays a secondary role. Here, we must watch out for the near-field components of the field, which represent reactive power and thus prevent the phase of the wave from varying in a linear fashion versus the distance. If we consider a simple dipole as our transmit antenna, this range will extend to about $d/\lambda \approx 1$ (Balanis, 1982).

Special care is needed to avoid intolerable interference with wave propagation (due to buildings, cables, vehicles, etc.). Clearing only obstacles from the first Fresnel zone are definitely not sufficient. The following analysis of a wave reflected by an obstacle by 90° with respect to the receive direction of the primary wave provides guidelines for the extent to which obstacles must be eliminated. In a system with a small aperture (antenna size $A \ll \lambda$), the reflected wave must not exceed an amplitude of approximately 2% of the direct wave so as not to cause more than 1° of bearing error in this scenario. Assuming equal intensities of the impinging field at the location of the DF antenna and the obstacle, the ratio between the amplitude $|E_r|$ of the reflected wave and the amplitude $|E_d|$

of the direct wave is determined by the distance r between the antenna and the obstacle and the obstacle's radar cross-section, RCS (Balanis, 1982):

$$\left|\frac{E_r}{E_d}\right|^2 = \frac{\text{RCS}}{4\pi r^2} \tag{2.32}$$

With a thin obstacle (wire, mast, etc.) of length and diameter $\ll 1$, we obtain a scattering cross-section, RCS, at its first resonance ($l = \lambda/2$) of

$$\text{RCS}_{\lambda/2} \cong \frac{\lambda^2 D_{\lambda/2}^2}{\pi} = 0.86\,\lambda^2 \tag{2.33}$$

$D_{\lambda/2}$ is the directivity of the scatterer in the case of a thin wire at half-wave resonance $D_{\lambda/2} = 1.64$. Considering Equations (2.32) and (2.33) and $|E_r/E_d|_{\max} = 0.02$, we must maintain a distance of

$$r_{\min} = \frac{0.26\lambda}{|E_r/E_d|_{\max}} \cong 13\lambda$$

to obtain not more than 1° of bearing error in small aperture systems.

DF Accuracy

The DF antenna is arranged in fixed angular positions, and the bearing is recorded. We must set the transmit power high enough so that the instrument noise does not influence the measurement. Now we can measure at different frequencies and with the smallest possible angular steps. For routine tests on antennas, we typically measure using 10° steps. The individual bearing errors $\Delta_{\phi i}$ are obtained from the difference between the bearing $\hat{\phi}_i$, which we read off, and the setting of the rotary table ϕ_{POSi} relative to the test transmitter.

$$\Delta_{\phi i} = \phi_{POSi} - \hat{\phi}_i \tag{2.34}$$

The total error is typically specified in the form of an RMS error. Of course, we must specify the sample type involved (e.g., azimuth only at constant frequency or azimuth and frequency in a single sample).

Sensitivity

Before beginning the actual measuring, we arrange a field strength–measuring antenna in place of the DF antenna and measure the transmitter power P_{T0} and field strength E_0 versus frequency. For determining the sensitivity of the direction finder, we can reduce the transmitter power to a value P_T, which results in a specified bearing fluctuation. Based on the ratio of transmit power to field strength E_0, we then obtain the sensitivity value as follows:

$$E_{sens} = E_0 \frac{P_T}{P_{T0}} \tag{2.35}$$

Influence of Polarization

Polarization measurement is analogous to bearing accuracy measurement. First, we determine the bearing for defined azimuth and elevation angles with the nominal polarization of the transmit antenna. For linear polarization, we then arrange the transmit antenna at an angle corresponding to the desired polarization angle. The polarization error is thus equal to the currently displayed bearing minus the bearing determined with nominal polarization. It is good practice to check the polarization at the DF antenna location as a function of the transmit antenna's offset position, and to prepare a compensation table if necessary.

Immunity to Multipath Propagation

For a defined generation of a secondary wave, we set up a second transmit antenna. Both antennas are connected to the test transmitter via a power splitter. We can vary the amplitude of the signal for the antenna, which simulates the second propagation path using a variable attenuator (Figure 2.27).

To perform tests for different phase angles of the secondary signal, we use a variable phase shifter (e.g., line sections of varying lengths) in addition to the attenuator. In this measurement, only the primary antenna is involved in direction

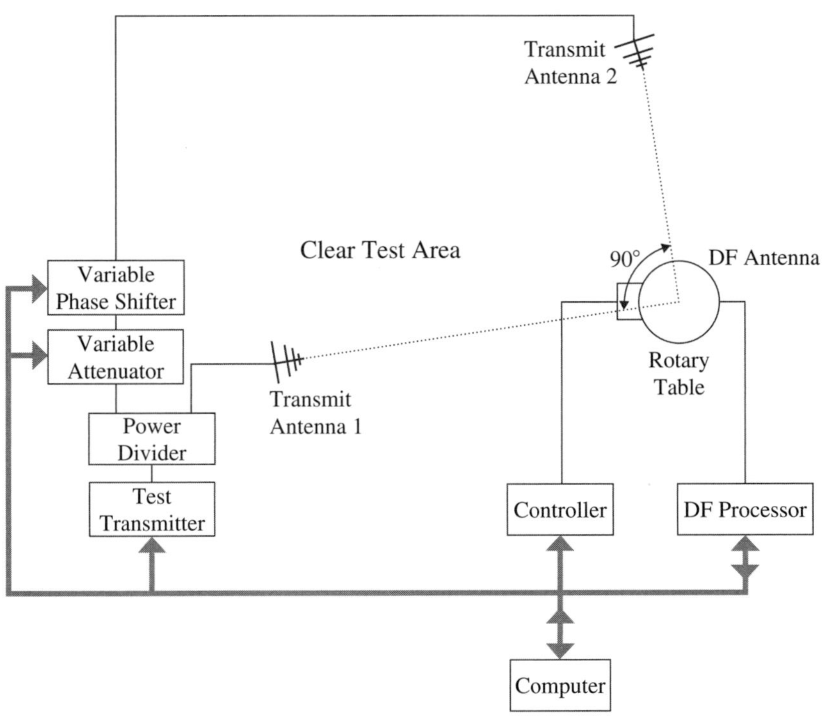

FIGURE 2.27 Setup for measuring immunity to multipath propagation.

finding at first. Then the secondary path is connected with a certain amplitude ratio and the phase of the signal is varied until the maximum bearing error is attained.

Angular Resolution

The test setup is the same as that used for simulation of multipath propagation. For testing with uncorrelated signals, the second antenna is supplied by a separate test generator. The bearing accuracy as well as certain other parameters can be measured as a function of the angular distance for the two transmit antennas.

2.11.2 Indoor Tests

Measurements not directly affected by the antenna properties are easier to make indoors. This is true in particular for measurements in which immunity to large interfering signals must be verified.

Antenna Simulators

If we also want to test the DF system indoors in DF mode, we must substitute an antenna simulator for the DF antenna array. The relative positions of the antenna elements with respect to the wave angle of incidence are replaced by different delay lines (Figure 2.28). The decoupling networks, calibration signal feeds, interface modules, and so on, from the original antenna are used.

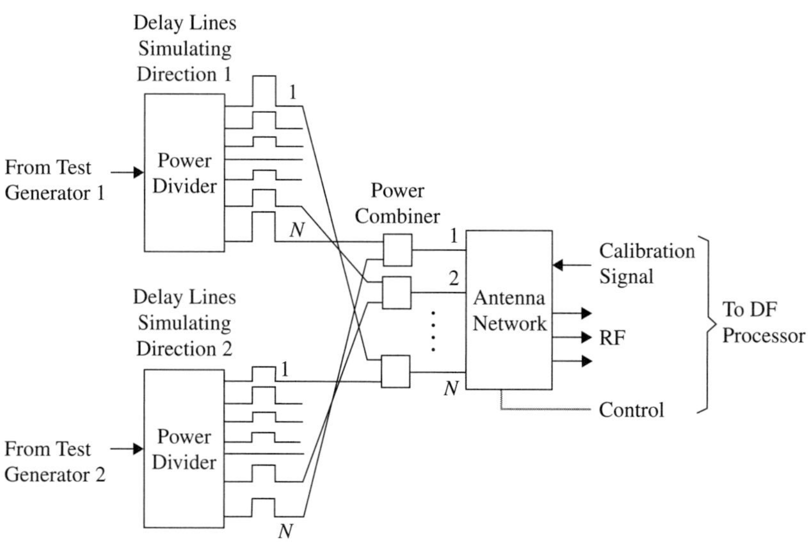

FIGURE 2.28 Antenna simulator for two different angles of wave incidence.

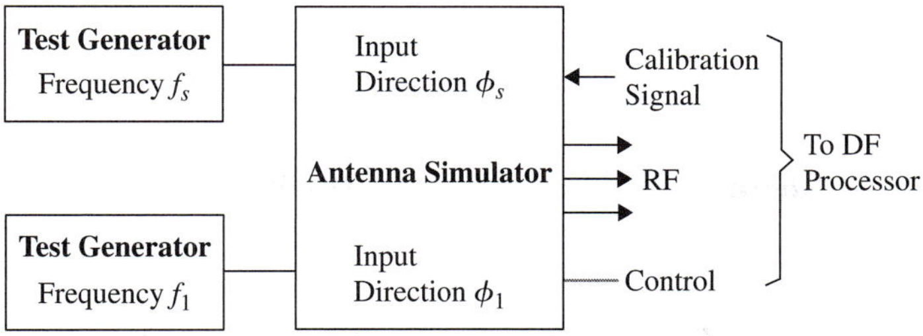

FIGURE 2.29 Setup for testing immunity against strong interfering signals.

Immunity to Strong Interfering Signals: Blocking, Reciprocal Mixing

Figure 2.29 shows the relevant test setup. A generator at the center frequency f_S simulates the desired signal with the wave angle of incidence ϕ_S. The second generator at the center frequency f_I provides the interfering signal with the wave angle of incidence ϕ_I. In the measurement, the result data for the desired signal is ascertained first (bearing, amplitude, signal-to-noise ratio, etc.). Next the interfering signal is connected and its power is increased until the maximum allowed influence on the result data is attained.

Inter-Modulation

The test setup from Figure 2.29 is also used here. Instead of the desired signal at f_S, however, a second interfering signal at f_{I2} is generated. Evaluation is carried out at frequency $f_S = |f_{I1} \pm f_{I2}|$ to measure the second-order inter-modulation, and at $f_S = |2f_{I1} - f_{I2}|$ or $f_S = |f_{I1} - 2f_{I2}|$ to measure the third-order inter-modulation.

Dynamic Range

The *lower limit* of the dynamic range is determined by the minimum power level of a desired signal for which certain performance data is fulfilled (e.g., a certain bearing fluctuation or a certain signal-to-noise ratio in the spectrum display). The *upper limit* is determined by the power of the signals that degrade the performance data of the desired signal by a defined amount. This measurement is performed analogously to the measurements previously described.

REFERENCES

Balanis, C.A., 1982. Antenna Theory. John Wiley & Sons.

Brandwood, D.H., 1994. Effect of phase errors on resolution of MUSIC. Signal Processing VII: Theories and Applications. European Association for Signal Processing.

Bussgang, J.J., Ehrman, L., Graham, J.W., 1974. Analysis of nonlinear systems with multiple inputs. Proc. IEEE 62 (8), 1088–1119.

Cheng, J., Ohira, T., 2006. ESPAR antenna signal processing for DOA estimation. In: Chandran, S. (Ed.), Advances in Direction-of-Arrival Estimation, pp. 395–417. Artech House.

Demmel, F., Unselt, U., 2002. New digital direction finder 0.5 MHz to 3000 MHz. News from Rohde & Schwarz 174, 47–49.

Friedlander, B., 1990. A sensitivity analysis of the MUSIC algorithm. IEEE Trans. Acoust. Speech Signal Process. 38 (10), 1740–1751.

Friedlander, B., Weiss, A.J., 1991. Direction finding in the presence of mutual coupling. IEEE Trans. Antennas Propag. 39 (3), 273–284.

Höring, C.H., 1998. Wahrscheinlichkeit der erfassung von frequenzsprungsignalen (Probability of intercept for frequency hopping signals by use of a scanning receiver). Frequenz 52 (7, 8), 134–140.

Hudson, J.E., 1981. Adaptive Array Principles. Peter Peregrinus.

ITU, 2007. Recommendation ITU-R P.372-9, Radio Noise, Publication of International Telecom Union.

Krolik, J., 1991. Focused wide-band array processing for spatial spectral estimation. In: Haykin, S. (Ed.), Advances in Spectrum Analysis and Array Processing, vol. 2, pp. 221–261. Prentice-Hall.

Manikas, A., 2004. Differential Geometry in Array Processing. Imperial College Press.

Pace, P.E., 2000. Advanced Techniques for Digital Receivers. Artech House.

Sheinvald, J., Wax, M., 1999. Direction finding with fewer receivers via time-varying preprocessing. IEEE Trans. Signal Process. 47 (1), 2–9.

Van Trees, H.L., 2002. Optimum Array Processing. John Wiley & Sons.

Wax, M., 1995. Model-based processing in sensor arrays. In: Haykin, S. (Ed.), Advances in Spectrum Analysis and Array Processing, vol. 3, pp. 1–47. Prentice-Hall.

Wax, M., Ziskind, I., 1989. On unique localization of multiple sources in passive sonar arrays. IEEE Trans. Acoust. Speech Signal Process. 37, 996–1000.

Worms, J.G., 2002. Direction finding with array sensors in broadband applications. Frequenz 56, 220–228.

Chapter 3

Calibration in Array Processing

Mats Viberg, Maria Lanne, Astrid Lundgren

3.1 INTRODUCTION

High-resolution direction-of-arrival (DOA) estimation has been an active area of research since the late 1970s. Its applications are wide ranging, including passive listening arrays, radar and sonar, and spatial (or space–time) characterization of wireless communication channels. Conventional beamforming techniques for direction estimation are limited by the aperture (or physical size) of the array. In contrast, parametric methods promise an unlimited resolution under ideal conditions (i.e., no noise or modeling errors). These methods take advantage of a precise mathematical model of the received array data, due for example to incoming plane waves. (See Krim and Viberg, 1996, for an overview of DOA estimation methods; and Trees, 2002, for details regarding array signal processing.) In practice, resolution and estimation accuracy is limited by noise as well as errors in the assumed data model. The focus of this chapter is thus on modeling errors and, in particular, calibration techniques to mitigate them.

The most natural and common approach is to measure the spatial response of the array in an anechoic chamber. These calibration measurements are then used to update the data model, either in the form of explicit unknown parameters or nonparametrically. Under favorable conditions it is also possible to estimate the response model together with the unknown directions—so-called auto-calibration. The purpose of this chapter is to give an overview of existing techniques and discuss their respective pros and cons. Some more recent developments and interpretations are also included.

It should be mentioned that calibration of a real-world antenna (or other sensor) array also involves hardware adjustments to compensate for temperature drift and the like. The methods considered here can be classified as *software calibration*, where errors are handled by adjusting the assumed data model rather

Note: This work was supported in part by the Swedish Foundation for Strategic Research within the Strategic Research Centre Charmant.

Copyright © 2009 by Academic Press, Inc. All rights of reproduction in any form reserved.

than correcting for it. An example of a hardware-based calibration procedure is given in Sakaguchi et al. (2002).

By array calibration we refer to the procedure of correcting the assumed array response model for errors due to several imperfections. These can include gain and phase errors in the receivers, imbalance between the I and Q channels, an incorrectly specified mutual coupling model, and uncertain sensor locations. The correction uses measured data. The least demanding approach from a data collection point of view is auto-calibration, where array response parameters are estimated jointly with the DOAs using the same data. Clearly, this requires a precise specification of the error model, which in turn cannot be "too complicated." Fundamental limitations of joint estimation of sensor positions and DOAs are presented in Levi and Messer (1990) and Rockah and Schultheiss (1987). Results for gain and phase estimation can be found in Astély et al. (1999), Paulraj and Kailath (1985), and Weiss and Friedlander (1990). A useful overview of auto-calibration is given in Li et al. (2003).

When calibration data is available, using sources at known positions, auto-calibration methods can of course be directly applied with fixed DOA parameters. Limitations and methods are discussed in various articles (e.g., Ng and Nehorai, 1995; Ng and See, 1996). Another interesting approach that does not require any specification of the array response model is the array interpolation, or manifold separation, approach (Doron and Doron, 1994a,b,c). The idea is to use a Fourier representation of the array response, which has the advantage of allowing simpler DOA estimation approaches developed especially for a uniform linear array (ULA) (e.g., Barabell, 1983; Roy and Kailath, 1989; Stoica and Sharman, 1990). A more recent contribution is Belloni et al. (2007), where the array interpolation technique is successfully applied in combination with the Root-MUSIC algorithm (Barabell, 1983). A different approach applicable also when the number of calibration DOAs is sparse is interpolation using a nominal model (Lanne et al., 2006; Ottersten et al., 1992b). A simple linear interpolation of a correction factor to the nominal model is often sufficient, although a smoothed interpolation can be beneficial, especially if the "true" correction is smooth or the calibration data is unreliable.

A natural question in this context is how errors in the calibration data propagate to errors in the DOA estimates, and how this effect compares to errors due to noise in finite samples. A theoretical study of accuracy requirements is provided in Porat and Friedlander (1997) for the case of parametric calibration. A general sensitivity analysis of DOA estimation algorithms is also a useful tool (e.g., Swindlehurst and Kailath, 1992, 1993; Viberg and Swindlehurst, 1994a), where the combined effects of modeling errors and finite samples are considered. Experimental results for antenna arrays using calibration measurements are presented in Dandekar et al. (2000), Gupta et al. (2003), Pettersson and Grahn (2003), and Pierre and Kaveh (1995), although these are mostly based on global models. Here we present a performance comparison between the various approaches based on simulated data using more general array error models.

Section 3.2 describes the data model under ideal as well as more realistic conditions. Next, selected methods for DOA estimation are reviewed in Section 3.3. Section 3.4 introduces the auto-calibration approach, and Section 3.5 discusses methods that rely on calibration data. Array interpolation techniques with and without availability of a nominal model are discussed in Section 3.6. The different calibration approaches are then compared in terms of applicability and performance in Section 3.7. The chapter concludes with Section 3.8.

3.2 DATA AND ERROR MODELS

This section presents the data models and the underlying assumptions of the chapter. We start with the ideal case and then move on to a more realistic model for an electromagnetic antenna array. Various sources of error are also pointed out, and their modeling is discussed.

3.2.1 Ideal Data Model

The underlying problem in this chapter is that of estimating the DOAs of incoming signals from distant sources (Figure 3.1). The different sensors will capture the same signals but with time delays that depend on the DOAs. Throughout this chapter, we will assume that the transmitted signals are narrowband with the same center frequency ω_c. After the front-end electronics (usually including down-converting to an intermediate frequency), the signal is digitized (analog-to-digital conversion, ADC); the direct down-conversion (DDC) block then takes the signal to baseband and extracts the real (in-phase) and imaginary (quadrature-phase) components. The resulting signal at sensor m with sample index n is denoted $x_m(n)$, for $m = 1, \ldots, M$. The sample index n corresponds to an arbitrary sampling instant t_n (in seconds), and all sensor elements are assumed to be sampled at the same time instant. A batch of N time samples is assumed available from each sensor. The resulting data is collected in the array output vector $\mathbf{x}(n) = [x_1(n), \ldots, x_M(n)]^T$, $n = 1, \ldots, N$. In the narrowband case, a time delay τ (seconds) results in a phase shift of $\omega_c \tau$ (radians) at the baseband, where ω_c is the center (or carrier) frequency. Thus, the noise-free array output from an ideal sensor array, due to a single signal source from the DOA θ, can be modeled by the linear relation

$$\mathbf{x}(n) = \mathbf{a}(\theta) s(n) \tag{3.1}$$

If we define the signal waveform $s(n)$ at some reference location (the array phase center), then the *array response vector* $\mathbf{a}(\theta)$ takes the form

$$\mathbf{a}(\theta) = [e^{j\omega_c \tau_1(\theta)}, \ldots, e^{j\omega_c \tau_M(\theta)}]^T \tag{3.2}$$

where $\tau_m(\theta)$ is the θ-dependent time delay of the signal at sensor m relative to the reference. Under the assumption of linearity, the super-position principle

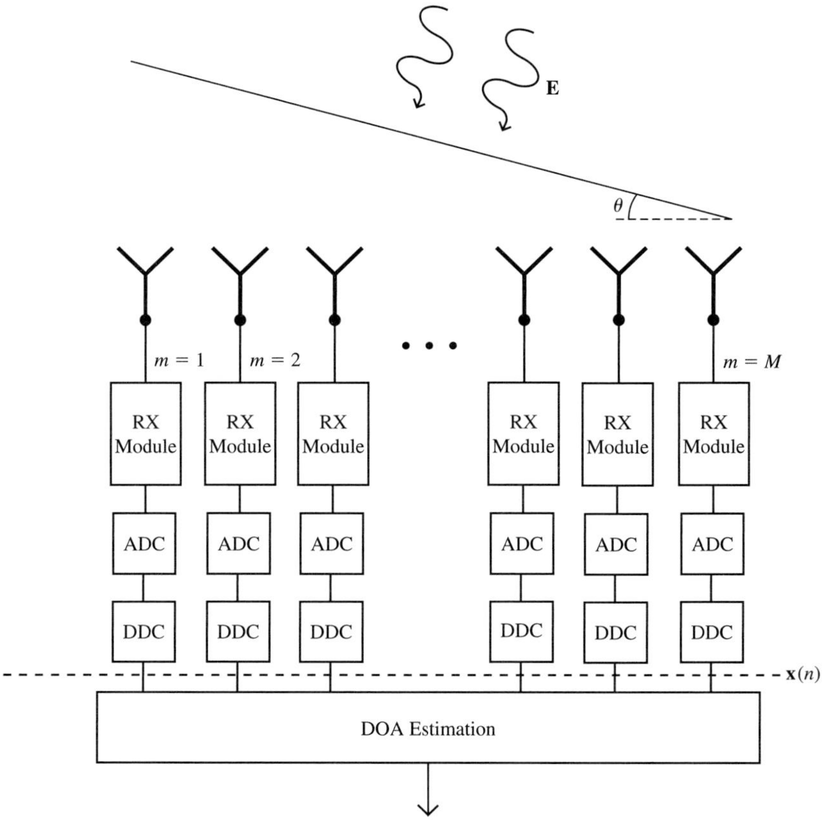

FIGURE 3.1 Plane waves from far-field sources arriving at an antenna array. The problem of interest is estimating the directions to the sources.

applies, and the sensor output due to Q far-field emitters can be written as

$$\mathbf{x}(n) = \sum_{q=1}^{Q} \mathbf{a}(\theta_q) s_q(n) + \mathbf{n}(n) \tag{3.3}$$

where θ_q are the DOAs of the signal sources, $s_q(n)$ are the corresponding signal waveforms, and $\mathbf{n}(n)$ represent an additive noise term. The noise represents components in the actual array output that cannot be well modeled by the simple relation (3.1)—for example, external noise sources that are not pointlike and thermal noise in the receivers. The model (3.3) can be put in compact matrix form by defining the array response matrix $\mathbf{A}(\boldsymbol{\theta}) = [\mathbf{a}(\theta_1), \ldots, \mathbf{a}(\theta_Q)]$, where $\boldsymbol{\theta} = [\theta_1, \ldots, \theta_Q]^T$ is the vector of DOA parameters, and the signal vector $\mathbf{s}(n) = [s_1(n), \ldots, s_Q(n)]^T$. The resulting model is

$$\mathbf{x}(n) = \mathbf{A}(\boldsymbol{\theta})\mathbf{s}(n) + \mathbf{n}(n), \quad n = 1, \ldots, N \tag{3.4}$$

or, more compactly,

$$\mathbf{X}=[\mathbf{x}(1),\ldots,\mathbf{x}(N)]=\mathbf{A}(\boldsymbol{\theta})\mathbf{S}+\mathbf{N} \tag{3.5}$$

Most methods use only second-order properties of the data. The array covariance matrix is defined as $\mathbf{R}=E[\mathbf{x}(n)\mathbf{x}^*(n)]$, where $E[\cdot]$ is the statistical expectation and $(\cdot)^*$ denotes the complex conjugate and transpose (the Hermitian operator). Assuming the signal and noise to be independent and the noise to be spatially white, $(E[\mathbf{n}(n)\mathbf{n}^*(n)]=\sigma^2\mathbf{I})$, the array covariance matrix takes the form

$$\mathbf{R}=\mathbf{A}(\boldsymbol{\theta})\mathbf{P}\mathbf{A}^*(\boldsymbol{\theta})+\sigma^2\mathbf{I} \tag{3.6}$$

where the signal waveforms are assumed stationary so that the signal covariance matrix $\mathbf{P}=E[\mathbf{s}(n)\mathbf{s}^*(n)]$ is independent of n. The array covariance matrix is estimated from the available data by the sample covariance matrix

$$\hat{\mathbf{R}}=\frac{1}{N}\sum_{n=1}^{N}\mathbf{x}(n)\mathbf{x}^*(n)=\frac{1}{N}\mathbf{X}\mathbf{X}^* \tag{3.7}$$

Since the chief interest here is modeling errors, we frequently assume that $N\rightarrow\infty$, so that the sample covariance matrix is a perfect estimate of $\mathbf{R}$. Note that this holds also in the case of quasi-stationary deterministic signals, where $\mathbf{P}$ then denotes the limiting sample covariance matrix of $\mathbf{s}(n)$.

The DOA estimation problem is now stated as follows: Given an observation of $\mathbf{X}$, modeled by Equation (3.5), determine the DOA parameters $\boldsymbol{\theta}$. An important partial problem is of course to determine the number of sources Q, which is the dimension of $\boldsymbol{\theta}$. Another problem of great interest, in particular in communication applications, is to estimate the signal waveforms in $\mathbf{S}$. However, the focus in this chapter is on DOA estimation. Methods for estimating Q are available (e.g., Wax and Kailath, 1985; Zhao et al., 1986), whereas Er and Ng (1994) and Ottersten et al. (1989) address the signal waveform estimation problem (see also the various publications on "smart antennas" for wireless communication (for example, Paulraj et al., 2003). A key issue in this context is that the functional form of the array response $\mathbf{a}(\theta)$, often termed the *array manifold*, must be accurately known. This is our main topic and various approaches to acquire this information are discussed. Perhaps the most natural start for an engineer is to revert to the underlying physics.

3.2.2 Real Antenna Array Modeling

Mutual coupling can have a significant influence on array performance in beamforming (Josefsson and Persson, 1999; Steyskal and Herd, 1990) and on DOA estimation if it is not correctly accounted for (Svantesson, 1998; Swindlehurst and Kailath, 1992). The strength of the mutual coupling is mostly determined by the element type (its radiation characteristics), the distance between the elements, and how the elements are oriented relative to one another.

Mutual coupling is commonly represented by its S-parameters, which are the elements of the scattering matrix **S**. Scattering matrix formulation is a well-known concept from circuit theory, which relates the output waves to the input waves of an N-port junction (Collin, 1992, chapter 4.7). The concept is easily applied to array antennas by assigning one port to each antenna element, as in Figure 3.2. If the applied excitation is called $\mathbf{v}^+$, the reflected waves are given by

$$\mathbf{v}^- = \mathbf{S}\mathbf{v}^+ \tag{3.8}$$

or

$$\begin{bmatrix} v_1^- \\ v_2^- \\ \vdots \\ v_M^- \end{bmatrix} = \begin{bmatrix} s_{11} & s_{12} & \cdots & s_{1M} \\ s_{21} & s_{22} & \cdots & s_{2M} \\ \vdots & \vdots & \ddots & \vdots \\ s_{M1} & s_{M2} & \cdots & s_{MM} \end{bmatrix} \begin{bmatrix} v_1^+ \\ v_2^+ \\ \vdots \\ v_M^+ \end{bmatrix} \tag{3.9}$$

In a transmitting array antenna, the backward-moving waves $\mathbf{v}^-$ are caused by mutual coupling from the surrounding elements and the mismatch in the element itself. Similarly, in a receiving array the backward-moving waves constitute the received voltages coming directly from the source as well as reradiated energy due to the mutual coupling. The received antenna voltages due to a far-field emitter from DOA θ can be modeled by

$$\mathbf{v}^- = \mathbf{C}\mathbf{a}(\theta) v_{isol}^-$$

where $\mathbf{a}(\theta)$ is the purely geometrical steering vector, v_{isol}^- is the received voltage for an isolated element, and $\mathbf{C}$ is the *mutual coupling* matrix, which describes

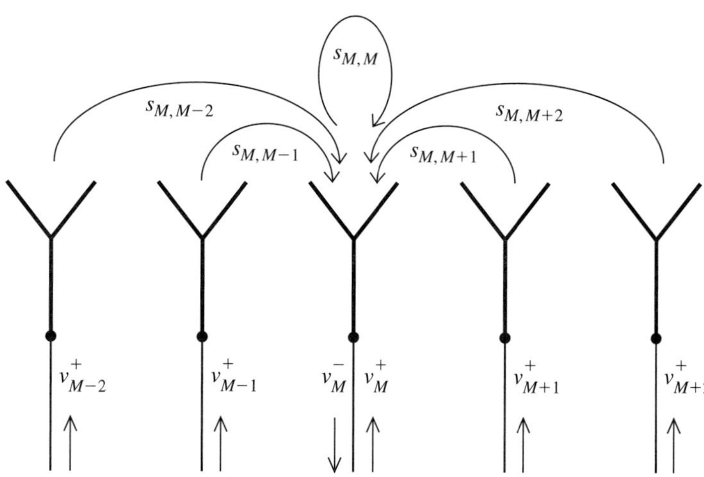

FIGURE 3.2 Mutual coupling and scattering parameters.

how the received signals change as a result of mutual coupling. Assuming an array of dipoles where the dipole self-impedance is equal to the characteristic impedance of the feeding lines, it can be shown that

$$\mathbf{C} = \mathbf{I} - \mathbf{S}$$

The corresponding derivation of the coupling matrix for an array of waveguide openings can be found in Steyskal and Herd (1990) and leads to

$$\mathbf{C} = \mathbf{I} + \mathbf{S}$$

If the mutual coupling is known, it can be included in the data model by changing the steering vectors such that

$$\mathbf{a}(\theta) \quad \rightarrow \quad \mathbf{a}_c(\theta) = \mathbf{C}\mathbf{a}(\theta) \tag{3.10}$$

If the mutual coupling matrix is unknown, or only approximately known, an error is introduced in the data model. In addition, there are uncertainties in the array response model caused by limited manufacturing accuracy and insufficient isolation between ideal components and nonideal components. It should also be mentioned that the incoming energy to a receiving array can include other unmodeled phenomena, such as near-field scattering and multipath propagation.

The main sources of array modeling errors are summarized as

- Mutual coupling unknown or misspecified
- Antenna element positions and orientations not perfectly known
- Gain and phase imbalances
- IQ imbalances in receivers (generally small)
- Near-field scattering due to platform or terrain
- Internal leakages in the system
- Quantization in phase shifters, attenuators, and ADCs
- Nonlinear components

For more information on array errors, see Allen (1964), Elliott (1958), Mailloux (2005), Pettersson (1992), and Swindlehurst and Kailath (1992).

The remedy for array modeling errors is *array calibration*. The calibration methods considered here are mostly based on calibration measurements. If the array manifold is assumed to be time-invariant, methods based on characterization at a measurement range are sufficient. On the other hand, to detect changes with time (and, for example, temperature), online calibration methods are needed. These include signal injection (e.g., Lanne and Drackner, 2003), mutual coupling–based measurements (Agrawal and Jablon, 2003; Aumann et al., 1989), and auto-calibration (see Section 3.4). Only characterization measurements using far-field sources at known positions and auto-calibration techniques are considered in this chapter.

3.3 DIRECTION-OF-ARRIVAL ESTIMATION

The purpose of this section is to present some of the most influential methods for DOA estimation, since this is the intended use of array calibration. The review presented here is far from complete (see Krim and Viberg (1996), Trees (2002), and the other chapters of this book for more details and methods). The DOA estimation problem is restated as follows: Given observations $\mathbf{X} = [\mathbf{x}(1), \ldots, \mathbf{x}(N)]$, determine the DOAs $\boldsymbol{\theta} = [\theta_1, \ldots, \theta_Q]$ of the incoming signals. The array response model $\mathbf{a}(\theta)$ is assumed to be a known function of θ over the DOA range of interest. The response may be computed from a detailed electromagnetic simulation, for example, or, as in this case, by a calibration procedure.

3.3.1 Classical Beamforming

Perhaps the most natural approach to DOA estimation is to coherently combine all sensor outputs, assuming a source at a hypothetical DOA θ, and measure the resulting power. The field of view is then scanned, and the peaks in the resulting spatial power spectrum estimate are taken as the DOA estimates. The coherent combination is $\mathbf{a}^*(\theta)\mathbf{x}(n)$ (a spatially matched filter), and the average power is given by

$$P_{BF}(\theta) = \frac{1}{N}\sum_{n=1}^{N} |\mathbf{a}^*(\theta)\mathbf{x}(n)|^2 \tag{3.11}$$

For an ULA, the propagation vectors take the form

$$\mathbf{a}(\theta) = [1, e^{j\phi}, \ldots, e^{j(M-1)\phi}]^T \tag{3.12}$$

where $\phi = \kappa\Delta \sin\theta$ is called the *electrical angle*. Here, $\kappa = \omega_c/c$ is the wavenumber, c is the speed of propagation, and Δ is the inter-element separation. The DOA θ is taken relative to the array broadside. Thus, Equation (3.11) is in this case the length-M periodogram, averaged over N realizations. Since the method only requires finding the Q highest peaks of the one-dimensional spectrum $P_{BF}(\theta)$, it is computationally attractive. However, beamforming shares the drawbacks of other Fourier-based techniques: limited resolution and possible masking of weak signals. For the ULA case, a single source gives rise to a mainlobe of width $\Delta\phi \approx 2\pi/M$, which is the classical Fourier resolution, and a maximum sidelobe 13 dB below the mainlobe. The sidelobes can be reduced by windowing, at the cost of poorer resolution, and the resolution can, in high-SNR scenarios, be improved by adaptive beamforming (Serebryakov, 1995). However, beamforming methods are in general unable to fully exploit the data model and achieve correct estimates (consistency) as $N \to \infty$.

3.3.2 Nonlinear Least Squares

Another very natural approach to parameter estimation is least squares (LS). In other words, the method tries to find the parameters that best match the observations in a least-squares sense. Given the model (3.5), with unknown parameters $\boldsymbol{\theta}$ and $\mathbf{S}$, the LS estimator is formulated as

$$\{\hat{\boldsymbol{\theta}}, \hat{\mathbf{S}}\} = \arg\min_{\boldsymbol{\theta},\mathbf{S}} \|\mathbf{X} - \mathbf{A}(\boldsymbol{\theta})\mathbf{S}\|_F^2 \tag{3.13}$$

where $\|\cdot\|_F$ is the Frobenius matrix norm. This is usually termed *nonlinear least squares* (NLLS), since the DOA parameters enter nonlinearly in the criterion function. It is noted that the method is identical to maximum likelihood (ML) if the signal waveforms $\mathbf{s}(n)$ are regarded as unknown deterministic parameters. Therefore, the term *deterministic ML* (DML), or *conditional ML* (CML), is also used. For a fixed (but unknown) value of $\boldsymbol{\theta}$, the criterion is linear in the signal matrix $\mathbf{S}$. Thus, it belongs to the class of separable NLLS problems (Golub and Pereyra, 2003). Suppressing the θ dependence of $\mathbf{A}$ for notational convenience, the LS solution is $\hat{\mathbf{S}}(\boldsymbol{\theta}) = (\mathbf{A}^*\mathbf{A})^{-1}\mathbf{A}^*\mathbf{X}$. Substituting this back into Equation (3.13), the NLLS estimate of $\boldsymbol{\theta}$ is found by

$$\hat{\boldsymbol{\theta}} = \arg\min_{\boldsymbol{\theta}} \mathrm{Tr}\{\mathbf{P}_{\mathbf{A}}^{\perp}\hat{\mathbf{R}}\} \tag{3.14}$$

where $\mathbf{P}_{\mathbf{A}}^{\perp} = \mathbf{A}(\mathbf{A}^*\mathbf{A})^{-1}\mathbf{A}^*$ is an orthogonal projection, and $\hat{\mathbf{R}} = \frac{1}{N}\mathbf{X}\mathbf{X}^*$ is the sample covariance matrix.

Note that (3.14) reduces to the classical beamforming estimate for $Q = 1$ and $\mathbf{A} = \mathbf{a}(\theta)$. However, for $Q > 1$, the NLLS method is generally superior. The resolution does not only depend on M, but can be arbitrarily good for high enough signal-to-noise ratio (SNR) or N. The drawback is that Equation (3.14) is a Q-dimensional nonlinear optimization, which is in general not convex. A popular implementation is based on solving for the parameters of one source at a time (Fessler and Hero, 1994; Fleury et al., 1999; Ziskind and Wax, 1988) in an iterative search. Once sufficiently good initial estimates have been found, it is more efficient to switch to a local optimization using a Newton-type method (Golub and Pereyra, 2003; Viberg et al., 1991).

We can alternatively model the signal waveforms as stochastic and Gaussian-distributed. The resulting maximum likelihood (ML) estimator can in some scenarios outperform NLLS, particularly when the signals are highly correlated. Interested readers are referred to Ottersten et al. (1992a) and Stoica and Nehorai (1990) for details.

3.3.3 Subspace Methods: The MUSIC Algorithm

Although the NLLS method can produce excellent statistical properties, its computational complexity motivates alternative approaches that are not only cheaper but superior to conventional beamforming. One important class of such approaches is *subspace methods*. The underlying observation is that if the signal matrix $\mathbf{S}$ has full rank, the noise-free data $\mathbf{X}=\mathbf{AS}$ is confined to the span of the columns of $\mathbf{A}$, denoted span$\{\mathbf{A}\}$, which is therefore termed the *signal subspace*. It is easy to show that the signal subspace is also spanned by the Q principal eigenvectors of the array covariance matrix $\mathbf{R}$, defined in Equation (3.6). Thus, the signal subspace is estimated from the available data by the Q principal eigenvectors of the sample covariance matrix

$$\hat{\mathbf{R}}=\sum_{k=1}^{M}\hat{\lambda}_k\hat{\mathbf{e}}_k\hat{\mathbf{e}}_k^*=\hat{\mathbf{E}}_s\hat{\Lambda}_s\hat{\mathbf{E}}_s^*+\hat{\mathbf{E}}_n\hat{\Lambda}_n\hat{\mathbf{E}}_n^* \tag{3.15}$$

where $\hat{\mathbf{E}}_s=[\hat{\mathbf{e}}_1,\ldots,\hat{\mathbf{e}}_Q]$ contains the Q principal "signal eigenvectors." Ideally, span$\{\mathbf{E}_s\}=$span$\{\mathbf{A}\}$, and because $\mathbf{E}_s$ and $\mathbf{E}_n$ are orthogonal, it holds that span$\{\mathbf{E}_n\}\perp$span$\{\mathbf{A}\}$. In the MUSIC algorithm (Schmidt, 1979, reprinted in Schmidt, 1986; see also Bienvenu and Kopp, 1979), this relation is exploited by forming a "pseudo-spectrum":

$$P_{MU}(\theta)=\frac{\|\mathbf{a}(\theta)\|^2}{\|\hat{\mathbf{E}}_n^*\mathbf{a}(\theta)\|^2} \tag{3.16}$$

The MUSIC DOA estimates are now the locations of the Q highest peaks of $P_{MU}(\theta)$. Since $\hat{\mathbf{E}}_s\rightarrow\mathbf{E}_s$ as $N\rightarrow\infty$ in a suitable statistical sense, the MUSIC spectrum can resolve arbitrarily close sources for large enough N (or SNR). Yet it only requires a one-dimensional search similar to the classical beamforming approach. The one additional burden is the eigendecomposition of $\hat{\mathbf{R}}$, which can be computed efficiently if $Q\ll M$ (Xu and Kailath, 1994; Yang, 1995). For this reason, the method and its variations (e.g., Gershman and Bohme, 1997; Lee and Wengrovitz, 1990; Xu and Kaveh, 1996) have become very popular since being introduced in the late 1970s. We reiterate that the signal matrix needs to have full rank for the MUSIC method to work (i.e., *coherent* sources cannot be handled). For ULAs, so-called spatial smoothing and forward-backward averaging techniques have been proposed to overcome this limitation (Friedlander and Weiss, 1992; Shan et al., 1985).

3.4 AUTO-CALIBRATION TECHNIQUES

The purpose of this section is to give a brief introduction to the auto-calibration approach. A different exposition including more methods is available in Li et al. (2003). Common to all of these techniques is that a parametric model of the array response, which includes some "array parameters," is exploited. Formally we have $\mathbf{a}=\mathbf{a}(\theta,\boldsymbol{\rho})$, where $\boldsymbol{\rho}=[\rho_1,\ldots,\rho_P]^T$ is a vector of P unknown parameters associated with the uncertainty in the response. As an example, consider the case of isotropic and identical antenna elements at unknown positions (x_m,y_m) in the xy-plane. The DOA parameter is the azimuth angle, taken clockwise relative to the y-axis. For simplicity, it is assumed that all sources are in the same plane as the sensors; that is, the elevation angle is ignored. In this case, the array response is modeled as

$$\mathbf{a}(\theta, \boldsymbol{\rho}) = [e^{j\kappa(x_1 \sin\theta + y_1 \cos\theta)}, \ldots, e^{j\kappa(x_M \sin\theta + y_M \cos\theta)}]^T \tag{3.17}$$

Thus, $P = 2M$ and $\boldsymbol{\rho} = [x_1, y_1, \ldots, x_M, y_M]^T$ (although, as will be pointed out later, not all of these parameters can be determined without additional knowledge). Another case of interest is calibration of gain and phase (i.e., complex gain). Again assuming a planar array, the array response is modeled by

$$\mathbf{a}(\theta, \boldsymbol{\rho}) = [\gamma_1 e^{j\kappa(x_1 \sin\theta + y_1 \cos\theta)}, \ldots, \gamma_M e^{j\kappa(x_M \sin\theta + y_M \cos\theta)}]^T \tag{3.18}$$

If the sensor locations are known, the $P = 2M$ real-valued array parameters can now be taken as

$$\boldsymbol{\rho} = [\Re\{\gamma_1\}, \Im\{\gamma_1\}, \ldots, \Re\{\gamma_M\}, \Im\{\gamma_M\}]^T$$

where $\Re\{\cdot\}$ and $\Im\{\cdot\}$ denote the real and imaginary parts, respectively. Equivalently, one can use the magnitudes and phases of the γ_ms as parameters. Also, in the case of auto-calibration with unknown complex gain, additional constraints are necessary to guarantee unique estimates.

A third case that has been extensively studied is estimation of mutual coupling parameters. From Equation (3.10), the model can be expressed as

$$\mathbf{a}(\theta, \boldsymbol{\rho}) = \mathbf{C}(\boldsymbol{\rho})\mathbf{a}_0(\theta) \tag{3.19}$$

where $\mathbf{C}(\boldsymbol{\rho})$ is the mutual coupling matrix (MCM) and $\mathbf{a}_0(\theta)$ is the "coupling-free" array response. The MCM is usually taken as a linear function of the array parameters $\boldsymbol{\rho}$, with a certain structure that depends on the sensor properties and the array geometry. Because of reciprocity, $\mathbf{C}(\boldsymbol{\rho})$ is always a symmetric matrix, and it is reasonable to assume that distant sensors are uncoupled so that $\mathbf{C}(\boldsymbol{\rho})$ is banded. For a ULA, the MCM is usually modeled as Toeplitz, which is a decent approximation for large arrays.

3.4.1 Identifiability

In the auto-calibration approach, also termed *self-calibration*, all parameters are estimated simultaneously. This can in principle be done using any of the available parametric methods for DOA estimation. However, a crucial question is identifiability of the parameters. In Rockah and Schultheiss (1987), it is shown that if the position of one sensor is fixed along with the direction to a second sensor, the positions can be estimated uniquely along with the DOAs in the model (3.17)—provided there are at least three sources present and the sensors are not collinear. More general results are given in Levi and Messer (1990), including the possibility of some sources at known locations. If the sensors have unknown DOA-independent complex gain factors, these can be determined together with the DOAs under fairly general conditions (Astély et al., 1999). However, the array needs to be nonlinear. If this is not the case, the DOAs can be determined only up to a rotational ambiguity (i.e., the differential DOAs are unique). A more general necessary condition for the case where $\mathbf{a}(\theta, \boldsymbol{\rho})$ is linear in $\boldsymbol{\rho}$ is given in Flieller et al. (1995). It is shown that to estimate P unknown array parameters requires that $P \leq Q(M-Q)$. This condition is generally not sufficient in all scenarios, but may serve as a useful guide for investigating a particular parameterization.

3.4.2 Algorithms

Most of the methods proposed in the literature fall into two categories. The first is based on an extension of the MUSIC algorithm to handle auto-calibration. The idea is to maximize the MUSIC pseudo-spectrum over all parameters simultaneously (a different approach is presented in Paulraj and Kailath, 1985). The resulting spectrum is

$$P_{MU}(\boldsymbol{\theta}, \boldsymbol{\rho}) = \sum_{q=1}^{Q} \frac{\|\mathbf{a}(\theta_q, \boldsymbol{\rho})\|^2}{\|\hat{\mathbf{E}}_n^* \mathbf{a}(\theta_q, \boldsymbol{\rho})\|^2} \tag{3.20}$$

The parameter estimates are found by performing a simultaneous maximization of $P_{MU}(\boldsymbol{\theta}, \boldsymbol{\rho})$ with respect to $\boldsymbol{\rho}$ and θ_q. In the original algorithm, from Weiss and Friedlander (1990), the normalization by $\|\mathbf{a}(\theta_q, \boldsymbol{\rho})\|^2$ is ignored and the "null spectrum" to be minimized is expressed as

$$\tilde{P}_{MU}(\boldsymbol{\theta}, \boldsymbol{\rho}) = \sum_{q=1}^{Q} \mathbf{a}^*(\theta_q, \boldsymbol{\rho}) \hat{\mathbf{E}}_n \hat{\mathbf{E}}_n^* \mathbf{a}(\theta_q, \boldsymbol{\rho}) \tag{3.21}$$

The advantage of this formulation is apparent for the case where $\mathbf{a}(\theta, \boldsymbol{\rho})$ is a linear function of $\boldsymbol{\rho}$ (i.e., for channel or mutual coupling errors). For fixed values of $\{\theta_q\}_{q=1}^{Q}$, Equation (3.21) is a quadratic function in $\boldsymbol{\rho}$, which can be minimized explicitly. To clarify this, note that a linear parameterization enables us to express the array response vector as

$$\mathbf{a}(\theta, \boldsymbol{\rho}) = \mathbf{M}(\theta) \boldsymbol{\rho} \tag{3.22}$$

for some $M \times P$ matrix–valued function $\mathbf{M}(\theta)$. Thus,

$$\tilde{P}_{MU}(\boldsymbol{\theta}, \boldsymbol{\rho}) = \sum_{q=1}^{Q} \boldsymbol{\rho}^* \mathbf{M}^*(\theta_q) \hat{\mathbf{E}}_n \hat{\mathbf{E}}_n^* \mathbf{M}(\theta_q) \boldsymbol{\rho} = \boldsymbol{\rho}^* \tilde{\mathbf{M}}(\boldsymbol{\theta}) \boldsymbol{\rho} \tag{3.23}$$

where

$$\tilde{\mathbf{M}}(\boldsymbol{\theta}) = \sum_{q=1}^{Q} \mathbf{M}^*(\theta_q) \hat{\mathbf{E}}_n \hat{\mathbf{E}}_n^* \mathbf{M}(\theta_q) \tag{3.24}$$

The minimization of Equation (3.23) w.r.t. $\boldsymbol{\rho}$ requires a constraint to avoid the trivial solution. Depending on the application, a norm constraint $\|\boldsymbol{\rho}\| = 1$ or a linear constraint, such as $\rho_1 = 1$, can be used. Given the constraint, the minimization of Equation (3.23) results in an updated value of $\boldsymbol{\rho}$. Next, for a fixed value of $\boldsymbol{\rho}$, the minimization of Equation (3.21) w.r.t. $\{\theta_q\}_{q=1}^{Q}$ reduces to the standard MUSIC algorithm. Thus, the joint minimization can be performed iteratively, updating $\boldsymbol{\rho}$ and $\boldsymbol{\theta}$ alternatingly until convergence. The iterations are started by estimating $\boldsymbol{\theta}$ using a "nominal value" for the array parameters $\boldsymbol{\rho}$. Since the joint optimization problem is nonlinear, the initial value needs to be sufficiently accurate to guarantee that the global optimum is found.

A similar approach is used to simultaneously estimate DOA and mutual coupling parameters (Friedlander and Weiss, 1991). Provided the DOA and array parameters are jointly identifiable, the same method is effective for calibrating any array parameters that enter linearly in the response vector.

Basic Weiss-Friedlander (W-F) auto-calibration is simple and computationally efficient for the case of linear array parameters. However, performance is limited by that of the MUSIC algorithm, and in particular the sources must be resolvable by MUSIC even when the nominal array parameters are used. A more direct approach is to estimate all parameters simultaneously in an NLLS or ML setting. The extension of Equation (3.14) to the case of unknown array parameters is conceptually straightforward (Weiss and Friedlander, 1989):

$$\{\hat{\boldsymbol{\theta}}, \hat{\rho}, \hat{\mathbf{S}}\} = \arg \min_{\boldsymbol{\theta}, \boldsymbol{\rho}, \mathbf{S}} \|\mathbf{X} - \mathbf{A}(\boldsymbol{\theta}, \rho)\mathbf{S}\|_F^2 \tag{3.25}$$

The criterion is still quadratic in $\mathbf{S}$, so the separable solution (3.14) is still applicable. Provided all parameters are identifiable, Equation (3.25) is expected to give the best possible performance. Yet the identifiability problem for auto-calibration methods is quite restrictive and limits their practical use. In short, the simultaneous parameter estimation of $\boldsymbol{\theta}$ and $\boldsymbol{\rho}$ is often severely ill-conditioned. This situation can to some extent be remedied if approximate values $\boldsymbol{\rho}_0$ of $\boldsymbol{\rho}$ are known. The array parameters can then be modeled as random with a certain a priori distribution. The simplest and most common approach is to assume a Gaussian distribution, where only the covariance matrix $\mathbf{C}_\rho = E[(\boldsymbol{\rho} - \boldsymbol{\rho}_0)(\boldsymbol{\rho} - \boldsymbol{\rho}_0)^T]$ (or $\mathbf{C}_\rho = E[(\boldsymbol{\rho} - \boldsymbol{\rho}_0)(\boldsymbol{\rho} - \boldsymbol{\rho}_0)^*]$, in the case of complex circularly symmetric error parameters) needs to be specified. With these assumptions, it is possible to put the joint estimation in a Bayesian framework, where $\boldsymbol{\rho}$ are treated as nuisance parameters. The simplest and most straightforward possibility is a joint maximum a posteriori (MAP) auto-calibration approach. Separating the criterion with respect to the noise variance σ^2 and the signal waveform matrix $\mathbf{S}$, the MAP criterion can be expressed as

$$\{\hat{\boldsymbol{\theta}}, \hat{\rho}\} = \arg \min_{\boldsymbol{\theta}, \boldsymbol{\rho}} NM \log \mathrm{Tr}\{\mathbf{P}_{\mathbf{A}}^{\perp} \hat{\mathbf{R}}\} + \frac{1}{2} (\boldsymbol{\rho} - \boldsymbol{\rho}_0)^T \mathbf{C}_\rho^{-1} (\boldsymbol{\rho} - \boldsymbol{\rho}_0) \tag{3.26}$$

It is interesting to interpret Equation (3.26) as a regularization of the supposedly ill-conditioned NLLS criterion. In theory, this means that the identifiability problem has been overcome. However, it should be stressed that for the case where joint modeling is nonunique without regularization, the array perturbations $\boldsymbol{\rho}-\boldsymbol{\rho}_0$ need to be "small enough." This is because the information from the data can only move $\boldsymbol{\rho}_0$ a small distance in the direction of the true array parameters, and thus can only reduce the effect of the modeling error on the estimation of $\boldsymbol{\theta}$ to a limited extent as compared to using the nominal value $\boldsymbol{\rho}_0$ as if it were true. In contrast, when the joint parameterization is unique, the size of the array perturbations is essentially irrelevant provided the true minimum of the respective criteria can be found. Alternative approaches to MAP-based auto-calibration are presented in Jansson et al. (1998), Viberg and Swindlehurst (1994b), and Wahlberg et al. (1991), using a stochastic model of the signal waveforms.

3.5 CALIBRATION USING SOURCES AT KNOWN POSITIONS

In the previous section we discussed methods that attempt to calibrate the array using the same data as used for DOA estimation. As was seen, this requires a parametric model with a known structure and relatively few parameters. This is not always a practical assumption. A useful remedy is to collect additional *calibration data*, normally in an anechoic chamber prior to the actual use of the antenna array. In most cases a single emitter is used, placed at several known positions. The calibration data is used to compute array response vectors for the different source directions. The calibration data for a wavefront coming from the DOA η_c is modeled by

$$\mathbf{x}_c(t)=\mathbf{a}_c s_c(t)+\mathbf{n}_c(t), \quad t=1,\ldots,N_c;\ c=1,\ldots,C \tag{3.27}$$

where $\mathbf{a}_c=\mathbf{a}(\eta_c)$ is the "true" array response vector at η_c, C is the number of emitter positions, and N_c is the number of data samples taken at position c. We distinguish between two cases of interest:

Coherent calibration. In this case $s_c(t)$ is known, and the array response vectors are estimated as

$$\hat{\mathbf{a}}_c=\frac{\sum_{t=1}^{N_c}\mathbf{x}_c(t)s_c^*(t)}{\sum_{t=1}^{N_c}|s_c(t)|^2} \tag{3.28}$$

This results in perfect calibration vectors for $N_c\to\infty$, even if the noise in the calibration data is not spatially white.

Noncoherent calibration. When $s_c(t)$ is unknown, for example because of synchronization problems, the array response is instead computed from the principal eigenvector of the sample covariance matrix

$$\hat{\mathbf{R}}_c=\frac{1}{N_c}\sum_{t=1}^{N_c}\mathbf{x}_c(t)\mathbf{x}_c^*(t)=\sum_{k=1}^{M}\hat{\lambda}_k\hat{\mathbf{e}}_k\hat{\mathbf{e}}_k^*\ \Rightarrow\ \hat{\mathbf{a}}_c\propto\hat{\mathbf{e}}_1 \tag{3.29}$$

In this case, the resulting estimates have a gain and phase ambiguity, which may need to be taken into account when the calibration vectors are used. Also, the noise needs to be spatially white to give consistent estimates. If this is not the case, data with no sources present can be used to determine the noise color, after which a prewhitening can be applied. For high enough SNR, the noise color can of course be ignored.

3.5.1 Parametric Methods

If a parametric model for the array parameters can be assumed, approaches similar to auto-calibration can be applied. However, since they are used with calibration data with sources at known positions, much more flexible error models can be handled. A straightforward approach is to apply an LS fit of the model to the calibration vectors. In the coherent case, where $\hat{\mathbf{a}}_c$ is computed according to Equation (3.28), this leads to

$$\hat{\rho} = \arg\min_{\hat{\rho}} \sum_{c=1}^{C} \|\hat{\mathbf{a}}_c - \mathbf{a}(\eta_c, \boldsymbol{\rho})\|^2 \tag{3.30}$$

For noncoherent calibration data, where $\hat{\mathbf{a}}_c$ is given by Equation (3.28), a complex gain factor must be included:

$$\{\hat{\rho}, \hat{\gamma}\} = \arg\min_{\boldsymbol{\rho}, \gamma} \sum_{c=1}^{C} \|\hat{\mathbf{a}}_c - \mathbf{a}(\eta_c, \boldsymbol{\rho})\gamma_c\|^2 \tag{3.31}$$

See Ng and Nehorai (1995) and Ng and See (1996) for extensions and performance analysis with application to sensor position uncertainties.

Parametric calibration is of course simplest when the array parameters enter the array response linearly. In this context, the linear model is often expressed as

$$\mathbf{a}_c = \mathbf{Q}\mathbf{a}(\eta_c) \tag{3.32}$$

When $\mathbf{Q}$ is independent of the DOA η, Equation (3.32) is a *global calibration model*. Two cases are of special interest (e.g., Pierre and Kaveh, 1995):

- $\mathbf{Q}$ is a full matrix, which can model, for example, channel and mutual coupling errors.
- $\mathbf{Q}$ is a diagonal matrix, which can correct for channel errors only.

In both cases the correction matrix is determined using an LS fit. Assuming coherent calibration, we have

$$\hat{\mathbf{Q}} = \arg\min_{\mathbf{Q}} \|\hat{\mathbf{A}}_c - \mathbf{Q}\mathbf{A}(\boldsymbol{\eta}_c)\|_F^2 \tag{3.33}$$

where $\hat{\mathbf{A}}_c = [\hat{\mathbf{a}}_1, \ldots, \hat{\mathbf{a}}_C]$ and $\mathbf{A}(\boldsymbol{\theta}_c) = [\mathbf{a}(\eta_1), \ldots, \mathbf{a}(\eta_C)]$. When $\mathbf{Q}$ is a full matrix, the solution is

$$\hat{\mathbf{Q}} = \hat{\mathbf{A}}_c\mathbf{A}^*(\boldsymbol{\eta}_c)(\mathbf{A}(\boldsymbol{\eta}_c)\mathbf{A}^*(\boldsymbol{\eta}_c))^{-1} \tag{3.34}$$

Note that $C \geq M$ is necessary to guarantee a unique solution. In the diagonal case, $\mathbf{Q} = \mathrm{diag}\{\mathbf{q}\}$. The elements of $\mathbf{q} = [q_1, \ldots, q_M]^T$ are estimated separately as

$$\hat{q}_m = \hat{\mathbf{A}}_{c,m} \mathbf{A}_m^*(\eta_c)(\mathbf{A}_m(\eta_c)\mathbf{A}_m^*(\eta_c))^{-1}, \quad m = 1, \ldots, M \tag{3.35}$$

where $\hat{\mathbf{A}}_{c,m}$ and $\mathbf{A}_m(\eta_c)$ denote the mth row of the respective matrix. These global calibration methods are simple and efficient. Their main drawback is that they cannot handle DOA-dependent errors due for example to uncertain element positions.

3.5.2 Interpolation of Calibration Data

Provided the calibration vectors are accurately measured, we can consider the array response to be essentially known at the calibration points. We can thus regard the calibration as an interpolation problem. The most direct approach is to interpolate the real and imaginary parts of the M elements of the measured calibration vectors $\{\hat{\mathbf{a}}_c(\eta_c)\}_{c=1}^C$ to any desired DOA θ using linear or spline interpolation. In general, this requires a very dense calibration grid since the array response is usually designed to vary rapidly with the DOA (necessary to enable high-resolution DOA estimation, which is the array's intended use). A more viable approach can be applied if a *nominal model* $\mathbf{a}(\theta)$ is available. Note that this is conceptually different from the parametric calibration case, since it is not necessary to specify how the nominal model depends on the array parameters. The "true" array response is now expressed as

$$\mathbf{a}_c = \mathbf{Q}(\eta_c)\mathbf{a}(\eta_c) \tag{3.36}$$

where $\mathbf{Q}(\eta) = \mathrm{diag}\{\mathbf{q}(\eta)\}$ is now a *local* (DOA-dependent) correction matrix. Using the calibration data, the correction $\mathbf{q}(\eta)$ can be determined at the DOAs $\{\eta_c\}_{c=1}^C$ as

$$\hat{\mathbf{q}}(\eta_c) = \hat{\mathbf{a}}_c./\mathbf{a}(\eta_c) \tag{3.37}$$

where ./ denotes element-wise division. The idea now is that the correction matrix will be a much smoother function of the DOA, provided that the nominal model captures most of the θ dependence. Thus, linear or spline-based interpolation can be used to interpolate the real and imaginary parts of $\hat{\mathbf{q}}(\cdot)$ instead of directly interpolating the calibration vectors. Given an interpolated $\hat{\mathbf{q}}(\theta)$ at a certain desired DOA θ, the array response is then computed as $\hat{\mathbf{a}}(\theta) = \mathrm{diag}\{\hat{\mathbf{q}}(\theta)\}\mathbf{a}(\theta)$. For a given calibration grid, this results in a (much) more accurate interpolation. We close this section by noting that it is of course possible to interpolate gain and phase rather than real and imaginary parts, although this in general does not change the result significantly. We also note that the same interpolation can in principle be used to calibrate over a multidimensional parameter space, such as azimuth, elevation, and frequency dependence. However, this is considerably more complicated, and may necessitate a more sparse calibration grid in order to keep the measurement costs manageable.

3.6 ARRAY INTERPOLATION TECHNIQUES

Interpolation of calibration data was introduced in Section 3.5.2. In this section we describe two more elaborate interpolation techniques. The first is based on a Fourier representation of the array response; the second uses a local polynomial approximation. Both methods assume the availability of calibration data in the form of Equation (3.28).

3.6.1 Fourier-Based Array Interpolation

This approach is based on the so-called Manifold Separation theorem (Belloni et al., 2007; Doron and Doron, 1994a), which says that any array can under mild conditions be represented by

$$\mathbf{a}_c(\eta) = \mathbf{G}\mathbf{v}(\eta) + \tilde{\mathbf{r}}(\eta) \tag{3.38}$$

where $\mathbf{G}$ is the $M \times \tilde{M}$, $\tilde{M} \geq M$ "sampling matrix," and (assuming $\tilde{M}$ to be odd)

$$\mathbf{v}(\eta) = \left[e^{-j\frac{\tilde{M}-1}{2}\eta}, e^{-j\frac{\tilde{M}-3}{2}\eta}, \ldots, 1, e^{j\eta}, \ldots, e^{j\frac{\tilde{M}-1}{2}\eta} \right]^T \tag{3.39}$$

is a Fourier vector. Note that η in Equation (3.39) is the "physical" DOA, and should not be interpreted as the electrical angle as in Equation (3.12). The error $\|\tilde{\mathbf{r}}(\eta)\|$ is shown to vanish "rapidly" as $\tilde{M} \to \infty$. The representation (3.38) is a truncated Fourier series representation of the array response, which is always a 2π-periodic function. In the antenna literature, the elements of the Fourier vector (3.39) are often termed *phase modes*. The idea behind the Fourier-based interpolation approach is that most arrays can be well approximated by a relatively small number of phase modes, since the corresponding terms in the sampling matrix decay super-exponentially (under some assumptions; Doron and Doron, 1994a). Given enough calibration vectors $\hat{\mathbf{A}}_c$, it is tempting to try to compute $\mathbf{G}$ by solving an LS problem similar to Equation (3.33):

$$\hat{\mathbf{G}} = \arg\min_{\mathbf{G}} \|\hat{\mathbf{A}}_c - \mathbf{G}\mathbf{V}(\boldsymbol{\eta}_c)\|_F^2 \tag{3.40}$$

In practice this can be problematic due to the potential ill-conditioning of $\mathbf{V}(\boldsymbol{\eta}_c)$ and the large dynamic range in $\mathbf{G}$. Instead, Belloni et al. (2007) proposes to compute the sampling matrix in two steps. First, an $M \times C$ matrix $\mathbf{G}_C$ is computed, thus using a maximum number of phase modes equal to the number of calibration vectors C. If the calibration vectors have been collected on a uniform grid covering the whole range $(0, 2\pi)$, $\mathbf{G}_C$ can be computed by an inverse discrete Fourier transform (IDFT) of $\hat{\mathbf{A}}_c$. This can be done efficiently using inverse fast Fourier transform (IFFT), and there is no risk of error amplification since the DFT vectors are orthogonal. The second step is to truncate the sampling matrix $\mathbf{G}_C$

to the desired size $M \times \tilde{M}$, for example by selecting the $\tilde{M}$ "middle" columns, which correspond to the lowest-order phase modes. Choosing the number of phase modes $\tilde{M}$ is crucial and can be guided by the Euclidean norm of the columns. In fact, if the size of the error in the calibration vectors is known, Belloni et al. (2007) suggest an "optimal" procedure for choosing $\tilde{M}$ based on matching the approximation and the calibration errors.

Besides the calibration aspect, an obvious advantage of Fourier-based interpolation is that the Fourier vectors resemble the array response for a ULA. Thus, the plethora of computationally efficient methods for the ULA case are directly applicable once the sampling matrix has been found. The difference with the ULA case is that $\mathbf{v}(\eta)$ depends directly on η and not on $(\sin\eta)$, as it would for a "physical" array. Since this does not change the "ULA structure" of the virtual response vector, one could ask if there are even better choices of nonlinear transformation $g(\eta)$ that could result in a better approximation for a given size $\tilde{M}$. An interesting contribution in this direction is given in Buhren et al. (2004), although it is not directly applicable to the calibration case considered here.

3.6.2 Interpolation Based on Local Polynomial Approximation

The global diagonal calibration (3.35) and the linear interpolation of correction vectors (3.37) represent two extreme cases of interpolation. In the former, the correction is assumed to be globally valid; in the latter, it is only valid between two calibration points. If the "true" correction functions are smooth, but not necessarily constant, it is natural to apply a method that takes several, but not all, neighboring points into account in the interpolation. In the absence of more structural information, these should be weighted according to their distance to the DOA θ of current interest. One way to achieve this smoothed interpolation is to model the correction vectors (3.37) using a set of basis functions $\{\phi_l(x)\}_{l=0}^{L}$ that are centered at the "query point" θ. The mth element of $\hat{q}(\eta_c)$ is thus expressed as

$$\hat{q}_m(\eta_c) = \sum_{l=0}^{L} \alpha_l \phi_l(\eta_c - \theta) = \boldsymbol{\phi}^T(\eta_c - \theta)\boldsymbol{\alpha} \tag{3.41}$$

It is beneficial to model the real and imaginary parts of $\hat{q}_m(\eta_c)$ separately, and thus we assume that $\hat{q}_m(\eta_c)$ is real-valued in Equation (3.41). Once the coefficients $\hat{\alpha}_l$ have been determined, the sought interpolator is obtained as $\hat{q}_m(\theta) = \boldsymbol{\phi}^T(0)\hat{\boldsymbol{\alpha}}$. The most commonly applied basis function is $\phi_l(x) = x^l$, in which case the approximation is called local polynomial approximation (LPA) (Fan and Gijbels, 1996; Katkovnic et al., 2006). See Lanne et al. (2006, 2007), Lundgren et al. (2007), and Ottersten et al. (1992b) for applications in array calibration. In Fan and Gijbels (1996) it is argued that an odd polynomial order L should be chosen, and in array calibration applications we have mostly used $L = 1$ (i.e., a locally linear approximation).

The polynomial coefficients $\boldsymbol{\alpha}$ are determined by a weighted regression

$$\hat{\boldsymbol{\alpha}} = \arg\min_{\alpha} \sum_{c=1}^{C} w_h(|\eta_c - \theta|)\, |\hat{q}_m(\eta_c) - \phi^T(\eta_c - \theta)\boldsymbol{\alpha}|^2 \tag{3.42}$$

Here $w_h(x)$ represents a weighting function or *smoothing kernel*, which should be a decreasing function of x. How fast $w_h(x)$ decays depends on the *smoothing parameter* (or bandwidth) h. Once the polynomial coefficients have been found from Equation (3.42), the interpolated value is computed from the postulated model (3.41). In the LPA case, the sought interpolation value is simply given as $\hat{q}_m(\theta) = \hat{\alpha}_0$. In fact, the higher-order coefficients give smoothed estimates of the derivatives: $\hat{q}'_m(\theta) = \hat{\alpha}_1$ and so on (Fan and Gijbels, 1996).

A common choice of weighting is the Gaussian kernel, $w\_h(x) \propto e^{-x^2/2h^2}$, in which case the smoothing parameter h is interpreted as a standard deviation. The actual shape of the kernel may not be so critical, but the choice of h will directly influence the amount of smoothing that is applied. When $h \to 0$ we expect the LPA to behave as a direct linear interpolation, whereas $h \to \infty$ means that a global model is used. By varying h we can thus "interpolate" between these extreme cases. The choice is a trade-off between bias and variance, and the optimal value depends on the reliability of the estimated correction values $\hat{q}_m(\eta_c)$, as well as the smoothness of the true function $q_m(\theta)$. Since the latter is typically not known, the bandwidth h must in practice be determined from the available data only. A technique often applied in the statistical literature is cross-validation. The data $\{\hat{q}_m(\eta_c)\}_{c=1}^{C}$ is then split into an estimation set—for example, $\{\hat{q}_m(\eta_1), \hat{q}_m(\eta_3), \ldots, \hat{q}_m(\eta_C - 1)\}$—and a validation set—for example, $\{\hat{q}_m(\eta_2), \hat{q}_m(\eta_4), \ldots, \hat{q}_m(\eta_C)\}$ (it might be wise to exclude the end point from the validation data to avoid data extrapolation). Estimates of the validation data are computed using interpolation of the estimation data, and the mean-square error (MSE) as compared to the actual validation data is computed. The optimal value of h is then the one that minimizes this MSE. With small data samples, the roles of the estimation and the data set can be reversed to obtain a new estimate of the MSE, and h is then chosen to minimize the average between the two.

One drawback with this approach is that the grid size in the estimation set is different from the one used in the original calibration data, and thus the calculated h may not be optimal when using the full data set. Another disadvantage is that it assumes that a single h can be used for the entire range of DOAs in the calibration data. Other ways to determine the bandwidth parameter, including adaptive scale methods, are discussed in Fan and Gijbels (1996), Katkovnik (1998), and Katkovnic et al. (2006).

It should be stressed that the basis function expansion (3.41) is only valid locally around the query point θ. Thus, a new set of coefficients $\boldsymbol{\alpha}$ must be computed for each DOA θ of interest. When evaluating, for example, the MUSIC spectrum at a dense set of DOAs, the LPA-based interpolation technique is thus computationally demanding, particularly when an optimal bandwidth h needs to be found for each real and imaginary part of each antenna element. However, if sufficient memory is available, the bulk of the computations can be done offline by computing new "artificial" calibration vectors on a finer grid than the original measured data. A simpler method like linear interpolation can then be used on the new "cleaned" calibration data. As we will see in the next section, the local interpolation technique can potentially yield superior performance in difficult scenarios, although the choice of h is both critical and difficult.

3.7 COMPARISON OF APPROACHES

The purpose of this section is to compare the applicability of the various calibration methods. This comparison is not always easy since the different approaches rely on their own assumptions. Table 3.1 classifies the techniques in terms of their dependence on a nominal model $\mathbf{a}(\theta)$ or a parametric model $\mathbf{a}(\theta, \boldsymbol{\rho})$ of the array response, as well as in terms of the need for calibration data.

Another aspect of interest is the computational complexity of the different calibration methods. Without going into detail, we note that the auto-calibration approach is in general the most demanding from this point of view because of the many parameters it involves. Also, interpolation using LPA requires substantial computation, but this is less critical if the computations are done offline, as explained in Section 3.6.2.

3.7.1 Simulation Setup

In the remainder of this section we present empirical results from computer simulations of the methods under various sources of error. The MUSIC algorithm (see Section 3.3) is applied using different calibration schemes and, in the first case, auto-calibration techniques as well. All simulations use a ULA of $M=10$ half-wavelength (nominal) separation. Two equi-powered signal sources are used, with the second DOA fixed at $\theta_2 = 12°$ with respect to array broadside, and the first varying between $\theta_1 = 5°$ and $\theta_1 = 11°$. The DOA estimation is free from finite sample errors; that is, an infinite number of samples ($N \to \infty$) is assumed. This is implemented by generating the array covariance matrix according to Equation (3.6), making the SNR irrelevant. The "true" array response used in the simulations is computed as

$$\mathbf{a}_c(\theta) = \mathbf{G}\,[e^{2\pi j(x_1 \sin\theta + y_1 \cos\theta)}, \ldots, e^{2\pi j(x_M \sin\theta + y_M \cos\theta)}]^T \tag{3.43}$$

TABLE 3.1 Classification of Calibration Techniques

	Without Calibration Data	With Calibration Data
No model	—	Fourier-based interpolation
Nominal model	Robust estimation	Interpolation of correction $\mathbf{q}(\eta)$
Parametric model	Auto-calibration	Parametric calibration

Thus, $\kappa = 2\pi$ is assumed, which means that the element positions are normalized to the wavelength.

The nominal array is an ideal ULA with half-wavelength element separation, so $x_m = (m-1)/2$ and $y_m = 0$. In the model (Equation 3.43), all signals are assumed to reside in the xy-plane, and the DOA is taken clockwise relative to the y-axis. Different errors are applied to the gain matrix (nominally $\mathbf{G} = \mathbf{I}$) and the element positions. In all cases, calibration data is collected on a uniform grid from the DOAs $\eta_1 = -40°$ to $\eta_C = 40°$. The grid size is varied between $\Delta\eta = 1°$ and $\Delta\eta = 20°$, which means that the number of calibration vectors varies between $C = 81$ and $C = 5$. All of these cases can be considered as relatively sparse calibration, which keeps data collection costs low. This is particularly apparent when the calibration needs to be done over several parameters, such as azimuth, elevation, and frequency. In all simulations, a coherent calibration is assumed, and the calibration vectors are generated as

$$\hat{\mathbf{a}}_c = \mathbf{a}_c + \tilde{\mathbf{a}}_c \tag{3.44}$$

where $\mathbf{a}_c$ is the true array response in the direction η_c and $\tilde{\mathbf{a}}_c$ is the zero-mean complex Gaussian distributed with covariance matrix $E[\tilde{\mathbf{a}}_c \tilde{\mathbf{a}}_c^*] = \sigma_a^2 \mathbf{I}$. The standard deviation is varied from $\sigma_a = 0.01$ to $\sigma_a = 0.1$. The former is a realistic measurement accuracy in an anechoic chamber, whereas the latter can account for perturbations in the operational data (e.g., due to temperature drift) that were not present during calibration data collection. In all figures that follow, the empirical RMS error (RMSE) is based on 500 independent realizations of the array perturbations for each point. Cases where an algorithm fails to resolve the sources (only one local extremum within $\pm\Delta\theta$ of the true value) or where the DOA estimation error is larger than half the DOA separation, $\Delta\theta$, are declared failures and are not included in the RMSE calculation. If the empirical failure rate exceeds 40%, the corresponding RMSE value is not included in the plot.

3.7.2 DOA-Independent Modeling Errors

In Example 3.1 we consider direction-independent errors only, meaning that the element positions are fixed at their nominal values. The calibration DOAs are $\eta = [-40°, -35°, \ldots, 40°]$. That is, $C = 17$ calibration vectors are generated. The standard deviation of the calibration vectors is $\sigma_a = 0.01$.

Example 3.1 *Channel Errors*

The first case considers channel errors only, so $\mathbf{G} = \mathbf{I} + \text{diag}(\tilde{\mathbf{g}})$, where the elements of $\tilde{\mathbf{g}}$ are zero-mean i.i.d., circularly symmetric, and distributed as $\tilde{g}_m \in \mathcal{N}(0, \sigma_g^2)$, where σ_g is varied from 0.02 to 0.22. The DOAs are fixed at $\boldsymbol{\theta} = [9°,\ 12°]^T$, which corresponds to approximately half-beamwidth separation.

We compare the following DOA estimation methods:

- MUSIC with no calibration.

- Auto-calibration using the W-F approach.
- Auto-calibration with the MAP method (3.26), using a correct statistical model of the array perturbations.
- MUSIC using linear interpolation of $\Re\{\hat{q}_m(\theta)\}$ and $\Im\{\hat{q}_m(\theta)\}$.
- MUSIC with a global full-correction matrix $\mathbf{Q}$, which is estimated using calibration data.
- MUSIC with LPA-based interpolation of $\mathbf{q}(\theta)$, with $L=1$ (locally linear model). A Gaussian window is used, where the window parameter h is optimized using cross-validation (CV) as described in Section 3.6.2. This is termed LPA CV.
- Similar to LPA CV, but optimal h is found using the true function $q_k(\theta)$. This "genie-aided" method, which is included as a reference, is called "LPA genie" in the plot.
- MUSIC with a global $\mathbf{Q}=\text{diag}\{\mathbf{q}\}$, which is estimated using calibration data. Note that this is the "correct" model in this example.

The W-F approach is included to illustrate the identifiability problem. With a nonlinear array, the method yields perfect gain and DOA estimates in all cases, since there are no finite sample errors. Note that the Fourier-based interpolation technique in Section 3.6.1 is omitted in this example. As will be seen in Example 3.3, the method is not applicable with such a sparse calibration grid ($\Delta\eta=5°$).

Figure 3.3 shows the average RMSE of the different estimates of θ_1 and θ_2 versus the standard deviation of the gain errors. It is seen that the uncalibrated MUSIC estimates deteriorate significantly as the size of the channel errors increases. When the standard deviation of the gain errors is larger than 0.14, the failure rate exceeds 40%, so these RMSE values are not included in the plots. The W-F is unable to improve on the uncalibrated case, as expected. The MAP method shows surprisingly good performance, especially for small perturbations. This clearly shows that a correct application of prior knowledge can mitigate the identifiability problem. However, MAP deteriorates rapidly as the size of the channel errors increases.

In contrast, the methods that use calibration data are insensitive to the size of the channel perturbations. It is also seen that linear interpolation performance is the worst in this scenario because of the inability to take advantage of the smoothness in $\mathbf{q}(\theta)$ (which is in fact constant in this case). Also, estimating a full-correction matrix $\mathbf{Q}$ from the calibration data results in higher variance of the estimates due to the large number of parameters, although the "true" model $\mathbf{Q}=\text{diag}\{\mathbf{q}\}$ is included as a special case.

Using the correct model (global $\mathbf{q}$) will of course result in the lowest DOA RMSE. The LPA interpolation reaches essentially the same RMSE if the optimal (which should be "large") smoothing parameter h is found (LPA genie), but the estimation of h from data by cross-validation (LPA CV) results in a notable performance penalty.

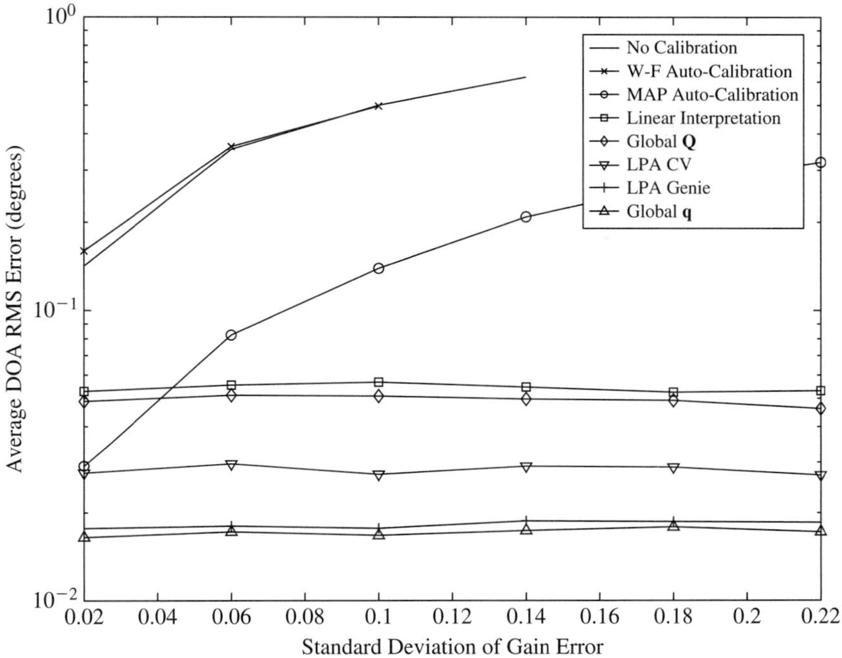

FIGURE 3.3 Empirical RMSE of a DOA estimate versus element gain perturbations size (channel errors only).

Example 3.2 *Mutual Coupling Errors*

In the second case we consider the effect of mutual coupling errors. Referring to Equation (3.43), the gain matrix is generated as $\mathbf{G} = \mathbf{I} + \tilde{\mathbf{G}}$, where the elements of $\tilde{\mathbf{G}}$ are random. The magnitude of the perturbations is scaled so that the diagonal elements are the largest and the error magnitude decreases across the subdiagonals. The elements of $\tilde{\mathbf{G}}$ are generated as independent complex Gaussian random variables: $\tilde{g}_{ij} \in \mathcal{N}(0, (0.1e^{-1.4\sqrt{|i-j|}})^2)$. The scaling is inspired by a mutual coupling model for a ULA of half-wavelength dipoles, the rationale being that the relative errors should be approximately constant. For simplicity, the nominal model $\mathbf{G}_0 = \mathbf{I}$ is used in place of a more realistic coupling model, keeping in mind that a known coupling does not affect the quality of the DOA estimates significantly (Svantesson, 1998). In this case, the DOA of the first source is varied from $\theta_1 = 5°$ to $\theta_1 = 11°$, whereas the second DOA is fixed at $\theta_2 = 12°$.

Figure 3.4 shows the empirical average RMSE of the MUSIC estimates of θ_1 and θ_2 using no calibration, global $\mathbf{q}$ calibration, linear interpolation of $\mathbf{q}(\theta)$, spline interpolation of $\mathbf{q}(\theta)$, LPA-based interpolation of $\mathbf{q}(\theta)$, and global $\mathbf{Q}$ calibration—the last being the "true" model. As expected, the global $\mathbf{Q}$ calibration performs best, whereas a diagonal $\mathbf{q}$ results in a substantial bias despite the errors to $\mathbf{G}$ being diagonal-dominant. The reference LPA genie achieves the

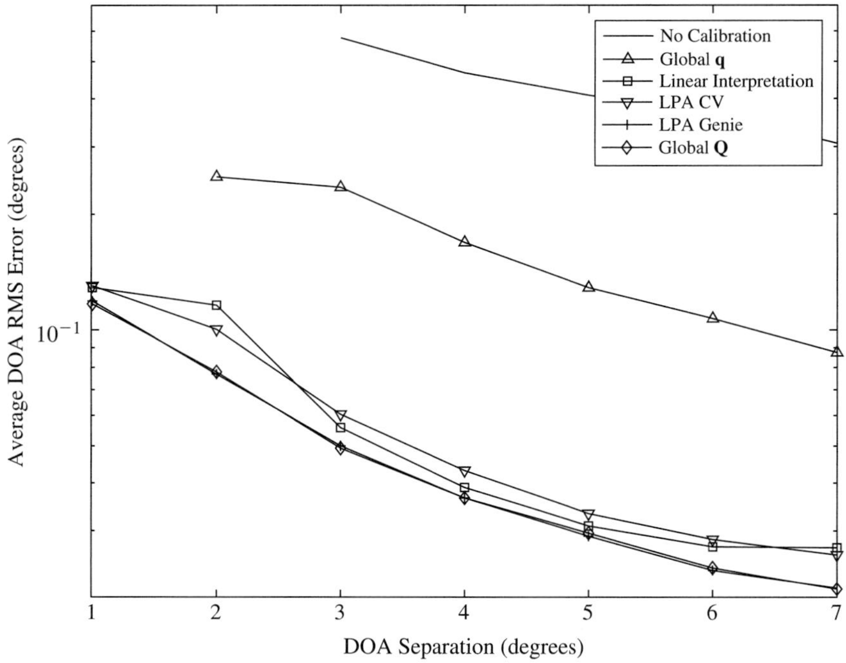

FIGURE 3.4 Empirical RMSE of a DOA estimate versus DOA separation (mutual coupling errors only).

same (optimal) performance as global $\mathbf{Q}$, whereas the LPA CV and the linear interpolation approaches are comparable, and their RMSE is slightly higher than that of global $\mathbf{Q}$ and LPA genie.

Examples 3.1 and 3.2 illustrate the performance of the various methods under DOA-independent (i.e., global) array perturbations. In both cases, as expected, the method that uses the correct model performs best. However, it is also clear that an interpolation-based approach, which makes no assumptions regarding the error model, can give almost the same performance. If an optimal smoothing parameter can be found, the LPA approach achieves the lowest RMSE. However, estimating h using CV results in a penalty, which means better methods are needed. We emphasize that the model-free interpolation must be applied to the correction data $\hat{\mathbf{q}}(\eta_c) = \hat{\mathbf{a}}_c./\mathbf{a}(\eta_c)$, not to the original calibration data $\hat{\mathbf{a}}_c$ directly. Although not shown here, the latter approach fails completely in the studied scenarios because of the relative sparseness of the calibration data, which leads to too high a variability in the true function $\mathbf{a}_c(\eta)$ between the sampling points $\{\eta_c\}_{c=1}^{C}$.

3.7.3 DOA-Dependent Modeling Errors

In the second case, both mutual coupling and element position errors are present. The gain matrix is generated as in Example 3.2. Referring to Equation (3.43), the x_m and y_m positions of the sensor elements are perturbed by i.i.d. $N(0, 0.05^2)$ random variables. The DOAs are fixed at $\boldsymbol{\theta} = [9°,\ 12°]^T$. The examples illustrate the importance of calibration data quality.

Example 3.3 *Varying Calibration Grid*

In this example, the calibration grid is varied from $\Delta\eta = 1°$ to $\Delta\eta = 20°$. The standard deviation of the calibration vectors is fixed at $\sigma_a = 0.01$. The uncalibrated MUSIC method fails completely in this scenario, as does the global $\mathbf{q}$ interpolation. The global $\mathbf{Q}$ approach yields reasonably good performance, but is omitted in the plots. Instead, we include results for manifold separation (Fourier-based interpolation). The method is implemented by solving the associated LS problem (3.40), although it is indeed highly ill-conditioned since the calibration data is not available over the whole field of view. The dimension of the virtual array is chosen as $\tilde{M} = 49$, which is an overall optimal choice for the smallest calibration grids $\Delta\eta \in \{1°, 2°, 3°\}$. Once the truncated sampling matrix $\hat{\mathbf{G}}$ $(M \times \tilde{M})$ is computed, it is used together with the Fourier vector (3.39) to interpolate the array response model to any desired DOA θ.

Figure 3.5 shows the empirical average RMSE versus the grid size. A number of effects can be seen. The first is that the amount of smoothing depends on

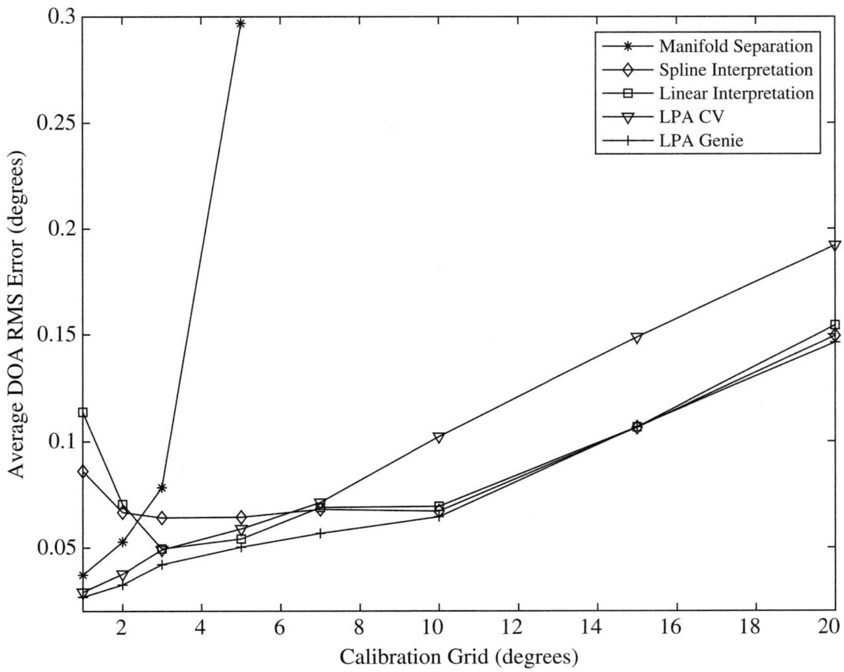

FIGURE 3.5 Empirical RMSE of a DOA estimate versus calibration data grid size (position and mutual coupling errors only).

the complexity of the model—that is, the variability of the true function to be interpolated. A sparse grid size means high variability, which favors the most local interpolation method (simple linear or spline interpolation). When more calibration data is available, there is less variation between the samples and more smoothing is necessary. Consequently, the LPA approaches outperform the others for finer interpolation grids. The second observation is that determining the optimal smoothing parameter h is increasingly difficult when less calibration data is available. For dense sampling, the LPA CV approach performs close to the LPA genie design, whereas for sparse calibration data, the optimal h is only found by the genie-aided optimization procedure, which cannot be implemented in a practical situation.

Figure 3.5 also illustrates that the manifold separation approach suffers from not taking advantage of the known (although poor) nominal response model. Thus, the method requires more calibration data and is useful only for $\Delta\eta \leq 3°$ in the chosen scenario. The other interpolation techniques also perform acceptably for $\Delta\eta = 20°$ or more, but this is only possible because of the known nominal array response model. Finally, there is no systematic difference between linear and spline interpolation in this case.

Example 3.4 *Varying Calibration Error Size*

In this example the calibration grid is fixed at $\Delta\eta = 5°$, whereas the standard deviation of the calibration vectors is varied from $\sigma_a = 0.01$ to $\sigma_a = 0.1$. The rest of the scenario is the same as in Example 3.3. Figure 3.6 displays the empirical average RMSE versus σ_a. It also illustrates how the quality of the calibration data affects the optimal amount of smoothing in the interpolation. For increasing variance of the calibration vectors, the LPA approach increasingly outperforms the simple interpolation approaches. It is also interesting that linear interpolation gives consistently better results than spline interpolation when calibration errors increase. One possible explanation may be that spline interpolation fits a slightly more complicated model to nonperfect data and thus suffers more from overfitting when the calibration measurements are noisy.

Examples 3.3 and 3.4 show that array interpolation using a nominal model can give highly accurate DOA estimates even if the calibration data is imperfect. Both cases illustrate the trade-offs involved in choosing the amount of smoothing to be applied. If the calibration data is accurate and the true function is highly variable (sparse sampling), a simple linear interpolation is a good choice. However, as the calibration data deteriorates or as the true function becomes smoother, a local interpolation technique like LPA can give superior performance, provided a good choice of smoothing parameter h is found.

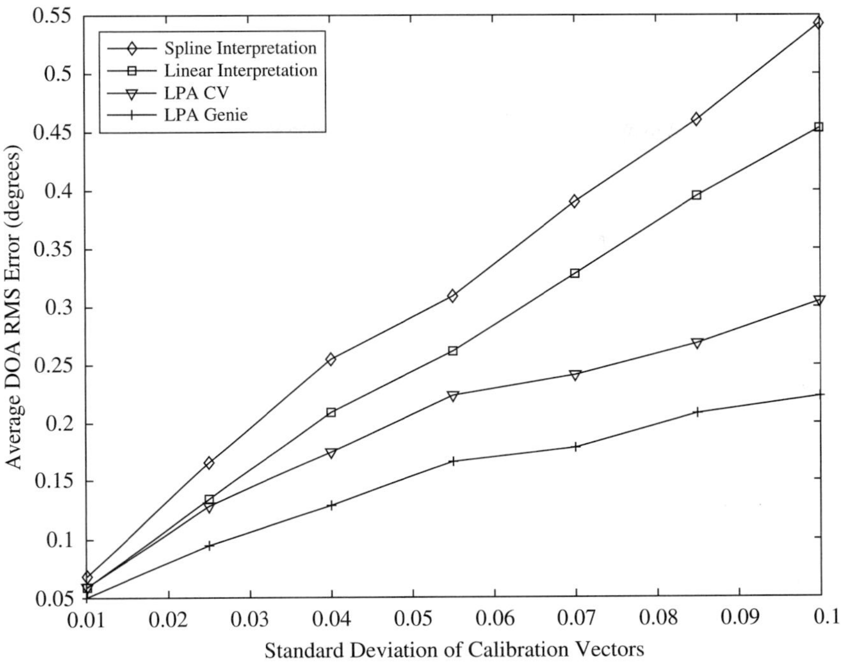

FIGURE 3.6 Empirical RMSE of a DOA estimate versus the standard deviation of calibration vectors (position and mutual coupling errors only).

3.8 CONCLUSION

This chapter presented methods for software calibration (or model update) applied to high-resolution DOA estimation. The methods are divided into categories depending on the involved assumptions. Auto-calibration techniques do not require any special calibration data, but a detailed error model must be specified. If the error model has many unknown parameters, the joint estimation of DOA and array parameters is likely to be ill-conditioned, leading to poor estimates or even nonuniqueness. The situation can to some extent be mitigated if prior information regarding the array parameters is available, in the form of nominal values and standard deviations. For large errors or in the absence of a specific error model, it is necessary to collect calibration data using sources at known positions. This is in fact a standard procedure before operational use of an antenna array. If a model is known, estimation of error parameters is straightforward when calibration data is available. A more realistic case is when a nominal model is available but the precise effect of the various error sources cannot be accurately quantified.

We demonstrated that interpolation of a correction factor for each sensor is viable for this situation. Such as approach is model based in that it exploits information regarding the nominal array but assumes no specific form for the correction. The two extreme cases of interpolation are simple linear interpolation, which uses no smoothing, and estimation of a global interpolation factor, which results in maximum smoothing. Local polynomial approximation (LPA) was introduced as a technique for "interpolating" between these extreme cases. The amount of smoothing is automatically adapted to the variability of the true function using cross-validation. Although this gives satisfactory results in most scenarios, performance is still limited by the choice of smoothing parameter h. Thus, more investigations regarding this aspect should be carried out, in particular trying "locally adaptive" methods where h is allowed to depend on the direction of arrival.

A final class of techniques discussed was Fourier-based array interpolation. The advantages of this approach are that no information regarding the array response is necessary, and that the Fourier-based array model allows computationally efficient DOA estimation. The drawback is that a relatively dense calibration grid is necessary to yield satisfactory results, to allow estimation of a sufficient number of Fourier coefficients (or phase modes).

The different calibration techniques were presented in their basic form, involving a single-direction parameter only and a scalar "wavefield." If a model can be specified, the extension of auto-calibration and parametric calibration techniques to multiple parameters per sources and/or polarization-sensitive antennas can be conceptually simple. However, practical applicability is limited by the increasing number of parameters. Also, interpolation of a correction factor is straightforward in principle, but computational complexity significantly increases.

It should be noted that one of the most investigated applications of LPA is image smoothing (Katkovnic et al., 2006), commonly referred to as moving least squares in the image-processing community (Bose and Ahuja, 2006). Image smoothing is a two-dimensional (2D) problem, corresponding to, for example, calibration over azimuth and elevation or azimuth and frequency. The increased complexity makes the search for computationally more efficient alternatives of great interest. An interesting alternative is manifold separation, if the implementation can be based on 2D FFT. The extension of the presented methods to a diversely polarized array, where the two polarization components (e.g., horizontal and vertical) can be measured separately, is straightforward, since the components act as separate arrays. Although it is more complicated to apply the techniques to circularly polarized antenna elements receiving signals of different polarizations, this is an interesting extension.

Besides those just mentioned, many other questions related to array calibration deserve further study. One interesting direction is to combine parametric and nonparametric techniques. A simple, but not necessarily correct, parametric model can capture most of the variability in the array model. What is left can then be corrected for using nonparametric interpolation. It is an open issue whether or not this would improve practical performance.

Regarding local polynomial approximation, one could question if choices of basis function other than polynomials can give better performance. In fact, using Fourier-based vectors would create a link between this approach and the Fourier-based array interpolation in Section 3.5.1. This combination could lead to a local Fourier-based model with potentially better performance, although at the expense of greater computational cost.

Finally, there is a need to develop tools for predicting performance of the different approaches, in terms of a statistical analysis for a given (known) model and methods for predicting performance using available data only. Robust testing based on the Bootstrap (Zoubir and Boashash, 1998), which is closely related to cross-validation, is a natural choice for this task.

Acknowledgement The authors would like to thank Professor Visa Koivunen of Helsinki University of Technology for clarifications regarding implementation of the manifold separation approach.

REFERENCES

Agrawal, A., Jablon, A., 2003. A calibration technique for active phased array antennas. In: IEEE International Symposium on Phased Array Systems and Technology, pp. 223–228.

Allen, J., 1964. Array antennas: new applications for an old technique. IEEE Spectr. 1 (11), 115–130.

Astély, D., Swindlehurst, L., Ottersten, B., 1999. Spatial signature estimation for uniform linear arrays with unknown receiver gains and phases. IEEE Trans. 47 (8), 2128–2138.

Aumann, H., Fenn, A., Willworth, F., 1989. Phased array antenna calibration and pattern prediction using mutual coupling measurements. IEEE Trans. Antennas Propag. 37 (7).

Barabell, A., 1983. Improving the resolution performance of eigenstructure-based direction-finding algorithms. In: Proc. ICASSP, pp. 336–339.

Belloni, F., Richter, A., Koivunen, V., 2007. DOA estimation via manifold separation for arbitrary array structures. IEEE Trans. Signal Process. 55 (10), 4800–4810.

Bienvenu, G., Kopp, L., 1979. Principle de la goniometrie passive adaptive. In: Proc. 7'eme Colloque GRESIT, 106/1–106/10.

Bose, N., Ahuja, N., 2006. Superresolution and noise filtering using moving least squares. IEEE Trans. Image Process. 15 (8), 2239–2248.

Buhren, M., Pesavento, M., Bohme, J., 2004. Virtual array design for array interpolation using differential geometry. In: Proc. ICASSP, pp. 229–232.

Collin, R., 1992. Foundations for Microwave Engineering, 2nd ed. McGraw-Hill.

Dandekar, K., Ling, H., Xu, G., 2000. Smart antenna array calibration procedure including amplitude and phase mismatch and mutual coupling effects. In: Proc. ICPWS, pp. 293–297.

Doron, M., Doron, E., 1994a. Wavefield modeling and array processing, part I—spatial sampling. IEEE Trans. Signal Process. 42 (10), 2549–2559.

Doron, M., Doron, E., 1994b. Wavefield modeling and array processing, part II—algorithms. IEEE Trans. Signal Process. 42 (10), 2560–2570.

Doron, M., Doron, E., 1994c. Wavefield modeling and array processing, part III—resolution capacity. IEEE Trans. Signal Process. 42 (10), 2571–2580.

Elliott, R., 1958. Mechanical and electrical tolerances for two-dimensional scanning antenna arrays. IRE Trans. Antennas Propag. AP-6, 114–120.

Er, M., Ng, B., 1994. A new approach to robust beamforming in the presence of steering vector errors. IEEE Trans. Signal Process. 42 (7), 1826–1829.

Fan, J., Gijbels, I., 1996. Local polynomial modeling and its application. Chapman and Hall.

Fessler, J., Hero, A., 1994. Space-alternating generalized expectation-maximization algorithm. IEEE Trans. Signal Process. 42 (10), 2664–2677.

Fleury, B., Tschudin, M., Heddergott, R., Dahlhaus, D., Pedersen, K.I., 1999. Channel parameter estimation in mobile radio environments using the SAGE algorithm. IEEE J. Sel. Areas Comm. 17 (3), 434–450.

Flieller, A., Ferreol, A., Larzabal, P., Clergeot, H., 1995. Robust bearing estimation in the presence of direction-dependent modelling errors: Identifiability and treatment. In: Proc. ICASSP, pp. 1884–1887.

Friedlander B., Weiss, A., 1991. Direction finding in the presence of mutual coupling. IEEE Trans. Antennas Propag. 39 (3), 273–284.

Friedlander, B., Weiss, A., 1992. Direction finding using spatial smoothing with interpolated arrays. IEEE Trans. Aerosp. Electron. Syst. 28 (4), 574–587.

Gershman, A., Bohme, J., 1997. Improved DOA estimation via pseudorandom resampling of spatial spectrum. IEEE Signal Process. Lett. 4 (2), 54–57.

Golub, G., Pereyra, V., 2003. Separable nonlinear least squares: the variable projection method and its applications. Inverse Prob. 19 (2), R1–R26.

Gupta, I., Baxter, J., Ellingson, S., Park, H.-G., Oh, H., Kyeong, M., 2003. An experimental study of antenna array calibration. IEEE Trans. Antennas Propag. 51 (3), 664–667.

Jansson, M., Swindlehurst, A., Wahlberg, B., 1998. Weighted subspace fitting for general array error models. IEEE Trans. Signal Process. 46 (9), 2484–2498.

Josefsson, L., Persson, P., 1999. Conformal array synthesis including mutual coupling. Electron. Lett. 35 (8), 625–627.

Katkovnik, V., 1998. On adaptive local polynomial approximation with varying bandwidth. In: Proc. ICASSP, pp. 2321–2324.

Katkovnic, V., Egiazarian, K., Astola, J., 2006. Local Approximation Techniques in Signal and Image Processing. SPIE Press.

Krim, H., Viberg, M., 1996. Two decades of array signal processing research: the parametric approach. IEEE Signal Process. Mag. 13 (4), 67–94.

Lanne, M., Drackner, P.N., 2003. T/R-module mismatch and antenna pattern characteristics. In: IEEE International Symposium on Phased Array Systems and Technology, pp. 229–234.

Lanne, M., Lundgren, A., Viberg, M., 2006. Calibrating an array antenna for scan dependent errors using a sparse grid. In: Proc. 40th Asilomar Conference on Signals, Systems and Computers.

Lanne, M., Viberg, M., Lundgren, A., 2007. Adaptive beamforming using calibration vectors with unknown gain and phase. In: IEEE AP-S, pp. 4208–4211.

Lee, H., Wengrovitz, M., 1990. Resolution threshold of beamspace MUSIC for two closely spaced emitters. IEEE Trans. Acoust. Speech Signal Process. 38 (9), 1545–1559.

Levi, M., Messer, H., 1990. Sufficient conditions for array calibration using sources of mixed types. In: Proc. ICASSP, pp. 2943–2946.

Li, Q., Gan, L., Ye, Z., 2003. An overview of self-calibration in sensor array processing. In: Proc. ISAPE'03, Beijing, Signal Process, pp. 279–282.

Lundgren, A., Lanne, M., Viberg, M., 2007. Two-step ESPRIT with compensation for modeling errors using a sparse calibration grid. In: Proc. ICASSP, pp. II-1045–II-1048.

Mailloux, R., 2005. Phased Array Antenna Handbook, 2nd ed. Artech House.

Ng, B., Nehorai, A., 1995. Active array sensor localization. Signal Process. 44, 309–327.

Ng, B., See, C.S., 1996. Sensor-array calibration using a maximum-likelihood approach. IEEE Trans. Antennas Propag. 44 (6), 827–835.

Ottersten, B., Roy, R., Kailath, T., 1989. Signal waveform estimation in sensor array processing. In: Proc. 23rd Asilomar Conference on Signals, Systems and Computers, pp. 787–791.

Ottersten, B., Viberg, M., Kailath, T., 1992a. Analysis of subspace fitting and ML techniques for parameter estimation from sensor array data. IEEE Trans. Signal Process. SP-40, 590–600.

Ottersten, B., Viberg, M., Wahlberg, B., 1992b. Robust source localization based on local array response modeling. In: Proc. ICASSP.

Paulraj, A., Kailath, T., 1985. Direction-of-arrival estimation by eigenstructure methods with unknown sensor gain and phase. In: Proc. IEEE ICASSP, pp. 17.7.1–17.7.4.

Paulraj, A., Nabar, R., Gore, D., 2003. Introduction to Space-Time Wireless Communications. Cambridge University Press.

Pettersson, L., 1992. Adaptive beamforming with imperfect arrays: pattern effects and their partial correction. In: International Symposium on Antennas (JINA92), pp. 621–624.

Pettersson, L., Grahn, P., 2003. Experiences using DOA methods with a small digital beamforming antenna. In: Proc. IEEE International Symposium on Phased Array Systems and Technology, pp. 164–169.

Pierre, J., Kaveh, M., 1995. Experimental evaluation of high-resolution direction-finding algorithms using a calibrated sensor array testbed. Digit. Signal Process. 5 (4), 243–254.

Porat, B., Friedlander, B., 1997. Accuracy requirements in off-line array calibration. IEEE Trans. Aerosp. Electron. Syst. 32 (2), 545–556.

Rockah, Y., Schultheiss, P.M., 1987. Array shape calibration using sources in unknown locations—part I: far-field sources. IEEE Trans. Acoust. Speech Signal Process. 35 (3), 286–299.

Roy, R., Kailath, T., 1989. ESPRIT—estimation of signal parameters via rotational invariance techniques. IEEE Trans. Acoust. Speech Signal Process. 37 (7), 984–995.

Sakaguchi, K., Kuroda, K., Takada, J.-I., Araki, K., 2002. Comprehensive calibration for MIMO systems. In: Proc. WPMC, vol. 2, pp. 440–443.

Schmidt, R., 1979. Multiple emitter location and signal parameter estimation. In: Proc. RADC Spectr. Estimation Workshop, Rome, NY, pp. 243–258.

Schmidt, R., 1986. Multiple emitter location and signal parameter estimation. IEEE Trans. Antennas Propag. 34 (5), 276–280. Reprinted from Proc. RADC Spectr. Estimation Workshop, Rome, NY, 1979.

Serebryakov, G., 1995. Direction-of-arrival estimation of correlated sources by adaptive beamforming. IEEE Trans. Signal Process. 43 (11), 2782–2787.

Shan, T., Wax, M., Kailath, T., 1985. On spatial smoothing for direction-of-arrival estimation of coherent signals. IEEE Trans. Acoust. Speech Signal Process. ASSP-33, 806–811.

Steyskal, H., Herd, J., 1990. Mutual coupling compensation in small array antennas. IEEE Trans. Antennas Propag. 38 (12), 1971–1975.

Stoica, P., Nehorai, A., 1990. Performance study of conditional and unconditional direction-of-arrival estimation. IEEE Trans. Acoust. Speech Signal Process. ASSP-38, 1783–1795.

Stoica, P., Sharman, K., 1990. A novel eigenanalysis method for direction estimation. IEE Proc. Radar Signal Process. 137 (1), 19–26.

Svantesson, T., 1998. The effects of mutual coupling using a linear array of thin dipoles of finite length. In: Proc. SSAP, pp. 232–235.

Swindlehurst, A., Kailath, T., 1992. A performance analysis of subspace-based methods in the presence of model errors—part I: the MUSIC algorithm. IEEE Trans. Signal Process. SP-40 (7), 1758–1774.

Swindlehurst, A., Kailath, T., 1993. A performance analysis of subspace-based methods in the presence of model errors: part 2—multidimensional algorithms. IEEE Trans. Signal Process. SP-41, 2882–2890.

Trees, H.V., 2002. Optimum Array Processing. John Wiley & Sons.

Viberg, M., Ottersten, B., Kailath, T., 1991. Detection and estimation in sensor arrays using weighted subspace fitting. IEEE Trans. Signal Process. SP-39 (11), 2436–2449.

Viberg, M., Swindlehurst, A., 1994a. Analysis of the combined effects of finite samples and model errors on array processing performance. IEEE Trans. Signal Process. 42 (10), 3073–3083.

Viberg, M., Swindlehurst, A., 1994b. A bayesian approach to auto-calibration for parametric array signal processing. IEEE Trans. Signal Process. 42, 3495–3507.

Wahlberg, B., Ottersten, B., Viberg, M., 1991. Robust signal parameter estimation in the presence of array perturbations. In: Proc. ICASSP 91 Conference, pp. 3277–3280.

Wax, M., Kailath, T., 1985. Detection of signals by information theoretic criteria. IEEE Trans. Acoust. Speech Signal Process. ASSP-33 (2), 387–392.

Weiss, A.J., Friedlander, B., 1989. Array shape calibration using sources in unknown locations—a maximum likelihood approach. IEEE Trans. Acoust. Speech Signal Process. 37 (12), 1958–1966.

Weiss, A.J., Friedlander, B., 1990. Eigenstructure methods for direction finding with sensor gain and phase uncertainties. Circ. Syst. Sig. Proc. 9 (3), 271–300.

Xu, G., Kailath, T., 1994. Fast subspace decomposition. IEEE Trans. Signal Process. 42 (3), 539–551.

Xu, W., Kaveh, M., 1996. Comparative study of the biases of MUSIC-like estimators. Signal Process. 50 (1–2), 39–56.

Yang, B., 1995. Projection approximation subspace tracking. IEEE Trans. Signal Process. 43 (1), 95–107.

Zhao, L., Krishnaiah, P., Bai, Z., 1986. On detection of the number of signals in the presence of white noise. J. Multivar. Anal. 20 (1), 1–25.

Ziskind, I., Wax, M., 1988. Maximum likelihood localization of multiple sources by alternating projection. IEEE Trans. Acoust. Speech Signal Process. ASSP-36, 1553–1560.

Zoubir, A., Boashash, B., 1998. The bootstrap and its application in signal processing. IEEE Signal Process. Mag. 15 (1), 56–76.

Chapter 4

Narrowband and Wideband DOA Estimation for Uniform and Nonuniform Linear Arrays

T. Engin Tuncer, T. Kaya Yasar, Benjamin Friedlander

4.1 INTRODUCTION

Direction-of-arrival (DOA) estimation for narrowband signals is well known in the literature. Recently, interest in nonuniform linear arrays has grown since they require fewer sensor and receiver structures and their performance can be increased via techniques such as array mapping. It is known that the DOA performance of nonuniform linear arrays can be improved when the data for the missing sensors is compensated (Tuncer et al., 2007a). Array mapping is an effective technique for this purpose, since it can improve the computational complexity and performance. In this chapter, different array-mapping techniques are considered and examples of DOA performance are outlined for the narrowband case.

Wideband direction of arrival has been explored for many years. Some signals are wideband, and so significant processing gain through wideband processing can be obtained. In wideband processing, unlike narrowband processing, several frequency bins carry different information regarding source DOA angles. This generates diversity and processing gain for DOA estimation. In this chapter, both well-known and new approaches are considered for wideband DOA estimation.

The main problem in wideband DOA estimation is how the information at each frequency can be combined to obtain the most accurate DOA information. Two main types of wideband processing are coherent and noncoherent processing. In coherent wideband processing, the covariance matrix at each frequency is combined coherently using a transformation and the final covariance matrix is used for DOA estimation. In noncoherent wideband processing, the DOA at each frequency is estimated separately and the results are combined noncoherently.

Classical and Modern Direction-of-Arrival Estimation
Copyright © 2009 by Academic Press, Inc. All rights of reproduction in any form reserved.

While several coherent wideband processing methods are discussed in the literature, the main philosophy behind them is the same: They all use a transformation matrix that maps the covariance matrices at different frequencies to a covariance matrix at the center frequency. The main differences among these methods involve the transformation matrix. The aforementioned philosophy for wideband processing was established in Hung and Kaveh (1988) and Wang and Kaveh (1985, 1987) and later named the coherent signal-subspace method (CSM). The transformation matrix is desired to be unitary in order to avoid focusing loss (Hung and Kaveh, 1988)—that is, the signal-to-noise (SNR) loss for the transformed system. The mapping matrices in Hung and Kaveh (1988) belong to the family of rotational signal subspaces (RSSs), and they do not have focusing loss.

The performance of the transformation matrix designed by the RSS technique is very good as long as the angular sector for direction of arrivals (DOAs) is small. The method has significant bias as the angular sector for the array mapping is increased. Another limitation of RSS is that it is configured for the design of square transformation matrices where the number of sensors or frequency bins for both the domain and range spaces is the same. In this chapter, we show that this limitation can be easily circumvented, allowing RSS to be used for the design of rectangular transformation matrices. This corresponds to generating different numbers of sensors or frequency bins for real and virtual arrays. The design of the square transformation matrix is further developed in Krolik and Swingler (1989, 1990) by considering, respectively, the steered covariance matrix and resampling in the spatial domain.

In Doron and Weiss (1992), a general class of transformation, namely signal subspace transformation (SST), is discussed and several properties are outlined. More specifically, the set of transformation matrices for SST covers nonunitary matrices that do not lead to focusing loss. A short review of these methods can be found in Sellone (2006). Another approach for array mapping is array interpolation (Bronez, 1988; Friedlander, 1993; Friedlander and Weiss, 1992; Hyberg et al., 2005), in which the transformation matrix can be designed in different ways (Doron et al., 1993; Friedlander and Weiss, 1993). In Friedlander and Weiss (1993), several interesting characteristics of coherent wideband processing using array interpolation are outlined. Weighted average of the signal subspace (WAVES) (Elio et al., 2001) is proposed to deal with various error sources. A Bayesian approach for the joint estimation of model order and DOA is presented in Ng et al. (2005).

Recently, the Wiener array interpolation technique, where the source and noise powers are estimated using a maximum likelihood method (Kaya Yasar and Tuncer, 2008; Tuncer et al., 2007a,b), was proposed. The use of source and noise power estimates makes the Wiener formulation a practically attractive approach. It was shown that its power estimation is accurate and that its performance is better than that of classical array interpolation, especially at low SNR (Kaya

Yasar and Tuncer, 2008). Another advantage is its improvement of the condition number of the inverse matrix for the array interpolation matrix. It also allows more sensors in the virtual array than in the real array. This feature of Wiener array interpolation is especially useful when fast algorithms like Root-MUSIC are used for nonuniform linear arrays.

Noncoherent wideband DOA estimation methods were developed before coherent techniques (Wax et al., 1982, 1984). While noncoherent wideband processing is a simple extension of narrowband processing, it is effective for DOA estimation. Its main problem is the computational complexity, since an eigendecomposition is needed for each frequency bin. This problem is worsened by the fact that noncoherent techniques require a search in the dense grid in order to avoid local minima. Computational complexity increases linearly as the number of frequency bins increases. Orthogonality of projected subspaces (TOPS) is presented in Yoon et al. (2006) as a way to achieve good performance at medium SNR levels. It is essentially a noncoherent method.

In this chapter, we focus on different array-mapping techniques for coherent wideband DOA estimation. Narrowband and wideband techniques are compared, and the advantages of both are outlined. One of the main problems in DOA estimation is the correlation of source signals. When the source signals are correlated, the algorithms' DOA performance degrades. The worst case is obtained when the source signals are coherent. For coherent sources, the CLEAN algorithm is adopted to improve performance. CLEAN is well known in astronomy (Clark, 1980) and radar signal processing (Deng, 2004) for obtaining clean images from those corrupted by multiple sources.

The advantages of the CLEAN algorithm are not well known in DOA estimation except for some limited research in beamforming (Stoica and Moses, 2005). The main reason is that CLEAN produces only marginal improvement for well-separated uncorrelated multiple sources. For correlated sources, its advantage becomes more obvious. The difference is more significant for coherent sources. In this chapter, CLEAN is adopted for both narrowband and wideband processing. Several simulations are done to show its performance in a variety of scenarios.

4.2 ARRAY MODELS

Linear array geometry can be constructed by using uniformly or nonuniformly spaced array elements. Nonuniform linear sensor arrays have variants depending on their covariance matrix lags. The following section describes the details of these arrays.

4.2.1 Uniform Linear Arrays

Sensors are placed uniformly on a straight line for a uniform linear array (ULA). The inter-element distance is usually less than a half-wavelength to avoid spatial ambiguity. Uniform linear arrays are one of the simplest forms of sensor arrays, and they have several useful properties. For example, forward-backward spatial smoothing (Pillai and Kwon, 1989) can only be applied to ULAs because of the Vandermonde structure of the array steering matrix. Furthermore, fast subspace algorithms like Root-MUSIC can be used in ULA, which increases computational efficiency. Usually it is desirable to increase the array aperture in order to improve resolution. This can be achieved by increasing the number of sensors when the inter-element distance is kept fixed for the ULA. However, adding elements is expensive and it increases system as well as computational complexity. One remedy for this problem is to keep the number of sensors the same and increase the distance between them. Thus, we obtain nonuniform linear arrays.

4.2.2 Nonuniform Linear Arrays

In nonuniform linear arrays (NLAs), sensors are usually placed at integer multiples of a unit distance $d < \lambda/2$, where λ is the wavelength. Assuming that the first sensor is the reference and is at the origin, the sensor displacements are $d_{NLA} = [0\ d_2\ \ldots\ d_M]$ for an NLA with M sensors. The treatment of the noninteger case can be found in Abramovich et al. (2000).

Since several sensors are placed with inter-element distances larger than a half-wavelength, there is a possibility of spatial ambiguity, where two different DOA angles generate the same covariance matrix, making it impossible to identify the true source DOA. The possibility of ambiguous DOA measurement is usually low as long as there is at least one pair of sensors with an inter-element distance of less than a half-wavelength.

Nonuniform linear arrays try to cover a large array aperture with a limited number of sensors. Therefore, efficient use of sensors is the priority. Since NLAs cover an aperture with fewer sensors, their performance is worse than that of ULAs with the same aperture. However, their performance is much better than ULAs with the same number of sensors. The target in DOA estimation with an NLA is to obtain the performance and advantages of the equivalent-aperture ULA. Root-MUSIC can still be applied, but its performance is usually worse than Spectral-MUSIC because missing sensors lead to large errors in polynomial root finding.

Co-array is used to characterize array structures. It can be defined as a symmetric function that represents the number of times each spatial correlation lag is contained for a given array structure. If we define $\mathbf{h}$ as the vector or sequence where the existence of a sensor at a certain position is denoted as one or zero, the co-array is obtained as the result of the following convolution operation:

$$c[n] = h[n] * h[-n] \tag{4.1}$$

For example, consider a 3-element NLA where $d_{NLA} = [0\ \ 1\ \ 3]$, $\mathbf{h}$ is given as $\mathbf{h} = [1\ \ 1\ \ 0\ \ 1]$, and $\mathbf{c} = [1\ \ 1\ \ 1\ \ 3\ \ 1\ \ 1\ \ 1]$. It does not have redundancy for the positive and negative covariance lags and has three zero lags. Arrays that have no redundancy and missing covariance lags are called perfect arrays.

There are different types of NLA, some of which are described next.

Nonredundant NLAs

Nonredundant arrays have co-arrays that are either zero or one except at the origin. In other words, there is only one element for any covariance lag. Certain covariance lags may not be present for these arrays and co-arrays may contain zeros. Table 4.1 presents a short list of nonredundant arrays. Those with $M \leq 4$ are called perfect arrays with full co-arrays. A more complete list can be found in Abramovich et al. (1999).

Nonredundant arrays use sensors very effectively, covering a large aperture. However, while this increases array resolution, it comes with a price. The array covariance matrix does not carry the same information as the equivalent ULA with the same aperture. Therefore, DOA performance of nonredundant arrays is worse than that of the same-aperture ULA. The missing sensors cause problems especially when the source signals are correlated or coherent.

Minimum Redundant NLAs

Minimum redundant arrays have no zeros or holes for their co-arrays. They have all the covariance lags, some of which may repeat. Therefore, a covariance lag may be represented by the correlation between more than one pair of sensors. Minimum redundant arrays have the largest aperture, although they have no gap in their co-arrays. Table 4.2 presents a short list of minimum redundant arrays.

TABLE 4.1 Nonredundant Arrays

M	Sensor Separation
2	1
3	1 2
4	1 3 2
5	1 3 5 2
6	1 3 6 2 5
7	1 3 6 8 5 2
8	1 3 5 6 7 10 2
9	1 4 7 13 2 8 6 3
10	1 5 4 13 3 8 7 12 2

TABLE 4.2 Minimum Redundant Arrays

M	Sensor Separation
2	1
3	1 2
4	1 3 2
5	1 3 3 2 , 3 4 1 1
6	1 3 1 6 2 , 1 5 3 2 2 , 1 1 4 4 3
7	1 3 6 2 3 2, 1 1 4 4 4 3 , 1 1 1 5 5 4, 1 1 6 4 2 3, 1 7 3 2 2 2
8	1 3 6 6 2 3 2 , 1 1 9 4 3 3 2

A more complete list can be found in Linebarger et al. (1993) and Van Trees (2002).

The covariance matrix of a minimum redundant array can be completed to make it similar to that of a ULA with the same aperture. Since all covariance lags are present, the covariance matrix can be completed effectively. There are different approaches for this purpose (Abramovich et al., 1998; Pillai and Haber, 1987). It turns out that it is possible to improve the DOA accuracy of minimum redundant arrays by covariance matrix completion.

Partly Filled NLAs

Partly filled NLAs have a uniform linear array portion in their array structure, making them a combination of ULAs and NLAs. This gives them certain advantages, especially for coherent sources (Tuncer et al., 2007a). Forward-backward spatial smoothing (Pillai and Kwon, 1989) can be used for the ULA part to obtain an initial estimate, which can be improved by subsequent processing. An example of a partly filled NLA is $d_{NLA} = [0 \;\; 1 \;\; 2 \;\; 3 \;\; 4 \;\; 8 \;\; 10 \;\; 11]$. The ULA part can be at the beginning or end or in the middle.

A partly filled NLA has many repeating covariance lags as opposed to minimum redundant or nonredundant arrays. This is an inefficient use of sensor elements, but it is advantageous for obtaining more accurate covariance matrices. In this respect, partly filled NLAs fill the gap between ULAs and NLAs. Furthermore, they can be used for coherent sources effectively since an initial DOA estimate can be obtained from the ULA part with forward-backward spatial smoothing. Sometimes partly filled and minimum redundant NLAs overlap. Examples of such arrays are presented in Table 4.3.

Another classification divides arrays as fully augmentable and partially augmentable. In fully augmentable arrays, all the covariance lags exist and

TABLE 4.3 Partly Filled and Minimum Redundant Arrays

M	Sensor Separation
7	1 1 1 5 5 4
12	1 1 1 (20) 5 4 4 4 4 3 3
13	1 1 1 (24) 5 4 4 4 4 4 3 3

their co-arrays do not have missing lags. Minimum redundant arrays are fully augmentable. Partially augmentable arrays have certain missing lags, and their co-arrays have holes. Nonredundant arrays are partially augmentable if the number of sensors is larger than four.

4.3 NARROWBAND DIRECTION-OF-ARRIVAL ESTIMATION

A narrowband model for array signal processing is the most frequently used framework because of its relative computational efficiency and simplicity. The following section describes this model and the DOA estimation techniques.

4.3.1 Narrowband Signal Model

In the narrowband model, it is assumed that the time-bandwidth product is small:

$$B\tau \ll 1 \tag{4.2}$$

where τ is the time that the wave propagates the array; n plane waves (or sources) impinge on a sensor array of M elements. The sensor array output, $\mathbf{y}_r(t)$, is an $M \times 1$ complex vector given as

$$\mathbf{y}_r(t) = \mathbf{A}_r(\theta)\mathbf{s}(t) + \mathbf{v}(t) \tag{4.3}$$

where the signal, $\mathbf{s}(t)$, and the noise, $\mathbf{v}(t)$, are uncorrelated. The following assumptions are required for subspace algorithms like MUSIC (Stoica and Nehorai, 1998).

Assumption 1. The number of sensors is greater than the number of sources (i.e., $M > n$), and the steering vectors in $\mathbf{A}_r(\theta)$ are linearly independent.

Assumption 2. Noise is zero mean and is both spatially and temporally white (i.e., $E\{\mathbf{v}(t)\} = 0$ and $E\{\mathbf{v}(t)\mathbf{v}^H(t)\} = \sigma_v^2\mathbf{I}$).

Assumption 3. The number of snapshots is greater than the number of sensors, $N \geq M$, and source signals are zero mean with a positive semi-definite covariance matrix, $\mathbf{R}_s = E\{\mathbf{s}(t)\mathbf{s}^H(t)\} \geq 0$.

In addition, sources are assumed to generate far-field planar waves in an isotropic and nondispersive medium. The sensors are identical, perfectly calibrated with no gain and phase mismatch as well as no mutual coupling.

$\mathbf{s}(t)$ is a vector of source signals

$$\mathbf{s}(t) = [s_1(t), \ldots, s_n(t)]^T \tag{4.4}$$

It is assumed that $\mathbf{A}_r(\theta)$ is the $M \times n$ steering matrix of the real array, which here corresponds to the physical array. When the virtual array is obtained by array interpolation, the steering matrix is denoted as $\mathbf{A}_v(\theta)$. $\mathbf{A}_r(\theta)$ is composed of steering vectors for each source:

$$\mathbf{A}_r(\theta) = [\mathbf{a}(\theta_1) \ldots \mathbf{a}(\theta_n)] \tag{4.5}$$

The steering vectors are given as

$$\mathbf{a}(\theta_i) = \left[1 \ \exp\left(j\frac{2\pi}{\lambda} d_2 \cos(\theta_i)\right) \ldots \exp\left(j\frac{2\pi}{\lambda} d_M \cos(\theta_i)\right) \right]^T \tag{4.6}$$

There are N snapshots or observations for $t = 1, 2, \ldots, N$. The covariance matrix for the array output is

$$E\left\{\mathbf{y}_r(t)\mathbf{y}_r^H(t)\right\} \triangleq \mathbf{R} = \mathbf{A}_r(\theta)\mathbf{R}_s\mathbf{A}_r^H(\theta) + \sigma^2\mathbf{I} \tag{4.7}$$

$$\mathbf{R}_s \triangleq E\left\{\mathbf{s}(t)\mathbf{s}^H(t)\right\} \tag{4.8}$$

The sample covariance matrix is the maximum-likelihood estimate of the theoretical covariance matrix and is given as

$$\hat{\mathbf{R}} = \frac{1}{N}\sum_{t=1}^{N} \mathbf{y}_r(t)\mathbf{y}_r^H(t) \tag{4.9}$$

It is a Gram matrix and is always positive semi-definite (Horn and Johnson, 1985). In practice, it becomes positive definite when the contribution of noise is considered.

4.3.2 Array Mapping for Narrowband Signals

Array mapping is an effective technique for both narrowband and wideband DOA estimation. It allows us to map an NLA to a virtual ULA with the same aperture, and it can be used to cohere covariance matrices for different frequencies. One strategy for DOA estimation is to first obtain an initial DOA estimate and then improve it iteratively. This strategy is sector independent and improves the accuracy of array mapping. Mapping accuracy, as well as the final DOA estimate, depends on the initial DOA estimate.

For coherent source signals the initial DOA estimation is an important problem for NLA. Different methods can be used to solve it. In this chapter, Toeplitz completion, the forward-backward approach, and forward-backward spatial smoothing are considered. It is also possible to use some special NLA structures to deal with multipath signals. The use of partly filled NLAs is promising since forward-backward spatial smoothing can be employed for their ULA part. Once an initial estimate is available, iterations improve DOA accuracy. Iterations do not monotonically decrease the error function, but only few iterations can lead to significant gains.

Different array-mapping techniques exist in the literature. One example, Davies's transformation (Davies, 1965), can be applied in a variety of cases, but it is sensitive to the array geometry and selection of parameters. Array interpolation (Bronez, 1988) is one of the most effective methods of array mapping. It is similar to simple sample interpolation in digital signal processing. The main difference between the two is the fact that array interpolation is model based. In fact, this is the power behind it.

In this section, classical array interpolation, Wiener array interpolation, and the coherent signal-subspace method with the rotational signal-subspace technique are considered. Furthermore, the initial DOA estimation for multipath signals is investigated using the forward-backward approach, Toeplitz completion, and forward-backward spatial smoothing.

Classical Array Interpolation

Array interpolation was introduced and developed in Bronez (1988), Doron and Weiss (1992), Doron et al. (1993), Friedlander (1993), and Friedlander and Weiss (1992). It allows us to obtain the sensor data in a virtual array to some accuracy depending on the geometries of the virtual and the real sensor arrays. In classical array interpolation, all sources of interest are assumed to be present within a certain angular sector. That is, DOA angle θ belongs to a sector $[\theta_b, \theta_f]$, which is uniformly divided by $\delta\theta$ intervals and its calibration angles are generated as $\bar{\theta} = \theta_b + i\delta\theta$, where $i = 0, 1, \ldots, (\theta_f - \theta_b)/\delta\theta$. The calibration steering matrix for the real $\mathbf{A}_r(\bar{\theta})$ and virtual $\mathbf{A}_v(\bar{\theta})$ arrays are constructed using these calibration angles, and the classical array interpolation matrix, $\mathbf{T}$, is obtained as

$$\mathbf{T} = \mathbf{A}_v(\bar{\theta})\mathbf{A}_r^H(\bar{\theta})(\mathbf{A}_r(\bar{\theta})\mathbf{A}_r^H(\bar{\theta}))^{-1} \tag{4.10}$$

Classical array interpolation (CAI) is an effective method, but it has certain limitations. The accuracy of the mapping matrix $\mathbf{T}$ increases as the interpolation sector decreases. However, the sector should be kept sufficiently large to cover a large region of interest. Any source outside the interpolation sector is not guaranteed to be identified correctly. The multiple sector approach in Friedlander (1993) tries to solve this problem. The bias and mean-square error (MSE) for the array interpolation can also be minimized to a certain extent (Hyberg et al., 2005). Another problem in classical array interpolation is the condition number of the matrix $\mathbf{A}_r(\bar{\theta})\mathbf{A}_r^H(\bar{\theta})$. This matrix may be ill-conditioned; certain techniques are used to improve its condition number (Friedlander, 1993).

Wiener Array Interpolation

Wiener array interpolation (WAI) was proposed in Tuncer et al. (2007a,b) as a way to overcome some of the problems of CAI. WAI is the approximation of the MSE optimum solution since the calibration angles are used to construct the array steering matrix. Its advantages include that it is more effective than CAI at low SNR, the condition number of the inverse matrix in the transformation matrix is significantly better, and it is possible to select the number of sensors in the virtual array to be larger than the number of sensors in the real array. Also, WAI performs better than CAI in wideband DOA estimation, especially when only a small number of frequency bins are available. The Wiener formulation in Tuncer et al. (2007a,b) assumes that the signal and source powers are known, which is unrealistic for a practical application. Fortunately, it is possible to accurately estimate the signal and noise powers with a maximum likelihood approach.

WAI can be carried out by considering the real array output, $\mathbf{y}_r(t)$, and the desired or virtual array output, $\mathbf{y}_v(t)$. It is assumed that there are M_v sensors in the virtual array and that the array geometry is known. The error between these outputs is

$$\mathbf{e} = \mathbf{y}_v - \mathbf{T}\mathbf{y}_r(t) \tag{4.11}$$

The MSE optimum solution for $M_v \times M$ matrix $\mathbf{T}$ is

$$\mathbf{T} = \mathbf{A}_v\mathbf{R}_s\mathbf{A}_r^H\left(\mathbf{A}_r\mathbf{R}_s\mathbf{A}_r^H + \mathbf{R}_v\right)^{-1} \tag{4.12}$$

If we assume $\mathbf{R}_v = \sigma_v^2\mathbf{I}$ and $\mathbf{R}_s = \sigma_s^2\mathbf{I}$ for uncorrelated source signals, we have

$$\mathbf{T} = \sigma_s^2\mathbf{A}_v\mathbf{A}_r^H\left(\sigma_s^2\mathbf{A}_r\mathbf{A}_r^H + \sigma_v^2\mathbf{I}\right)^{-1} \tag{4.13}$$

The noise component in $\left(\sigma_s^2\mathbf{A}_r\mathbf{A}_r^H + \sigma_v^2\mathbf{I}\right)^{-1}$ improves the condition number and allows more sensors in the virtual array than in the real array. In order to have a sector-independent array interpolation, we need an initial estimate of the source angles. It is possible to obtain initial estimates by Toeplitz completion in the case of NLA. Assuming that a good initial estimate is available, DOA accuracy can be improved significantly by subsequent processing.

Once we have an initial DOA estimate, $\hat{\boldsymbol{\theta}} = [\hat{\theta}_1, \ldots, \hat{\theta}_n]$, we construct $\mathbf{T}$ by considering narrow sectors for each $\hat{\theta}_i$. These small sectors are divided with calibration angles $\tilde{\theta}_{i,j} \in [\hat{\theta}_i - \theta_\epsilon, \hat{\theta}_i + \theta_\epsilon]$ and $\tilde{\theta}_{i,j} = \hat{\theta}_i - \theta_\epsilon + j\delta\theta$, where $j = 0, 1, \ldots, 2\theta_\epsilon/\delta\theta$. $\mathbf{A}_v$ and $\mathbf{A}_r$ are constructed from the calibration angles $\tilde{\theta}_{i,j}$. For simplicity, $\tilde{\theta}$ is dropped from the steering matrix expressions.

To apply the WAI as in Equation (4.13), we need to find the signal and noise powers. These parameters can be found using the stochastic maximum likelihood estimation. Noise variance is estimated by projecting the covariance matrix to a space orthogonal to the signal space. The orthogonal projection matrix, $\mathbf{P_h}$, is

$$\mathbf{A_h} = (\mathbf{A}_r^H \mathbf{A}_r)^{-1} \mathbf{A}_r^H \tag{4.14}$$

$$\mathbf{P_h} = \mathbf{I} - \mathbf{A}_r \mathbf{A_h} \tag{4.15}$$

Then the noise variance is found by projecting the covariance matrix as

$$\sigma_v^2 = \frac{1}{M-n} tr(\mathbf{P_h} \hat{\mathbf{R}}) \tag{4.16}$$

Once the noise variance is found, the signal covariance matrix, $\mathbf{P_{SML}}$, can be obtained by first removing the noise contribution and then using the following expression:

$$\mathbf{P_{SML}} = \mathbf{A_h} (\hat{\mathbf{R}} - \sigma_v^2 \mathbf{I}) \mathbf{A_h}^H \tag{4.17}$$

In general, signal variance can be different for different sources. The source signal variances can be obtained as

$$diag(\sigma_{s_1}^2, \sigma_{s_2}^2, \ldots, \sigma_{s_n}^2) = diag(\mathbf{P_{SML}}) \tag{4.18}$$

The SNR for each source can be obtained as

$$\mathrm{SNR}_i = 10 \log 10(\sigma_{s_i}^2 / \sigma_v^2), \qquad i = 1, \ldots, n \tag{4.19}$$

WAI can be carried out in the following steps:

Step 1. Given the sensor outputs, $\mathbf{y}_r$, obtain the sample covariance matrix $\hat{\mathbf{R}}$ from Equation (4.9). Use Toeplitz completion for NLA. Fill the missing covariance lags using spline interpolation if necessary, and find the complete Hermitian symmetric Toeplitz covariance matrix, $\mathbf{R}_{TC}$.

Step 2. Use $\mathbf{R}_{TC}$ in the Root-MUSIC algorithm and find the initial DOA estimate $\hat{\boldsymbol{\theta}}$.

Step 3. Given $\hat{\boldsymbol{\theta}}$, construct $\mathbf{T}$ from Equation (4.13).

Step 4. Find $\mathbf{R}_F = \mathbf{T}\hat{\mathbf{R}}\mathbf{T}^H$ and use Root-MUSIC to find the true DOA estimate.

Step 5. Iterate steps 3 and 4 by updating the initial estimate.

It turns out that iterations can significantly improve DOA accuracy, with a price of increased computational complexity.

Rotational Signal Subspace Method

CSM is proposed for wideband DOA estimation (Wang and Kaveh, 1985). Its application requires the design of a transformation matrix, where the RSS technique is presented in Hung and Kaveh (1988) for this purpose. The main idea is to keep the same SNR after the application of the transformation matrix. This is also called focusing design, and the resulting mapping is the focusing matrix. The focusing matrix can easily be obtained with the desired property if the mapping matrix is constrained to be unitary. The advantages of such a design are twofold. As already mentioned, there is no SNR loss after the transformation; nor is the noise covariance matrix changed. One important assumption for subspace techniques such as MUSIC is that noise should be white or at least approximately so. Unitary transformation does not change the noise statistics, and MUSIC is as effective as the original data. In RSS, the transformation matrix is designed as the solution to

$$\mathbf{T} = \underset{\mathbf{T}}{argmin} ||\mathbf{T}\mathbf{A}_r - \mathbf{A}_v||_F^2, \quad s.t. \; \mathbf{T}^H\mathbf{T} = \mathbf{I}_M \tag{4.20}$$

where $M \times K$ matrix $\mathbf{A}_r$ stands for the real array steering matrix and $M_v \times K$ matrix $\mathbf{A}_v$ is the array steering matrix for the virtual array. It is assumed that K calibration angles are used in the neighborhood of the initial DOA estimates and that $M_v = M$.

The solution to $M \times M$ matrix $\mathbf{T}$ is given in Hung and Kaveh (1988) as

$$\mathbf{T} = \mathbf{V}\mathbf{U}^H \tag{4.21}$$

where $M \times M$ unitary matrices $\mathbf{V}$ and $\mathbf{U}$ are obtained from the singular value decomposition,

$$\mathbf{A}_r\mathbf{A}_v^H = \mathbf{U}\Sigma\mathbf{V}^H \tag{4.22}$$

When the number of sources is less than the number of sensors, the solution is not unique. However, this is not an important problem and the performance of the transformation matrix is good.

Generalized RSS Method

In Hung and Kaveh (1988), the RSS method is presented for the case where the number of sensors in real and virtual arrays is equal. It is possible to generalize the RSS method (generalized RSS, GRSS) for the case where the number of sensors in a virtual array is larger than the number of real sensors (i.e., $M_v > M$). This can be easily achieved with a small modification of the transformation matrix $\mathbf{T}$. In this case, the solution to $M_v \times M$ matrix $\mathbf{T}$ is

$$\mathbf{T} = \mathbf{V}\hat{\Sigma}\mathbf{U}^H \tag{4.23}$$

where $M_v \times M$ matrix $\hat{\Sigma} = [\mathbf{I}_{M\times M} \;\; \mathbf{0}]^T$, $M_v \times M_v$, and $M \times M$ unitary matrices $\mathbf{V}$ and $\mathbf{U}$ are obtained from the singular value decomposition,

$$\mathbf{A}_r\mathbf{A}_v^H = \mathbf{U}\Sigma\mathbf{V}^H \tag{4.24}$$

Note that the formulation in Equation (4.23) can be used to map arrays with $M_v > M$ as well as $M_v = M$. $\mathbf{T}$ satisfies all the desired properties of the focusing matrices (Hung and Kaveh, 1988).

4.3.3 Initial DOA Estimation for Coherent Sources

Source signals are coherent for the multipath reflections. The DOA estimation becomes problematic for coherent sources, and preprocessing is usually needed. When the source signals are coherent, the covariance matrix becomes rank deficient and it is not possible to find the source DOA. Certain methods can be used at the preprocessing stage to improve the rank, but their performance and computational complexity vary. These methods can be applied to uniform linear arrays; however, array mapping can be employed for NLA to obtain an equivalent virtual ULA to take advantage of them. In this section, forward-backward, Toeplitz completion, and forward-backward spatial smoothing techniques are considered.

Forward-Backward Approach

Forward-backward approach (FBA) is a simple and efficient technique used in a variety of applications, including linear prediction. The modified covariance matrix is obtained from the sample covariance matrix as

$$\tilde{\mathbf{R}} = \frac{1}{2}(\hat{\mathbf{R}} + \mathbf{J}\hat{\mathbf{R}}^T\mathbf{J}) \tag{4.25}$$

where $\mathbf{J}$ is an off-diagonal matrix,

$$\mathbf{J} = \begin{bmatrix} 0 & & 1 \\ & \cdot^{\cdot^{\cdot}} & \\ 1 & & 0 \end{bmatrix} \tag{4.26}$$

$\tilde{\mathbf{R}}$ is a persymmetric matrix like the theoretical covariance matrix, which is also Toeplitz.

FBA performance is good for coherent sources, but DOA-dependent errors are observed in general. Furthermore, this technique is usually applicable when the number of sources is less than three, $n < 3$. With several sources, FBA is not effective.

Toeplitz Completion

Toeplitz completion (TC) is a well-known method for improving the covariance matrix of NLA. In Pillai and Haber (1987), it is shown that TC generates an $M_v \times M_v$ covariance matrix for an $M < M_v$-element minimum redundant array, which is identical to the covariance matrix of a ULA of M_v elements when true statistics are available. If there is only a sample covariance matrix, TC generates a matrix that results in some degradations in DOA performance (Pillai and Haber, 1987). In Pillai and Haber (1987) the covariance matrix obtained by the TC is shown to be nonpositive semi-definite. This point is also expressed in Abramovich et al. (1998). In Abramovich et al. (1998), it is shown that the covariance matrix obtained by TC performs well for low SNR and can be used as a good initial estimate for high SNR for uncorrelated sources. TC performance for coherent sources is not considered in the literature.

Toeplitz completion is a simple and computationally efficient way to deal with coherent sources, especially to obtain an initial DOA estimate. In the case of ULAs, it corresponds to averaging the elements along each diagonal of the covariance matrix. For NLAs, it requires augmentation for the missing elements. Consider an NLA with M sensors. We want to obtain an $M_v \times M_v$ $(M_v > M)$ covariance matrix, $\mathbf{R}_{TC}$, corresponding to the full virtual array with the same aperture as the NLA. Let $\mathbf{r}_{TC}$ be the first row of this Hermitian symmetric Toeplitz matrix. Once $\mathbf{r}_{TC}$ is known, $\mathbf{R}_{TC}$ can be easily constructed by Toeplitz completion. An $M_v \times M_v$ sparse sample covariance matrix, $\hat{\mathbf{R}}_{va}$, is obtained from the sensor outputs by assuming that the outputs of missing sensors are zero. Then the elements of $\mathbf{r}_{TC}$ are obtained:

$$r_{TC}(k) = \frac{\sum_{i=1}^{M_v-(k-1)} \hat{\mathbf{R}}_{va}(i, i+(k-1))}{c(M_v+k-1)} \qquad k=1,\ldots,M_v \tag{4.27}$$

assuming that the co-array $c(k)$ has all nonzero elements. If the co-array has zero lags, $r_{TC}(k)=0$ for that lag. TC is more effective for minimum redundant arrays since they do not have any zero elements for the co-array. When $c(k)$ has missing covariance lags, interpolation techniques can be employed to fill them.

Forward-Backward Spatial Smoothing

FBA and TC methods do not change the size of the covariance matrix. Therefore, there is no loss in array resolution. Forward-backward spatial smoothing (FBSS) uses the subarrays, and the resulting covariance matrix size is less than the original covariance matrix size. This may be a disadvantage, but FBSS is still the best method to deal with the coherent sources, especially when ULAs are considered. When NLAs are used, TC gives better estimates. Applying FBSS after TC makes no difference.

FBSS (Pillai and Kwon, 1989) is more efficient than spatial smoothing (Shan et al., 1985) in the sense that it requires fewer sensors. The linear array is divided into subarrays, and the corresponding covariance matrices are summed to obtain a high-rank matrix. There are certain conditions on the number of array elements and subarray size. The number of sensors, M, and the number of sensors in the subarray, L, should satisfy

$$M \geq \lfloor L+n/2-1 \rfloor \tag{4.28}$$

where $\lfloor\rfloor$ rounds to the nearest integer, and n is the number of coherent sources. The subarray size, L, satisfies

$$n < L \leq \lfloor M-n/2+1 \rfloor \tag{4.29}$$

The minimum value for L is $L_{min}=n+1$ and $M_{min}=\lfloor 3n/2 \rfloor$. The FBSS covariance matrix, $\mathbf{R}_{fb}$, can be obtained from the sample covariance matrix,

$\hat{\mathbf{R}}$, for ULAs or from the completed covariance matrix for NLAs. First, the forward-backward matrix $\tilde{\mathbf{R}}$ is obtained:

$$\tilde{\mathbf{R}} = \frac{1}{2}(\hat{\mathbf{R}} + \mathbf{J}\hat{\mathbf{R}}^T\mathbf{J}) \tag{4.30}$$

Then

$$\mathbf{R}_{fb} = \frac{1}{P}\sum_{k=1}^{P}\mathbf{Z}_k\tilde{\mathbf{R}}\mathbf{Z}_k^T \tag{4.31}$$

where $P = M - L + 1$ is the number of subarrays and

$$\mathbf{Z}_k = [\mathbf{0}_{L\times(k-1)}\ \ \mathbf{I}_{L\times L}\ \ \mathbf{0}_{L\times M_v-(L+k-1)}] \tag{4.32}$$

Comparison of Methods for Initial DOA Estimation

Here, FBA, TC, and FBSS are compared for narrowband coherent sources in the case of ULA, minimum redundant NLA, and partly filled NLA, respectively. The number of snapshots is set to $N = 256$, and $\text{SNR} = 10$ dB. There are 500 trials for each experiment. The results with ULA are shown in Figure 4.1. ULA has ten sensors. In Figure 4.1(a) there are two coherent sources, one fixed at 70°, the other swept between 10° and 170°. An experiment with three coherent sources is shown Figure 4.1(b), where two sources at 70° and 95° are fixed and the third is swept. From Figure 4.1, it is seen that the FBA returns DOA-dependent estimates for two sources, where for certain DOA angles performance is unacceptable. FBA does not perform well for three coherent sources. TC has an acceptable performance for ULA in the case of a coherent source; it returns

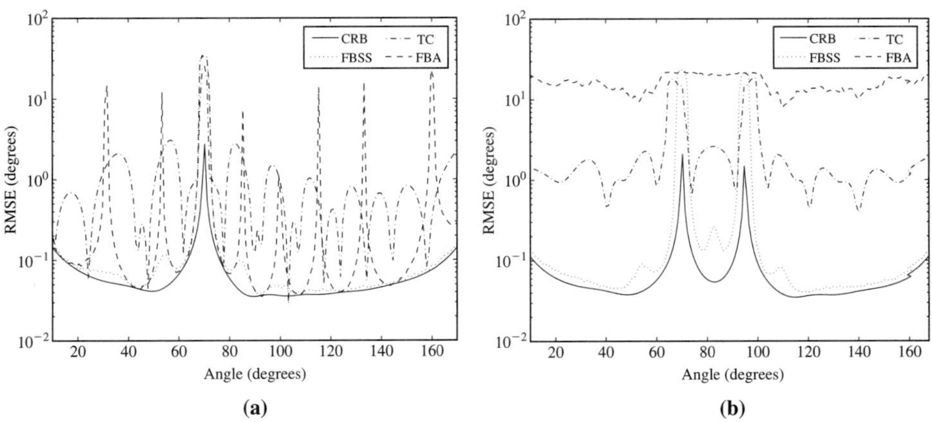

FIGURE 4.1 Initial DOA estimation for coherent sources by TC, FBA, and FBSS for ULA. (a) One source is fixed at 70°; the second is swept. (b) Two sources are fixed at 70° and 95° and the third is swept. SNR = 10 dB.

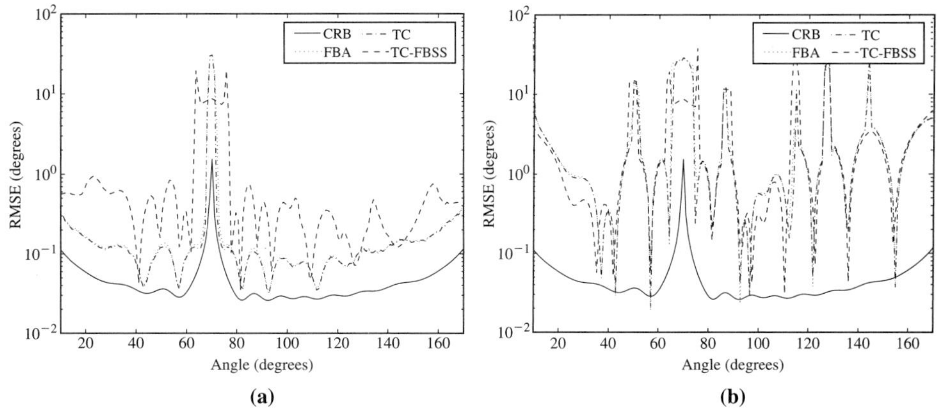

FIGURE 4.2 Initial DOA estimation for uncorrelated and coherent sources by TC, FBA, and TC-FBSS for NLA. One source is fixed at 70°; the second is swept. (a) Uncorrelated sources. (b) Coherent sources. SNR = 10 dB.

biased estimates but the accuracy is satisfactory for an initial estimate. FBSS has the best performance, and it approaches the CRB.

The second experiment is with a minimum redundant NLA with sensor positions $d_{NLA} = [0\ 1\ 4\ 5\ 11\ 13]$. FBSS cannot be directly applied to obtain an initial DOA estimate. It turns out that applying it after TC (TC-FBSS) does not alter DOA estimation accuracy. Figures 4.2(a) and 4.2(b) show the DOA estimates for different methods for uncorrelated and coherent sources, respectively. There are two sources: one fixed at 70° and the other swept. Figure 4.2(a) shows that TC and TC-FBSS have similar performance and that TC performs better than FBA. When the sources are coherent, as in Figure 4.2(b), TC, TC-FBSS, and FBA do not perform well for certain DOA angles, which means that initial DOA estimation is a problem in NLA.

One solution for initial NLA estimation is to use partly filled NLA and employ FBSS for the ULA part. Figure 4.3 shows the DOA estimate of $d_{NLA} = [0\ 1\ 2\ 3\ 8\ 13\ 17]$ when only the first four elements are used. FBSS is applied to the covariance matrix of four sensors and an initial DOA estimate is found. There are two coherent sources: one fixed at 70°. As seen from Figure 4.3, partly filled NLA has an advantage compared to other types of NLA when the source signals are coherent.

CLEAN Algorithm for Coherent Sources

DOA estimation in multipath signals is challenging. The covariance matrix is rank deficient and does not reflect the multiple-source DOA angles. The CLEAN algorithm is well known in astronomy (Clark, 1980; Schwarz, 1978) and radar (Abramovich, 1979; Deng, 2004; Tsao and Steinberg, 1988), especially for reducing sidelobe artifacts. Different versions of CLEAN exist in the literature, but

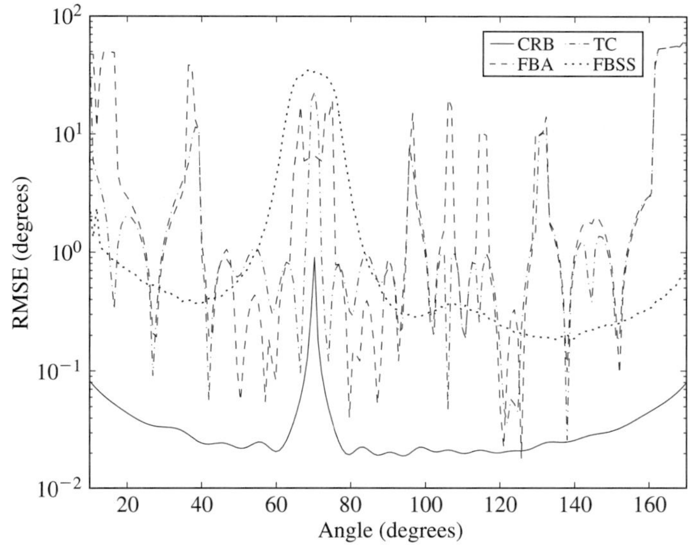

FIGURE 4.3 Initial DOA estimation for two coherent sources by a partly filled NLA. One source is fixed at 70°; the second is swept. SNR = 10 dB.

the main idea is the same: namely, to remove the strongest signals from the observed data successively. Some of these algorithms operate over the observed signals (Tsao and Steinberg, 1988); others operate over the covariance matrix or the spatial spectrum (Stoica and Moses, 2005). Their performance and behavior are different in general. CLEAN is seldom used in DOA estimation since it can offer only marginal improvement for uncorrelated sources. Furthermore, it is believed to be suited only for uncorrelated sources since its operation is over the covariance matrix (Stoica and Moses, 2005). Here, we present the CLEAN algorithm that operates over sensor signals. This algorithm works well for correlated and coherent sources. In fact, as correlation between sources increases, improvement becomes significant and the best result is achieved when the sources are coherent.

In order to apply CLEAN, initial estimates for the signal parameters should be available. Since these estimates are not perfect, the cancellation of signal components cannot be perfect and there are residuals. However, CLEAN can still perform well for coherent sources as long as the residuals are sufficiently small.

Initial DOA estimates can be obtained in different ways. In our case, TC or FBSS techniques are used. Sensor positions are assumed to be known. Therefore, the array steering matrix can be constructed once the initial DOA estimates are available. The CLEAN algorithm for narrowband DOA estimation can be summarized as follows:

Step 1. Given the NLA array, obtain the sample covariance matrix and apply TC.

Step 2. Use the covariance matrix obtained in step 1 for initial DOA estimation by employing the Root-MUSIC algorithm.

Step 3. Given the initial DOA estimates $\bar{\theta} = [\bar{\theta}_1, \ldots, \bar{\theta}_n]$, apply CLEAN as follows: For each $\bar{\theta}_i, i = 1, \ldots, n$:

1. Construct the $M \times n$ array steering matrix $\bar{\mathbf{A}}_r$ and find source signal $\bar{\mathbf{s}} = [\bar{\mathbf{A}}_r^H \bar{\mathbf{A}}_r]^{-1} \bar{\mathbf{A}}_r^H \mathbf{y}_r$.
2. Construct the $M \times (n-1)$ steering matrix $\hat{\mathbf{A}}_{r,i}$. $\hat{\mathbf{A}}_{r,i}$ is obtained from $\bar{\mathbf{A}}_r$ by deleting the ith column. Also obtain the source $\hat{\mathbf{s}}_i$ from $\bar{\mathbf{s}}$ by deleting the ith row.
3. Find the cleaned signal $\mathbf{y}_{r,i} = \mathbf{y}_r - \hat{\mathbf{A}}_{r,i}\hat{\mathbf{s}}_i$ and the corresponding covariance matrix $\mathbf{R}_i = \frac{1}{N}\sum_{m=1}^{N} \mathbf{y}_{r,i}\mathbf{y}_{r,i}^H$.
4. Find the array interpolation matrix $\mathbf{T}_i$ from Equation (4.23) by using $\tilde{\mathbf{A}}_{r,i}$ and $\tilde{\mathbf{A}}_{v,i}$ for the real and virtual arrays, respectively, by considering K calibration angles in the neighborhood of $\bar{\theta}_i$.
5. Obtain $\mathbf{R}_{T,i} = \mathbf{T}_i \mathbf{R}_i \mathbf{T}_i^H$.
6. Use $\mathbf{R}_{T,i}$ in the Root-MUSIC algorithm to find the θ_i estimate.

Step 4. Repeat step 3, updating the initial estimate for better results.

CLEAN is not optimal in any sense, but it can perform as well as MUSIC for uncorrelated sources. Note that the signal estimate in step 3.1 is a maximum likelihood estimate, and its behavior due to several error sources can be found in Weiss and Friedlander (1993). The cleaning operation in step 3.3 is the orthogonal projection onto the null space of $\hat{\mathbf{A}}_{r,i}^H$. Therefore, the measurements are projected onto a subspace orthogonal to the signal components except for one. Because the sample covariance matrix at step 3.3 is always positive semi-definite (Horn and Johnson, 1985), the problem of obtaining a positive semi-definite covariance matrix in CLEAN algorithms operating over the covariance matrices (Stoica and Moses, 2005) is not observed in our case. Thus, a sample-based CLEAN algorithm statistically makes sense. Note that the sample covariance matrix at step 3.3 is different from the covariance matrix obtained in CLEAN algorithms that operate over the covariance matrices.

In general, there is not much CLEAN can offer for ULA since FBSS performs sufficiently well for coherent sources. In this respect, CLEAN is an effective algorithm for NLA only. It can improve NLA DOA performance significantly. To show its performance, three experiments are performed with a minimum redundant NLA with sensor placement $d_{NLA} = [0\ \ 1\ \ 4\ \ 5\ \ 11\ \ 13]$. SNR is set to 10dB, and two uncorrelated sources at 75° and 80° are considered. In Figures 4.4(a) and 4.4(b), the CLEAN spectrum for each source is shown. Initial estimates for the NLA, obtained by Toeplitz completion, are 75.1187° and 79.9340°, respectively. As can be seen from Figures 4.4(a) and 4.4(b), CLEAN can successfully isolate the source spectra.

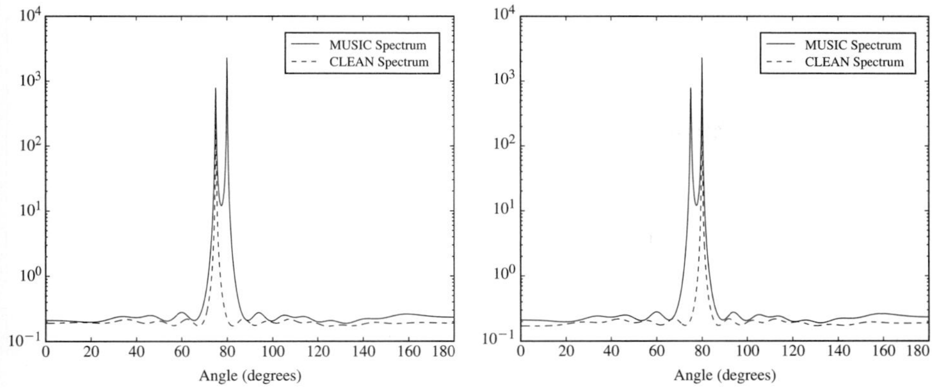

(a) (b)

FIGURE 4.4 MUSIC and estimated CLEAN spectrum for two narrowband uncorrelated sources at 75° and 80°, respectively. (a) CLEAN spectrum for the source at 75°. (b) CLEAN spectrum for the source at 80°. SNR = 10 dB.

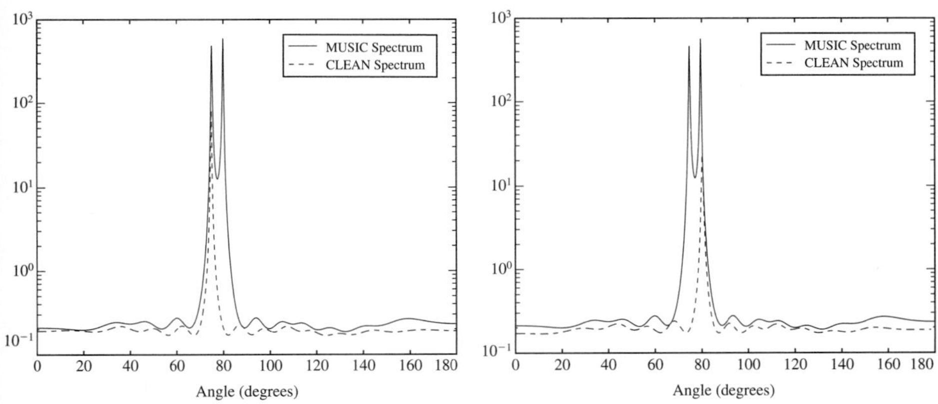

(a) (b)

FIGURE 4.5 MUSIC and estimated CLEAN spectrum for uncorrelated sources at 75° and 80°. Initial estimates are intentionally given as 73° and 82°. (a) CLEAN spectrum for the source at 75°. (b) CLEAN spectrum for the source at 80°. SNR = 10 dB.

In Figure 4.5, the robustness of the CLEAN algorithm is shown when the initial estimates are mistakenly given as 73° and 82°, respectively. After three iterations, the resulting DOA angles are 74.8804° and 80.8160°. As can be seen, CLEAN can isolate the sources successfully even when the initial estimates are not accurate.

In Figure 4.6, the same experiment is repeated for the coherent sources at 75° and 80°, respectively. The initial estimates, obtained by Toeplitz completion, are 74.9292° and 79.9747°, respectively. The MUSIC spectrum does not carry any

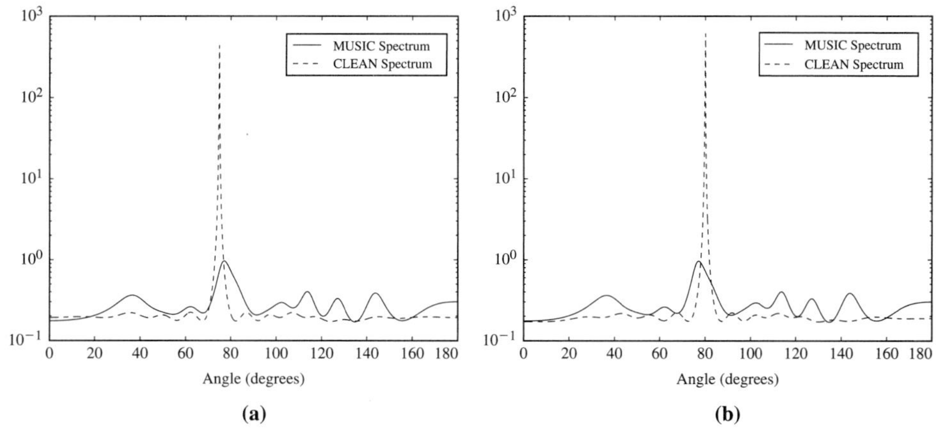

FIGURE 4.6 MUSIC and estimated CLEAN spectrum for two coherent sources at 75° and 80°. (a) CLEAN spectrum for the source at 75°. (b) CLEAN spectrum for the source at 80°. SNR = 10 dB.

information about the two sources, whereas the CLEAN spectrum clearly shows the source DOAs for both.

4.3.4 Performance Comparisons

The performance of different algorithms is evaluated for narrowband DOA estimation to observe the advantages and disadvantages of each: WAI with the estimated signal and noise powers; Wiener array interpolation when the true signal and noise powers are used (WAI-TRUE) is the generalized rotational signal-subspace (GRSS) method; and CLEAN, described previously. A partly filled NLA is used with $d_{NLA} = [0 \;\; 1 \;\; 2 \;\; 3 \;\; 8 \;\; 13 \;\; 17]$. In the experiments, the number of snapshots is $N = 256$, and the number of trials is 500.

In Figure 4.7, both uncorrelated and coherent cases are considered when there are two sources. One source is fixed at 70°, and the other is swept between 10° and 170°. An initial estimate is obtained from the first four elements by FBSS. Each array-mapping algorithm has three iterations. Figure 4.7(a) shows the performance of the algorithms when the sources are uncorrelated. CLEAN and GRSS have the best performance, followed by WAI-TRUE and WAI. The closely spaced sources are resolved better by GRSS. WAI is more sensitive to the initial estimate, and in fact performs better when the initial estimate is obtained via TC. Figure 4.7(b) shows DOA performance when the two sources are coherent. All the algorithms apply FBSS when the final covariance matrix is obtained. CLEAN and GRSS perform well, followed by WAI-TRUE and WAI.

The same experiment is repeated for two fixed- and equal-power sources at 70° and 85°, respectively. In Figure 4.8(a) the sources are uncorrelated and DOA performance is observed with respect to SNR. Figure 4.8(b) shows the

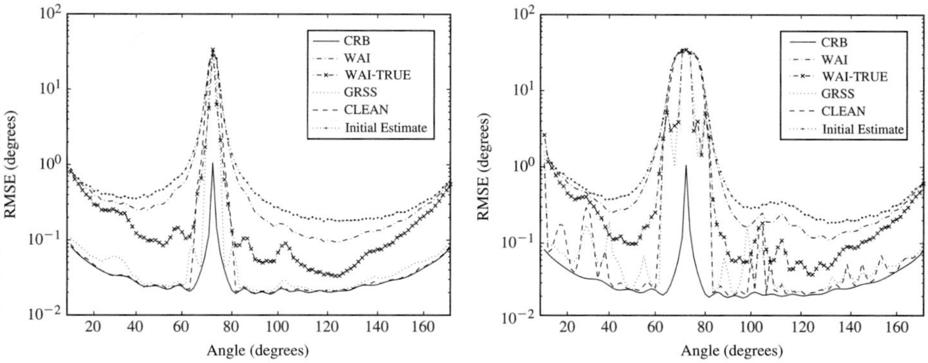

(a) (b)

FIGURE 4.7 DOA performance for two sources: (a) uncorrelated; (b) coherent (SNR = 10 dB).

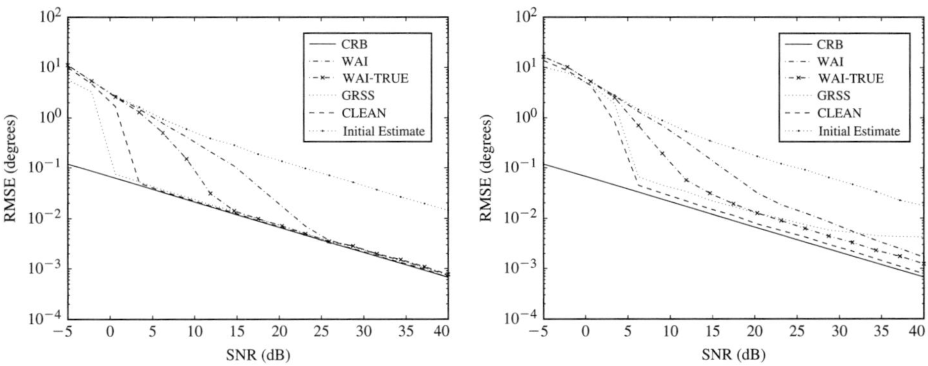

(a) (b)

FIGURE 4.8 SNR performance for two sources at 70° and 85°: (a) uncorrelated; (b) coherent.

case when the sources are coherent. It turns out that the performance of GRSS shows a flooring effect as SNR is increased. As the number of iterations for array mapping increases, this effect decreases. CLEAN, WAI, and WAI-TRUE are more robust to source coherence as SNR increases.

The performance of CLEAN and GRSS significantly improves as the number of iterations increases. Figure 4.9 shows the performance of the array-mapping algorithms with only one iteration. Figure 4.9(a) can be compared with Figure 4.7(a) and Figure 4.9(b) can be compared with Figure 4.8(a) to see the difference for one and three iterations, respectively. It seems that CLEAN performs slightly better than GRSS when there is only a single iteration. In general, for the resolution of closely spaced sources CLEAN is not as good as GRSS. Also, its computational complexity is higher than GRSS's.

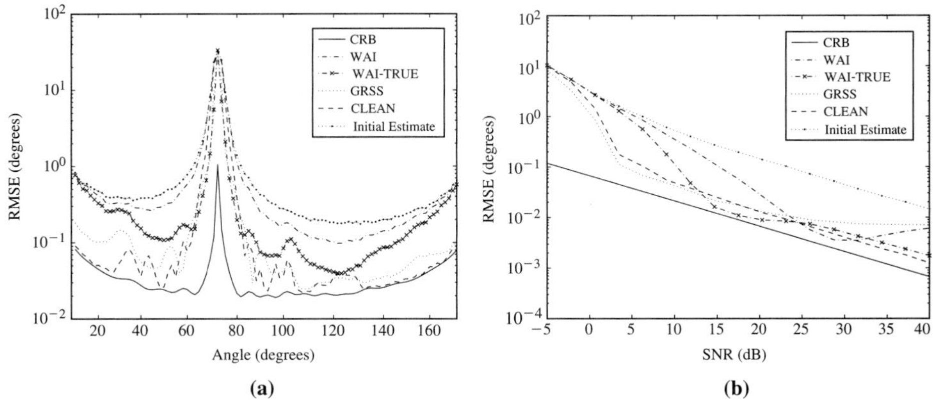

FIGURE 4.9 DOA performance for two uncorrelated sources when the array-mapping algorithms have a single iteration. (a) One source is fixed at 70° and the second is swept (SNR = 10 dB). (b) Two sources are fixed at 70° and 85°, SNR performance is observed.

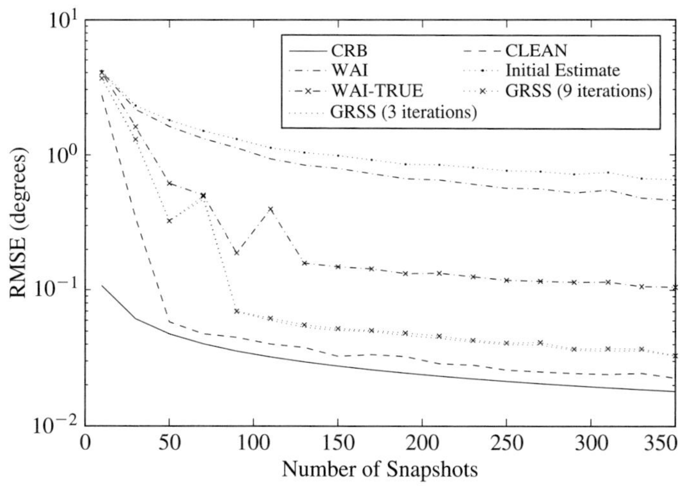

FIGURE 4.10 Effect of the number of snapshots for two narrowband coherent sources at 70° and 85°, respectively. SNR = 10 dB.

In Figure 4.10, the performance of the algorithms is shown when the number of snapshots is changed. The two coherent sources are placed at 70° and 85°, respectively. In this case, the number of iterations for each algorithm is 3 except for the GRSS, which is considered for 3 and 9 iterations. The figure shows that increasing the number of iterations beyond a certain value does not improve the performance of GRSS for the narrowband coherent signal case.

It turns out that the RSS method in Hung and Kaveh (1988) is very effective for array mapping, especially in the case of uncorrelated source signals when

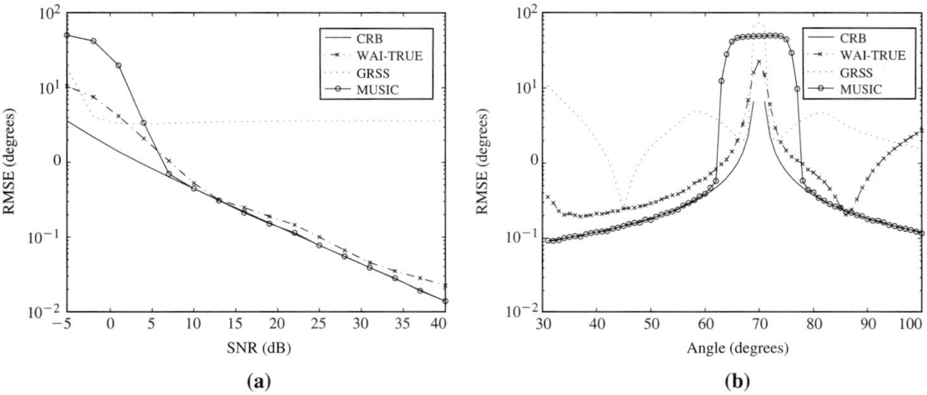

FIGURE 4.11 Array mapping from a 7-element circular array to a virtual ULA. (a) Two uncorrelated sources at 70° and 85°. (b) One source is fixed at 70°, the second is swept. SNR = 15 dB.

the angular sector for DOAs is small. As the angular sector for array mapping is increased, it has significant bias. When the sensor array is circular, an initial DOA estimate is not available and the angular sector should be kept sufficiently large. In such cases, WAI can perform significantly better than RSS. To see the performance loss for RSS, a 7-element circular array with $\lambda/2$ inter-element spacing is mapped to a virtual ULA with the same number of elements, where the elements are spaced by $\lambda/5$. A 50° sector $\tilde{\theta} \in [40°, 90°]$ is selected for the array interpolation.

Figure 4.11(a) shows the SNR performance when there are two uncorrelated sources at 70° and 85°, respectively. In Figure 4.11(b), one source at 70° is fixed and the second is swept between 30° and 100°. As can be seen, WAI is more effective than RSS for this case. RSS has a significant bias when the angular sector is large.

An interesting observation is that the Spectral-MUSIC algorithm applied on the original circular array performs worse than WAI when the source DOAs are close to each other. This is mainly due to the fact that in Wiener array interpolation Root-MUSIC is used. When the error has a significant radial component, Spectral-MUSIC fails to show two distinct peaks, whereas Root-MUSIC is unaffected. This phenomenon is analyzed in Rao and Hari (1989) and practical cases are reported in Covarruhias-Rosales and Olague (2004) and Teague (2002).

4.4 WIDEBAND DIRECTION-OF-ARRIVAL ESTIMATION

In practice, most signals are wideband. Narrowband processing of such signals ignores the useful information available for parameter estimation. Wideband processing in this context allows better parameter estimation because of the processing gain. Wideband processing gain (WBPG) can be defined as the ratio of the variance of DOA estimation for one frequency to the variance of DOA estimation for L frequencies:

$$WBPG \triangleq \frac{\sigma^2_{\theta,1}}{\sigma^2_{\theta,L}} \tag{4.33}$$

When the sensor outputs for different frequencies are uncorrelated, a Fisher information matrix (FIM) can be written as the sum of the narrowband terms:

$$\mathbf{FIM} = \sum_{j=1}^{L} \mathbf{F}(w_j) \tag{4.34}$$

Assuming that the statistics for each frequency are the same,

$$\mathbf{FIM}^{-1} = \frac{1}{L}\mathbf{F}^{-1}(w) \tag{4.35}$$

Therefore, $WBPG \leq L$ in wideband DOA estimation, which indicates that there is significant gain for DOA estimation when wideband processing is employed. Figure 4.12 shows the Cramer-Rao bound for a 10-element ULA in the case of a single source as the number of frequencies, L, in wideband processing is changed.

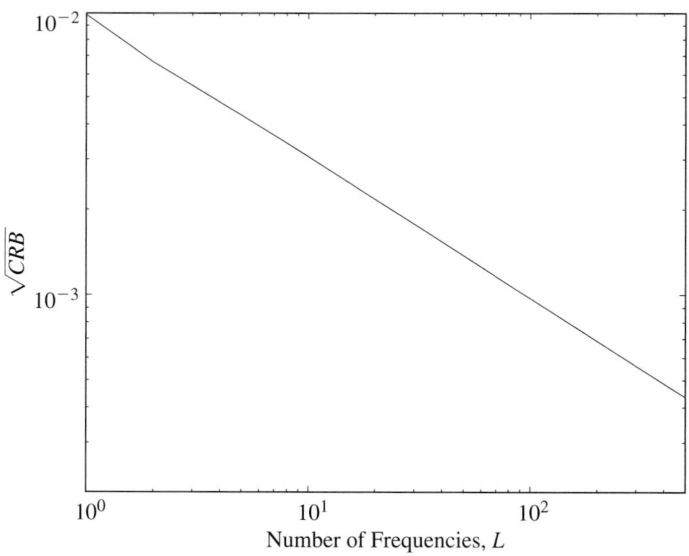

FIGURE 4.12 CRB versus number of frequencies used in wideband processing. SNR = 20 dB, 10-element ULA.

Wideband DOA estimation can be performed in different ways, which can be divided into two main groups: coherent and noncoherent.

4.4.1 Wideband Signal Model

In the wideband signal model, the time-bandwidth product in Equation (4.2) is not small. In this case, sensor signals are usually treated in the frequency domain. The received signal for the ith sensor can be written as

$$y_{r_i}(t) = \sum_{k=1}^{n} s_k(t - \tau_{ki}) + v_i(t) \tag{4.36}$$

where $s_k(t)$ and $v_i(t)$ are the source and noise signals. Noise is assumed to be both spatially and temporally white and uncorrelated with the signal. Equation (4.44) can be written in the frequency domain as

$$Y_{r_i}(\omega) = \sum_{k=1}^{n} e^{-j\omega\tau_{ki}} S_k(\omega) + V_i(\omega) \tag{4.37}$$

where $\tau_{ki} = \frac{d_i}{c} \cos\theta_k$, and c is the propagation velocity. In matrix vector form, this equation becomes

$$\mathbf{Y}_r(\omega) = \mathbf{A}_r(\omega)\mathbf{S}(\omega) + \mathbf{V}(\omega) \tag{4.38}$$

It is assumed that there are L frequency bins, $w_j, j = 1, \ldots, L$, for the bandwidth under consideration. The $M \times n$ matrix $\mathbf{A}_r(\omega)$ is the real array steering matrix at frequency ω. The $M \times M$ sample covariance matrix is

$$\hat{\mathbf{R}}(\omega) = \frac{1}{N} \sum_{m=1}^{N} \mathbf{Y}_r(\omega, m)\mathbf{Y}_r^H(\omega, m) \tag{4.39}$$

where N is the number of snapshots.

4.4.2 Signal Correlation and Coherence

Signal correlation for wideband signals is defined slightly differently than for the narrowband case (Friedlander and Weiss, 1993). Consider two signals $s_1(t)$ and $s_2(t)$ for simplicity. Let $s_1(t)$ be a white signal and

$$s_2(t) = \alpha s_1(t - \tau) + g(t) \tag{4.40}$$

where $g(t)$ is another white signal, and α is a constant factor. In the frequency domain,

$$S_2(w)=\alpha e^{-jw\tau}S_1(w)+G(w) \tag{4.41}$$

If we assume that $s_1(t)$ and $s_2(t)$ have the same power (i.e., $E\{|S_1(w)|^2\}=E\{|S_2(w)|^2\}$), the signal power for $g(t)$ is

$$E\{|G(w)|^2\}=(1-\alpha^2)E\{|S_1(w)|^2\} \tag{4.42}$$

The correlation coefficient between $S_1(w)$ and $S_2(w)$ is defined as

$$\rho(w)=\frac{E\{S_1(w)S_2^*(w)\}}{\sqrt{E\{|S_1(w)|^2\}E\{|S_2(w)|^2\}}}=\alpha e^{jw\tau} \tag{4.43}$$

Let $B=w_L-w_1$ be the signal bandwidth in radians per second. The delay-bandwidth product (DBP) is defined as

$$DBP=\frac{B\tau}{2\pi} \tag{4.44}$$

If we assume a nonzero bandwidth, $DBP=0$ has different effects depending on the correlation coefficient $\rho(w)$. If $\rho(w)=1$, $s_1(t)=s_2(t)$ and multipath signals arrive at the sensors at the same time.

This is the most problematic case for DOA estimation because the covariance matrix becomes rank deficient. $DBP=1$ is the mildest case of correlation. In this case, multipath signals arrive at different times. It is known that wideband processing can handle correlated sources efficiently provided that the *DBP* is sufficiently large (Friedlander and Weiss, 1993). For such cases, the average of the focused covariance matrices in coherent wideband processing generates an effect similar to the average of the covariance matrices for subarrays in FBSS for the narrowband case. When $DBP=0$, coherent wideband processing alone is not sufficient to estimate DOA angles accurately.

4.4.3 Noncoherent Wideband DOA Estimation

The idea in noncoherent wideband DOA estimation is simple. The DOA information at every frequency bin is separately processed and the final results combined. This is a simple approach, but there are different ways of implementing it. Trivial noncoherent DOA estimation finds the DOA estimates at each frequency bin and takes their average. It turns out that it is possible to obtain better accuracy than this trivial approach. We consider two variants of noncoherent wideband DOA estimation: NCO1 and NCO2, which use Spectral-MUSIC as a tool. The narrowband spectra are summed to obtain a final spectrum of estimation. Since this summation uses different frequency bins without considering subspace matching, these methods are considered noncoherent. The NCO1 estimate is given as

$$\hat{\theta} = \underset{\theta}{argmin} \sum_{j=0}^{L-1} \mathbf{a}_\theta(w_j)\mathbf{G}(w_j)\mathbf{G}^H(w_j)\mathbf{a}_\theta^H(w_j) \tag{4.45}$$

where L is the number of frequency bins, $\mathbf{G}(w_j)$ is the noise subspace at frequency w_j, and $\mathbf{a}_\theta(w_j)$ is the array steering vector. Similarly, the NCO2 estimate is

$$\hat{\theta} = \underset{\theta}{argmax} \sum_{j=0}^{L-1} \frac{1}{\mathbf{a}_\theta(w_j)\mathbf{G}(w_j)\mathbf{G}^H(w_j)\mathbf{a}_\theta^H(w_j)} \tag{4.46}$$

It turns out that NCO1 and NCO2 perform similarly even though their formulations are quite different. In the following sections of this chapter, we consider NCO1 as the noncoherent wideband DOA estimation method.

Noncoherent methods perform well with uncorrelated sources. However, a search in a dense grid should be performed to avoid the local minima of the multidimensional surface of θ parameters (4.45), which makes the computational complexity of noncoherent methods high. Noncoherent methods also perform poorly for coherent sources—just like the MUSIC algorithm.

4.4.4 Coherent Wideband DOA Estimation

The idea in coherent wideband DOA estimation is to coherently combine the covariance matrices at each frequency bin in order to apply narrowband techniques directly over a single covariance matrix. This is achieved by transforming the covariance matrices at each frequency into a covariance matrix at a focusing frequency, such as the center frequency. Thus, coherent methods find the transformation matrices for each frequency, perform the transformation, and sum the covariance matrices to obtain the final and focused covariance matrix. Then they apply narrowband DOA estimation techniques. Clearly, coherent methods are computationally more efficient than noncoherent techniques. It turns out that their performance can be better than that of coherent methods.

Wiener Array Interpolation for Coherent Wideband Processing

WAI is discussed in detail in Section 4.3.2. It is possible to extend the WAI formulation to wideband processing. The main difference between the narrowband and wideband cases is that there is a different transformation matrix for each frequency:

$$\mathbf{T}(w) = \sigma_s^2\mathbf{A}_v(w_c)\mathbf{A}_r^H(w)\left(\sigma_s^2\mathbf{A}_r(w)\mathbf{A}_r^H(w) + \sigma_v^2\mathbf{I}\right)^{-1} \tag{4.47}$$

In addition, as is obvious in Equation (4.47), two mappings are performed simultaneously in the case of NLA. The first is for the transformation of the covariance matrix at frequency w to frequency w_c. The second is for obtaining a virtual array response from the real array. The application of WAI here is similar to its application in the narrowband case. The algorithm steps are as follows:

Step 1. Given the sensor outputs, $\mathbf{Y}_r(w_c)$, at center frequency w_c, obtain the sample covariance matrix $\hat{\mathbf{R}}(w_c)$. Use Toeplitz completion for NLA. Fill the missing covariance lags using spline interpolation if necessary and find the complete Hermitian symmetric Toeplitz covariance matrix, $\mathbf{R}_{TC}(w_c)$.

Step 2. Use $\mathbf{R}_{TC}(w_c)$ in Root-MUSIC and find the initial DOA estimate $\hat{\boldsymbol{\theta}}$.

Step 3. Given $\hat{\boldsymbol{\theta}}$, construct $\mathbf{T}(w_j)$ from Equation (4.47) for $j=1,\ldots,L$.

Step 4. Find $\mathbf{R}_F=\frac{1}{L}\sum_{j=1}^{L}\mathbf{T}(w_j)\hat{\mathbf{R}}(w_j)\mathbf{T}^H(w_j)$ and use Root-MUSIC to find the true DOA estimate.

Step 5. Iterate steps 3 and 4 by updating the initial estimates.

GRSS for Coherent Wideband Processing

A coherent subspace method (Wang and Kaveh, 1985) was originally proposed for wideband processing. In Wang and Kaveh (1987), CSM performance was investigated analytically. The RSS technique was presented in Hung and Kaveh (1988) as an effective method for the design of the mapping matrix. In Section 4.3.2, RSS was discussed for the narrowband case. Wideband extension can be easily done. In this case, the formulation is frequency dependent:

$$\mathbf{T}(\omega)=\underset{\mathbf{T}(\omega)}{argmin}||\mathbf{T}(\omega)\mathbf{A}_r(\omega)-\mathbf{A}_v(\omega_c)||_F^2,\quad s.t.\ \mathbf{T}^H(\omega)\mathbf{T}(\omega)=\mathbf{I}_M \tag{4.48}$$

where $M\times K$ matrix $\mathbf{A}_r(\omega)$ stands for the real array steering matrix at frequency ω, and $M_v\times K$ matrix $\mathbf{A}_v(\omega_c)$ is the array steering matrix for the virtual array at center frequency. It is assumed that K calibration angles are used in the neighborhood of the initial DOA estimates. The solution for $M_v\times M$ $\mathbf{T}(\omega)$ is

$$\mathbf{T}(\omega)=\mathbf{V}\hat{\Sigma}\mathbf{U}^H \tag{4.49}$$

where $M_v\times M$ matrix $\hat{\Sigma}=[\mathbf{I}_{M\times M}\ \mathbf{0}]^T$, and $M_v\times M_v$ and $M\times M$ unitary matrices $\mathbf{V}$ and $\mathbf{U}$ are obtained from the singular value decomposition:

$$\mathbf{A}_r(\omega)\mathbf{A}_v^H(\omega_c)=\mathbf{U}\Sigma\mathbf{V}^H \tag{4.50}$$

The formulation in Equation (4.49) can be used to map arrays with $M_v\geq M$.

4.4.5 Coherent Wideband Processing for Multipath Signals

Multipath signals generate different signal scenarios, as described in Section 4.4.2. When $DBP=0$, additional processing is required to correct the rank of the covariance matrix. For ULA, FBSS is an effective technique for this additional processing; it is applied to the final covariance matrix obtained in coherent wideband processing. For NLA, it is not possible to use FBSS directly. Furthermore, the initial DOA estimation is a difficult problem for coherent source signals in this case. Here CLEAN is presented to deal with it. Although intended for NLA, CLEAN is effective for ULA as well.

CLEAN Algorithm for Coherent Wideband DOA Estimation

CLEAN was discussed in detail for narrowband processing in Section 4.3.3. It is possible to use the same approach in wideband processing with small modifications, following these steps:

Step 1. Given an NLA, obtain the sample covariance matrix at ω_c and apply Toeplitz completion.

Step 2. Use the covariance matrix obtained in step 1 for initial DOA estimation by employing Root-MUSIC.

Step 3. Given the initial DOA estimates $\bar{\theta} = [\bar{\theta}_1, \ldots, \bar{\theta}_n]$, apply CLEAN as follows: Take the frequency as $\omega_j, j = 1, \ldots, L$. For each $\bar{\theta}_i, i = 1, \ldots, n$:

1. Construct the $M \times n$ array steering matrix $\bar{\mathbf{A}}_r(\omega_j)$ and find $\bar{\mathbf{S}}(\omega_j) = [\bar{\mathbf{A}}_r^H(\omega_j)\bar{\mathbf{A}}_r(\omega_j)]^{-1}\bar{\mathbf{A}}_r^H(\omega_j)\mathbf{Y}_r(\omega_j)$.
2. Construct the $M \times (n-1)$ steering matrix $\hat{\mathbf{A}}_r^i(\omega_j)$. $\hat{\mathbf{A}}_r^i(\omega_j)$ is obtained from $\bar{\mathbf{A}}_r^i(\omega_j)$ by deleting the ith column. Also obtain the source $\hat{\mathbf{S}}^i(\omega_j)$ from $\bar{\mathbf{S}}(\omega_j)$ by deleting the ith row.
3. Find the cleaned signal $\mathbf{Y}_r^i(\omega_j) = \mathbf{Y}_r(\omega_j) - \hat{\mathbf{A}}_r^i(\omega_j)\hat{\mathbf{S}}^i(\omega_j)$ and the corresponding covariance matrix $\mathbf{R}^i(\omega_j) = \frac{1}{N}\sum_{m=1}^{N}\mathbf{Y}_r^i(\omega_j, m)\mathbf{Y}_r^{i\,H}(\omega_j, m)$.
4. Find the array interpolation matrix $\mathbf{T}^i(\omega_j)$ from Equation (4.49) using $\tilde{\mathbf{A}}_r^i(\omega_j)$ and $\tilde{\mathbf{A}}_v^i(\omega_j)$ for the real and virtual arrays, respectively, considering K calibration angles in the neighborhood of $\bar{\theta}_i$.
5. Obtain $\mathbf{R}_T^i(\omega_j) = \mathbf{T}^i(\omega_j)\mathbf{R}^i(\omega_j)\mathbf{T}^i(\omega_j)^H$ and $\mathbf{R}_F^i = \frac{1}{L}\sum_{j=1}^{L}\mathbf{R}_T^i(\omega_j)$.
6. Use $\mathbf{R}_F^i$ in Root-MUSIC to find the θ_i estimate.

Step 4. Repeat step 3 by updating the initial estimate for better results.

The CLEAN algorithm is very effective for coherent (or multipath) sources. The main reason for this is that the accuracy of array mapping increases significantly when a single source is considered. Since the main source of error for NLA is the array mapping, CLEAN performs well compared to the other algorithms. Note that FBSS can be applied to the final covariance matrix after array mapping to improve performance for coherent sources.

4.4.6 Performance Comparisons

Comparison of the algorithms for the wideband case is similar to that for narrowband simulations. The partly filled NLA has a sensor placement of

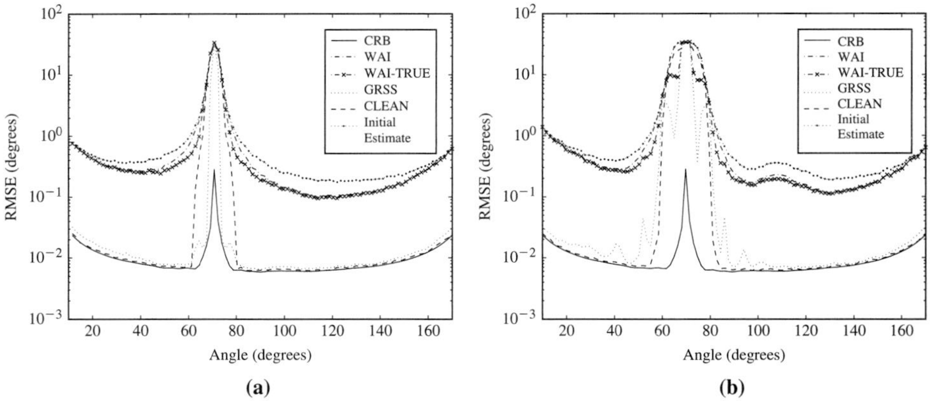

FIGURE 4.13 DOA performance for two wideband sources: (a) uncorrelated; (b) coherent. SNR = 10 dB.

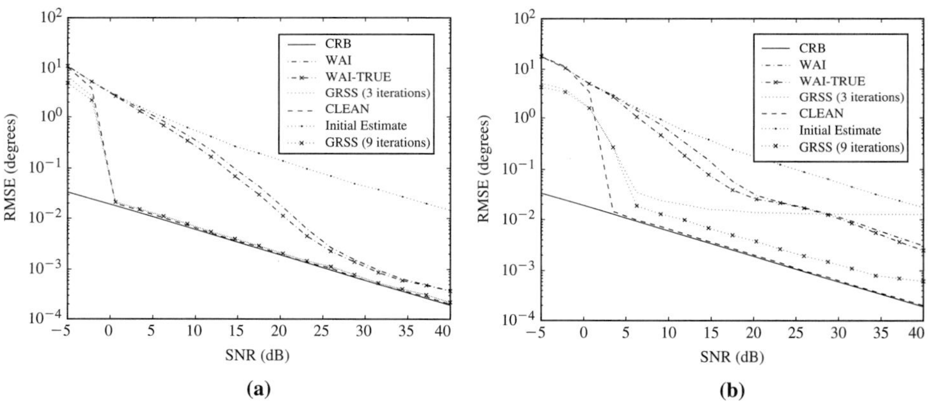

FIGURE 4.14 SNR performance for two wideband sources: (a) uncorrelated; (b) coherent. SNR = 10 dB.

$d_{NLA} = [0 \quad 1 \quad 2 \quad 3 \quad 8 \quad 13 \quad 17]$. The ratio of the signal bandwidth to the center frequency is $B/f_c = 2/3$. Eleven frequency bins are used for wideband processing. Figure 4.13 shows the performance for two sources: one fixed at 70°, and the other swept from 10° to 170°. In Figure 4.13(a) source signals are uncorrelated. CLEAN and GRSS have very good performance for these. When the sources are coherent with $DBP = 0$, CLEAN shows slightly better performance than GRSS, although GRSS can resolve closely spaced sources better. In these experiments array mapping has three iterations for all algorithms.

Figure 4.14 shows the SNR performance of the algorithms. There are two sources at 70° and 85°, respectively. Figure 4.14(a) shows the case when the source signals are uncorrelated. GRSS-3-ITER and GRSS-9-ITER are the GRSS methods where array-mapping iterations are 3 and 9, respectively. Iterations are

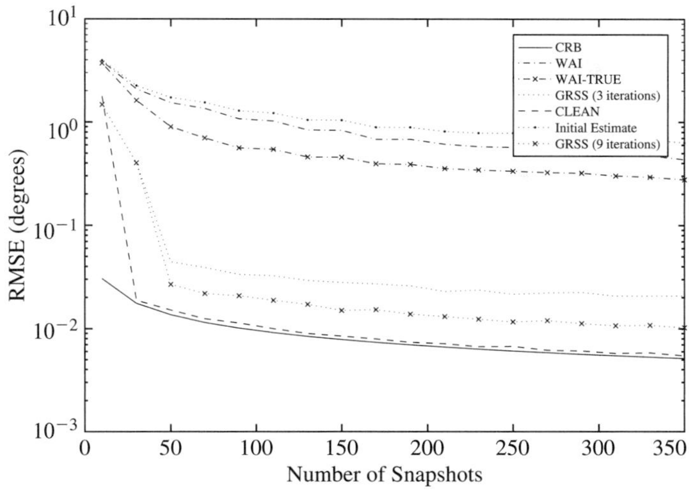

FIGURE 4.15 Performance for two wideband coherent sources in terms of number of snapshots. SNR = 10 dB.

increased for GRSS to show the gain with iterations as well as to show that CLEAN may still outperform GRSS significantly for certain cases. When the sources are uncorrelated, a small number of iterations are sufficient and GRSS and CLEAN perform well. When the sources are coherent with $DBP = 0$, CLEAN performs significantly better. As the number of iterations increases, GRSS gets closer to CLEAN. In general, WAI and WAI-TRUE have similar performance, and the estimated noise and source powers are sufficiently accurate.

Figure 4.15 shows the performance of the algorithms in terms of the number of snapshots for the same two coherent sources. CLEAN performs better in comparison.

Additional comparisons are done with noncoherent and TOPS algorithms. Since these algorithms are computationally intense, the number of trials is decreased to 100. In Figure 4.16(a), two wideband sources at 70° and 85° are fixed and the third is swept between 10° and 170°. Source signals are uncorrelated and the same NLA is used. The initial estimate is obtained by TC. CLEAN, GRSS, and noncoherent algorithms perform well, while TOPS does not come close to the Cramer–Rao bound (CRB). In Figure 4.16(b) three uncorrelated sources are placed at 70°, 85°, and 95°, respectively. A similar behavior for the algorithm's performance is observed.

In Figure 4.17, a ULA with ten sensors is used. An initial estimate is obtained directly from the sensor data by the Root-MUSIC algorithm using only a single-frequency covariance matrix. There are three uncorrelated wideband sources at 70°, 85°, and 95°, respectively. While CLEAN, GRSS, and the noncoherent algorithms perform well, TOPS does not come close to the same performance.

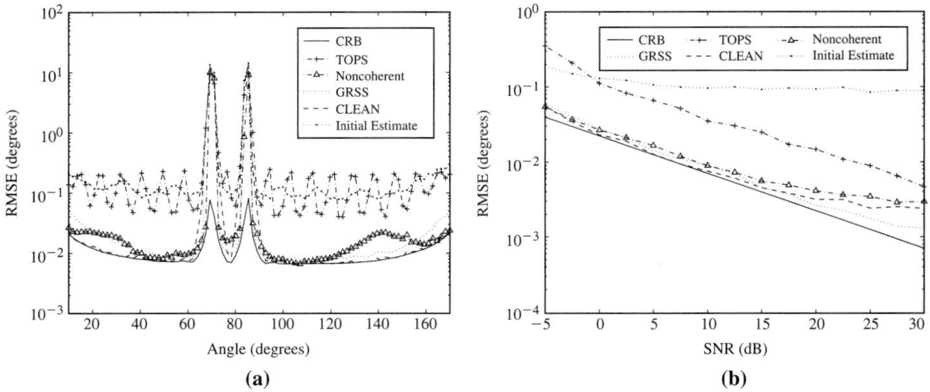

FIGURE 4.16 Performance for NLA for three wideband uncorrelated sources. (a) Two sources at 70° and 85° are fixed; the third source is swep (SNR = 10 dB). (b) Sources are at 70°, 85°, and 95°, respectively.

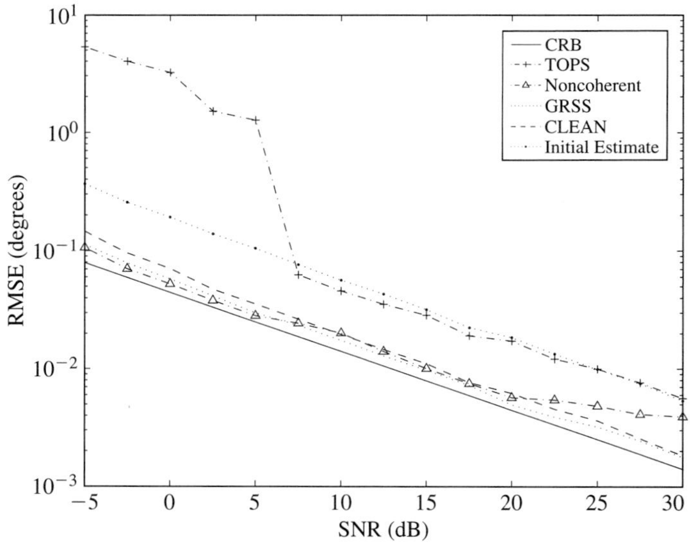

FIGURE 4.17 Performance for three wideband uncorrelated sources for a ULA with ten sensors.

Figure 4.18 shows the DBP characteristics for a ULA with ten sensors. There are two sources at 70° and 85°, respectively. $SNR = 10$ dB, and the initial estimate is obtained from a single frequency using FBSS. Neither algorithm uses FBSS for the final covariance matrix. CLEAN performs better than the others since it treats single sources. GRSS performance improves as the DBP increases. Noncoherent and TOPS seem to be insensitive to DBP, which is reasonable. They do not perform well for coherent sources.

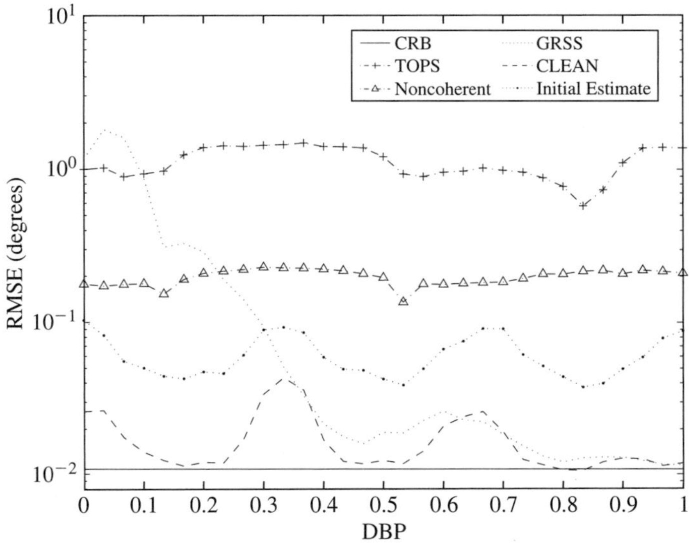

FIGURE 4.18 DBP performance for two coherent wideband sources for a ULA with ten sensors.

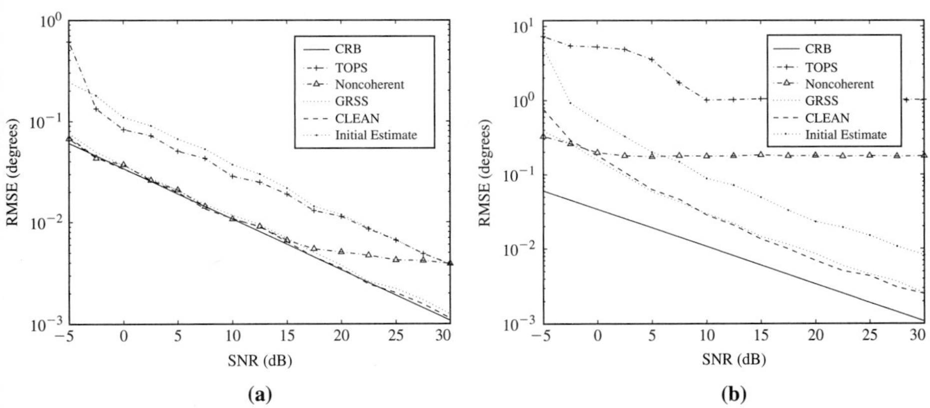

FIGURE 4.19 ULA performance for two wideband sources at 70° and 85°: (a) uncorrelated sources; (b) coherent sources (DBP = 0).

Figure 4.19 shows the performance of the algorithms for uncorrelated and coherent sources in the case of a ULA. In Figure 4.19(a) two sources at 70° and 85° are uncorrelated. An initial estimate is obtained from the ULA by the Root-MUSIC algorithm for a single frequency. In Figure 4.19(b) sources are coherent with $DBP = 0$. An initial estimate is obtained using FBSS for a single frequency. All the algorithms use FBSS for improving their covariance matrices. CLEAN and GRSS perform well for both uncorrelated and coherent cases. Noncoherent and TOPS algorithms do not perform satisfactorily.

4.5 CONCLUSION

NLAs have certain advantages and problems for DOA estimation. They cover a large array aperture with fewer sensors, and they require fewer matched channels or receivers. Their disadvantage is that coherent sources cannot be easily handled. Furthermore, they need additional computation to compensate for and augment the missing sensor information. The completion of missing sensor data is required to improve accuracy. Array mapping is an effective method of augmenting the NLA covariance matrix. Array-mapping accuracy can be improved significantly if an initial DOA estimate is used and then the estimates are improved iteratively. For this reason initial DOA estimation is a key problem for NLA, but it can be easily solved for uncorrelated sources. Toeplitz completion can be used directly for this purpose. Initial DOA estimation for coherent sources is not an easy task. While TC and FBA return good estimates for certain DOA angles, they do not perform well for others.

A promising approach for coherent signals is to use partly filled NLA. Initial DOA estimates can be obtained via FBSS for the ULA part of this array. Then array mapping can generate a covariance matrix corresponding to a full array with the same aperture. Different array-mapping techniques exist in the literature. Classical array interpolation is well known, but it has certain limitations. Wiener array interpolation performs well, but it produces focusing loss for NLA. It performs better for circular arrays compared to alternative techniques where a large angular sector for array mapping is used. Array mapping with the RSS and CLEAN algorithms works well for a variety of cases as long as the angular sector is small. In this respect, RSS and CLEAN are good candidates for DOA estimation for NLA.

Wideband DOA estimation has a processing gain proportional to the number of frequency bins that carry information about the source signal. In this respect, it can improve DOA accuracy compared to narrowband processing. The main problem is computational complexity. Noncoherent techniques generally perform well for uncorrelated sources, but computational complexity increases with the number of frequency bins. Coherent wideband processing is an effective and computationally efficient alternative.

When array mapping is used to cohere the covariance matrices, narrowband techniques can be applied on a single covariance matrix. GRSS and CLEAN methods perform well even with coherent source signals. They allow FBSS to be used in the final covariance matrix, and as a result they can be employed for both uncorrelated and coherent wideband sources. They also perform better than the noncoherent method. These techniques can be used for both ULA and NLA. Since array mapping is applied more effectively when an initial DOA estimate is used, partly filled NLA can be used for wideband DOA estimation when there are multipath signals.

Acknowledgement T. Tuncer and T. Kaya Yasar were supported by the Scientific and Research Council of Turkey, (TÜBİTAK), through grant 104E045.

REFERENCES

Abramovich, Y.I., 1979. A recursion method for the resolution of wideband signals having substantially different intensities. Radio Eng. Electron. Phys. 23 (8), 66–70.

Abramovich, Y.I., Gray, D.A., Gorokhov, A.Y., Spencer, N.K., 1998. Positive-definite Toeplitz completion in DOA estimation for nonuniform linear antenna arrays—part I: fully augmentable arrays. IEEE Trans. Signal Process. 46 (9), 2458–2471.

Abramovich, Y.I., Spencer, N.K., Gorokhov, A.Y., 1999. Positive-definite Toeplitz completion in DOA estimation for nonuniform linear antenna arrays-part II: partially augmentable arrays. IEEE Trans. Signal Process. 47 (6), 1502–1521.

Abramovich, Y.I., Spencer, N.K., Gorokhov, A.Y., 2000. DOA estimation for noninteger linear antenna arrays with more uncorrelated sources than sensors. IEEE Trans. Signal Process. 48 (4), 943–955.

Bronez, T.P., 1988. Sector interpolation of non-uniform arrays for efficient high-resolution bearing estimation. Int. Conf. Acoust. Speech Signal Process. 5, 2885–2888.

Clark, B.G., 1980. An efficient implementation of the algorithm CLEAN. Astron. Astrophys. 89, 377–378.

Covarruhias-Rosales, D.H., Olague, J.G.A., 2004. Improving resolution on DOA estimation in a multipath macrocell environment using eigenstructure-based methods. Symposium on Trends in Communications, SympoTIC-2004, 39–42.

Davies, D.E.N., 1965. A transformation between the phasing techniques required for linear and circular aerial arrays. Proc. IEE. 112, 2041–2045.

Deng, H., 2004. Effective CLEAN algorithms for performance-enhanced detection of binary coding radar signals. IEEE Trans. Signal Process. 52 (1), 72–78.

Doron, M.A., Doron, E., Weiss, A.J., 1993. Coherent wideband processing for arbitrary array geometry. IEEE Trans. Signal Process. 41 (1), 414–417.

Doron, M.A., Weiss, A.J., 1992. On focusing matrices for wide-band array processing. IEEE Trans. Signal Process. 40 (6), 1295–1302.

Elio, D., Claudio, D., Parisi, R., 2001. WAVES: Weighted average of signal subspace for robust wideband direction finding. IEEE Trans. Signal Process. 49 (10), 2179–2191.

Friedlander, B., 1993. The Root-MUSIC algorithm for direction finding with interpolated arrays. Signal Process. 30 (1), 15–29.

Friedlander, B., Weiss, A.J., 1992. Direction finding using spatial smoothing with interpolated arrays. IEEE Trans. Aerosp. Electron. Syst. 28 (2), 574–587.

Friedlander, B., Weiss, A.J., 1993. Direction finding for wideband signals using an interpolated array. IEEE Trans. Signal Process. 41 (4), 1618–1634.

Horn, R.A., Johnson, C.R., 1985. Matrix Analysis. Cambridge University Press.

Hung, H., Kaveh, M., 1988. Focusing matrices for coherent signal-subspace processing. IEEE Trans. Signal Process. 36 (8), 1272–1281.

Hyberg, P., Jansson, M., Ottersten, B., 2005. Array interpolation and DOA MSE reduction. IEEE Trans. Signal Process. 53 (12), 4464–4471.

Kaya Yasar, T., Tuncer, T.E., 2008. Wideband DOA estimation for nonuniform linear arrays with Wiener array interpolation. Fifth IEEE Sensor Array and Multichannel Signal Processing Workshop, SAM-2008, pp. 207–211.

Krolik, J., Swingler, D., 1989. Multiple broad-band source location using steered covariance matrices. IEEE Trans. Signal Process. 37 (10), 1481–1494.

Krolik, J., Swingler, D., 1990. Focused wide-band array processing by spatial resampling. IEEE Trans. Signal Process. 38 (2), 356–360.

Linebarger, D.A., Sudborough, I.H., Tollis, I.G., 1993. Difference bases and sparse arrays. IEEE Trans. Inf. Theory. 39 (2), 716–721.

Ng, W., Reilly, J.P., Kirubarajan, T., Larocque, J., 2005. Wideband array signal processing using MCMC methods. IEEE Trans. Signal Process. 53 (2), 411–426.

Pillai, S.U., Haber, F., 1987. Statistical Analysis of a high resolution spatial spectrum estimator utilizing an augmented covariance matrix. IEEE Trans. Signal Process. 35 (11), 1517–1523.

Pillai, S.U., Kwon, B.H., 1989. Forward/backward spatial smoothing techniques for coherent signal identification. IEEE Trans. Acoust. Speech Signal Process. [see also IEEE Trans. Signal Process.] 37 (1), 8–15.

Rao, B.D., Hari, K.V.S., 1989. Performance analysis of root-MUSIC. IEEE Trans. Signal Process. 37 (12), 1939–1949.

Schwarz, U.J., 1978. Mathematical-statistical description of the iterative beam removing technique (Method CLEAN). Astron. Astrophys. 65, 345–356.

Sellone, F., 2006. Robust auto-focusing wideband DOA estimation. Signal Process. 86, 17–37.

Shan, T.-J., Wax, M., Kailath, T., 1985. On spatial smoothing for direction-of-arrival estimation of coherent signals. IEEE Trans. Acoust. Speech Signal Process. [see also IEEE Trans. Signal Process.] 33 (4), 806–811.

Stoica, P., Nehorai, A., 1998. MUSIC, maximum likelihood and Cramer-Rao bound. IEEE Trans. Signal Process. 37 (5), 720–741.

Stoica, P., Moses, R., 2005. Spectral Analysis of Signals. Prentice-Hall.

Teague, C.C., 2002. Root-MUSIC direction finding applied to multifrequency coastal radar. Geoscience and Remote Sensing Symposium, IGARSS-2002, 1896–1898.

Tsao, J., Steinberg, B.D., 1988. Reduction of sidelobe and speckle artifacts in microwave imaging: the CLEAN technique. IEEE Trans. Antennas Propag. 36 (4), 543–556.

Tuncer, T., Kaya Yasar, T., Friedlander, B., 2007a. DOA estimation for nonuniform linear arrays by using array interpolation. Radio Sci. 42 (4), 4002-I–4002-II.

Tuncer, T.E., Kaya Yasar, T., Friedlander, B., 2007b. Partly-filled nonuniform linear arrays for DOA estimation in multipath signals. Int. Conf. Acoust. Speech Signal Process. 2, II-1053–II-1056.

Van Trees, H.L., 2002. Optimum Array Processing in Detection, Estimation and Modulation Theory. John Wiley & Sons.

Wang, H., Kaveh, M., 1985. Coherent signal-subspace processing for the detection and estimation of angles of arrival of multiple wide-band sources. IEEE Trans. Signal Process. 33 (4), 823–831.

Wang, H., Kaveh, M., 1987. On the performance of signal-subspace processing—part II: coherent wide-band systems. IEEE Trans. Signal Process. 35 (11), 1583–1591.

Wax, M., Shan, T.J., Kailath, T., 1982. Source location and spectral density estimation of multiple sources. In: Proc. 16th Asilomar Conference on Circuits, Systems and Computers.

Wax, M., Shan, T., Kailath, T., 1984. Spatio-temporal spectral analysis by eigenstructure methods. IEEE Trans. Signal Process. 32 (4), 817–827.

Weiss, A.J., Friedlander, B., 1993. Maximum likelihood signal estimation for polarization sensitive arrays. IEEE Trans. Antennas Propag. 41 (7), 918–925.

Yoon, Y., Kaplan, L.M., McClellan, J.H., 2006. TOPS: new DOA estimator for wideband signals. IEEE Trans. Signal Process. 54 (6), 1977–1989.

Chapter 5

Search-Free DOA Estimation Algorithms for Nonuniform Sensor Arrays

Michael Rübsamen, Alex B. Gershman

5.1 INTRODUCTION

Over the last four decades, a large number of methods for estimating the directions of arrival (DOA) of multiple narrowband far-field signal sources has been proposed. Important characteristics of these methods are their DOA estimation mean-square error (MSE) performance, computational complexity, and required array geometry specifications. Many techniques are applicable only to quite specific array geometries, such as uniform linear arrays (ULAs) or uniform circular arrays (UCAs). However, it is often unsuitable in practical scenarios to restrict the array geometry to a certain class. Therefore, methods that can be applied to arbitrary nonuniform array geometries are in great demand.

Among the known DOA estimation methods applicable to arbitrary arrays, those based on the maximum likelihood (ML) principle have been shown to achieve the best MSE performance (Böhme, 1986; Jaffer, 1988; Schweppe, 1968; Wax and Kailath, 1983; Ziskind and Wax, 1988). However, their computational complexity is usually prohibitively high because a multidimensional search is required to find the global maximum of the nonconvex likelihood function.

Several computationally efficient high-resolution DOA estimation methods applicable to arbitrary arrays have been proposed based on the concept of signal and noise subspaces. One of the most popular subspace-based techniques is the Multiple-Signal Classification (MUSIC) algorithm (Bienvenu and Kopp, 1980; Schmidt, 1979). Compared to ML techniques, MUSIC leads to a spectral search over a reduced parameter space and therefore offers dramatically lower computational complexity. However, the complexity of the spectral search in the conventional MUSIC algorithm may still be too high for real-time applications.

Copyright © 2009 by Academic Press, Inc. All rights of reproduction in any form reserved.

The lowest computational complexities in subspace-based techniques are achieved by so-called *search-free* DOA estimation methods. We provide a detailed overview of search-free techniques in this chapter, with a particular emphasis on nonuniform sensor arrays.

The oldest and perhaps the most popular search-free DOA estimation techniques are Root-MUSIC (Barabell, 1983; Rao and Hari, 1989) and ESPRIT (Estimation of Signal Parameters via Rotational Invariance Technique; Paulraj et al., 1986; Roy and Kailath, 1989; Roy et al.,, 1986). However, the latter two can only be applied to arrays with very specific geometries. In particular, Root-MUSIC applies only to ULAs and nonuniform arrays (NUAs) of which the sensors lie on a uniform grid, whereas ESPRIT requires arrays that consist of two identical and identically oriented subarrays. Several useful extensions of Root-MUSIC and ESPRIT to more general classes of array geometries have been proposed, such as UCA Root-MUSIC, UCA ESPRIT (Mathews and Zoltowski, 1994), Rank Reduction Estimator (RARE; Pesavento et al., 2002), and generalized ESPRIT (Gao and Gershman, 2005).

There has been a growing interest in developing search-free DOA estimation techniques for arbitrary array geometries. The array interpolation approach (Friedlander, 1993; Friedlander and Weiss, 1992, 1993; Weiss et al., 1995) employs the idea of approximating any actual NUA by a virtual ULA. Then any ULA-based search-free method can be applied to the virtual ULA observations, including the standard Root-MUSIC algorithm. The manifold separation (MS) approach (Belloni et al., 2007; Doron and Doron, 1994a,b,c) extends Root-MUSIC to arbitrary array geometries by modeling the received wavefield via an orthogonal expansion that approximates the true array steering vector as a product of a matrix depending only on the array parameters and a Vandermonde vector depending only on the angle. The Vandermonde structure of the latter vector is then exploited to obtain a polynomial that can be rooted to estimate the source DOAs.

Another recent approach to search-free DOA estimation for arbitrary arrays is Fourier Domain (FD) Root-MUSIC (Rübsamen and Gershman, 2008, 2009). FD Root-MUSIC exploits the fact that the null-spectrum MUSIC function is periodic in angle. More specifically, it uses the truncated Fourier series expansion of this periodic function to reformulate the DOA estimation problem in terms of polynomial rooting rather than spectral search. FD Root-MUSIC is related to the MS technique, but it has been shown to achieve substantial performance improvements in the asymptotic domain at no additional cost. Further refinement of this technique is the FD-Weighted Least-Squares (FDWLS) Root-MUSIC algorithm (Rübsamen and Gershman, 2009), which uses a weighted least-squares approximation of the MUSIC null-spectrum function to improve the DOA estimation performance.

This chapter is organized as follows. In Section 5.2 we introduce our notation and provide some necessary background on the DOA estimation problem.

Then in Section 5.3 we focus on search-free DOA estimation methods applicable to specific array geometries. First, classic search-free subspace-based techniques are discussed such as Root-MUSIC and ESPRIT. Next, more recent search-free direction-finding methods are considered that extend the traditional Root-MUSIC and ESPRIT approaches to several useful classes of nonuniform sensor arrays. These methods include the UCA extensions of Root-MUSIC, the generalized ESPRIT methods, and the Root-RARE techniques. Section 5.4 is devoted to search-free DOA estimation methods applicable to arbitrary array geometries; we discuss the interpolated Root-MUSIC technique, the MS method, and the family of FD Root-MUSIC algorithms.

Simulation results are presented in Section 5.5, where performance of the search-free methods for arbitrary arrays is compared.

5.2 BACKGROUND

We assume L narrowband far-field signal sources that impinge on an array of N omnidirectional sensors. The unknown DOAs of the signals are $(\theta_1, \phi_1), \ldots, (\theta_L, \phi_L)$, where θ_L and ϕ_L are the azimuth and elevation angles of the Lth source, respectively. The $N \times 1$ array snapshot vector at time k can be modeled as

$$\boldsymbol{x}(k) = \boldsymbol{A}(\boldsymbol{\theta})\boldsymbol{s}(k) + \boldsymbol{n}(k) \tag{5.1}$$

where $\boldsymbol{\theta} = [\theta_1, \phi_1, \ldots, \theta_L, \phi_L]^T$ is the $2L \times 1$ vector of signal DOAs,

$$\boldsymbol{A}(\boldsymbol{\theta}) = [\boldsymbol{a}(\theta_1, \phi_1), \ldots, \boldsymbol{a}(\theta_L, \phi_L)] \tag{5.2}$$

is the $N \times L$ signal direction matrix, $\boldsymbol{s}(k)$ is the $L \times 1$ vector of signal waveforms, $\boldsymbol{n}(k)$ is the $N \times 1$ vector of sensor noise, and $(\cdot)^T$ is the transpose.

The $N \times 1$ steering vector is given by

$$\boldsymbol{a}(\theta, \phi) = \begin{bmatrix} \exp\left\{j\frac{2\pi}{\lambda}(x_1 \sin\theta \sin\phi + y_1 \cos\theta \sin\phi + z_1 \cos\phi)\right\} \\ \vdots \\ \exp\left\{j\frac{2\pi}{\lambda}(x_N \sin\theta \sin\phi + y_N \cos\theta \sin\phi + z_N \cos\phi)\right\} \end{bmatrix} \tag{5.3}$$

where λ is the signal wavelength, $j = \sqrt{-1}$, and $\{x_i, y_i, z_i\}$ are the coordinates of the ith array sensor. In the sequel we assume, unless specified otherwise, that the sensor array manifold is known.

If the signal and noise vectors are zero mean, and if the noise is spatially white and statistically independent of the source signals, then the $N \times N$ array covariance matrix can be written as

$$\boldsymbol{R}_x = \mathrm{E}\{\boldsymbol{x}(k)\boldsymbol{x}^H(k)\} = \boldsymbol{A}\boldsymbol{R}_s\boldsymbol{A}^H + \sigma^2\boldsymbol{I}_N \tag{5.4}$$

where $\boldsymbol{R}_s = \mathrm{E}\{\boldsymbol{s}(k)\boldsymbol{s}^H(k)\}$ is the source covariance matrix, σ^2 is the sensor noise variance, $\boldsymbol{I}_N$ is the $N \times N$ identity matrix, $\mathrm{E}\{\cdot\}$ denotes the statistical expectation, and $(\cdot)^H$ denotes the Hermitian transpose.

The eigendecomposition of $\boldsymbol{R}_x$ yields

$$\boldsymbol{R}_x = \sum_{k=1}^{N} \lambda_k \boldsymbol{e}_k \boldsymbol{e}_k^H \tag{5.5}$$

where λ_k and $\boldsymbol{e}_k$ $(k = 1, \ldots, N)$ are the eigenvalues and corresponding eigenvectors of $\boldsymbol{R}_x$. Let the eigenvalues λ_k be sorted in nonascending order. Then the matrices

$$\boldsymbol{E}_S \triangleq [\boldsymbol{e}_1, \ldots, \boldsymbol{e}_L], \quad \boldsymbol{E}_N \triangleq [\boldsymbol{e}_{L+1}, \ldots, \boldsymbol{e}_N] \tag{5.6}$$

contain the L signal-subspace and $N-L$ noise-subspace eigenvectors, respectively.

In practical situations, the exact array covariance matrix $\boldsymbol{R}_x$ is unavailable, so its sample estimate

$$\hat{\boldsymbol{R}}_x = \frac{1}{K} \sum_{k=1}^{K} \boldsymbol{x}(k)\boldsymbol{x}^H(k) \tag{5.7}$$

is used, where K is the number of snapshots.

The eigendecomposition of the sample covariance matrix (5.7) yields

$$\hat{\boldsymbol{R}}_x = \hat{\boldsymbol{E}}_S \hat{\boldsymbol{\Lambda}}_S \hat{\boldsymbol{E}}_S^H + \hat{\boldsymbol{E}}_N \hat{\boldsymbol{\Lambda}}_N \hat{\boldsymbol{E}}_N^H \tag{5.8}$$

where the sample eigenvalues are again sorted in nonascending order ($\hat{\lambda}_1 \geq \hat{\lambda}_2 \geq \ldots \geq \hat{\lambda}_N$), and the matrices $\hat{\boldsymbol{E}}_S \triangleq [\hat{\boldsymbol{e}}_1, \ldots, \hat{\boldsymbol{e}}_L]$ and $\hat{\boldsymbol{E}}_N \triangleq [\hat{\boldsymbol{e}}_{L+1}, \ldots, \hat{\boldsymbol{e}}_N]$ contain in their columns the signal- and noise-subspace eigenvectors of $\hat{\boldsymbol{R}}_x$, respectively. Correspondingly, the diagonal matrices $\hat{\boldsymbol{\Lambda}}_S \triangleq \mathrm{diag}\{\hat{\lambda}_1, \ldots, \hat{\lambda}_L\}$ and $\hat{\boldsymbol{\Lambda}}_N \triangleq \mathrm{diag}\{\hat{\lambda}_{L+1}, \ldots, \hat{\lambda}_N\}$ are built from the signal- and noise-subspace eigenvalues of $\hat{\boldsymbol{R}}_x$, respectively.

The conventional MUSIC null-spectrum function (Bienvenu and Kopp, 1980; Schmidt, 1979) can be expressed as

$$f(\theta, \phi) = \boldsymbol{a}^H(\theta, \phi)\hat{\boldsymbol{E}}_N \hat{\boldsymbol{E}}_N^H \boldsymbol{a}(\theta, \phi) = \|\hat{\boldsymbol{E}}_N^H \boldsymbol{a}(\theta, \phi)\|^2 \tag{5.9}$$

where $\|\cdot\|$ is the vector-2 norm. The Spectral-MUSIC technique estimates the signal DOAs from the minima of this function by searching over ϕ and θ using a fine grid. The computational complexity of this spectral search step is typically substantially higher than that of the eigendecomposition step because, as a rule, $J \gg N$, where J is the total number of spectral points. Note that for each spectral point, the product of $\hat{\boldsymbol{E}}_N^H$ and $\boldsymbol{a}(\theta, \phi)$ (or, alternatively, of $\hat{\boldsymbol{E}}_S^H$ and $\boldsymbol{a}(\theta, \phi)$) has to be computed. The statistical performance of MUSIC has been studied by Kaveh and Barabell (1986) and Stoica and Nehorai (1989), where it has been shown that at large values of K and N, and for uncorrelated signals, this algorithm asymptotically achieves the deterministic Cramer-Rao bound (CRB).

5.3 SEARCH-FREE METHODS FOR SPECIFIC ARRAY STRUCTURES

Historically, search-free DOA estimation methods were originally proposed for specific array geometries (Barabell, 1983; Paulraj et al., 1986). Search-free techniques applicable to arbitrary array geometries were developed later, often based on extensions of the earlier methods (Belloni et al., 2007; Doron and Doron, 1994a,b,c; Rübsamen and Gershman, 2008, 2009). In this section, we present an overview of search-free DOA estimation methods that require specific array geometries.

In the rest of this chapter, we will consider the one-dimensional DOA estimation problem, assuming that the sensors and sources are located in the xy-plane of the coordinate system. In this case, we have to estimate only the vector of the source azimuth angles $\boldsymbol{\theta} = [\theta_1, \ldots, \theta_L]^T$.

5.3.1 Root-MUSIC

As mentioned before, the Root-MUSIC technique (Barabell, 1983) is applicable to ULAs or to linear arrays with sensors that lie on a uniform grid. For the sake of simplicity, we consider the ULA case in this section. Let the array sensors be located on the x-axis of the coordinate system and centered with respect to the origin. Then, with a slight abuse of notation, the steering vector (5.3) can be written as

$$\boldsymbol{a}(\theta) = \begin{bmatrix} \exp\left\{-j\frac{N-1}{2}\frac{2\pi}{\lambda}\, d_x \sin\theta\right\} \\ \exp\left\{-j\frac{N-3}{2}\frac{2\pi}{\lambda}\, d_x \sin\theta\right\} \\ \vdots \\ \exp\left\{j\frac{N-1}{2}\frac{2\pi}{\lambda}\, d_x \sin\theta\right\} \end{bmatrix} = \begin{bmatrix} z^{-\frac{N-1}{2}} \\ z^{-\frac{N-3}{2}} \\ \vdots \\ z^{\frac{N-1}{2}} \end{bmatrix} \triangleq \boldsymbol{a}(z) \tag{5.10}$$

where $z \triangleq \exp\{j(2\pi/\lambda)\, d_x \sin\theta\}$, and d_x is the array inter-element spacing.

Using Equation (5.10), the MUSIC null-spectrum function can be written as

$$\begin{aligned} f(\theta) &= \boldsymbol{a}^H(\theta)\hat{\boldsymbol{E}}_N\hat{\boldsymbol{E}}_N^H\boldsymbol{a}(\theta) \\ &= \boldsymbol{a}^T(1/z)\hat{\boldsymbol{E}}_N\hat{\boldsymbol{E}}_N^H\boldsymbol{a}(z) \triangleq f(z) \end{aligned} \tag{5.11}$$

The polynomial $f(z)$ has $2(N-1)$ roots, which form conjugate reciprocal pairs. That is, if any z_0 is a root of $f(z)$, then $1/z_0^*$ is its root as well, where $(\cdot)^*$ denotes the complex conjugate. In the noise-free case, the polynomial $f(z)$ has L pairs of roots $z_i = \exp\{j(2\pi/\lambda)\, d_x \sin\theta_i\}$, $i = 1, \ldots, L$, and there are $2(N-L-1)$

additional "noise" roots. In the presence of noise, the root locations are distorted, but the signal DOAs can be estimated from the roots of $f(z)$ that lie closest to the unit circle. Because of the root conjugate reciprocity property, the roots inside the unit circle contain all information about the signal DOAs. Thus, the Root-MUSIC algorithm computes all roots of $f(z)$ and estimates the signal DOAs from the L largest-magnitude roots inside the unit circle.

Several modifications of Root-MUSIC have been proposed that offer improved performance and/or reduced computational complexity (see Pesavento et al., 2000 and Zoltowski et al., 1993 and references therein).

Clearly, Root-MUSIC is also applicable to any linear array with sensor displacements that are integer multiples of a common baseline d_x. This enables its application to sparse linear arrays.

5.3.2 ESPRIT

The ESPRIT algorithm was proposed by Paulraj et al. (1986) and further developed and studied by Ottersten et al. (1991), Roy and Kailath (1989), Roy et al., (1986). It is applicable to sensor arrays that consist of two identical and identically oriented subarrays, with one being a shifted "replica" of the other, where the displacement vector between the subarrays has to be known, whereas the geometry of each subarray may be unknown.

The steering vectors for the two subarrays can be written as

$$\boldsymbol{a}_1(\theta) = \boldsymbol{J}_1 \boldsymbol{a}(\theta), \quad \boldsymbol{a}_2(\theta) = \boldsymbol{J}_2 \boldsymbol{a}(\theta) \tag{5.12}$$

where $\boldsymbol{J}_1$ and $\boldsymbol{J}_2$ are the selection matrices having in each row a single entry equal to one and all the others zero. Without any loss of generality, we assume that the two identical and identically oriented subarrays are displaced along the x-axis of the coordinate system. Then the steering vectors for the two subarrays are related as

$$\boldsymbol{a}_1(\theta) = \boldsymbol{a}_2(\theta) e^{j(2\pi/\lambda)\, d_x \sin\theta} \tag{5.13}$$

where d_x now denotes the displacement between the two subarrays. Similarly, the two subarray steering matrices

$$\boldsymbol{A}_1(\boldsymbol{\theta}) = \boldsymbol{J}_1 \boldsymbol{A}(\boldsymbol{\theta}), \quad \boldsymbol{A}_2(\boldsymbol{\theta}) = \boldsymbol{J}_2 \boldsymbol{A}(\boldsymbol{\theta}) \tag{5.14}$$

are related as

$$\boldsymbol{A}_1(\boldsymbol{\theta}) = \boldsymbol{A}_2(\boldsymbol{\theta}) \boldsymbol{Q}(\boldsymbol{\theta}) \tag{5.15}$$

where

$$\boldsymbol{Q}(\boldsymbol{\theta}) \triangleq \mathrm{diag}\left\{ e^{j(2\pi/\lambda)\, d_x \sin\theta_1}, \ldots, e^{j(2\pi/\lambda)\, d_x \sin\theta_L} \right\} \tag{5.16}$$

Note that the signal DOAs can be obtained from the diagonal elements of $\boldsymbol{Q}(\boldsymbol{\theta})$.

For incoherent signal sources, the columns of $\boldsymbol{E}_S$ and $\boldsymbol{A}(\boldsymbol{\theta})$ span the same subspace. Therefore,

$$\boldsymbol{E}_S = \boldsymbol{A}(\boldsymbol{\theta}) \boldsymbol{T} \tag{5.17}$$

where $\boldsymbol{T}$ is nonsingular. From Equations (5.15) and (5.17), it can be readily shown that

$$\boldsymbol{E}_{S1} = \boldsymbol{J}_1\boldsymbol{E}_S = \boldsymbol{J}_1\boldsymbol{A}\boldsymbol{T} = \boldsymbol{A}_1\boldsymbol{T} = \boldsymbol{A}_2\boldsymbol{Q}\boldsymbol{T} = \boldsymbol{E}_{S2}\boldsymbol{T}^{-1}\boldsymbol{Q}\boldsymbol{T} \tag{5.18}$$

where $\boldsymbol{E}_{S2} = \boldsymbol{J}_2\boldsymbol{E}_S$. Clearly, the matrices $\boldsymbol{\Psi} = \boldsymbol{T}^{-1}\boldsymbol{Q}\boldsymbol{T}$ and $\boldsymbol{Q}$ have the same eigenvalues. Therefore, the ESPRIT algorithm proceeds in two steps. First, an estimate of $\boldsymbol{\Psi}$ is computed. Second, the signal DOAs are estimated from the eigenvalues of the latter matrix.

One simple way to estimate $\boldsymbol{\Psi}$ is to use the least-squares (LS) approach:

$$\hat{\boldsymbol{\Psi}} = \arg\min_{\boldsymbol{\Psi}} \|\hat{\boldsymbol{E}}_{S1} - \hat{\boldsymbol{E}}_{S2}\boldsymbol{\Psi}\|_F = \hat{\boldsymbol{E}}_{S2}^{\dagger}\hat{\boldsymbol{E}}_{S1} \tag{5.19}$$

where $\|\cdot\|_F$ denotes the Frobenius norm, $\hat{\boldsymbol{E}}_{S1} = \boldsymbol{J}_1\hat{\boldsymbol{E}}_S$, $\hat{\boldsymbol{E}}_{S2} = \boldsymbol{J}_2\hat{\boldsymbol{E}}_S$, and $(\cdot)^\dagger$ stands for the pseudo-inverse.

The minimization in Equation (5.19) can be written as

$$\min_{\boldsymbol{\Psi}} \left\|\hat{\boldsymbol{E}}_{S1} - \hat{\boldsymbol{E}}_{S2}\boldsymbol{\Psi}\right\|_F = \min_{\boldsymbol{\Psi},\boldsymbol{Z}_1} \left\|\boldsymbol{Z}_1\right\|_F \quad \text{s.t.} \quad \hat{\boldsymbol{E}}_{S1} + \boldsymbol{Z}_1 = \hat{\boldsymbol{E}}_{S2}\boldsymbol{\Psi} \tag{5.20}$$

Thus, the matrix $\hat{\boldsymbol{E}}_{S1}$ is considered to contain errors, whereas $\hat{\boldsymbol{E}}_{S2}$ is interpreted to be error-free. In practice, both $\hat{\boldsymbol{E}}_{S1}$ and $\hat{\boldsymbol{E}}_{S2}$ contain errors, so improved DOA estimates can be obtained by the total least-squares (TLS) approach (Ottersten et al., 1991; Roy and Kailath, 1989), which solves

$$\min_{\boldsymbol{\Psi},\boldsymbol{Z}_1,\boldsymbol{Z}_2} \|[\boldsymbol{Z}_1, \boldsymbol{Z}_2]\|_F \quad \text{s.t.} \quad \hat{\boldsymbol{E}}_{S1} + \boldsymbol{Z}_1 = \left(\hat{\boldsymbol{E}}_{S2} + \boldsymbol{Z}_2\right)\boldsymbol{\Psi} \tag{5.21}$$

The solution to Equation (5.21) can be written as

$$\hat{\boldsymbol{\Psi}} = -\boldsymbol{M}_{22}\boldsymbol{M}_{12}^{-1} \tag{5.22}$$

where the $L \times L$ matrices $\boldsymbol{M}_{12}$ and $\boldsymbol{M}_{22}$ follow from the eigenvalue decomposition of the matrix

$$\begin{pmatrix} \hat{\boldsymbol{E}}_{S1}^H \\ \hat{\boldsymbol{E}}_{S2}^H \end{pmatrix} \left(\hat{\boldsymbol{E}}_{S1}\ \hat{\boldsymbol{E}}_{S2}\right) = \begin{pmatrix} \boldsymbol{M}_{11} & \boldsymbol{M}_{12} \\ \boldsymbol{M}_{21} & \boldsymbol{M}_{22} \end{pmatrix} \tilde{\boldsymbol{\Lambda}} \begin{pmatrix} \boldsymbol{M}_{11}^H & \boldsymbol{M}_{21}^H \\ \boldsymbol{M}_{12}^H & \boldsymbol{M}_{22}^H \end{pmatrix} \tag{5.23}$$

In Equation (5.23) the eigenvalues are sorted in nonascending order, that is $\tilde{\boldsymbol{\Lambda}} = \operatorname{diag}\left\{\tilde{\lambda}_1, \ldots, \tilde{\lambda}_{2L}\right\}$, with $\tilde{\lambda}_1 \geq \tilde{\lambda}_2 \geq \ldots \geq \tilde{\lambda}_{2L}$.

It has been shown that, for many cases of practical interest, the asymptotic performance of TLS-ESPRIT is close to the CRB for fully calibrated sensor arrays (Ottersten et al., 1991). For centro-symmetric sensor arrays (e.g., ULAs) forward-backward (FB) averaging can be applied to achieve lower computational complexity and improved DOA estimation performance (Haardt and Nossek, 1995).

The concept of ESPRIT can also be extended to sensor arrays that consist of more than two identical and identically oriented subarrays (Sidiropoulos et al., 2000; Swindlehurst et al., 1992).

5.3.3 UCA Root-MUSIC

UCAs are often preferred to ULAs because of ULA direction-of-arrival ambiguities and strongly varying beamshapes. Although for UCAs neither Root-MUSIC nor ESPRIT can be directly applied, search-free DOA estimation algorithms can be obtained using the phase mode excitation theory (Mathews and Zoltowski, 1994).

Let the nth UCA sensor be located at

$$x_n = r\sin\left(2\pi\frac{n-1}{N}\right), \quad y_n = r\cos\left(2\pi\frac{n-1}{N}\right), \quad z_n = 0 \tag{5.24}$$

where r denotes the array radius. Substituting Equation (5.24) into Equation (5.3) yields

$$\boldsymbol{a}(\theta) = \left[e^{j\frac{2\pi}{\lambda}r\cos(\theta)}, e^{j\frac{2\pi}{\lambda}r\cos(2\pi\frac{1}{N}-\theta)}, \ldots, e^{j\frac{2\pi}{\lambda}r\cos(2\pi\frac{N-1}{N}-\theta)}\right]^T \tag{5.25}$$

The excitation of the mth phase mode

$$\boldsymbol{w}_m = \frac{1}{N}\left[1, e^{-j2\pi\frac{m}{N}}, \ldots, e^{-j2\pi\frac{m(N-1)}{N}}\right]^T \tag{5.26}$$

by a signal with steering vector $\boldsymbol{a}(\theta)$ is given by

$$\boldsymbol{w}_m^H\boldsymbol{a}(\theta) = \frac{1}{N}\sum_{n=0}^{N-1} e^{j2\pi\frac{mn}{N}} e^{j\frac{2\pi}{\lambda}r\cos(2\pi\frac{n}{N}-\theta)} \approx j^m J_m\left(\frac{2\pi}{\lambda}r\right)e^{jm\theta} \tag{5.27}$$

if $|m| < N/2$. In Equation (5.27), $J_m(\cdot)$ denotes the Bessel function of the first kind of order m and the approximation becomes arbitrarily tight for large N. It is a common rule of thumb that $J_m(\gamma) \approx 0$ if $|m| > \gamma$; that is, essentially only phase modes with $|m| < 2\pi r/\lambda$ are excited. For $m = -M+1, \ldots, M-1$, where $M < \min\{N/2, 2\pi r/\lambda\}$, we obtain from Equation (5.27) that

$$\boldsymbol{W}^H\boldsymbol{a}(\theta) \approx \boldsymbol{v}(\theta) \tag{5.28}$$

where the lth row of the $2M-1 \times N$ matrix $\boldsymbol{W}^H$ is given by

$$\left(\boldsymbol{W}^H\right)_{l,:} = \frac{\boldsymbol{w}_{-M+l}^H}{j^{-M+l}J_{-M+l}(2\pi r/\lambda)} \tag{5.29}$$

and the $2M-1 \times 1$ Vandermonde vector $\boldsymbol{v}(\theta)$ is

$$\boldsymbol{v}(\theta) = \begin{bmatrix} e^{j(-M+1)\theta} \\ e^{j(-M+2)\theta} \\ \vdots \\ e^{j(M-1)\theta} \end{bmatrix} = \begin{bmatrix} z^{-M+1} \\ z^{-M+2} \\ \vdots \\ z^{M-1} \end{bmatrix} \triangleq \boldsymbol{v}(z) \tag{5.30}$$

Note that in Equation (5.30) we use the definition $z \triangleq \exp\{j\theta\}$, which is different from that used in Section 5.3.1. The Vandermonde structure of $\boldsymbol{v}(\theta)$ enables direct application of Root-MUSIC, ESPRIT, and FB averaging techniques. In what follows, we focus on UCA Root-MUSIC.

From Equations (5.4) and (5.28), it follows that

$$\boldsymbol{W}^H \boldsymbol{R}_x \boldsymbol{W} \approx \boldsymbol{V}(\boldsymbol{\theta}) \boldsymbol{R}_s \boldsymbol{V}^H(\boldsymbol{\theta}) + \sigma^2 \boldsymbol{W}^H \boldsymbol{W} \tag{5.31}$$

where $\boldsymbol{V}(\boldsymbol{\theta}) = [\boldsymbol{v}(\theta_1), \ldots, \boldsymbol{v}(\theta_L)]$. In the finite sample case, after noise prewhitening, the eigendecomposition of the following matrix can be used:

$$\begin{aligned}(\boldsymbol{W}^H \boldsymbol{W})^{-1/2} \boldsymbol{W}^H \hat{\boldsymbol{R}}_x \boldsymbol{W} (\boldsymbol{W}^H \boldsymbol{W})^{-1/2} = {} & \hat{\boldsymbol{E}}_{C,S} \hat{\boldsymbol{\Lambda}}_{C,S} \hat{\boldsymbol{E}}_{C,S}^H \\ & + \hat{\boldsymbol{E}}_{C,N} \hat{\boldsymbol{\Lambda}}_{C,N} \hat{\boldsymbol{E}}_{C,N}^H\end{aligned} \tag{5.32}$$

where the columns of $\hat{\boldsymbol{E}}_{C,N}$ span the noise subspace. The UCA Root-MUSIC polynomial (Mathews and Zoltowski, 1994) can then be defined as

$$f_C(z) = \boldsymbol{v}^T(1/z) (\boldsymbol{W}^H \boldsymbol{W})^{-1/2} \hat{\boldsymbol{E}}_{C,N} \hat{\boldsymbol{E}}_{C,N}^H (\boldsymbol{W}^H \boldsymbol{W})^{-1/2} \boldsymbol{v}(z) \tag{5.33}$$

As in Root-MUSIC, the DOAs can be estimated from the largest-magnitude roots of $f_C(z)$ lying inside the unit circle.

It is important to note that the approximation in Equation (5.28) leads to biased DOA estimates. The bias can be reduced by increasing the number of sensors N (Mathews and Zoltowski, 1994).

5.3.4 Root-RARE

The search-free Root-RARE estimator (Pesavento et al., 2002) is a generalization of the Root-MUSIC algorithm for partly calibrated sensor arrays. This algorithm is applicable to sensor arrays that consist of several linear, identically oriented, calibrated subarrays with inter-element spacings of each subarray being integer multiples of a common baseline. Therefore, RARE is applicable to an array composed of several colinear Root-MUSIC-type arrays with inter-subarray displacements that can be unknown. Figure 5.1 depicts an example array geometry where the crosses represent the sensor locations and the lines indicate the apertures of calibrated subarrays.

Let us consider a sensor array in the xy-plane of the coordinate system that consists of M linear identically oriented subarrays. Without loss of generality, we assume that all of the subarrays are parallel to the x-axis. The mth subarray consists of N_m sensors, and the unknown xy-coordinates of its first sensor

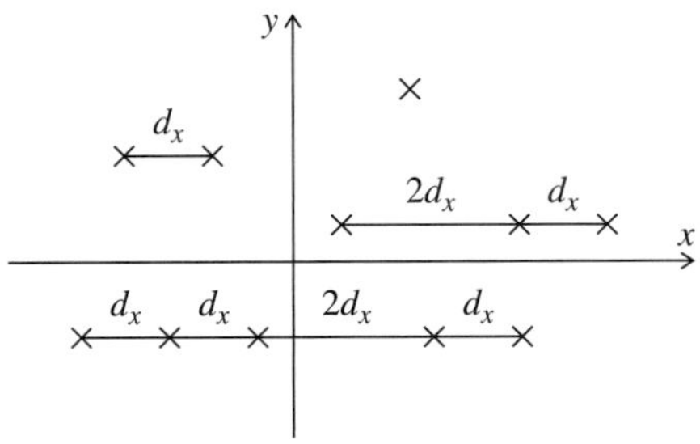

FIGURE 5.1 Example array geometry for Root-RARE.

are $(\alpha_{x,m}, \alpha_{y,m})$. Thus, the location of the rth sensor of the mth subarray can be expressed as $(\alpha_{x,m} + \kappa_{m,r} d_x, \alpha_{y,m})$, where $\kappa_{m,r}$ is the integer multiple of the common baseline d_x. The $N_m \times 1$ steering vector of the mth subarray is

$$\boldsymbol{a}_m(\theta, \alpha_{x,m}, \alpha_{y,m}) = \boldsymbol{b}_m(\theta) e^{j\frac{2\pi}{\lambda}(\alpha_{x,m}\sin\theta + \alpha_{y,m}\cos\theta)} \tag{5.34}$$

where

$$\boldsymbol{b}_m(\theta) = \begin{bmatrix} 1 \\ e^{j\kappa_{m,2}\frac{2\pi}{\lambda} d_x \sin\theta} \\ \vdots \\ e^{j\kappa_{m,N_m}\frac{2\pi}{\lambda} d_x \sin\theta} \end{bmatrix} = \begin{bmatrix} 1 \\ z^{\kappa_{m,2}} \\ \vdots \\ z^{\kappa_{m,N_m}} \end{bmatrix} \triangleq \boldsymbol{b}_m(z) \tag{5.35}$$

and where we return to the notation $z \triangleq \exp\{j(2\pi/\lambda) d_x \sin\theta\}$.

Equation (5.34) leads to the following expression for the steering vector of the whole array:

$$\boldsymbol{a}(\theta, \boldsymbol{\alpha}) = \boldsymbol{B}(\theta)\boldsymbol{h}(\theta, \boldsymbol{\alpha}) \tag{5.36}$$

where $\boldsymbol{\alpha} = \left[\alpha_{x,1}, \alpha_{y,1}, \ldots, \alpha_{x,M}, \alpha_{y,M}\right]^T$ stacks all the unknown subarray locations,

$$\boldsymbol{h}_m(\theta, \boldsymbol{\alpha}) = \begin{bmatrix} e^{j\frac{2\pi}{\lambda}(\alpha_{x,1}\sin\theta + \alpha_{y,1}\cos\theta)} \\ \vdots \\ e^{j\frac{2\pi}{\lambda}(\alpha_{x,M}\sin\theta + \alpha_{y,M}\cos\theta)} \end{bmatrix} \tag{5.37}$$

and the $N \times M$ matrix $\boldsymbol{B}(\theta)$ is defined as

$$\begin{aligned} \boldsymbol{B}(\theta) &\triangleq \begin{bmatrix} \boldsymbol{b}_1(\theta) & \boldsymbol{0}_{N_1\times 1} & \cdots & \boldsymbol{0}_{N_1\times 1} \\ \boldsymbol{0}_{N_2\times 1} & \boldsymbol{b}_2(\theta) & \cdots & \boldsymbol{0}_{N_2\times 1} \\ \vdots & \vdots & \ddots & \\ \boldsymbol{0}_{N_M\times 1} & \boldsymbol{0}_{N_M\times 1} & & \boldsymbol{b}_M(\theta) \end{bmatrix} \\ &= \begin{bmatrix} \boldsymbol{b}_1(z) & \boldsymbol{0}_{N_1\times 1} & \cdots & \boldsymbol{0}_{N_1\times 1} \\ \boldsymbol{0}_{N_2\times 1} & \boldsymbol{b}_2(z) & \cdots & \boldsymbol{0}_{N_2\times 1} \\ \vdots & \vdots & \ddots & \\ \boldsymbol{0}_{N_M\times 1} & \boldsymbol{0}_{N_M\times 1} & & \boldsymbol{b}_M(z) \end{bmatrix} \triangleq \boldsymbol{B}(z) \end{aligned} \tag{5.38}$$

where $\boldsymbol{0}_{k\times l}$ denotes a $k \times l$ matrix of zeros. The steering vectors of the true signal DOAs are orthogonal to the noise subspace, so it follows from Equation (5.36) that

$$\|\boldsymbol{E}_N^H \boldsymbol{a}(\theta_l, \boldsymbol{\alpha})\|^2 = \boldsymbol{h}^H(\theta_l, \boldsymbol{\alpha})\boldsymbol{B}^H(\theta_l)\boldsymbol{E}_N\boldsymbol{E}_N^H\boldsymbol{B}(\theta_l)\boldsymbol{h}(\theta_l, \boldsymbol{\alpha}) = 0, \quad l = 1, \ldots, L. \tag{5.39}$$

The RARE algorithm exploits the fact that the matrix

$$\boldsymbol{Z}(\theta) \triangleq \boldsymbol{B}(\theta)^H \boldsymbol{E}_N \boldsymbol{E}_N^H \boldsymbol{B}(\theta) \tag{5.40}$$

does not depend on $\boldsymbol{\alpha}$ and that, according to Equation (5.39), this matrix is rank deficient for $\theta = \theta_l, l = 1, \ldots, L$. The RARE null spectrum (Pesavento et al.,

2002) can be defined as

$$f_{\text{RARE}}(\theta) \triangleq \det(\boldsymbol{B}^H(\theta)\hat{\boldsymbol{E}}_N\hat{\boldsymbol{E}}_N^H\boldsymbol{B}(\theta)) \tag{5.41}$$

so that the source DOAs can be estimated from the minima of this function. Similar to Root-MUSIC, the spectral search can be replaced by rooting the polynomial

$$f_{\text{RARE}}(z) \triangleq \det(\boldsymbol{B}^T(1/z)\hat{\boldsymbol{E}}_N\hat{\boldsymbol{E}}_N^H\boldsymbol{B}(z)) \tag{5.42}$$

It has been shown by Pesavento et al. (2002) that, similar to the Root-MUSIC case, the roots of $f_{\text{RARE}}(z)$ form conjugate reciprocal pairs. Consequently, the source DOAs can be estimated from the largest-magnitude roots lying inside the unit circle.

A modification of the RARE technique that is robust to subarray misorientations has been proposed by Abd Elkader et al. (2006). Several other search-free approaches have been developed for subarray-based arrays by Swindlehurst et al. (2007).

5.3.5 Generalized ESPRIT

The generalized ESPRIT algorithm developed by Gao and Gershman (2005) is an extension of the standard ESPRIT algorithm for a broader class of array geometries. In contrast to ESPRIT (which requires identical displacements between any pair of corresponding sensors of the first and second subarray), generalized ESPRIT accounts for different collinear displacement vectors the lengths of which must be integer multiples of a common baseline. Figure 5.2 shows an example array geometry for the generalized ESPRIT algorithm.

As in the conventional ESPRIT case, let all the array sensors lie in the xy-plane, and let the displacements be parallel to the x-axis of the coordinate system. Using the same notations as in Section 5.3.2, the steering matrices of the two subarrays can be written as

$$\boldsymbol{A}_1(\boldsymbol{\theta}) = [\boldsymbol{a}(\theta_1), \ldots, \boldsymbol{a}(\theta_L)], \quad \boldsymbol{A}_2(\boldsymbol{\theta}) = [\boldsymbol{\Gamma}(\theta_1)\boldsymbol{a}(\theta_1), \ldots, \boldsymbol{\Gamma}(\theta_L)\boldsymbol{a}(\theta_L)] \tag{5.43}$$

where

$$\boldsymbol{\Gamma}(\theta) \triangleq \text{diag}\left\{e^{j\frac{2\pi}{\lambda}\kappa_1 d_x \sin\theta}, \ldots, e^{j\frac{2\pi}{\lambda}\kappa_M d_x \sin\theta}\right\} \tag{5.44}$$

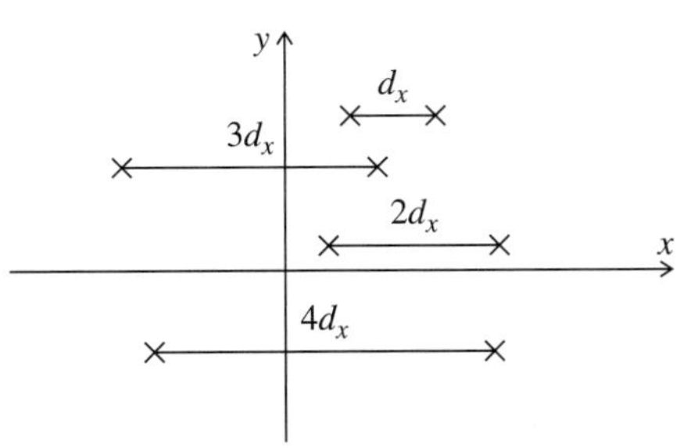

FIGURE 5.2 Example array geometry for generalized ESPRIT.

M is the number of sensor pairs, and κ_m, $m=1,\ldots,M$ are the integer multiples of the baseline d_x such that $\kappa_m d_x$ is the displacement between the two sensors of the mth pair. Thus, we obtain

$$\boldsymbol{E}_{S2}-\boldsymbol{\Gamma}(\theta)\boldsymbol{E}_{S1}=\boldsymbol{A}_2\boldsymbol{T}-\boldsymbol{\Gamma}(\theta)\boldsymbol{A}_1\boldsymbol{T}=\tilde{\boldsymbol{Q}}(\theta)\boldsymbol{T} \tag{5.45}$$

where

$$\tilde{\boldsymbol{Q}}(\theta)=[(\boldsymbol{\Gamma}(\theta_1)-\boldsymbol{\Gamma}(\theta))\boldsymbol{a}(\theta_1),\ldots,(\boldsymbol{\Gamma}(\theta_L)-\boldsymbol{\Gamma}(\theta))\boldsymbol{a}(\theta_L)] \tag{5.46}$$

$\boldsymbol{E}_{S1}=\boldsymbol{J}_1\boldsymbol{E}_S$, and $\boldsymbol{E}_{S2}=\boldsymbol{J}_2\boldsymbol{E}_S$.

The generalized ESPRIT algorithm is based on the observation that $\tilde{\boldsymbol{Q}}(\theta)$ becomes singular for any $\theta=\theta_l$, $l=1,\ldots,L$. Therefore, its null spectrum can be defined as

$$f_{\text{GESPRIT}}(\theta)\triangleq\det\left\{(\hat{\boldsymbol{E}}_{S2}-\boldsymbol{\Gamma}(\theta)\hat{\boldsymbol{E}}_{S1})^H(\hat{\boldsymbol{E}}_{S2}-\boldsymbol{\Gamma}(\theta)\hat{\boldsymbol{E}}_{S1})\right\} \tag{5.47}$$

By defining

$$\boldsymbol{\Gamma}(z)\triangleq\operatorname{diag}\left\{z^{\kappa_1},\ldots,z^{\kappa_M}\right\} \tag{5.48}$$

with $z=\exp\{j(2\pi/\lambda)d_x\sin\theta\}$, the following polynomial can be obtained from Equation (5.47):

$$f_{\text{GESPRIT}}(z)\triangleq\det\left\{\left(\hat{\boldsymbol{E}}_{S2}^H-\hat{\boldsymbol{E}}_{S1}^H\boldsymbol{\Gamma}(1/z)\right)\left(\hat{\boldsymbol{E}}_{S2}-\boldsymbol{\Gamma}(z)\hat{\boldsymbol{E}}_{S1}\right)\right\} \tag{5.49}$$

The source DOAs can be estimated from the roots of $f_{\text{GESPRIT}}(z)$ in the same way as for the standard Root-MUSIC technique.

Note that our definition of the generalized ESPRIT null-spectrum function is somewhat different from that used in Gao and Gershman (2005). The null-spectrum in Equation (5.47) is real-valued and positive, and the roots of $f_{\text{GESPRIT}}(z)$ appear in conjugate reciprocal pairs.

5.4 SEARCH-FREE METHODS FOR ARBITRARY ARRAYS

In this section, we provide an overview of search-free DOA estimation methods that can be applied to arrays of arbitrary geometry.

5.4.1 Interpolated Root-MUSIC

The essence of the array interpolation technique (Friedlander, 1993; Friedlander and Weiss, 1992, 1993; Weiss et al., 1995) is to approximate the steering vector of the actual sensor array as

$$\boldsymbol{a}(\theta)\simeq\boldsymbol{G}_1\boldsymbol{g}_1(\theta) \tag{5.50}$$

where $\boldsymbol{g}_1(\theta)$ is the $M_1\times 1$ steering vector of a virtual ULA, M_1 is the number of virtual sensors, and $\boldsymbol{G}_1$ is the $N\times M_1$ array interpolation matrix designed to minimize the interpolation error. For example, if the virtual sensors are aligned along the x-axis of the coordinate system and centered with respect to its origin, the "virtual" steering vector is described by Equation (5.10), with N replaced by M_1 and with d_x being the inter-element spacing of the virtual ULA.

The MUSIC null-spectrum function (5.9) can thus be approximated as

$$f(\theta) \simeq \boldsymbol{g}_1^H(\theta)\boldsymbol{G}_1^H\hat{\boldsymbol{E}}_N\hat{\boldsymbol{E}}_N^H\boldsymbol{G}_1\boldsymbol{g}_1(\theta) \triangleq f_1(\theta) \tag{5.51}$$

which can be expressed in terms of $z = \exp\{j(2\pi/\lambda)d_x \sin\theta\}$ as

$$f_1(z) = \boldsymbol{g}_1^T(1/z)\boldsymbol{G}_1^H\hat{\boldsymbol{E}}_N\hat{\boldsymbol{E}}_N^H\boldsymbol{G}_1\boldsymbol{g}_1(z) \tag{5.52}$$

This is a polynomial of degree $2(M_1 - 1)$ the roots of which appear in conjugate reciprocal pairs. As with Root-MUSIC, the signal DOAs can be estimated from the largest-magnitude roots located inside the unit circle.

Because the approximation in Equation (5.50) is typically inaccurate for the whole array angular field of view, angular sectors have to be defined and such an approximation has to be used separately for each sector. The number of sectors, the number of virtual sensors, and their locations are the design parameters of the interpolated Root-MUSIC method.

5.4.2 Manifold Separation

Another elegant Root-MUSIC method for arbitrary arrays was proposed in Doron and Doron (1994a,b,c) and further analyzed in Belloni et al. (2007). In Doron and Doron (1994a), it was shown that for any arbitrary array the steering vector can be approximated as

$$\boldsymbol{a}(\theta) \simeq \boldsymbol{G}_2\boldsymbol{g}_2(\theta) \tag{5.53}$$

where $\boldsymbol{G}_2$ is an $N \times M_2$ matrix that depends only on the array parameters and

$$\boldsymbol{g}_2(\theta) = \left[e^{-j\frac{M_2-1}{2}\theta}, \ldots, e^{j\frac{M_2-1}{2}\theta}\right]^T$$

is an $M_2 \times 1$ Vandermonde vector that depends only on θ and M_2. The elements of $\boldsymbol{g}_2(\theta)$ form a subset of the Fourier basis, and the parameter M_2 characterizes the accuracy of the approximation in Equation (5.53). Given the completeness of the Fourier basis, Equation (5.53) becomes exact when $M_2 \to \infty$, and the accuracy of the approximation improves when increasing M_2. Thus, sector-wise processing as for the array interpolation method is no longer needed.

Using the results of Doron and Doron (1994a), it was proposed by Belloni et al. (2007) and Doron and Doron (1994b) to use Equation (5.53) with some finite M_2 to approximate the MUSIC null-spectrum function as

$$\begin{aligned} f(\theta) &\simeq \boldsymbol{g}_2^H(\theta)\boldsymbol{G}_2^H\hat{\boldsymbol{E}}_N\hat{\boldsymbol{E}}_N^H\boldsymbol{G}_2\boldsymbol{g}_2(\theta) \triangleq f_2(\theta) \\ &= \boldsymbol{g}_2^T(1/z)\boldsymbol{G}_2^H\hat{\boldsymbol{E}}_N\hat{\boldsymbol{E}}_N^H\boldsymbol{G}_2\boldsymbol{g}_2(z) \triangleq f_2(z) \end{aligned} \tag{5.54}$$

where, in contrast to Equation (5.52), $z \triangleq \exp\{j\theta\}$ and the degree of the polynomial is $2M_2 - 2$. It was suggested in Doron and Doron (1994b) that the signal

DOAs be obtained from the largest-magnitude roots of $f_2(z)$ located inside the unit circle, in a way similar to that used in the standard Root-MUSIC algorithm.

Several methods have been proposed to compute the matrix $\boldsymbol{G}_2$, including the least-squares method and another technique that determines each element of $\boldsymbol{G}_2$ via the inverse discrete Fourier transform (IDFT) of different components of the steering vector $\boldsymbol{a}(\theta)$ taken at different angles. The LS technique is optimal in the sense that it minimizes the manifold approximation error.

The parameter M_2 should be large enough to obtain an acceptable DOA estimation performance. The following rule of thumb was suggested for choosing M_2 (Doron and Doron, 1994a):

$$M_2 \approx 8\pi r/\lambda \tag{5.55}$$

where r is the largest distance between the array sensors and the origin of the coordinate system.

5.4.3 Fourier Domain Root-MUSIC

The FD Root-MUSIC technique of Rübsamen and Gershman (2008, 2009) uses a different approach to obtain a polynomial that approximates the MUSIC null-spectrum function. It exploits the fact that the MUSIC null-spectrum function (5.9) is periodic in θ with the period 2π. Therefore, the latter function can be expressed using its Fourier series expansion as

$$f(\theta) = \sum_{m=-\infty}^{\infty} F_m e^{jm\theta} \tag{5.56}$$

where the Fourier coefficients are given by

$$F_m = \frac{1}{2\pi} \int_{-\pi}^{\pi} f(\theta) e^{-jm\theta} d\theta \tag{5.57}$$

Truncating the Fourier series in Equation (5.56) to $2M_3 - 1$ points, the function $f(\theta)$ can be approximated as

$$\begin{aligned} f(\theta) &\simeq \sum_{m=-(M_3-1)}^{M_3-1} F_m e^{jm\theta} \triangleq \tilde{f}(\theta) \\ &= \sum_{m=-(M_3-1)}^{M_3-1} F_m z^m \triangleq \tilde{f}(z) \end{aligned} \tag{5.58}$$

where the notation $z \triangleq \exp\{j\theta\}$ is used. It can be seen that, if $M_2 = M_3$, the two polynomials in Equations (5.54) and (5.58) differ only in their polynomial coefficients.

It was shown in Rübsamen and Gershman (2009) that the polynomial coefficients (5.57) provide a better approximation of the original MUSIC null spectrum than does the MS technique. Therefore, it may be expected that FD Root-MUSIC shows improved DOA estimation performance as compared to the MS technique. This conjecture was validated in Rübsamen and Gershman (2009) by computer simulations.

The Fourier series coefficients F_m $(m=-(M_3-1),\ldots,M_3-1)$ can be approximately obtained using the discrete Fourier transform (DFT) as

$$F_m \simeq \frac{1}{2\pi} \sum_{l=-(M_3-1)}^{M_3-1} f(l\,\Delta\theta)\, e^{-jml\,\Delta\theta}\,\Delta\theta \triangleq \hat{F}_m \tag{5.59}$$

where $\Delta\theta = 2\pi/(2M_3-1)$. With such a DFT approximation of $\tilde{f}(z)$, the final expression for the FD Root-MUSIC polynomial can be written as

$$\begin{aligned} \tilde{f}(z) &\simeq \sum_{m=-(M_3-1)}^{M_3-1} \hat{F}_m z^m \triangleq f_3(z) \\ &= \sum_{m=-(M_3-1)}^{M_3-1} \hat{F}_m e^{jm\theta} \triangleq f_3(\theta) \end{aligned} \tag{5.60}$$

The DFT coefficients $\hat{F}_m$ differ from the Fourier series coefficients F_m because of aliasing introduced by sampling the MUSIC null-spectrum function $f(\theta)$. The aliasing effect can be reduced by increasing M_3, but this leads to a higher computational complexity, a detailed discussion of which is presented in Rübsamen and Gershman (2009).

The functions $f_1(\theta)$ and $f_2(\theta)$ are non-negative by definition, which is not the case for $f_3(\theta)$. That is, $f_3(\theta)$ may have some values that are slightly below zero in some of its minima. Because of the two sign changes of $f_3(\theta)$, whenever $f_3(\theta)$ takes a negative value we obtain two roots that lie exactly on the unit circle very close to each other. Taking into account that $\hat{F}_m^* = \hat{F}_{-m}$, it can be shown again that the roots of Equation (5.60) satisfy the conjugate reciprocity property. That is, if z_0 is a root, then $1/z_0^*$ also nullifies $f_3(z)$. However, this property becomes trivial for roots that lie exactly on the unit circle and therefore do not appear in conjugate reciprocal pairs.

Based on this observation, a procedure different from Root-MUSIC was proposed in Rübsamen and Gershman (2009) to estimate the signal DOAs from the $2(M_3-1)$ roots of the polynomial $f_3(z)$. The signal roots of $f_3(z)$ can be divided into two groups. The first group contains the roots that appear in pairs, lying exactly at the unit circle very close to each other. As discussed before, these root pairs correspond to negative minimum values of $f_3(\theta)$, and there are no conjugate reciprocal roots caused by roots of this type. The second group contains the roots that do not belong to the unit circle. These appear in conjugate reciprocal pairs and are the same as the roots of the standard Root-MUSIC polynomial.

A simple procedure to estimate the source DOAs from the roots of $f_3(z)$ can therefore be summarized as follows:

Step 1. Take the root closest to the unit circle and determine if it belongs to the first or second group by checking whether its conjugate reciprocal value is another root.

Step 2. If this root belongs to the first group, estimate the source DOA from the average of this root and its closest neighbor, and drop both. Go to step 4.

Step 3. If this root belongs to the second group, use it to estimate the source DOA and then drop both this root and its corresponding conjugate reciprocal root.

Step 4. If fewer than L DOAs have been estimated, return to step 1. Otherwise, stop.

It is worth noting that the $\hat{F}_m$ can be obtained in a computationally efficient way using the fast Fourier transform (FFT) algorithm.

5.4.4 Fourier Domain–Weighted Least-Squares Root-MUSIC

To estimate the signal DOAs, it is important to have a close approximation of the MUSIC null-spectrum function in the vicinity of the true DOAs; that is, in the areas where the null-spectrum function has its minima. This idea was exploited in Rübsamen and Gershman (2009) to obtain the following weighted least-squares approximation of the MUSIC null-spectrum function:

$$\min_{\{H_m\}} \sum_{l=1}^{Q} w(\theta_l) \left| f(\theta_l) - \sum_{m=-(M_4-1)}^{M_4-1} H_m e^{jm\theta_l} \right|^2 \quad \text{s.t.} \;\; H_m^* = H_{-m} \; \forall\, m = 0, \ldots, M_4 - 1 \tag{5.61}$$

where $w(\theta_l)$ are the weight coefficients, Q is the number of MUSIC null-spectrum samples, and $2M_4 - 1$ is the number of basis functions $\exp\{jm\theta\}$ that are used to approximate the null-spectrum samples. The weighted least-squares approximation in Equation (5.61) requires that $Q > 2M_4 - 1$. A natural choice of weight coefficients to stress low values of $f(\theta)$ is given by

$$w(\theta_l) = \frac{1}{f(\theta_l)}, \qquad l = 1, \ldots, Q \tag{5.62}$$

The constraints in Equation (5.61) can be used to guarantee that the roots of the resulting FDWLS Root-MUSIC polynomial

$$f_4(z) \triangleq \sum_{m=-(M_4-1)}^{M_4-1} H_{*m} z^m = \sum_{m=-(M_4-1)}^{M_4-1} H_{*m} e^{jm\theta} \triangleq f_4(\theta) \tag{5.63}$$

obey the conjugate reciprocity property, where H_{*m} $(m = -(M_4 - 1), \ldots, M_4 - 1)$ are the optimal coefficients that solve Equation (5.61). Eliminating the equality constraints in Equation (5.61) leads to the following equivalent unconstrained optimization problem:

$$\min_{\{H_m\}} \sum_{l=1}^{Q} w(\theta_l) \left(f(\theta_l) - H_0 - 2 \sum_{m=1}^{M_4-1} \Re\left(H_m e^{jm\theta_l}\right) \right)^2 \tag{5.64}$$

and its solution is given by

$$\boldsymbol{u} = \left(\boldsymbol{B}^T \boldsymbol{D} \boldsymbol{B}\right)^{-1} \boldsymbol{B}^T \boldsymbol{D} \boldsymbol{f} \tag{5.65}$$

where

$$\boldsymbol{f} \triangleq \left[f(\theta_1), \ldots, f(\theta_Q)\right]^T \tag{5.66}$$

$$\boldsymbol{u} \triangleq \left[H_0, \Re\left(H_1\right), \ldots, \Re\left(H_{M_4-1}\right), \Im\left(H_1\right), \ldots, \Im\left(H_{M_4-1}\right)\right]^T \tag{5.67}$$

$$\boldsymbol{D} \triangleq \operatorname{diag}\{w(\theta_1), \ldots, w(\theta_Q)\} \tag{5.68}$$

$$\boldsymbol{B} \triangleq \left[\mathbf{1}_Q, 2\Re(\tilde{\boldsymbol{B}}), -2\Im(\tilde{\boldsymbol{B}})\right] \tag{5.69}$$

$[\tilde{\boldsymbol{B}}]_{l,m} \triangleq e^{jm\theta_l}$ ($l = 1, \ldots, Q$; $m = 1, \ldots, M_4$), $\mathbf{1}_Q$ is a $Q \times 1$ vector of ones, and $\Im(\cdot)$ denotes the imaginary part of a complex value.

Equation (5.65) determines the coefficients of the FDWLS Root-MUSIC polynomial (5.63). The same procedure as in FD Root-MUSIC must be used to estimate the signal DOAs from the polynomial's roots. Rübsamen and Gershman (2009) showed that for $M_3 = M_4$ the performace of FDWLS Root-MUSIC is substantially better than that of FD Root-MUSIC.

5.5 SIMULATION RESULTS

In our simulation examples, we focus on the search-free techniques for arbitrary arrays and compare the DOA estimation performances of the interpolated Root-MUSIC, MS, FD Root-MUSIC, and FDWLS Root-MUSIC algorithms. As the LS method provides the smallest possible approximation errors, it has been used in the interpolated Root-MUSIC and MS techniques to compute the matrices $\boldsymbol{G}_1$ and $\boldsymbol{G}_2$, respectively. The interpolated Root-MUSIC technique has been applied to sectors of width 60°. Virtual arrays with the same number of sensors as the real array have been selected, and their apertures have been chosen to be orthogonal to the center directions of the sectors. Throughout our simulations, we assume $M_2 = M_3 = M_4 = M$, which guarantees that the asymptotic complexity of FD Root-MUSIC is no higher than that of the MS technique. The number of null-spectrum samples Q in FDWLS Root-MUSIC is twice that of Fourier basis functions.

In our first example, a randomly generated (although fixed throughout all simulation runs) network user address (NUA) of $N = 6$ sensors is used. Its sensor locations are depicted in Figure 5.3. Two closely spaced equal-power signal sources are assumed to impinge on the array from the directions $\theta_1 = 15°$ and

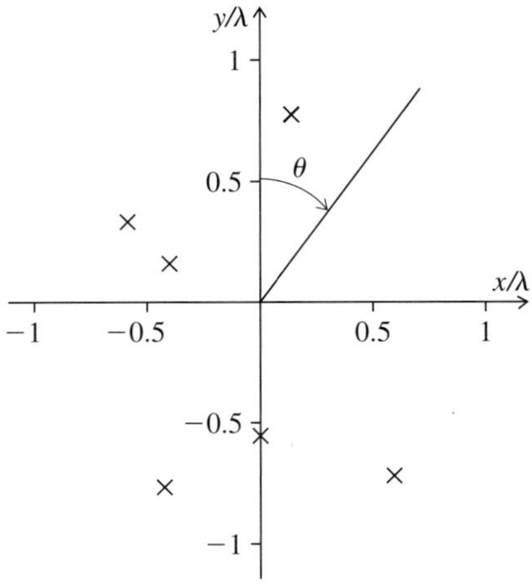

FIGURE 5.3 Array geometry of Example 1.

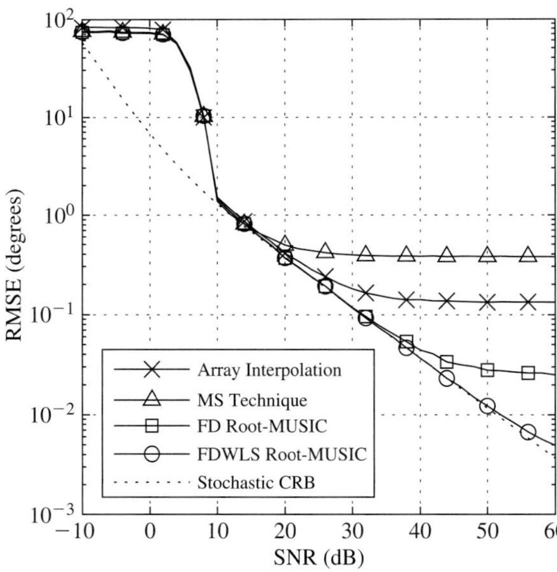

FIGURE 5.4 DOA estimation RMSEs versus SNR for $K = 100$, $\theta_1 = 15°$, $\theta_2 = 20°$, and $M = 19$. The array geometry corresponds to Figure 5.3.

$\theta_2 = 20°$. The parameters $K = 100$ and $M = 19$ are taken in this example. All simulation results have been averaged over 1000 independent Monte Carlo runs and over the sources.

Figure 5.4 displays the DOA estimation root mean-square errors (RMSEs) of the interpolated Root-MUSIC, MS, FD Root-MUSIC, and FDWLS Root-MUSIC techniques. The stochastic CRB is also shown. The DOA estimation methods tested saturate in their performance at high signal-to-noise ratios (SNRs), due to the approximation errors in, Equation (5.50) for interpolated Root-MUSIC Equation (5.53) for MS, Equations (5.58) and (5.60) for FD Root-MUSIC, and Equation (5.61) for FDWLS Root-MUSIC. It can be observed that the performance saturation is less pronounced in the case of FD Root-MUSIC and FDWLS Root-MUSIC, which in particular substantially outperform the MS technique in the high SNR region. As expected, the performance of FDWLS Root-MUSIC is the best among these methods.

In our second example, the RMSE performance of the same methods versus the angular separation between the sources is examined. To demonstrate that the search-free methods tested have an improved resolution threshold as compared to Spectral-MUSIC for closely spaced signals (Rao and Hari, 1989), we also add the Spectral-MUSIC performance curve in Figure 5.5. In this figure, the DOA of the first source is varied, whereas the DOA of the second source is fixed and equal to $\theta_2 = 20°$. It has been assumed that SNR = 20 dB and all other parameters are as in the previous example. As FDWLS Root-MUSIC has essentially the same performance as FD Root-MUSIC for the chosen set of parameters, we only plot the curve for the latter.

Figure 5.5 shows that the rooting-based methods have quite similar performance in the threshold domain, where they substantially outperform Spectral-MUSIC. Furthermore, FD Root-MUSIC outperforms MS in the asymptotic domain (i.e., in the case of widely spaced signals).

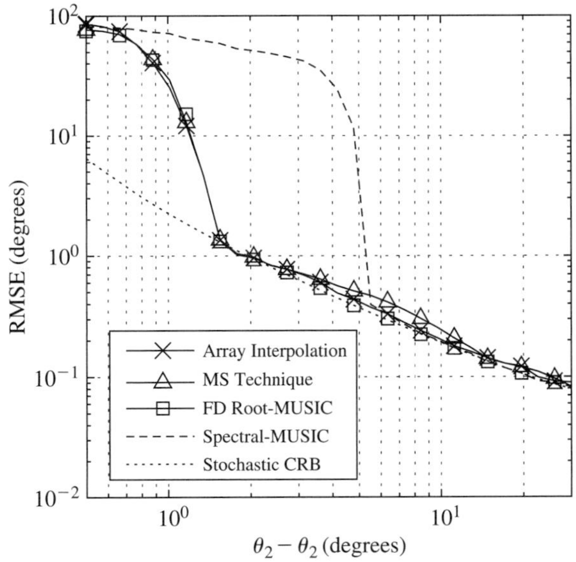

FIGURE 5.5 DOA estimation RMSEs versus $\theta_2 - \theta_1$ for $K = 100$, SNR = 20 dB, $\theta_2 = 20°$, and $M = 19$. The array geometry corresponds to Figure 5.3.

In our third example, we extend the previous examples now considering the case when different (randomly generated) array geometries are used in each simulation run. In this example, the locations of $N = 8$ array sensors are drawn uniformly from the interior of a circle with radius λ. The performances of the FD Root-MUSIC and MS techniques are compared for SNR $= 20$ dB and $K = 100$ snapshots. The angular spacing between the two sources is assumed to be equal to 30°. In Figure 5.6, a scatter plot for the DOA estimation RMSEs of the MS and FD Root-MUSIC techniques is shown for $M = 15$. Each point on this plot corresponds to one random realization of the array geometry.

Figure 5.6 demonstrates the substantial performance improvements achieved by FD Root-MUSIC as compared to the MS technique. Note that the cases where FD Root-MUSIC outperforms the MS technique correspond to the points located above the diagonal line. The geometric mean of the RMSE improvement in this example is approximately given by a factor of 2.9.

In the last example, we study the impact of the parameter $M = M_2 = M_3 = M_4$ on the performance of the methods tested. We consider a random array geometry drawn in each run uniformly from the interior of a circle of radius 1.25λ. The other array and source parameters are chosen as before. In Figure 5.7, the DOA estimation RMSEs are plotted versus the value of M. All curves in this figure have

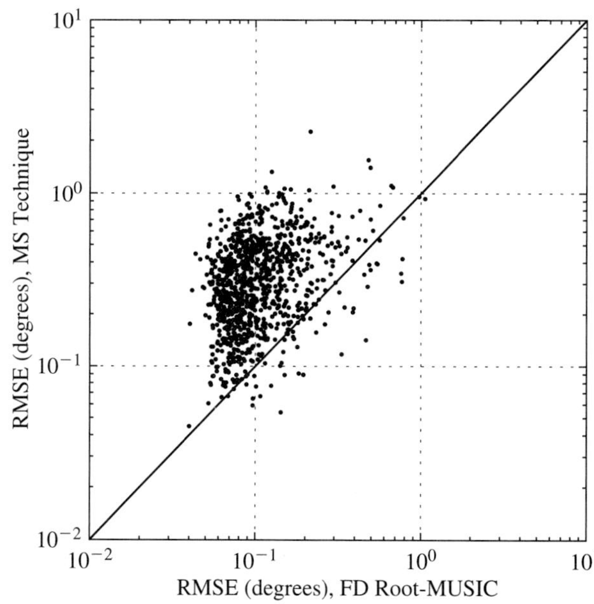

FIGURE 5.6 Scatter plot of the RMSEs of the MS and FD Root-MUSIC techniques for $N = 8$, $K = 100$, SNR $= 20$ dB, $\theta_2 - \theta_1 = 30°$, and $M = 15$. The array geometry is randomly generated in each simulation run.

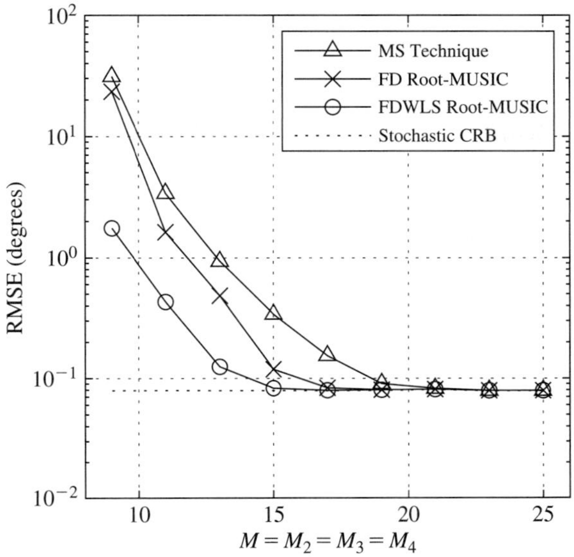

FIGURE 5.7 DOA estimation RMSEs versus M for $K = 100$, SNR = 20 dB, and $\theta_2 - \theta_1 = 30°$. The array geometry is randomly generated in each simulation run.

been averaged over 1000 realizations of the random array geometry. Clearly, FD and FDWLS Root-MUSIC require reduced polynomial orders as compared to the MS technique to achieve the same performance.

5.6 CONCLUSION

An overview of search-free DOA estimation methods was presented with a particular emphasis on nonuniform sensor arrays. First, the traditional Root-MUSIC and ESPRIT techniques were discussed. Then, more recent search-free direction-finding methods were introduced that extend the classic Root-MUSIC and ESPRIT approaches to more general classes of nonuniform arrays. In the final sections of the chapter, we discussed search-free approaches that are applicable to arbitrary array geometries. Simulation results were presented to compare the performance of the latter techniques.

Our results demonstrate that the family of Fourier Domain Root-MUSIC techniques represents a computationally attractive alternative to existing search-free DOA estimation methods for arbitrary nonuniform arrays.

Acknowledgement This work was supported by the European Research Council (ERC) under Advanced Investigator Grant scheme, Project No. 227477.

REFERENCES

Abd Elkader, S., Gershman, A.B., Wong, K.M., 2006. Rank reduction direction-of-arrival estimators with improved robustness against subarray orientation errors. IEEE Trans. Signal Process. 54 (5), 1951–1955.

Barabell, A., 1983. Improving the resolution performance of eigenstructure-based direction-finding algorithms. In: Proc. ICASSP, pp. 336–339.

Belloni, F., Richter, A., Koivunen, V., 2007. DOA estimation via manifold separation for arbitrary array structures. IEEE Trans. Signal Process. 55, 4800–4810.

Bienvenu, G., Kopp, L., 1980. Adaptivity to background noise spatial coherence for high resolution passive methods. In: Proc. ICASSP, pp. 307–310.

Böhme, J.F., 1986. Separated estimation of wave parameters and spectral parameters by maximum likelihood. In: Proc. ICASSP, pp. 2819–2822.

Doron, M.A., Doron, E., 1994a. Wavefield modeling and array processing, part I—spatial sampling. IEEE Trans. Signal Process. 42, 2549–2559.

Doron, M.A., Doron, E., 1994b. Wavefield modeling and array processing, part II—algorithms. IEEE Trans. Signal Process. 42, 2560–2570.

Doron, M.A., Doron, E., 1994c. Wavefield modeling and array processing, part III—resolution capacity. IEEE Trans. Signal Process. 42, 2571–2580.

Friedlander, B., 1993. The Root-MUSIC algorithm for direction finding with interpolated arrays. Signal Process. 30, 15–25.

Friedlander, B., Weiss, A.J., 1992. Direction finding using spatial smoothing with interpolated arrays. IEEE Trans. Aerosp. Electron. Syst. 28 (4), 574–587.

Friedlander, B., Weiss, A.J., 1993. Direction finding for wide-band signals using an interpolated array. IEEE Trans. Signal Process. 41 (4), 1618–1634.

Gao, F., Gershman, A.B., 2005. A generalized ESPRIT approach to direction-of-arrival estimation. IEEE Signal Process. Lett. 12 (3), 254–257.

Haardt, M., Nossek, J.A., 1995. Unitary ESPRIT: how to obtain increased estimation accuracy with a reduced computational burden. IEEE Trans. Signal Process. 43 (5), 1232–1242.

Jaffer, A.G., 1988. Maximum likelihood direction finding of stochastic sources: a separable solution. In: Proc. ICASSP, pp. 2893–2896.

Kaveh, M., Barabell, A.J., 1986. The statistical performance of MUSIC and the minimum-norm algorithms in resolving plane waves in noise. IEEE Trans. Acoust. Speech Signal Process. 34 (4), 331–340.

Mathews, C.P., Zoltowski, M.D., 1994. Eigenstructure techniques for 2-D angle estimation with uniform circular arrays. IEEE Trans. Signal Process. 42 (9), 2395–2407.

Ottersten, B., Viberg, M., Kailath, T., 1991. Performance analysis of the total least squares ESPRIT algorithm. IEEE Trans. Signal Process. 39 (5), 1122–1135.

Paulraj, A., Roy, R., Kailath, T., 1986. A subspace rotation approach to signal parameter estimation. Proc. IEEE 74 (11), 1044–1046.

Pesavento, M., Gershman, A.B., Haardt, M., 2000. Unitary Root-MUSIC with a real-valued eigendecomposition: a theoretical and experimental performance study. IEEE Trans. Signal Process. 48 (5), 1306–1314.

Pesavento, M., Gershman, A.B., Wong, K.M., 2002. Direction finding using partly calibrated sensor arrays composed of multiple subarrays. IEEE Trans. Signal Process. 50 (9), 2103–2115.

Rao, B.D., Hari, K.V.S., 1989. Performance analysis of Root-MUSIC. IEEE Trans. Acoust. Speech Signal Process. 37 (12), 1939–1949.

Roy, R., Kailath, T., 1989. ESPRIT: estimation of signal parameters via rotational invariance techniques. IEEE Trans. Acoust. Speech Signal Process. 37 (7), 984–995.

Roy, R., Paulraj, A., Kailath, T., 1986. ESPRIT—a subspace rotation approach to estimation of parameters of cisoids in noise. IEEE Trans. Acoust. Speech Signal Process. 34 (10), 1340–1342.

Rübsamen, M., Gershman, A.B., 2008. Root-MUSIC based direction-of-arrival estimation methods for arbitrary non-uniform arrays. In: Proc. ICASSP, pp. 2317–2320.

Rübsamen, M., Gershman, A.B., 2009. Direction-of-arrival estimation for non-uniform sensor arrays: From manifold separation to Fourier domain MUSIC methods. IEEE Trans. Signal Process. 57 (2), 588–599.

Schmidt, R.O., 1979. Multiple emitter location and signal parameter estimation. In: Proc. RADC Spectral Estimation Workshop, pp. 243–258.

Schweppe, F., 1968. Sensor array data processing for multiple signal sources. IEEE Trans. Inf. Theory 4 (3), 294–305.

Sidiropoulos, N.D., Bro, R., Giannakis, G.B., 2000. Parallel factor analysis in sensor array processing. IEEE Trans. Signal Process. 48 (8), 2377–2388.

Stoica, P., Nehorai, A., 1989. MUSIC, maximum likelihood and Cramér-Rao bound. IEEE Trans. Acoust. Speech Signal Process. 37 (5), 720–741.

Swindlehurst, A.L., Ottersten, B., Roy, R., Kailath, T., 1992. Multiple invariance ESPRIT. IEEE Trans. Signal Process. 40, 867–881.

Swindlehurst, A.L., Stoica, P., Jansson, M., 2007. Exploiting arrays with multiple invariances using MUSIC and MODE. IEEE Trans. Signal Process. 49, 2511–2521.

Wax, M., Kailath, T., 1983. Optimum localization of multiple sources by passive arrays. IEEE Trans. Acoust. Speech Signal Process. 31 (10), 1210–1218.

Weiss, A.J., Friedlander, B., Stoica, P., 1995. Direction-of-arrival estimation using MODE with interpolated arrays. IEEE Trans. Signal Process. 43 (1), 296–300.

Ziskind, I., Wax, M., 1988. Maximum likelihood localization of multiple sources by alternating projection. IEEE Trans. Acoust. Speech Signal Process. 36 (10), 1553–1560.

Zoltowski, M.D., Kautz, G.M., Silverstein, S.D., 1993. Beamspace Root-MUSIC. IEEE Trans. Signal Process. 41 (1), 344–364.

Chapter 6

Spatial Time-Frequency Distributions and DOA Estimation

Moeness Amin, Yimin Zhang

6.1 INTRODUCTION

This chapter discusses the direction-of-arrival (DOA) estimation of far-field sources producing nonstationary signals, particularly those in the form of frequency-modulated (FM) waveforms, using a class of methods based on the quadratic (bilinear) spatial time-frequency distribution (STFD) framework. Nonstationary signals are encountered in various applications in communications, radar systems, and biomedicine. For example, many modern radar systems use linear FM (LFM, or "chirp") signals to achieve pulse compression. Maneuvering or rotation due to targets or surrounding environments may generate time-varying Doppler frequency with well-defined Doppler frequency signatures (Boashash, 2003; Qian and Chen, 1996). FM signals are also among the popularly used intentional jammer sources for blocking communications in a wide-frequency spectrum (Amin, 1997).

For many decades, time-frequency (t-f) signal representations, such as the Wigner-Ville distribution (WVD) and spectrogram, were only applied to analyze nonstationary signals incident on single-sensor receivers. The objective was to characterize signals in the t-f domain, leading to proper signal characterization, separation, classification, and cancellation. These offerings were subsequently enhanced by new quadratic time-frequency distributions (TFDs), which have led to improved multicomponent signal power localizations in the t-f domain. As many applications of signal processing in the areas of communication, radar, acoustics, and biomedicine call for the use of multisensor receivers, it has become important to consider TFDs in the context of array processing. This is achieved within the framework of STFD (Belouchrani and Amin, 1998, 1999; Zhang et al., 2001).

Copyright © 2009 by Academic Press, Inc. All rights of reproduction in any form reserved.

The STFD framework was first developed by Belouchrani and Amin (1998) for the blind separation of narrowband nonstationary signals. It was shown that the STFD matrix is related to the source TFD matrix by the spatial mixing matrix in a manner similar to the commonly used formula in narrowband array-processing problems that use second-order statistics to relate the sensor spatial covariance and source covariance matrices. The framework was then used for DOA estimations of nonstationary signals using *t-f*-based DOA estimation approaches, such as *t-f* MUSIC, *t-f* Root-MUSIC, *t-f* maximum likelihood (*t-f* ML), and *t-f* ESPRIT (Amin and Zhang, 2000; Belouchrani and Amin, 1999; Gershman and Amin, 2000; Hassanien et al., 2002; Sekihara et al., 1999; Wang and Wu, 2002; Zhang et al., 2000). These techniques have shown improved performance compared to their conventional DOA counterparts.

Thorough supporting analyses of the advantages of STFD-based DOA estimation techniques and the robustness of the signal and noise subspaces estimated from STFD matrices were provided by Zhang et al. (2001). It was shown that, by constructing an STFD matrix from the selected *t-f* points of highly localized signal energy, the corresponding signal and noise subspace estimates become more robust to noise than their counterparts obtained using the data covariance matrix as a result of signal-to-noise ratio (SNR) enhancement. In addition, source masking and filtering in the *t-f* domain allow separation and thus consideration of individual sources or subsets of sources in the field of view. Source elimination, rendered through the selection of specific *t-f* regions corresponding only to a subset of signal arrivals, further increases the SNR and reduces mutual interference between signals, yielding improved subspace robustness. With such discriminatory capability, the receiver can process more signals than sensors and provide improved DOA estimation. In some applications, particularly when acoustic signals are involved, the TFD of multiple signals can be approximately disjoint or orthogonal; that is, only one signal is active in the *t-f* plane at a given *t-f* point. Therefore, the signals' DOAs can be estimated using only two receive sensors (Rickard and Dietrich, 2000). Accordingly, DOA estimation for nonstationary signals that are well localized in the *t-f* domain or characterized by their instantaneous frequencies (IFs) should be performed using *t-f* methods.

The merits of *t-f* DOA estimation can only be materialized through the selection of appropriate *t-f* points in the construction of STFD matrices. While in some scenarios the selection of peak *t-f* points may be relatively easy, the problem may become more challenging in other situations—for example, when the signals are highly contaminated by noise. Utilization of the spatial diversity inherent in the STFD matrix can enhance the *t-f* signature of the signals of interest. A basic example is simple averaging of the TFDs over all the receive sensors (Mu et al., 2003). Analysis of TFDs across different sensors also helps identify auto-term and cross-term points (Linh-Trung et al., 2005; Zhang and Amin, 2006). The separation of the auto-term *t-f* points from cross-term *t-f* points in general is a less-critical issue in DOA estimation when compared to blind-source separation applications, because in performing DOA estimation the STFD matrices only need to meet the full-rank requirement (Amin and Zhang, 2000). Nevertheless, the capability of separating auto-term from cross-term points often helps the selection of appropriate *t-f* regions corresponding to a subset of sources for source discrimination.

This chapter also reviews relevant advances in different aspects of STFD. One is the augmentation of the polarization dimension, yielding spatial polarimetric time-frequency distributions (SPTFD). SPTFD methods include polarimetric time-frequency MUSIC (PTF-MUSIC) and polarimetric time-frequency ESPRIT (PTF-ESPRIT) (Mu and Amin, 2000; Obeidat et al., 2004; Zhang et al., 2006a), which take advantage of the distinction of signal signatures between the sources in both the *t-f* and polarization domains for the improvement of DOA estimation performance. In this way they outperform their respective counterparts that only use source diversity in terms of *t-f* or polarization.

The advantage of using polarization dimensionality becomes particularly evident when the sources have distinct polarizations but are closely spaced and have similar waveforms. Application of PTF-MUSIC for the tracking of moving sources with time-varying polarizations was documented in Zhang et al. (2006b). The exploitation of spatial joint variable distributions (SJVDs), such as the spatial ambiguity function (SAF) that employs the Doppler and lag variables, permits estimation of DOA information for nonstationary signals that can be represented using their ambiguity domain characteristics (Amin et al., 2000).

Another important extension of STFD and SJVD DOA methods is the consideration of wideband signals where the narrowband assumption does not hold (Gershman and Amin, 2000; Ma and Goh, 2006; Obeidat et al., 2004).

Note that this chapter focuses on the DOA estimation approaches of nonstationary signals based on the quadratic STFD framework. In addition to quadratic TFDs, a nonstationary signal can also be described using linear transforms, such as the short-time Fourier transform (STFT) and the wavelet transform. Accordingly, DOA estimation can be performed based on linear *t-f* representations, providing that power localization and enhanced SNR are achieved. However, the STFT has a well-known shortcoming of trading off the resolutions in the time and frequency domains. Multiresolution analysis, on the other hand, is not most effective when dealing with signals characterized by their IF laws.

6.2 TIME-FREQUENCY DISTRIBUTION

Cohen's (1989, 1995) class of TFDs of a narrowband signal $x(t)$ is defined as

$$D_{xx}(t,f) = \iint \phi(t-u,\tau)\, x\left(u+\frac{\tau}{2}\right) x^*\left(u-\frac{\tau}{2}\right) e^{-j2\pi f\tau}\, du\, d\tau \qquad (6.1)$$

where t and f represent the time and frequency indexes, respectively, $\phi(t,\tau)$ is the *t-f* kernel, τ is the time-lag variable, and * denotes complex conjugation. In this chapter, all the integrals are from $-\infty$ to ∞.

The cross-term TFD of two signals $x_i(t)$ and $x_k(t)$ is defined by

$$D_{x_i x_k}(t,f) = \iint \phi(t-u,\tau) x_i\left(u+\frac{\tau}{2}\right) x_k^*\left(u-\frac{\tau}{2}\right) e^{-j2\pi f\tau} du d\tau \tag{6.2}$$

In practice, TFDs are often evaluated using their discrete-time forms (Boashash and Putland, 2003). To use integer time delay τ, we rewrite Equation (6.1) as

$$D_{xx}(t,f) = 2\iint \phi(t-u,2\tau) x(u+\tau) x^*(u-\tau) e^{-j4\pi f\tau} du d\tau \tag{6.3}$$

The discrete form of the auto-term TFD corresponding to Equation (6.3) is often expressed as

$$D_{xx}(t,f) = \sum_{u=-\infty}^{\infty} \sum_{\tau=-\infty}^{\infty} \phi(t-u,\tau) x(u+\tau) x^*(u-\tau) e^{-j4\pi f\tau} \tag{6.4}$$

which excludes the constant of two and a scaling factor in τ for expressional convenience. Similarly, the cross-term TFD corresponding to Equation (6.2) is expressed as

$$D_{x_i x_k}(t,f) = \sum_{u=-\infty}^{\infty} \sum_{\tau=-\infty}^{\infty} \phi(t-u,\tau) x_i(u+\tau) x_k^*(u-\tau) e^{-j4\pi f\tau} \tag{6.5}$$

The TFD maps one-dimensional (1D) signals in the time domain into two-dimensional (2D) signal representations in the *t-f* domain. The TFD property of concentrating the input signal energy around its IF while spreading the noise energy over the entire *t-f* domain increases the effective SNR and proves valuable in DOA estimation. For a single-component LFM signal, pseudo–Wigner-Ville distribution (PWVD) can achieve SNR improvement up to the window length (Zhang et al., 2001). This enhancement is dominantly determined by window size, but is less sensitive to kernel type (Mu and Amin, 2000). When all *t-f* points are selected within a 3-dB bandwidth from the peaks, SNR improvement remains proportional to window length (Xia and Chen, 1999). Such observations are valid for a general class of signals provided that the third-order derivative of the waveform phase is negligible or, equivalently, that the waveforms can be approximated by an LFM within each sliding-window interval.

The properties of a TFD can be characterized by simple constraints on the kernel. Different kernels can be used to generate TFDs with prescribed, desirable properties. WVD is often regarded as the basic or prototype quadratic TFD, because the other quadratic TFDs can be described as filtered versions of it. WVD is known to provide the best *t-f* resolution for single-component LFM

TABLE 6.1 Time-Frequency Kernels

Distribution	Kernel $\phi(t,\tau)$
Wigner-Ville	$\delta(t)$
Pseudo–Wigner-Ville	$\delta(t)w(\tau)$
Choi-Williams	$\frac{\sqrt{\pi\sigma}}{\lvert\tau\rvert}\exp(-\pi^2\sigma t^2/\tau^2)$
Zhao-Atlas-Marks	$w(\tau)\text{rect}\left(\frac{t}{2\tau/a}\right)$

signals, but it yields high cross-terms when the frequency law is not linear or when multicomponent signals are considered.

Various reduced-interference kernels have been developed to lower cross-term interference. Table 6.1 shows some commonly used kernel functions (Boashash, 2003), where $\delta(t)$ is a Dirac delta function, $\text{rect}(t)$ is a rectangular window function, $w(\tau)$ is an arbitrary window function, and σ and a are scalars. TFD examples that use these kernels are illustrated in Examples 6.1 and 6.2.

Example 6.1

Figures 6.1(a) and 6.1(b) show the real and imaginary parts of an analytic LFM signal, respectively, expressed as $x(t) = \exp[j2\pi(0.1t + 0.001176t^2)]$, $t = 0, \ldots, 255$. The start and end frequencies of the LFM signals are, respectively, 0.1 and 0.4. Fourier transform of the signal yields a spectrum spreading over the normalized frequency band [0.1, 0.4], as shown in Figure 6.1(c). The WVD of the waveform, shown in Figure 6.1(d), depicts high energy concentration of the instantaneous narrowband signal with a linearly time-varying IF signature.

Example 6.2

Figure 6.2 shows the t-f representations of two time-limited parallel LFM signals using different kernels. The WVD provides sharp auto-term signatures, whereas the cross-terms are evidently present in the middle of the two auto-term signatures. With PWVD, the cross-terms in the time domain are mitigated, whereas those in the frequency domain remain. Both the Choi-Williams distribution and the Zhao-Atlas-Marks distribution provide much reduced cross-term presence. The three non-WVD distributions yield much wider auto-term signatures in the t-f domain.

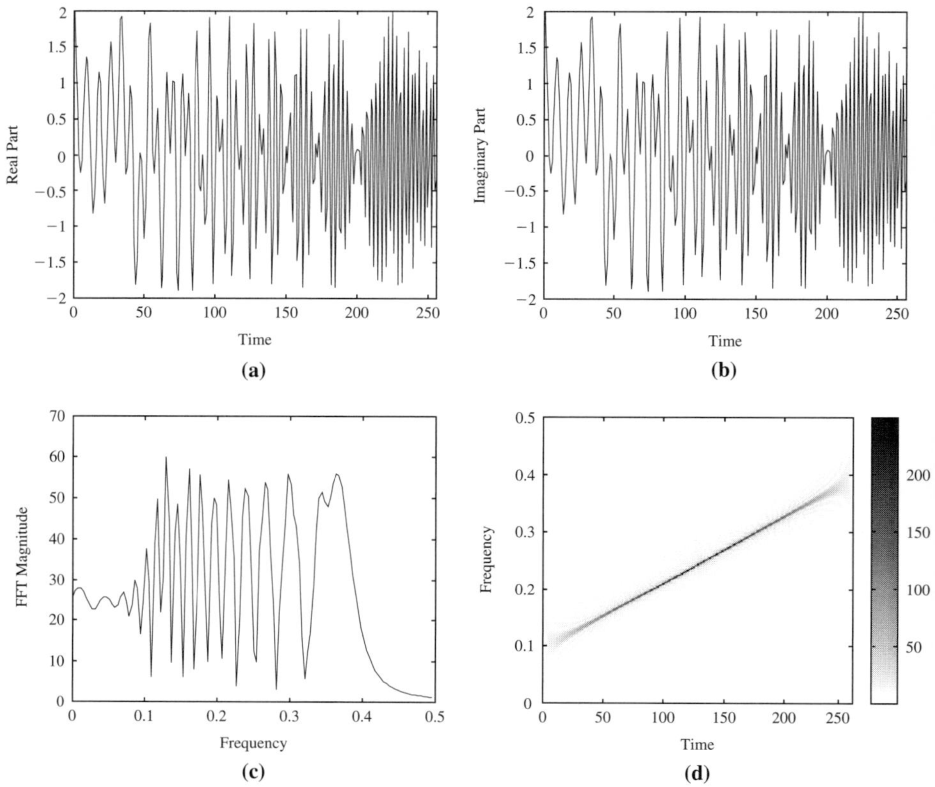

FIGURE 6.1 Waveform and Wigner-Ville distribution of an LFM signal: (a) real part of waveform; (b) imaginary part of waveform; (c) FFT magnitude; and (d) Wigner-Ville distribution.

6.3 SPATIAL TIME-FREQUENCY DISTRIBUTION

This section introduces the concept of STFD and then analyzes the robustness of the subspace estimated from STFD matrices. The fact that STFD matrices provide higher robustness compared to the corresponding covariance matrices is the foundation of extending subspace-based DOA estimation techniques in the STFD platform for improved DOA estimation. As examples, we introduce the *t-f* MUSIC and *t-f* ML techniques and demonstrate their performance advantages over conventional DOA estimation techniques.

6.3.1 Definitions

Consider n narrowband nonstationary signals impinging on an array consisting of m single-polarized sensors. For simplicity, we assume a 1D DOA estimation problem (e.g., only the azimuth angle is considered), but the extension to a 2D problem (i.e., both the azimuth and elevation angles are considered) is

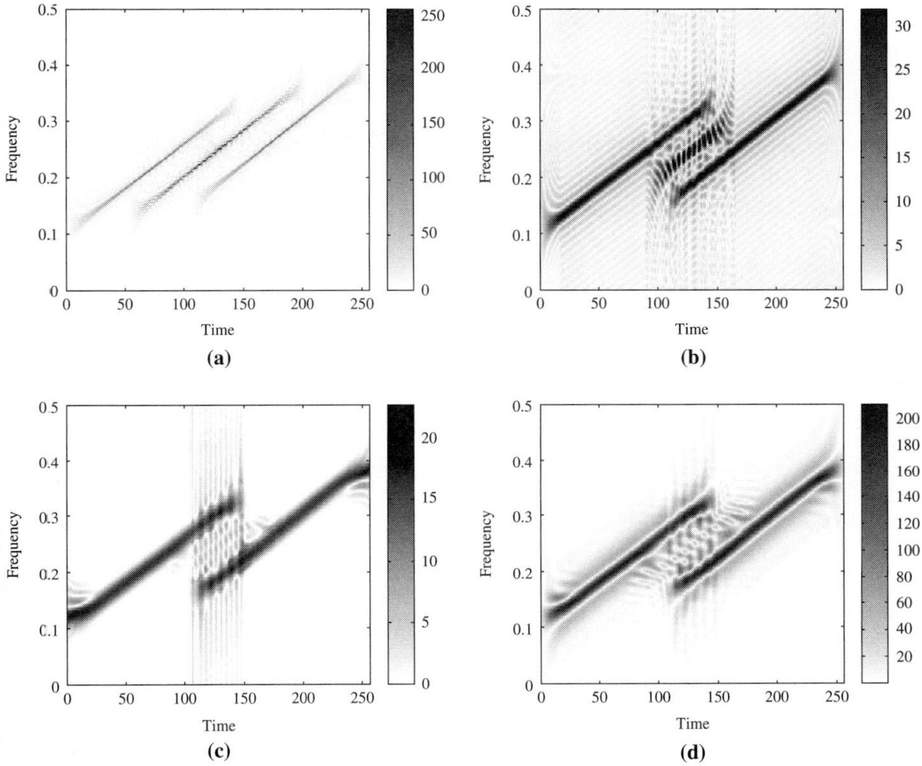

FIGURE 6.2 TFDs of two LFM signals corresponding to different kernels: (a) Wigner-Ville distribution; (b) pseudo–Wigner-Ville distribution; (c) Choi-Williams distribution; and (d) Zhao-Atlas-Marks distribution.

straightforward. The $m \times 1$ received data vector $\mathbf{x}(t)$ and the $n \times 1$ source signal vector $\mathbf{s}(t)$ are related by

$$\mathbf{x}(t) = \mathbf{A}(\boldsymbol{\Theta})\mathbf{s}(t) + \mathbf{n}(t) \tag{6.6}$$

where the $m \times n$ matrix $\mathbf{A}(\boldsymbol{\Theta}) = [\mathbf{a}(\theta_1), \mathbf{a}(\theta_2), \ldots, \mathbf{a}(\theta_n)]$ is the mixing matrix that holds the steering vectors of the n signals, $\boldsymbol{\Theta} = [\theta_1, \theta_2, \ldots, \theta_n]$, and $\mathbf{a}(\theta_q)$ is the steering vector for the qth source, $s_q(t)$, that arrives from direction θ_q. Each element of $\mathbf{s}(t) = [s_1(t), s_2(t), \ldots, s_n(t)]^T$ is assumed to be a mono-component signal, where the superscript T denotes the transpose of a matrix or a vector.

Because of the signal mixing occurring at each sensor, the elements of $\mathbf{x}(t)$ become multicomponent signals. $\mathbf{n}(t)$ is an $m \times 1$ additive noise vector that consists of independent and identically distributed (i.i.d.) zero-mean, white and complex Gaussian distributed processes with variance $\sigma^2\mathbf{I}$, where $\mathbf{I}$ denotes any identity matrix of a proper dimension. The noise elements are assumed to be independent of the signals, which are assumed to be deterministic.

The STFD matrix of vector $\mathbf{x}(t)$ (Belouchrani and Amin, 1998) is expressed as

$$\mathbf{D_{xx}}(t,f) = \iint \phi(t-u,\tau)\,\mathbf{x}\left(u+\frac{\tau}{2}\right)\mathbf{x}^H\left(u-\frac{\tau}{2}\right)e^{-j2\pi f\tau}du\,d\tau \tag{6.7}$$

where the superscript H denotes conjugate transpose (Hermitian) operation, and the (i,k)-th element of $\mathbf{D_{xx}}(t,f)$ is given in Equation (6.2) for $i,k=1,2,\ldots,m$. The noise-free STFD matrix is obtained by substituting Equation (6.6) into Equation (6.7), resulting in

$$\mathbf{D_{xx}}(t,f) = \mathbf{A}(\boldsymbol{\Theta})\mathbf{D_{ss}}(t,f)\mathbf{A}^H(\boldsymbol{\Theta}) \tag{6.8}$$

where $\mathbf{D_{ss}}(t,f)$ is the TFD matrix of $\mathbf{s}(t)$, which consists of auto-source TFDs as the diagonal elements and cross-source TFDs as the off-diagonal elements. In the presence of noise, the expected value of $\mathbf{D_{xx}}(t,f)$ becomes

$$E\left[\mathbf{D_{xx}}(t,f)\right] = \mathbf{A}(\boldsymbol{\Theta})\mathbf{D_{ss}}(t,f)\mathbf{A}^H(\boldsymbol{\Theta}) + \sigma^2\mathbf{I} \tag{6.9}$$

where $E[\cdot]$ denotes statistical expectation.

Equation (6.9) relates the STFD matrix to the source TFD matrix in a manner similar to the formula that is commonly used in narrowband array-processing problems to relate the source covariance and sensor spatial covariance matrices. It is clear, therefore, that the two subspaces spanned by the principle eigenvectors of $\mathbf{D_{xx}}(t,f)$ and the columns of $\mathbf{A}(\boldsymbol{\Theta})$ are identical. As will be discussed, the construction of the STFD matrix from the *t-f* points of highly localized signal energy allows the corresponding signal and noise subspace estimates to become more robust to noise than their counterparts obtained using the data covariance matrix (Belouchrani and Amin, 1998; Zhang et al., 2001). Further, source elimination, rendered through the selection of specific *t-f* regions, improves DOA estimation (Zhang et al., 2001).

6.3.2 Utilization of Multiple Time-Frequency Points

Let $\mathbf{S}^{tf}$ and $\mathbf{G}^{tf}$, respectively, span the signal and noise subspaces corresponding to a *t-f* region. A simple way to estimate $\mathbf{S}^{tf}$ and $\mathbf{G}^{tf}$ is through eigendecomposition on a single STFD matrix $\mathbf{D_{xx}}(t,f)$ obtained at a *t-f* point. However, subspace estimation from an STFD matrix is usually not robust and may have a singularity problem. Multiple *t-f* points that share the same spatial signature can be used to avoid such a problem. Joint block diagonalization (JBD) of the combined set of K *t-f* points, $\mathbf{D_{xx}}(t_i,f_i)$, $i=1,\ldots,K$, was introduced in Belouchrani and Amin (1999). The JBD is achieved by maximization, under unitary transform, of the following criterion (Belouchrani et al., 1997):

$$C\left(\mathbf{E}^{tf}\right) = \sum_{k}^{K}\sum_{i,l}^{m}\left|\mathbf{e}_i^H\mathbf{D_{xx}}(t_k,f_k)\,\mathbf{e}_l\right|^2 \tag{6.10}$$

over the set of unitary matrices, where $\mathbf{E}^{tf} = [\mathbf{S}^{tf}\mathbf{G}^{tf}]$, and $\mathbf{e}_i$ is the ith column of $\mathbf{E}^{tf}$. In Belouchrani et al. (1997), an efficient Jacobi-like algorithm for solving Equation (6.10) was presented.

Alternatively, STFD matrices evaluated at multiple t-f points can be incorporated by eigendecomposition of an STFD matrix to be obtained by averaging them (Zhang et al., 2001). This approach is simpler than JBD, particularly when the number of t-f points is large.

6.3.3 SNR Enhancement and Source Discrimination

To understand the properties of STFDs, we consider PWVD applied to FM signals. This consideration is motivated by the fact that these FM signals have constant magnitude and clear t-f signatures characterized by their IFs. They can be modeled as

$$\mathbf{s}(t) = [s_1(t), \ldots, s_n(t)]^T = \left[D_1 e^{j\psi_1(t)}, \ldots, D_n e^{j\psi_n(t)}\right]^T \tag{6.11}$$

where D_q and $\psi_q(t)$ are the fixed amplitude and time-varying phase of the qth source signal. For each sampling time t, the IF of $d_q(t)$ is expressed as $f_q(t) = d\psi_q(t)/(2\pi dt)$.

Using the PWVD, we construct the spatial pseudo–Wigner-Ville distribution (SPWVD) matrix, which, for a vector signal $\mathbf{x}(t)$, is expressed in discrete format as

$$\mathbf{D}_{\mathbf{xx}}(t,f) = \sum_{\tau=-(L-1)/2}^{(L-1)/2} x(t+\tau)x^*(t-\tau)e^{-j4\pi f\tau} \tag{6.12}$$

where the window size L is chosen as an odd integer. An STFD matrix can be constructed by averaging $\mathbf{D}_{\mathbf{xx}}(t,f)$ at the peak auto-term t-f points of n_o selected FM signals:

$$\hat{\mathbf{D}} = \frac{1}{n_o N'} \sum_{q=1}^{n_o} \sum_{i=1}^{N'} \mathbf{D}_{\mathbf{xx}}\left(t_i, f_{q,i}(t_i)\right) \tag{6.13}$$

where $f_{q,i}(t_i)$ is the IF law of the qth signal at the ith time sample, and $N' = N - L + 1$ is the number of t-f points that are not computed from padded zeros, with N denoting the total number of observation samples. We assume $N \gg n$ and $N \gg m$. In the case where the IF law of the FM signals is approximated by a linear behavior over each kernel window (i.e., the third- and higher-order derivatives of the phase of the FM signals are negligible over the time period $[t-L+1, t+L-1]$ for all values of t), the expectation of $\hat{\mathbf{D}}$ is given by Zhang et al. (2001) as

$$\begin{aligned}\mathbf{D} = E\left[\hat{\mathbf{D}}\right] &= \frac{1}{n_o N'} \sum_{q=1}^{n_o} \sum_{i=1}^{N'} E\left[\mathbf{D}_{\mathbf{xx}}\left(t_i, f_{q,i}\left(t_i\right)\right)\right] \\ &= \frac{L}{n_o} \mathbf{A}\left(\boldsymbol{\Theta}^o\right) \mathbf{R}_{\mathbf{ss}}^o \mathbf{A}\left(\boldsymbol{\Theta}^o\right) + \sigma^2 \mathbf{I}\end{aligned} \tag{6.14}$$

where $\mathbf{R}_{\mathbf{ss}}^o = \text{diag}\left[D_1^2, \ldots, D_{n_o}^2\right]$ and $\mathbf{A}(\boldsymbol{\Theta}^o) = \left[\mathbf{a}(\theta_1), \ldots, \mathbf{a}\left(\theta_{n_o}\right)\right]$, respectively, represent the signal covariance matrix and the mixing matrix formulated by only considering the n_o signals selected from the total number of n signal arrivals.

It becomes clear from the preceding expression that the expected STFD matrix is similar to a covariance matrix constructed as if only the n_o selected sources are present, and the SNR is enhanced by a factor of $G = L/n_o$, which is proportional to the window length L and inversely proportional to the number of sources contributing to matrix $\mathbf{D}$. Therefore, when $L \gg n_o$ is satisfied, DOA estimation of the selected n_o sources can be achieved based on the expected STFD matrix $\mathbf{D}$ with improved performance.

With the use of STFD matrices, array processing can be performed for a subclass of the impinging signals with specific t-f signatures. In this respect, t-f DOA estimation techniques have implicit spatial filtering, removing the undesired signals from consideration. From the SNR perspective, it is desirable to choose a small subset of signals. Ideally, it is best to set $n_o = 1$. That is, select the sets of N' t-f points that belong to only a single signal at a time. It is also important to note that, with the ability to construct the STFD matrix from one or few signal arrivals, the well-known $m > n$ condition on DOA estimation using arrays can be relaxed to $m > n_o$. In other words, we can perform DOA estimation with fewer array sensors than the total number of impinging signals. Further, from the angular resolution perspective, closely spaced sources with different t-f signatures can be resolved by constructing two separate STFDs, each corresponding to one source. It was demonstrated in Rickard and Dietrich (2000) that when the TFD of multiple signals can be considered approximately disjoint or orthogonal (i.e., only one signal is active in the t-f plane at a given t-f point), the signals' DOAs can be estimated using only two receive sensors. The drawback to using STFD matrices corresponding to a subclass of signal arrivals is of course the need for repeated computations if the DOA information on all source signals is of interest.

6.3.4 Subspace Analysis for FM Signals

To analyze the subspace robustness, we first present the case of FM signals using the conventional covariance matrix approach. In this case, it is required that the number of sensors be greater than the number of sources (i.e., $m > n$) and that matrix $\mathbf{A}(\boldsymbol{\Theta})$ be full column rank. Then the covariance matrix of $\mathbf{x}(t)$ is given by

$$\mathbf{R}_{\mathbf{xx}} = E\left[\mathbf{x}(t)\mathbf{x}^H(t)\right] = \mathbf{A}(\boldsymbol{\Theta})\mathbf{R}_{\mathbf{ss}}\mathbf{A}(\boldsymbol{\Theta}) + \sigma^2\mathbf{I} \tag{6.15}$$

where $\mathbf{R}_{\mathbf{ss}} = E[\mathbf{s}(t)\mathbf{s}^H(t)]$ is the source covariance matrix.

Let $\lambda_1 > \lambda_2 > \ldots > \lambda_n > \lambda_{n+1} = \lambda_{n+2} = \ldots = \lambda_m = \sigma^2$ denote the eigenvalues of $\mathbf{R_{xx}}$. The unit-norm eigenvectors associated with $\lambda_1, \ldots, \lambda_n$ constitute the columns of the signal subspace matrix $\mathbf{S} = [\mathbf{s}_1, \ldots, \mathbf{s}_n]$, and those corresponding to $\lambda_{n+1}, \ldots, \lambda_m$ make up the noise subspace matrix $\mathbf{G} = [\mathbf{g}_1, \ldots, \mathbf{g}_{m-n}]$. When $\mathbf{R_{xx}}$ is estimated from the available data samples $\mathbf{x}(i), i = 1, 2, \ldots, N$, the estimated covariance matrix is given by $\hat{\mathbf{R}}_{\mathbf{xx}} = \frac{1}{N}\sum_{i=1}^{N}\mathbf{x}(i)\mathbf{x}^H(i)$. Let $\{\hat{\mathbf{s}}_1, \ldots, \hat{\mathbf{s}}_n, \hat{\mathbf{g}}_1, \ldots, \hat{\mathbf{g}}_{m-n}\}$ denote the unit-norm eigenvectors of $\hat{\mathbf{R}}_{\mathbf{xx}}$, arranged in descending order of associated eigenvalues, and denote $\hat{\mathbf{S}} = [\hat{\mathbf{s}}_1, \ldots, \hat{\mathbf{s}}_n]$ and $\hat{\mathbf{G}} = [\hat{\mathbf{g}}_1, \ldots, \hat{\mathbf{g}}_{m-n}]$. For mutually uncorrelated signals, the orthogonal projections of $\hat{\mathbf{g}}_i$ onto the column space of $\mathbf{S}$ are asymptotically (for large N) jointly Gaussian distributed with zero means and covariance matrices (Zhang et al., 2001) given by

$$E\left[\left(\mathbf{S}\mathbf{S}^H\hat{\mathbf{g}}_i\right)\left(\mathbf{S}\mathbf{S}^H\hat{\mathbf{g}}_j\right)^H\right] = \frac{\sigma^2}{N}\left[\sum_{k=1}^{n}\frac{\lambda_k}{\left(\sigma^2 - \lambda_k\right)^2}\mathbf{s}_k\mathbf{s}_k^H\right]\delta_{i,j} \stackrel{\text{def}}{=} \frac{1}{N}\mathbf{U}\delta_{i,j} \tag{6.16}$$

$$E\left[\left(\mathbf{S}\mathbf{S}^H\hat{\mathbf{g}}_i\right)\left(\mathbf{S}\mathbf{S}^H\hat{\mathbf{g}}_j\right)^T\right] = 0, \quad \text{for all } i, j \tag{6.17}$$

where $\delta_{i,j}$ is the Kronecker delta function. Both equations are identical to those derived for independent stochastic signals (Stoica and Nehorai, 1989).

Now we consider the subspace of STFD matrix $\mathbf{D}$ and its estimate $\hat{\mathbf{D}}$. The results are compared with those obtained from $\mathbf{R}^o_{\mathbf{xx}} = \mathbf{A}^o\mathbf{R}^o_{\mathbf{dd}}(\mathbf{A}^o)^H + \sigma^2\mathbf{I}$ and its estimate $\hat{\mathbf{R}}^o_{\mathbf{xx}}$. Let $\lambda^o_1 > \lambda^o_2 > \ldots > \lambda^o_{n_0} > \lambda^o_{n_0+1} = \lambda^o_{n_0+1} = \ldots = \lambda^o_m = \sigma^2$ denote the eigenvalues of $\mathbf{R}^o_{\mathbf{xx}}$, and let its signal subspace and noise subspace be represented, respectively, by $\mathbf{S}^o = [\mathbf{s}^o_1, \ldots, \mathbf{s}^o_{n_o}]$ and $\mathbf{G}^o = [\mathbf{g}^o_1, \ldots, \mathbf{g}^o_{m-n_o}]$. We also denote $\lambda^{tf}_1 > \lambda^{tf}_2 > \ldots > \lambda^{tf}_{n_0} > \lambda^{tf}_{n_0+1} = \lambda^{tf}_{n_0+1} = \ldots = \lambda^{tf}_m = \left(\sigma^{tf}\right)^2$ as the eigenvalues of $\mathbf{D}$, and $\mathbf{S}^{tf} = \left[\mathbf{s}^{tf}_1, \ldots, \mathbf{s}^{tf}_{n_o}\right]$ and $\mathbf{G}^{tf} = \left[\mathbf{g}^{tf}_1, \ldots, \mathbf{g}^{tf}_{m-n_o}\right]$ as the corresponding signal and noise subspaces. Then

- The signal and noise subspaces of $\mathbf{S}^{tf}$ and $\mathbf{G}^{tf}$ are respectively the same as those of $\mathbf{S}^o$ and $\mathbf{G}^o$.
- The eigenvalues have the following relationship:

$$\lambda^{tf}_i = \begin{cases} \frac{L}{n_o}\left(\lambda^o_i - \sigma^2\right) + \sigma^2 = \frac{L}{n_o}\lambda^o_i + \left(1 - \frac{L}{n_o}\right)\sigma^2, & i \le n_o \\ \left(\sigma^{tf}\right)^2 = \sigma^2, & n_o < i \le m \end{cases} \tag{6.18}$$

That is, the largest n_o eigenvalues are amplified by a factor of L/n_o when the STFD matrix is used.

The robustness of the noise subspace obtained from a finite number of data samples is described as follows (Zhang et al., 2001). If the third- and higher-order derivatives of the phase of the FM signals are negligible over the time period $[t-L+1, t+L-1]$ for all values of t, then the orthogonal projections of $\{\hat{\mathbf{g}}^{tf}_i\}$ onto the column space of $\mathbf{S}^{tf}$ are asymptotically (for $N \gg L$) jointly Gaussian distributed, with zero means and covariance matrices given by

$$E\left[\left(\mathbf{S}^{tf}(\mathbf{S}^{tf})^H\hat{\mathbf{g}}_i^{tf}\right)\left(\mathbf{S}^{tf}(\mathbf{S}^{tf})^H\hat{\mathbf{g}}_j^{tf}\right)^H\right] = \frac{\sigma^2 L}{n_o N'}\left[\sum_{k=1}^{n_o}\frac{\lambda_k^{tf}}{\left(\sigma^2-\lambda_k^{tf}\right)^2}\mathbf{s}_k^{tf}\left(\mathbf{s}_k^{tf}\right)^H\right]\delta_{i,j} \overset{\text{def}}{=} \frac{1}{N'}\mathbf{U}^{tf}\delta_{i,j} \tag{6.19}$$

$$E\left[\left(\mathbf{S}^{tf}(\mathbf{S}^{tf})^H\hat{\mathbf{g}}_i^{tf}\right)\left(\mathbf{S}^{tf}(\mathbf{S}^{tf})^H\hat{\mathbf{g}}_j^{tf}\right)^T\right] = 0, \quad \text{for all } i,j \tag{6.20}$$

Two important observations are in order. First, if the signals are both localizable and separable in the t-f domain, reducing the number of signals from n to n_o greatly reduces the estimation error, specifically when the signals are closely spaced. Second, regarding SNR enhancements, the preceding equations show that error reductions using STFDs are more pronounced for the cases of low SNR and/or closely spaced signals. When $\lambda_k^o \gg \sigma^2$ for all $k = 1, 2, \ldots, n_o$, the results are almost independent of L (assume $N \gg L$ so that $N' = N - L + 1 \approx N$), and therefore there is no obvious improvement in using the STFD over conventional array processing. On the other hand, when some of the eigenvalues are close to $\sigma^2 (\lambda_k^o \approx \sigma^2$ for some $k = 1, 2, \ldots, n_o)$, which is the case for weak or closely spaced signals, the result of Equation (6.19) is reduced by a factor of up to $G = L/n_o$. This factor represents, in essence, the gain achieved from STFD processing.

6.4 TIME-FREQUENCY DOA ESTIMATION TECHNIQUES

As we addressed earlier, various DOA estimation techniques can be applied in the STFD framework. In this section, we introduce time-frequency MUSIC and time-frequency maximum likelihood techniques as examples.

6.4.1 Time-Frequency MUSIC

We first recall that the DOAs are estimated in the MUSIC technique by determining the n values of θ for which the following spatial pseudo-power spectrum (abbreviated as "spatial spectrum" hereafter) is maximized (Schmidt, 1986),

$$f_{\text{MU}}(\theta) = \left[\mathbf{a}^H(\theta)\hat{\mathbf{G}}\hat{\mathbf{G}}^H\mathbf{a}(\theta)\right]^{-1} = \left[\mathbf{a}^H(\theta)\left(\mathbf{I}-\hat{\mathbf{S}}\hat{\mathbf{S}}^H\right)\mathbf{a}(\theta)\right]^{-1} \tag{6.21}$$

where $\mathbf{a}(\theta)$ is the steering vector corresponding to θ. The variance of the estimates in the MUSIC technique (Stoica and Nehorai, 1989), assuming white noise processes, is given by

$$E\left(\hat{\omega}_i - \omega_i\right)^2 = \frac{1}{2N} \times \frac{\mathbf{a}^H(\theta_i)\mathbf{U}\mathbf{a}(\theta_i)}{h(\theta_i)} \tag{6.22}$$

where ω_i is the spatial frequency associated with DOA θ_i, and $\hat{\omega}_i$ is its estimate obtained from MUSIC. The spatial frequency is defined as $\omega_i = \frac{2\pi d \sin(\theta_i)}{\lambda}$ for a uniform linear array (ULA), where λ is the wavelength of the impinging wavefront, and d is the inter-element spacing between two adjacent sensors. Moreover, $\mathbf{U}$ is defined in Equation (6.16), and

$$h(\theta_i) = \mathbf{d}^H(\theta_i)\mathbf{G}\mathbf{G}^H\mathbf{d}(\theta_i), \text{ with } \mathbf{d}(\theta_i) = d\mathbf{a}(\theta_i)/d\omega \tag{6.23}$$

The t-f MUSIC algorithm is very similar to MUSIC. With n_o signals selected, the DOAs are determined by locating the n_o peaks of the spatial spectrum defined from the TFD of the n_o selected signals:

$$\begin{aligned} f_{\mathrm{MU}}^{tf}(\theta) &= \left[\mathbf{a}^H(\theta)\hat{\mathbf{G}}^{tf}\left(\hat{\mathbf{G}}^{tf}\right)^H\mathbf{a}(\theta)\right]^{-1} \\ &= \left[\mathbf{a}^H(\theta)\left(\mathbf{I} - \hat{\mathbf{S}}^{tf}\left(\hat{\mathbf{S}}^{tf}\right)^H\right)\mathbf{a}(\theta)\right]^{-1} \end{aligned} \tag{6.24}$$

The variance of the DOA estimates based on t-f MUSIC (Zhang et al., 2001) is obtained as

$$E\left(\hat{\omega}_i^{tf} - \omega_i\right)^2 = \frac{1}{2N} \cdot \frac{\mathbf{a}^H(\theta_i)\mathbf{U}^{tf}\mathbf{a}(\theta_i)}{h^{tf}(\theta_i)} \tag{6.25}$$

where $\hat{\omega}_i^{tf}$ is its estimate of ω_i obtained from t-f MUSIC, $\mathbf{U}^{tf}$ is defined in Equation (6.19), and

$$h^{tf}(\theta_i) = \mathbf{d}^H(\theta_i)\mathbf{G}^{tf}\left(\mathbf{G}^{tf}\right)^H\mathbf{d}(\theta_i) \tag{6.26}$$

Note that $h^{tf}(\theta_i) = h(\theta_i)$ if $n_o = n$.

Example 6.3

Consider a ULA consisting of eight sensors spaced by a half-wavelength. Two LFM signals are emitted by two sources from DOAs of $\theta_1 = -10°$ and $\theta_2 = 10°$, respectively. The start and end frequencies of the first source are $f_{s1} = 0$ and $f_{e1} = 0.5$, whereas the corresponding two frequencies for the other source are $f_{s2} = 0.5$ and $f_{e2} = 0$. An observation period of 1024 samples is assumed.

Figure 6.3 displays the root mean-square error (RMSE) of the estimated DOA $\hat{\theta}_1$ versus the input SNR. The curves in the figure show the theoretical and experimental results of conventional MUSIC and t-f MUSIC (for two cases of $L = 33$ and $L = 129$). The Cramer-Rao bound (CRB) is also shown for comparison.

Both signals were selected when performing t-f MUSIC ($n_o = n = 2$). Simulation results were averaged over 100 independent trials. The advantages of t-f MUSIC in low-SNR cases are evident. The experimental results deviate from the theoretical results for low SNR, since we only considered the lowest order of

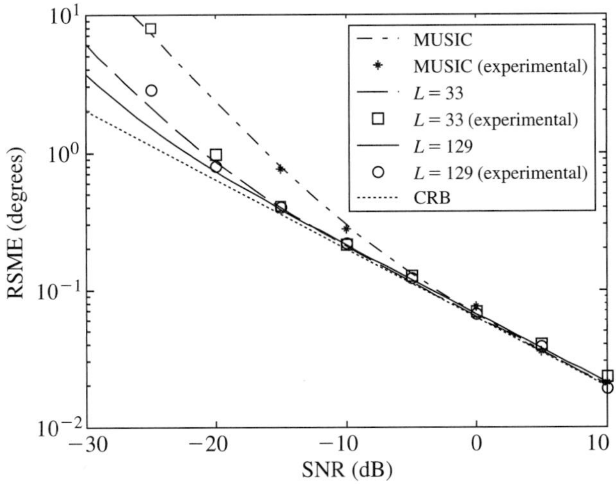

FIGURE 6.3 RMSE performance of DOA estimates using *t-f* MUSIC and MUSIC versus input SNR.

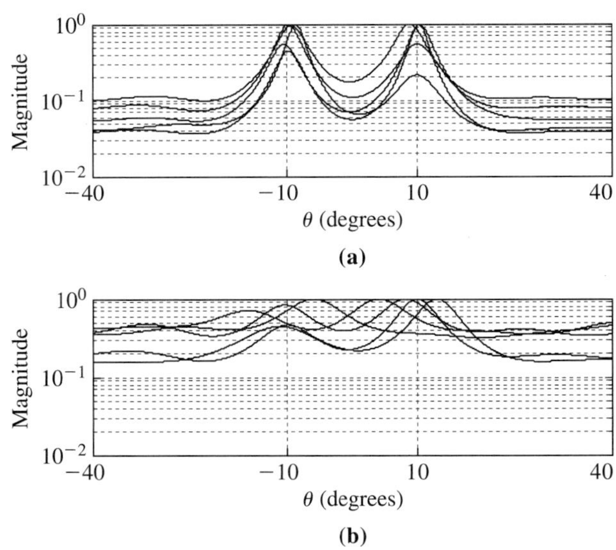

FIGURE 6.4 Estimated spatial spectra using (a) *t-f* MUSIC and (b) MUSIC.

the coefficients of the perturbation expansion in deriving the latter (Zhang et al., 2001). Figure 6.4 shows estimated spatial spectra at SNR = −20 dB based on *t-f* MUSIC ($L = 129$) and conventional MUSIC. The *t-f* MUSIC spectral peaks are clearly resolved in all trials.

Example 6.4

Figure 6.5 shows examples of the estimated spatial spectrum based on *t-f* MUSIC and conventional MUSIC where the angle separation is small ($\theta_1 = -2°, \theta_2 = 2°$). L is chosen to be 129 and the input SNR is -5 dB. Two *t-f* MUSIC algorithms are performed using two sets of *t-f* points; each set belongs to the *t-f* signature of one source ($n_o = 1$). It is evident that the two signals cannot be resolved when conventional MUSIC is applied. However, by using their distinct *t-f* signatures and applying *t-f* MUSIC separately for each signal, the two signals become clearly separated and moderate DOA estimation performance is achieved. It is noted that there is a small bias in the estimates of *t-f* MUSIC due to the sidelobe leakage of one signal's TFD when the other signal is selected.

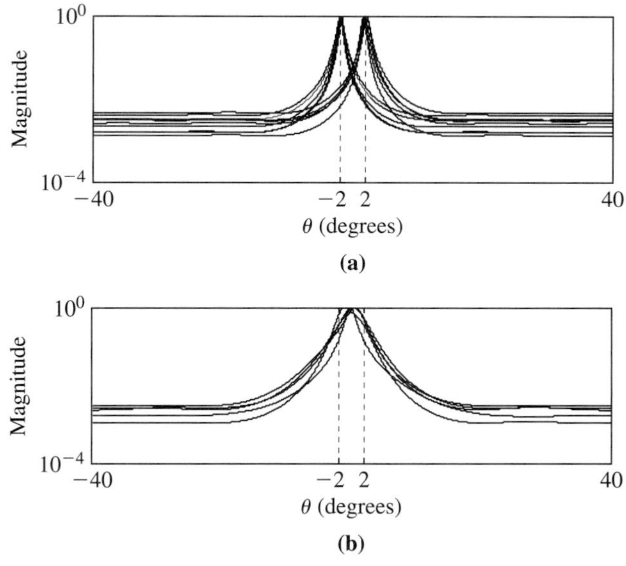

FIGURE 6.5 Estimated spatial spectra using (a) *t-f* MUSIC and (b) MUSIC for closely spaced signals.

6.4.2 Time-Frequency Maximum Likelihood

Compared to *t-f* MUSIC, *t-f* maximum likelihood (ML) methods have advantages in dealing with coherent nonstationary sources (Zhang et al., 2000). Before we introduce *t-f* ML, we first review conventional ML methods (Ziskind and Wax, 1988).

When the noise elements are assumed to be i.i.d. complex Gaussian variables with zero means, the joint density function of the sampled data vectors

$\mathbf{x}_1, \mathbf{x}_2, \ldots, \mathbf{x}_N$ is given by

$$f(\mathbf{x}(1), \ldots, \mathbf{x}(N)) = \prod_{i=1}^{N} \frac{1}{\pi^m \sigma^{2m}} \exp\left(-\frac{1}{\sigma^2}[\mathbf{x}(i) - \mathbf{A}\mathbf{d}(i)]^H [\mathbf{x}(i) - \mathbf{A}\mathbf{d}(i)]\right) \tag{6.27}$$

The concentrated likelihood function (Ziskind and Wax, 1988) is obtained as

$$F_{\text{ML}}(\boldsymbol{\Theta}) = \text{tr}\left\{\left[\mathbf{I} - \hat{\mathbf{A}}\left(\hat{\mathbf{A}}^H\hat{\mathbf{A}}\right)^{-1}\hat{\mathbf{A}}^H\right]\hat{\mathbf{R}}_{\mathbf{xx}}\right\} \tag{6.28}$$

where tr(.) denotes the trace of a matrix, and $\hat{\mathbf{A}}$ is the estimate of $\mathbf{A}$. The ML estimate of $\boldsymbol{\Theta}$ is obtained as the minimizer of this expression. Let ω_i and $\hat{\omega}_i$, respectively, denote the spatial frequency and its ML estimate associated with θ_i. The estimation error $(\hat{\omega}_i - \omega_i)$ is thus asymptotically (for large N) jointly Gaussian distributed with zero means and the following covariance matrix (Stoica and Nehorai, 1989):

$$E\left[(\hat{\omega}_i - \omega_i)^2\right] = \frac{1}{2N}\left[\text{Re}(\mathbf{H} \circ \mathbf{R}_{\mathbf{dd}}^T)\right]^{-1} \text{Re}\left[\mathbf{H} \circ (\mathbf{R}_{\mathbf{dd}}\mathbf{A}^H\mathbf{U}\mathbf{A}\mathbf{R}_{\mathbf{dd}})^T\right]\left[\text{Re}(\mathbf{H} \circ \mathbf{R}_{\mathbf{dd}}^T)\right]^{-1} \tag{6.29}$$

where $\circ$ denotes the Hadamard product, Re(.) denotes real part, and

$$\mathbf{H} = \mathbf{C}^H\left[\mathbf{I} - \mathbf{A}(\mathbf{A}^H\mathbf{A})^{-1}\mathbf{A}^H\right]\mathbf{C}, \text{ with } \mathbf{C} = d\mathbf{A}/d\omega \tag{6.30}$$

Now we consider the t-f ML method. As discussed in the previous section, we select $n_o \leq n$ signals in the t-f domain. The concentrated likelihood function defined from the STFD matrix is obtained by replacing $\hat{\mathbf{R}}_{\mathbf{xx}}$ by $\hat{\mathbf{D}}$:

$$F_{\text{ML}}^{tf}(\Theta) = \text{tr}\left[\mathbf{I} - \hat{\mathbf{A}}^o\left((\hat{\mathbf{A}}^o)^H\hat{\mathbf{A}}^o\right)^{-1}(\hat{\mathbf{A}}^o)^H\right]\hat{\mathbf{D}} \tag{6.31}$$

where $\hat{\mathbf{A}}^o$ is the estimate of $\mathbf{A}^o$. Therefore, the estimation error $\left(\hat{\omega}_i^{tf} - \omega_i\right)$ associated with the t-f ML method is asymptotically (for $N \gg L$) jointly Gaussian distributed with zero means and the covariance matrix (Zhang et al., 2000)

$$\begin{aligned}
&E\left[\left(\hat{\omega}_i^{tf} - \omega_i\right)^2\right] \\
&= \frac{\sigma^2}{2N'}\left[\text{Re}(\mathbf{H}^o \circ \mathbf{D}_{\mathbf{dd}}^T)\right]^{-1} \text{Re}\left[\mathbf{H}^o \circ \left(\mathbf{D}_{\mathbf{dd}}(\mathbf{A}^o)^H\mathbf{U}^{tf}\mathbf{A}^o\mathbf{D}_{\mathbf{dd}}\right)^T\right]\left[\text{Re}(\mathbf{H}^o \circ \mathbf{D}_{\mathbf{dd}}^T)\right]^{-1} \\
&= \frac{\sigma^2}{2N'}\left[\text{Re}\left(\mathbf{H}^o \circ (\mathbf{R}_{\mathbf{dd}}^o)^T\right)\right]^{-1} \text{Re}\left[\mathbf{H}^o \circ \left(\mathbf{R}_{\mathbf{dd}}^o(\mathbf{A}^o)^H\mathbf{U}^{tf}\mathbf{A}^o\mathbf{R}_{\mathbf{dd}}^o\right)^T\right]\left[\text{Re}\left((\mathbf{H}^o \circ \mathbf{R}_{\mathbf{dd}}^o)^T\right)\right]^{-1}
\end{aligned} \tag{6.32}$$

where

$$\mathbf{H}^o = (\mathbf{C}^o)^H \left[\mathbf{I} - \mathbf{A}^o \left((\mathbf{A}^o)^H \mathbf{A}^o \right)^{-1} (\mathbf{A}^o)^H \right] \mathbf{C}^o, \text{ with } \mathbf{C}^o = d\mathbf{A}^o / d\omega \quad (6.33)$$

In the case of $n_o = n$, $\mathbf{H}^o = \mathbf{H}$, and $\mathbf{C}^o = \mathbf{C}$.

The signal localization in the *t-f* domain enables us to select fewer signal arrivals. This fact not only is important in improving estimation performance, particularly when the signals are closely spaced, but it also reduces the dimension of the optimization problem solved by the maximum likelihood algorithm, and subsequently reduces the computational requirement.

Example 6.5

To demonstrate the advantages of *t-f* ML over conventional ML, consider a ULA consisting of eight sensors separated by a half-wavelength. Two FM signals arrive from $\theta_1 = -10°$ and $\theta_2 = 10°$ with the IFs $f_1(t) = 0.2 + 0.1t/N + 0.2\sin(2\pi t/N)$ and $f_2(t) = 0.2 + 0.1t/N + 0.2\sin(2\pi t/N + \pi/2)$, $t = 1, \ldots, N$. The input SNR of both signals is -20 dB, and the number of snapshots used in the simulation is $N = 1024$. $L = 129$ is used for *t-f* ML. Figure 6.6 shows the (θ_1, θ_2) pairs that minimize the respective likelihood function of the *t-f* ML and ML methods for 20 independent trials. It is evident that *t-f* ML provides much improved DOA estimation over conventional ML.

Example 6.6

In this example, *t-f* ML and *t-f* MUSIC are compared for coherent sources. The ULA considered in Example 6.5 is also used here. The two coherent FM signals now have a common IF of $f_{1,2}(t) = 0.2 + 0.1t/N + 0.2\sin(2\pi t/N)$, $t = 1, \ldots, N$, with a $\pi/2$ phase difference. The signals arrive at $\theta_1 = -2°$ and $\theta_2 = 2°$. The input SNR of both signals is 5 dB and the number of snapshots is $N = 1024$. Figure 6.7 shows the (θ_1, θ_2) estimates of *t-f* ML and the estimated spatial spectra of *t-f* MUSIC for five independent trials. It is clear that *t-f* ML can separate the two signals, whereas *t-f* MUSIC cannot.

6.5 POLARIMETRIC TIME-FREQUENCY DOA ESTIMATION

Polarization has been incorporated in array processing for improved estimation of various signal parameters, such as DOA and time of arrival (TOA) (Ferrara and Parks, 1983; Li and Compton, 1991; Yamada et al., 1997). The polarimetric signature may contain valuable information that the single-polarization

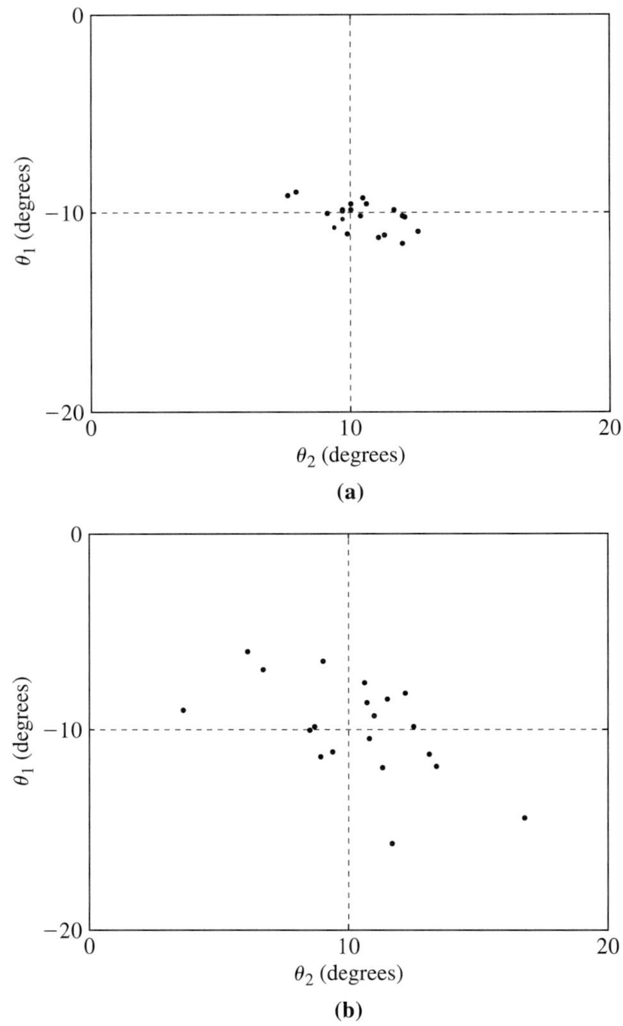

FIGURE 6.6 Estimation of θ_1 and θ_2 obtained from (a) *t-f* ML and (b) ML.

Doppler frequency signature fails to provide. The spatial polarimetric time-frequency technique for multisensor receivers uses not only the time-varying Doppler frequency signatures but also the polarization signatures, whether they are stationary or time-varying (Zhang et al., 2004, 2006a). Signal polarization information empowers the STFDs to achieve enhanced DOA estimation resolution, as it retains the integrity of eigenstructure methods and improves the robustness of the respective signal and noise subspaces under low SNR and highly correlated signal environments.

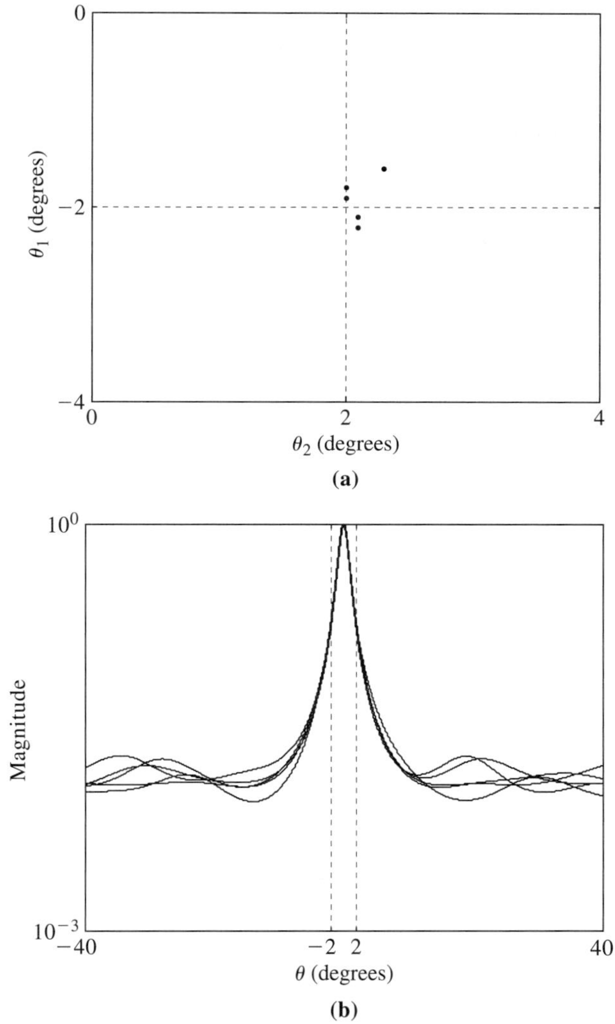

FIGURE 6.7 Contour plots of (a) the *t-f* ML likelihood function and (b) the spatial spectra of *t-f* MUSIC.

In this section, we introduce SPTFD and demonstrate its effectiveness by polarimetric time-frequency ESPRIT (PTF-ESPRIT) (Obeidat et al., 2003; Zhang et al., 2004). A MUSIC-like method for noncoherent and coherent signals is introduced separately in Zhang et al. (2003a,b, 2006a).

6.5.1 Spatial Polarimetric Time-Frequency Distribution

To utilize the polarization information, an electromagnetic (EM) vector sensor can be configured to hold up to six collocated and orthogonally oriented EM sensors and thus receive up to six EM components (three electric and three magnetic) (Gong et al., 2009; Nehorai and Paldi, 1994). For an active radar or sonar system, the polarization channels can be further extended into combinations of the transmit and receive polarizations, such as vertical–vertical (VV) and vertical–horizontal (VH). In this section, however, we only consider passive dual-polarization scenarios for representational simplicity. The extension to higher polarization dimensions is straightforward.

The received signal with dual polarizations can be expressed as

$$\underline{x}(t) = \left[x^{[p]}(t) \;\; x^{[q]}(t)\right]^T \tag{6.34}$$

where $(.)^{[p]}$ and $(.)^{[q]}$, respectively, denote two orthogonal polarizations. They can be, for example, vertical and horizontal or right and left circular.

When an array consisting of m dual-polarized sensors is considered, the data vector for each polarization i is expressed as

$$\mathbf{x}^{[i]}(t) = \mathbf{A}(\Theta)\mathbf{s}^{[i]}(t) + \mathbf{n}^{[i]}(t) \tag{6.35}$$

where i represents either p or q. The polarization components of the kth signal can be expressed as

$$s_k^{[p]}(t) = s_k(t)\cos(\gamma_k) \quad \text{and} \quad s_k^{[q]}(t) = s_k(t)\sin(\gamma_k)\, e^{j\phi_k} \tag{6.36}$$

where $\tan(\gamma_k) = |s_k^{[q]}(t)/s_k^{[p]}(t)|$ and ϕ_k denote the magnitude ratio and phase difference between the two components, respectively.

We are now in a position to tie together the polarization, spatial, and t-f properties of the signals incident on the sensor array. The following vector can be constructed for both polarizations:

$$\begin{aligned}\mathbf{x}(t) = \begin{bmatrix}\mathbf{x}^{[p]}(t)\\ \mathbf{x}^{[q]}(t)\end{bmatrix} &= \begin{bmatrix}\mathbf{A}(\boldsymbol{\Theta}) & \mathbf{0}\\ \mathbf{0} & \mathbf{A}(\boldsymbol{\Theta})\end{bmatrix}\begin{bmatrix}\mathbf{s}^{[p]}(t)\\ \mathbf{s}^{[q]}(t)\end{bmatrix} + \begin{bmatrix}\mathbf{n}^{[p]}(t)\\ \mathbf{n}^{[q]}(t)\end{bmatrix}\\ &= \mathbf{B}(\boldsymbol{\Theta})\mathbf{s}(t) + \mathbf{n}(t)\end{aligned} \tag{6.37}$$

where $\mathbf{B}(\boldsymbol{\Theta}) = \begin{bmatrix}\mathbf{A}(\boldsymbol{\Theta}) & \mathbf{0}\\ \mathbf{0} & \mathbf{A}(\boldsymbol{\Theta})\end{bmatrix}$ is block-diagonal, and $\mathbf{s}^{[i]}(t), i = p, q$ are the source signal vectors for polarization i. The STFD of the dual-polarization vector, $\mathbf{x}(t)$, can therefore be defined as

$$\begin{aligned}\mathbf{D_{xx}}(t,f) &= \iint \phi(t-u,\tau)\mathbf{x}(t+\tau/2)\mathbf{x}^H(t-\tau/2)e^{-j2\pi f\tau}du\,d\tau\\ &= \begin{bmatrix}\mathbf{A}(\boldsymbol{\Theta}) & \mathbf{0}\\ \mathbf{0} & \mathbf{A}(\boldsymbol{\Theta})\end{bmatrix}\begin{bmatrix}\mathbf{D}_{\mathbf{s}^{[p]}\mathbf{s}^{[p]}}(t,f) & \mathbf{D}_{\mathbf{s}^{[p]}\mathbf{s}^{[q]}}(t,f)\\ \mathbf{D}_{\mathbf{s}^{[q]}\mathbf{s}^{[p]}}(t,f) & \mathbf{D}_{\mathbf{s}^{[q]}\mathbf{s}^{[q]}}(t,f)\end{bmatrix}\begin{bmatrix}\mathbf{A}(\boldsymbol{\Theta}) & \mathbf{0}\\ \mathbf{0} & \mathbf{A}(\boldsymbol{\Theta})\end{bmatrix}^H\end{aligned} \tag{6.38}$$

In Equation (6.38), $\mathbf{D_{xx}}(t,f)$ is referred to as the spatial polarimetric time-frequency distribution (SPTFD) matrix. This distribution serves as a general framework within which DOA estimation techniques can be applied to take advantage of diversity provided by the polarimetric and *t-f* signatures of the sources (Obeidat et al., 2003; Zhang et al., 2003a,b, 2006a). In the next section, PTF-ESPRIT (Obeidat et al., 2003; Zhang et al., 2004) is shown as an application example of SPTFD.

6.5.2 Polarimetric Time-Frequency ESPRIT

To achieve the rotational invariance required to apply the ESPRIT technique, we consider a ULA with m identical cross-dipoles. As illustrated in Figure 6.8, the array is divided into two overlapping subarrays, each consisting of $m-1$ elements. Let the first subarray be composed of the leftmost $m-1$ cross-dipoles, and the second subarray be composed of the rightmost $m-1$ cross-dipoles. The array response matrices of the two subarrays are denoted as $\mathbf{A}_1(\boldsymbol{\Theta})$ and $\mathbf{A}_2(\boldsymbol{\Theta})$, respectively. Accordingly,

$$\begin{bmatrix} \mathbf{A}_2(\boldsymbol{\Theta}) & \mathbf{0} \\ \mathbf{0} & \mathbf{A}_2(\boldsymbol{\Theta}) \end{bmatrix} = \begin{bmatrix} \mathbf{A}_1(\boldsymbol{\Theta}) & \mathbf{0} \\ \mathbf{0} & \mathbf{A}_1(\boldsymbol{\Theta}) \end{bmatrix} \begin{bmatrix} \boldsymbol{\Phi} & \mathbf{0} \\ \mathbf{0} & \boldsymbol{\Phi} \end{bmatrix} \tag{6.39}$$

where the rotation operator $\boldsymbol{\Phi}$ can be described as

$$\boldsymbol{\Phi} = \mathrm{diag}\left[e^{-j2\pi\frac{d}{\lambda}\sin(\theta_1)}, e^{-j2\pi\frac{d}{\lambda}\sin(\theta_2)}, \ldots, e^{-j2\pi\frac{d}{\lambda}\sin(\theta_n)}\right] \tag{6.40}$$

By performing JBD on the SPTFD matrices $\mathbf{D_{xx}}(t,f)$ over multiple (t, f) points where the energy of the signal arrivals is concentrated, we obtain the signal and noise subspaces, represented as matrices $\mathbf{S}^{tf}$ and $\mathbf{G}^{tf}$, respectively. The signal eigenvectors, which make up the columns of $\mathbf{S}^{tf}$, span the signal subspace such that there exists a transformation matrix $\mathbf{T}$ that satisfies

$$\mathbf{S}^{tf} = \begin{bmatrix} \mathbf{A}(\boldsymbol{\Theta}) & \mathbf{0} \\ \mathbf{0} & \mathbf{A}(\boldsymbol{\Theta}) \end{bmatrix} \mathbf{T} \tag{6.41}$$

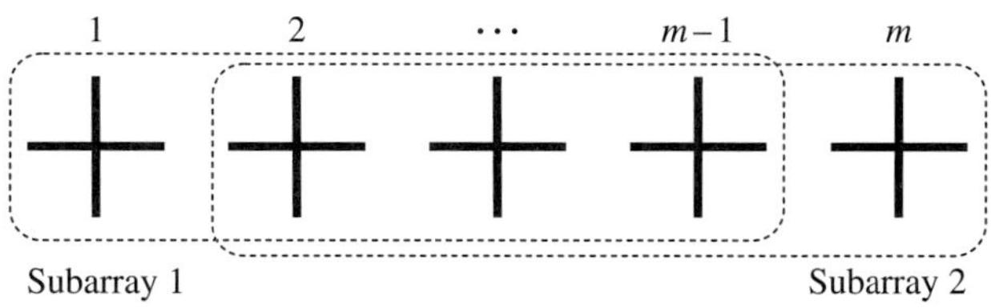

FIGURE 6.8 Partitioning of a ULA into two subarrays.

Applying the same transformation matrix $\mathbf{T}$ to the steering matrices of the two subarrays, we define the following matrices:

$$\mathbf{S}_1^{tf} = \begin{bmatrix} \mathbf{A}_1(\Theta) & \mathbf{0} \\ \mathbf{0} & \mathbf{A}_1(\Theta) \end{bmatrix} \mathbf{T} \quad \text{and} \quad \mathbf{S}_2^{tf} = \begin{bmatrix} \mathbf{A}_2(\Theta) & \mathbf{0} \\ \mathbf{0} & \mathbf{A}_2(\Theta) \end{bmatrix} \mathbf{T} \tag{6.42}$$

which are related by

$$\mathbf{S}_2^{tf} = \mathbf{S}_1^{tf} \mathbf{T}^{-1} \begin{bmatrix} \mathbf{\Phi} & \mathbf{0} \\ \mathbf{0} & \mathbf{\Phi} \end{bmatrix} \mathbf{T} = \mathbf{S}_1^{tf} \mathbf{\Psi} \tag{6.43}$$

where the eigenvalues of $\mathbf{\Psi}$ are $e^{-j2\pi \frac{d}{\lambda} \sin(\theta_i)}$, and $i = 1, 2, \ldots, n$. $\mathbf{\Psi}$ can be solved using the least-squares or the total least-squares approach (Roy and Kailath, 1989). When least squares is applied, the solution is given by

$$\mathbf{\Psi} = \left[\left(\mathbf{S}_1^{tf} \right)^H \mathbf{S}_1^{tf} \right]^{-1} \left(\mathbf{S}_1^{tf} \right)^H \mathbf{S}_2^{tf} \tag{6.44}$$

Example 6.7

We consider a ULA consisting of four cross-dipoles separated by a half-wavelength. Each comprises one vertical and one horizontal dipole. To demonstrate the advantages of the SPTFD framework and the PTF-ESPRIT algorithm, we consider two sources (sources 1 and 2) with high-order FM waveforms, separable in the *t-f* domain, in the presence of an undesired sinusoidal signal (source 3). The DOA of the three sources is, respectively, $\theta = -3°, 3°$, and $5°$; the amplitude ratio between the two polarization components is represented by $\gamma = 45°, 45°$, and $20°$; and the phase difference between the two polarization components for the three sources is $\phi = 0°$, $180°$, and $0°$, respectively. The normalized frequency of the sinusoid is 0.1. All signals have the same signal power with input SNR = 5 dB, and the data length is 256 samples. The PWVD with a rectangular window of length 65 is used for *t-f* signal representations.

For both *t-f* ESPRIT and PTF-ESPRIT, the array-averaged PWVD is first used to identify the auto-term regions. The search-free PTF-ESPRIT provides a DOA estimation that is compared to that of conventional ESPRIT, polarimetric ESPRIT, and *t-f* ESPRIT. Because the source signatures can be separated in the *t-f* domain in this example, only the *t-f* points on the signal signature of a single source are considered for the construction of the STFD and SPTFD matrices. PTF-ESPRIT outperforms other ESPRIT-based methods by taking advantage of such source-discriminatory capability, in addition to SNR enhancement and polarimetric selection.

Figure 6.9 shows the RMSE performance of PTF-ESPRIT and other ESPRIT-based methods versus the input SNR, where the least-squares approach is used for all methods. Each result is obtained from 100 independent trials. For conventional and *t-f* ESPRIT, only the vertical polarization components of the source signals

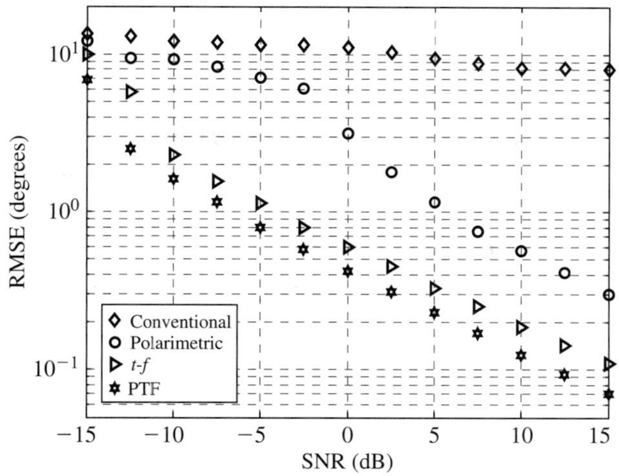

FIGURE 6.9 RMSE performance of ESPRIT algorithms versus input SNR.

are used. For the *t-f* and PTF-ESPRIT methods, which can discriminate sources based on their TFDs, only the first source signal is selected in STFD and SPTFD matrix construction, respectively. The RMSE performance is evaluated for the first source signal. It is evident that PTF-ESPRIT outperforms all other methods. Polarimetric ESPRIT provides satisfactory DOA estimation only when input SNR is high, whereas conventional ESPRIT fails to do so for all input SNR levels simulated. In constrast, PTF-ESPRIT provides 1° RMSE when input SNR is at a low level of about −7 dB.

6.6 THE SPATIAL AMBIGUITY FUNCTION AND APPLICATIONS TO DOA ESTIMATION

In addition to the *t-f* domain, a nonstationary signal can be analyzed in other joint variable domains, such as time lag (t, τ), Doppler lag (ν, τ), and Doppler frequency (ν, f). These are related by 1D or 2D Fourier transforms (Boashash, 2003).

Ambiguity domains that describe a signal using the Doppler and lag variables are found to be useful in constructing the SAF for DOA estimation. The ambiguity function of a signal $x(t)$ is defined as

$$D_{xx}(\nu, \tau) = \int x\left(t + \frac{\tau}{2}\right) x^*\left(t - \frac{\tau}{2}\right) e^{j2\pi\nu t} dt \tag{6.45}$$

where ν and τ represent the Doppler frequency and the lag, respectively. The ambiguity function between two signals of $x_1(t)$ and $x_2(t)$ can be similarly defined.

Thus, for a vector signal $\mathbf{x}(t)$ received at an array as expressed in Equation (6.6), the SAF matrix (Amin et al., 2000) can be defined as

$$\mathbf{D}_{\mathbf{xx}}(\nu,\tau) = \int \mathbf{x}\left(t+\frac{\tau}{2}\right)\mathbf{x}^H\left(t-\frac{\tau}{2}\right)e^{j2\pi\nu t}dt \tag{6.46}$$

The ambiguity function has several inherent advantages. First, its auto-terms of a signal are positioned near or at the origin ($\nu=0$, $\tau=0$). As a result, selection of auto-term points becomes relatively easy. For example, while in the t-f domain an LFM signal is characterized by chirp rate and initial frequency; LFM signals with the same chirp rate but different initial frequencies share the same trajectory in the ambiguity plane. Second, the auto-terms of all narrowband signals, regardless of their frequencies and phases, fall on the time lag axis ($\nu=0$), while those of the wideband signals fall on a different (ν,τ) region or spread over the entire ambiguity domain.

6.6.1 Ambiguity-Domain MUSIC

We introduce now, as an application example of SAF, ambiguity-domain MUSIC (AD-MUSIC). As in t-f MUSIC, matrix $\mathbf{E}=[\mathbf{S},\mathbf{G}]$, which spans the signal and noise subspaces of the SAF matrix, $\mathbf{D}_{\mathbf{xx}}(\nu,\tau)$, can be obtained by performing JBD on $\mathbf{D}_{\mathbf{xx}}(\nu,\tau)$ at different (ν,τ) points. Once the noise subspace $\mathbf{G}$ of $\mathbf{D}_{\mathbf{xx}}(\nu,\tau)$, which is constructed from n_o signals, is estimated as $\hat{\mathbf{G}}$, the AD-MUSIC technique estimates the DOAs by finding the n_o largest peaks of the localization function

$$f_{\mathrm{AD}}(\theta) = \left[\mathbf{a}^H(\theta)\hat{\mathbf{G}}\hat{\mathbf{G}}^H\mathbf{a}(\theta)\right]^{-1} \tag{6.47}$$

Example 6.8

Consider the scenario of a ULA consisting of four elements separated by a half-wavelength, where one LFM signal and two sinusoidal signals are received. All three signals have the same input SNR of 20 dB. The DOA of the LFM signal is 15°, and the DOAs of the two sinusoidal signals are 0° and 10°, respectively. The data record has 128 samples. As shown in Figure 6.10, while the ambiguity function of the LFM signal sweeps the ambiguity domain with contributions at the origin, the exact auto-term ambiguity function of the narrowband signals is zero for nonzero frequency lags and may have nonzero values only along the vertical axis $\nu=0$. In the figure, the two vertical lines off the lag axis represent the cross-terms between the sinusoidal components.

Suppose that we are interested only in the DOAs of the sinusoidal signals. For this purpose, we select 24 points on the lag axis excluding the origin, and in so doing, emphasize the narrowband components. Figure 6.11 shows the two estimated spatial spectra, respectively corresponding to conventional MUSIC

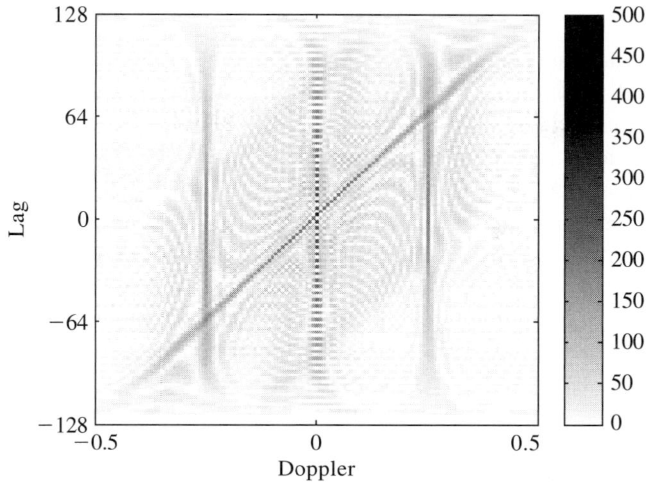

FIGURE 6.10 Ambiguity function of an LFM signal and two sinusoidal signals.

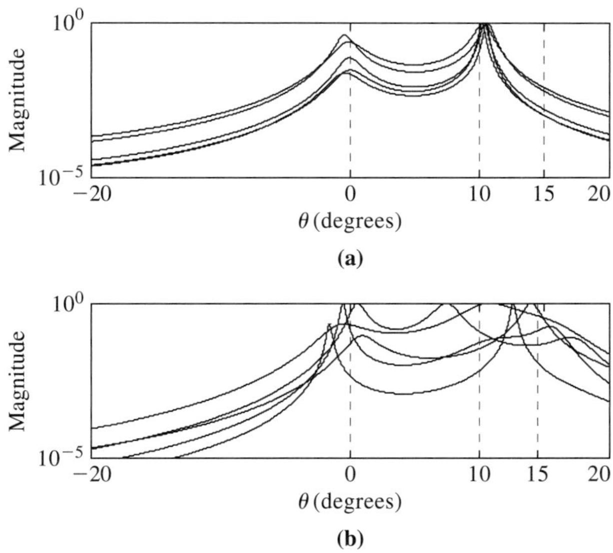

FIGURE 6.11 Estimated spatial spectra using (a) AD-MUSIC and (b) conventional MUSIC.

and AD-MUSIC, of five independent trials. There are two dominant eigenvalues for the case of AD-MUSIC since the LFM signal has been dropped through the selection of ambiguity domain points. It is clear that AD-MUSIC resolves the two sinusoidal signals while conventional MUSIC cannot separate three signals.

6.7 WIDEBAND DOA ESTIMATION

The discussion in this chapter so far has focused on the DOA estimation of narrowband nonstationary signals. However, in many applications the signals are wideband. In this case, the DOA estimator should consider the fact that the steering vector is now frequency-dependent.

To estimate the DOA for a general class of wideband signals, conventional techniques use Fourier transform to decompose these signals into a set of narrowband components, which can then be processed either incoherently or coherently. The incoherent approaches are relatively simple and estimate the DOA from the average of the spatial spectra corresponding to different frequency bins. However, coherent approaches are often preferred because of their superior performance. A popularly used technique, coherent signal-subspace (CSS) processing, was proposed by Wang and Kaveh (1985) and was further developed in several papers (see Friedlander and Weiss, 1993, and references therein). The fundamental concept of CSS is to use a set of focusing matrices that map the steering vector at different frequencies into a steering vector at a reference frequency before they can be coherently combined.

Several *t-f* and ambiguity domain DOA estimation methods have been developed for the estimation of wideband LFM signals (Gershman and Amin, 2000; Ma and Goh, 2006; Wang and Xia, 2000). Wang and Xia (2000) employed *t-f* analysis to estimate chirp rates and iteratively compensated the signal chirp structure. A good estimate of the signal DOAs was required to initialize this iterative processing. By assuming that the wideband signals are separable in the *t-f* domain and that their IFs do not change rapidly, Gershman and Amin (2000) used a sufficiently short sliding window to construct the STFD matrices so as to preserve the narrowband structure of the array manifold. The focusing matrices were then applied to the STFD matrices at selected *t-f* points corresponding to the source *t-f* signatures. Ma and Goh (2006) considered the ambiguity domain for the DOA estimation of wideband LFM signals with chirp rates that were assumed to be known. Multiple chirps with identical chirp rates were allowed in this technique. Incoherent and coherent processing techniques were also compared (Ma and Goh, 2006).

In essence, STFD wideband DOA estimation methods can incorporate CSS processing for nonstationary signals, discussed previously. The LFM signal properties and the *t-f* signal representations offered may be used in several ways:

- Decomposition of LFM signals into a spectrum of frequency bins is inherently performed in *t-f* analysis.
- For LFM signals that have distinct characteristics in the *t-f* domain, DOA estimation can be performed on individual sources.
- LFM signals are instantaneous narrowband, allowing the focusing matrices to be applied to *t-f* points.

6.8 TIME-FREQUENCY POINTS

Auto-term and cross-term time-frequency points represent different aspects of the contribution of signals in the bilinear transform and they behave differently in DOA estimation. This section considers the behavior of cross-terms in DOA estimation and the proper selection of auto-term and cross-term points.

6.8.1 Auto-Term and Cross-Term Points

As discussed in Section 6.2, cross-terms are a by-product of the TFD's bilinearity. Although different kernels have different ways of mitigating cross-terms (Cohen's, 1995; Jeong and Williams, 1992), their complete removal is nevertheless very difficult to achieve.

There are two types of cross-terms in the underlying DOA estimation problems. The first type is due to interactions between the components of the same source signal. These cross-terms always reside, along with the auto-terms, on the main diagonal of the source TFD matrix. They share the same steering vector as that of the auto-terms and thus can be similarly treated. The other type of cross-term is generated from the interactions between two signal components belonging to two different sources. These are associated with cross-TFD of the source signals and, at any given time-frequency point, they constitute the off-diagonal entries of the source TFD matrices. Here we consider the second type.

To understand the role of cross-terms in DOA estimation, it is important to compare them to the cross-correlation between signals in conventional array processing, with properties that are well studied. When cross-terms are present at the selected t-f point, they appear as off-diagonal elements in the source TFD matrix. On the other hand, when signals are correlated, the off-diagonal elements of the covariance matrix of the source signals represent the cross-correlation between two source signals. DOA estimation problems can usually be solved when the signals are partially correlated, provided that the full-rank property of the source signals' covariance matrix is maintained.

The cross-correlation terms and the cross-term TFDs have an analogous form and similar function. That is, cross-term TFDs can be exploited in DOA estimation as long as the full-rank subspace of the STFD matrix is achievable (Amin and Zhang, 2000). It is noted that the covariance matrix is obtained as a result of statistical or ensemble averages, whereas the STFD matrix is defined at a (t,f) point and its value usually varies with respect to time t and frequency f. When multiple (t,f) points are incorporated, the effect of a cross-term may be reduced, since the cross-term usually oscillates with respect to time.

6.8.2 Selection of Time-Frequency Points

The selection of t-f points is relatively straightforward when nonstationary signals are present with clear IF signature and moderate-input SNR. Most TFDs result in high values in the vicinity of the same true IF signatures as their auto-terms. A problem may arise, however, when the signals have a low SNR and their IF signature in the *t-f* domain is not clear. In addition, we may want to separate auto-term from cross-term points. For example, including cross-terms between selected and other sources may compromise the source discrimination capability in DOA estimation.

A relatively simple way of enhancing auto-term TFDs and suppressing cross-term TFDs is through computing the TFDs corresponding to the signals obtained at each individual receive sensor and then averaging the TFDs across the array sensors (Mu et al., 2003). Averaging introduces a weighing function in the *t-f* domain that decreases the noise levels, reduces the interactions of the source signals, and mitigates the cross-terms. This is achieved independently of the temporal characteristics of the source signals and without causing any smearing of the signal terms.

The TFD of the noise-free signal received at the ith array sensor, $x_i(t) = \sum_{k=1}^{n} a_i(\theta_k)s_k(t)$, where $a_i(\theta_k)$ is the ith element of steering vector $\mathbf{a}(\theta_k)$, is expressed as

$$D_{x_i x_i}(t,f) = \sum_{k=1}^{n}\sum_{l=1}^{n} a_i(\theta_k)\, a_i^*(\theta_l) D_{s_k s_l}(t,f) \tag{6.48}$$

The averaging of $D_{x_i x_i}(t,f)$ for $i = 1,\ldots,m$ yields

$$\overline{D}_{\mathbf{xx}}(t,f) = \frac{1}{m}\sum_{i=1}^{m} D_{x_i x_i}(t,f) = \sum_{k=1}^{n}\sum_{l=1}^{n} \beta_{k,l} D_{s_k s_l}(t,f) \tag{6.49}$$

where $\beta_{k,l} = \mathbf{a}_k^H \mathbf{a}_l / m$ is the spatial correlation coefficient between the kth and the lth sources. The average of the TFDs over different array sensors is the trace of the corresponding STFD matrix, $\mathbf{D_{xx}}(t,f)$, up to the normalization factor, m. With the introduction of the spatial signature between two source signals, it becomes clear that $\beta_{k,l}$ is equal to unity for the same source signal (i.e., $k = l$, corresponding to the auto-term *t-f* points) and that its magnitude is smaller than unity for two different source signals (i.e., $k \neq l$, corresponding to the cross-term *t-f* points).

In the presence of additive noise, the received signal becomes

$$x_i(t) = \sum_{k=1}^{n} a_i(\theta_k)s_k(t) + n_i(t) \tag{6.50}$$

and $n_i(t)$ is the additive Gaussian noise. The TFD of $x_i(t)$ is expressed as

$$\begin{aligned} D_{x_i x_i}(t,f) = &\sum_{k=1}^{n}\sum_{l=1}^{n} a_i(\theta_k)\, a_i^*(\theta_l)\, D_{s_k s_l}(t,f) \\ &+ 2\mathrm{Re}\left[\sum_{k=1}^{n} a_i(\theta_k)\, D_{s_k n_i}(t,f)\right] + D_{n_i n_i}(t,f) \end{aligned} \tag{6.51}$$

When the noise elements are assumed independent of each other and of the signals, averaging over the array elements will reduce the variance of both the second and third terms at the right side of Equation (6.51) by a factor of m.

To make the selection of auto-term points meaningful, it is often desirable that the selected t-f points have enough significance in the t-f domain. This can be accomplished by judging whether the TFD strength at a t-f point exceeds a certain preset level. The TFD strength indicator may be chosen as the sum of the absolute value of the auto-sensor TFDs, $\sum_{i=1}^{m} |D_{x_i x_i}(t,f)|$, or, alternatively, as the norm of the STFD matrix, norm$[\mathbf{D_{xx}}(t,f)]$. Then the auto-term t-f points can be selected by comparing the ratio between the average of the auto-sensor TFDs (or, equivalently, the trace of the STFD matrix) and the TFD strength indicator (Belouchrani et al., 2001; Linh-Trung et al., 2005; Zhang and Amin, 2006). On the other hand, a t-f point can be considered as a cross-term point when the above ratio is below a certain threshold.

Although the array averaging or trace of the STFD matrix is simple, it may identify some false auto-term t-f points when the spatial correlation between sources is high (e.g., the sources have close DOAs). In this case, auto-term selection can be improved by prewhitening the received data to orthogonalize the source spatial signatures before TFD averaging or matrix tracing is applied (Belouchrani et al., 2001; Linh-Trung et al., 2005; Zhang and Amin, 2006). When the condition $m > n$ is satisfied, a convenient way to do this is to compute the prewhitening matrix (Belouchrani and Amin, 1998) as

$$\mathbf{W} = \left(\mathbf{\Lambda}_n - \sigma^2 \mathbf{I}_n\right)^{-1/2} \mathbf{S}^H \tag{6.52}$$

where $\mathbf{S}$ is the signal subspace and $\mathbf{\Lambda}_n$ is the diagonal matrix containing the eigenvalues corresponding to $\mathbf{S}$. This yields in the $n \times 1$ prewhitened data vector

$$\mathbf{z}(t) = \mathbf{W}\mathbf{x}(t) \tag{6.53}$$

which, in the absence of noise, has the $n \times n$ identity matrix as its covariance matrix. The prewhitened matrix can be derived from an STFD matrix to take advantage of SNR enhancement and source discriminations (Zhang and Amin, 2006).

Example 6.9

Consider an eight-sensor ULA with inter-element spacing of a half-wavelength. Three LFM signals, $s_1(t)$, $s_2(t)$, and $s_3(t)$, arrive at the array with angle of arrival (AOAs) of $-25°$, $0°$, and $25°$, with the respective start and end frequencies given by $(0.45, 0.25)$, $(0.33, 0.13)$, and $(0.25, 0.05)$. The length of the signal sequence is set to 256. In the t-f plane, the source signals have parallel signatures. The cross-term of $s_1(t)$ and $s_2(t)$ also forms a structure with a frequency that changes linearly from 0.35 to 0.15 and therefore lies close to the t-f signature of $s_3(t)$.

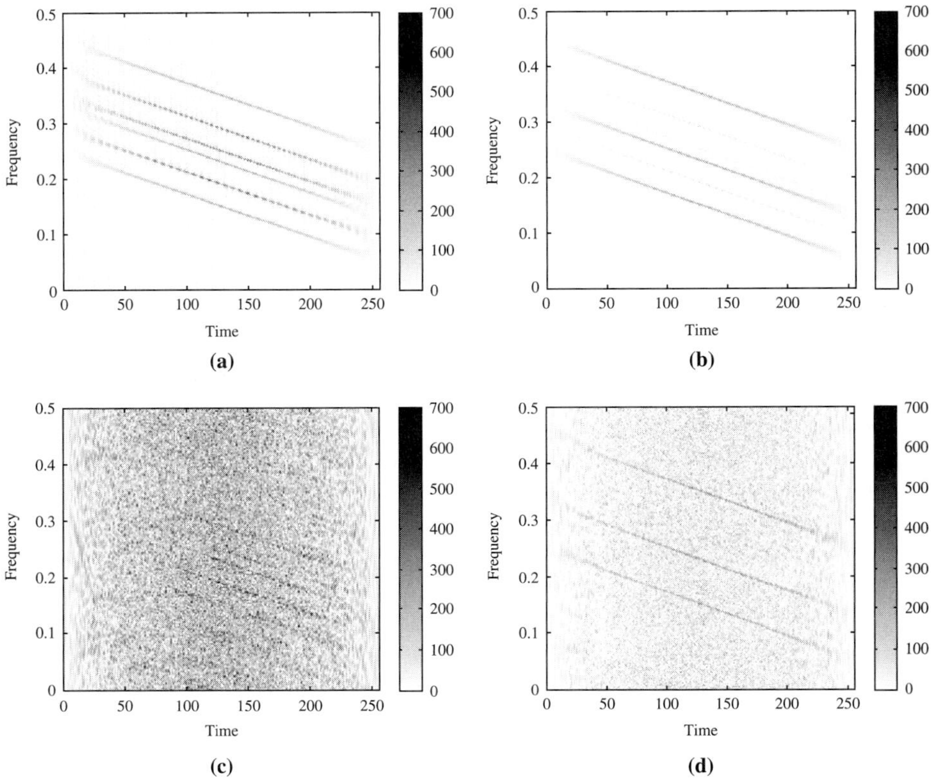

FIGURE 6.12 WVD with and without array averaging: (a) WVD in the noise-free case at the first element; (b) array-averaged WVD in the noise-free case; (c) WVD of the corrupted signal at the first element; and (d) array-averaged WVD of the corrupted signals.

Figure 6.12(a) depicts the WVD of the signals at the first sensor in a noise-free environment. With averaging of the WVD over the eight sensors, substantial cross-term reduction is achieved because the spatial signatures of the sources are weakly correlated. Figure 6.12(b) shows the corresponding array-averaged WVD. As a result, averaging WVDs across the array significantly reduces the cross-terms, whereas the three signals' auto-terms remain intact.

Next, we add spatially and temporally white Gaussian noise to the data at each array sensor so that the input SNR is -10 dB. Figures 6.12(c) and 6.12(d) depict both the reference sensor WVD and the array-averaged WVD. It is evident that the noise obscures both the signal auto-terms and the cross-terms of the WVD at a single sensor, so it is difficult to find appropriate t-f regions that contain high energy of the signal arrivals. Upon averaging, both noise and cross-terms are sufficiently reduced to clearly manifest the individual source t-f signature, and the signals can be individually selected for DOA estimation.

6.9 CONCLUSION

In this chapter, we reviewed the spatial time-frequency distribution concept and various STFD-based DOA estimation techniques. STFD provides a very flexible platform on which to perform effective DOA estimation for nonstationary signals. This is because the similar structures of STFD and data covariance matrices allow the use of virtually all subspace-based DOA estimation techniques. Compared to conventional DOA estimation techniques, those based on STFD provide significant performance improvement because of they are capable of enhancing signal-to-noise ratio and source discrimination in the time-frequency domain.

Extensions of the STFD concept to spatial polarimetric time-frequency distribution (SPTFD) and wideband signal applications were also addressed. We included various examples of STFD- and SPTFD-based DOA estimation techniques built on the theory underlying the MUSIC, maximum likelihood, and ESPRIT methods.

REFERENCES

Amin, M.G., 1997. Interference mitigation in spread spectrum communication systems using time-frequency distributions. IEEE Trans. Signal Process. 45 (1), 90–101.

Amin, M.G., Belouchrani, A., Zhang, Y., 2000. The spatial ambiguity function and its applications. IEEE Signal Process. Lett. 7 (6), 138–140.

Amin, M.G., Zhang, Y., 2000. Direction finding based on spatial time-frequency distribution matrices. Digital Signal Process. 10 (4), 325–339.

Belouchrani, A., Abed-Meraim, K., Amin, M.G., Zoubir, A., 2001. Joint antidiagonalization for blind source separation. In: Proc. IEEE International Conference on Acoustics, Speech, and Signal Processing, pp. 2196–2199.

Belouchrani, A., Amin, M.G., 1998. Blind source separation based on time-frequency signal representations. IEEE Trans. Signal Process. 46 (11), 2888–2897.

Belouchrani, A., Amin, M.G., 1999. Time-frequency MUSIC. IEEE Signal Process. Lett. 6 (5), 109–110.

Belouchrani, A., Amin, M.G., Abed-Meraim, K., 1997. Direction finding in correlated noise fields based on joint block-diagonalization of spatio-temporal correlation matrices. IEEE Signal Process. Lett. 4 (9), 266–268.

Boashash, B., 2003. Theory of quadratic TFDs. In: Boashash, B. (Ed.), Time-Frequency Signal Analysis and Processing, pp. 59–81 Elsevier.

Boashash, B., Putland, G.R., 2003. Discrete time-frequency distributions. In: Boashash, B. (Ed.), Time-Frequency Signal Analysis and Processing, pp. 232–241. Elsevier.

Choi, H.I., Williams, W.J., 1989. Improved time-frequency representation of multicomponent signals using exponential kernels. IEEE Trans. Acoust. Speech Signal Process. ASSP-37 (6), 862–871.

Cohen, L., 1989. Time-frequency distributions—a review. Proc. IEEE 77 (7), 941–981.

Cohen, L., 1995. Time-Frequency Analysis. Prentice-Hall.

Ferrara, E.R., Parks, T.M., 1983. Direction finding with an array of antennas having diverse polarizations. IEEE Trans. Antennas Propag. 31 (3), 231–236.

Friedlander, B., Weiss, J., 1993. Direction finding for wide-band signals using an interpolated array. IEEE Trans. Signal Process. 41 (4), 1618–1634.

Gershman, A., Amin, M.G., 2000. Wideband direction-of-arrival estimation of multiple chirp signals using spatial time-frequency distributions. IEEE Signal Process. Lett. 7 (6), 152–155.

Gong, X., Liu, Z., Xu, Y., Ahmad, M.I., 2009. Direction-of-arrival estimation via twofold mode-projection. Signal Process. 89 (5), 831–842.

Hassanien, A., Gershman, A.B., Amin, M.G., 2002. Time-frequency ESPRIT for direction-of-arrival estimation of chirp signals. IEEE Sensor Array and Multichannel Signal Processing Workshop, pp. 337–341.

Jeong, J., Williams, W.J., 1992. Kernel design for reduced interference distributions. IEEE Trans. Signal Process. 42, 402–412.

Li, J., Compton, R.T., 1991. Angle estimation using a polarization sensitive array. IEEE Trans. Antennas Propag. 3 (10), 1539–1543.

Linh-Trung, N., Belouchrani, A., Abed-Meraim, K., Boashash, B., 2005. Separating more sources than sensors using time-frequency distributions. EURASIP J. Appl. Signal Process. 2005 (17), 2828–2847.

Ma, N., Goh, J.T., 2006. Ambiguity-function-based techniques to estimate DOA of broadband chirp signals. IEEE Trans. Signal Process. 54 (5), 1826–1839.

Mu, W., Amin, M.G., 2000. SNR analysis of time-frequency distributions. In: Proc. IEEE International Conference on Acoustics, Speech, and Signal Processing, pp. II645–II648.

Mu, W., Amin, M.G., Zhang, Y., 2003. Bilinear signal synthesis in array processing. IEEE Trans. Signal Process. 51 (1), 90–100.

Nehorai, A., Paldi, E., 1994. Vector-sensor array processing for electromagnetic source localization. IEEE Trans. Signal Process. 42 (2), 376–398.

Obeidat, B.A., Zhang, Y., Amin, M.G., 2003. Polarimetric time-frequency ESPRIT. In: Proc. Asilomar Conference on Signals, Systems, and Computers.

Obeidat, B.A., Zhang, Y., Amin, M.G., 2004. DOA and polarization estimation for wideband sources. In: Proc. Asilomar Conference on Signals, Systems, and Computers.

Qian, S., Chen, D., 1996. Joint Time-Frequency Analysis—Methods and Applications. Prentice-Hall.

Rickard, S., Dietrich, F., 2000. DOA estimation of many W-disjoint orthogonal sources from two mixtures using DUET. IEEE Workshop on Statistical Signal and Array Processing, pp. 311–314.

Roy, R., Kailath, T., 1989. ESPRIT-estimation of signal parameters via rotational invariance techniques. IEEE Trans. Acoust. Speech Signal Process. ASSP-37 (7), 984–995.

Schmidt, R.O., 1986. Multiple emitter location and signal parameter estimation. IEEE Trans. Antennas Propag. AP-34 (3), 276–280.

Sekihara, K., Nagarajan, S., Poeppel, D., Miyashita, Y., 1999. Time-frequency MEG-MUSIC algorithm. IEEE Trans. Med. Imaging 18 (1), 92–97.

Stoica, P., Nehorai, A., 1989. MUSIC, maximum likelihood and Cramer-Rao bound. IEEE Trans. Acoust. Speech Signal Process. ASSP-37 (5), 720–741.

Wang, G., Xia, X.-G., 2000. Iterative algorithm for direction of arrival estimation with wideband chirp signals. IEE Proc. Radar Sonar Navig. 147 (5), 233–238.

Wang, H., Kaveh, M., 1985. Coherent signal-subspace processing for the detection and estimation of angles of arrival of multiple wideband sources. IEEE Trans. Acoust. Speech Signal Process. ASSP-33, 823–831.

Wang, Q., Wu, C., 2002. A high reliability DOA estimation method—TF-ESPRIT method. In: Proc. International Conference on Signal Processing, pp. 374–377.

Xia, X.-G., Chen, V., 1999. A quantitative SNR analysis for the pseudo Wigner-Ville distribution. IEEE Trans. Signal Process. 47 (10), 2891–2894.

Yamada, H., Onishi, K., Yamaguchi, Y., 1997. Polarimetric superresolution technique for 2-D radar target imaging. Proc. SPIE 3120, 317–326.

Zhang, Y., Amin, M.G., 2006. Blind separation of nonstationary sources based on spatial time-frequency distributions. EURASIP J. Appl. Signal Process. 64785, 1–13.

Zhang, Y., Amin, M.G., Obeidat, B.A., 2004. Polarimetric array processing for nonstationary signals. In: Chandran, S. (Ed.), Adaptive Antenna Arrays: Trends and Applications, pp. 205–218. Springer-Verlag.

Zhang, Y., Mu, W., Amin, M.G., 2000. Time-frequency maximum likelihood methods for direction finding. J. Franklin Inst. 337 (4), 483–497.

Zhang, Y., Mu, W., Amin, M.G., 2001. Subspace analysis of spatial time-frequency distribution matrices. IEEE Trans. Signal Process. 49 (4), 747–759.

Zhang, Y., Amin, M.G., Obeidat, B.A., 2003a. Direction finding using spatial polarimetric time-frequency distributions. In: Proc. International Symposium on Signal Processing and Its Applications.

Zhang, Y., Obeidat, B.A., Amin, M.G., 2003b. Polarimetric time-frequency MUSIC in coherent signal environment. In: Proc. IEEE Workshop on Statistical Signal Processing.

Zhang, Y., Obeidat, B.A., Amin, M.G., 2006a. Spatial polarimetric time-frequency distributions for direction-of-arrival estimations. IEEE Trans. Signal Process. 54 (4), 1327–1340.

Zhang, Y., Obeidat, B., Amin, M.G., 2006b. Polarimetric time-frequency MUSIC for direction finding of moving sources with time-varying polarizations. In: Chandran, S. (Ed.), Advances in Direction-of-Arrival Estimation, pp. 103–121. Artech House.

Zhao, Y., Atlas, L.E., Marks, R.J., 1990. The use of cone-shaped kernels for generalized time-frequency representations of non-stationary signals. IEEE Trans. Acoust. Speech Signal Process. 38 (6), 1084–1091.

Ziskind, I., Wax, M., 1988. Maximum likelihood localization of multiple sources by alternating projection. IEEE Trans. Acoust. Speech Signal Process. ASSP-36 (10), 1553–1560.

Chapter 7

DOA Estimation in the Small-Sample Threshold Region

Yuri I. Abramovich, Ben A. Johnson, Xavier Mestre

7.1 INTRODUCTION

Methods and fundamental limitations of detection/estimation techniques for multiple closely spaced emitters in noise under small-sample support conditions have always been treated as an important but difficult analytical problem. Yet most of the theoretical results from estimation of the number of emitters (detection) and their direction of arrival (DOA) in M-element antenna arrays still rely on large-sample assumptions. In particular, well-known attempts to analyze resolution and detection limits (Kaveh and Barabell, 1986; Lee and Li, 1993; Zhang, 1995) have been derived under the traditional asymptotic assumption regarding the number of independent identically distributed (i.i.d.) training samples T:

$$M = \text{const},\ T \to \infty\ (M/T \to 0) \tag{7.1}$$

In some cases, the signal-to-noise ratio (SNR) has been alternatively assumed to be asymptotically large, though the two asymptotic assumptions

$$M = \text{const},\ \text{SNR} = \text{const},\ T \to \infty \tag{7.2}$$

Note: This chapter is partially based on "MUSIC, GMUSIC, Maximum Likelihood Performance Breakdown" by B. A. Johnson, Y. I. Abramovich, and X. Mestre, which appeared in *IEEE Transactions on Signal Processing* in August, © 2008 IEEE (Johnson et al., 2008), as well as "GLRT-Based Detection-Estimation for Under-Sampled Training Conditions" by Y. I. Abramovich and B. A. Johnson, which appeared in *IEEE Transactions on Signal Processing* in August, © 2008 IEEE (Abramovich and Johnson, 2008). The work was conducted under RLM/DSTO Collaboration 290905.

Copyright © 2009 by Academic Press, Inc. All rights of reproduction in any form reserved.

and

$$M = \text{const},\ T = \text{const},\ \text{SNR} \to \infty \tag{7.3}$$

are not always interchangeable. Athleys's observation (Athley, 2002), that for a small T and SNR $\to \infty$ the DOA estimates are not efficient (they do not reach the Cramer-Rao bound), was analytically reinforced recently by Renaux et al. (2004) for the traditional Gaussian model with m independent sources ($m < M$).

Therefore, if Equations (7.2) and (7.3) are not interchangeable even for sufficiently large $T \to \infty$, SNR$\to \infty$, where the performance of DOA detection/estimation techniques is high, one can expect even more profound distinctions between large-sample training conditions ($T \gg M$) and low-sample training conditions ($T \leq M$), where detection/estimation procedures reach their limits. In fact, in the low-sample support case, we are in the circumstance where not only do the actual detection/estimation performance of various techniques approach fundamental limits, but the traditional asymptotic ($T \to \infty$, SNR$\to \infty$) analysis techniques lose their ability to adequately describe this performance. This severely complicates analysis of "threshold" phenomena as the detection/estimation techniques break down and produce large errors that depart from the Cramer-Rao bound (CRB).

The question with respect to asymptotic assumptions (Equations 7.2 and 7.3)—namely, how large T and/or SNR should be so that the asymptotic analysis accurately describes a considered scenario—is frequently asked. An interesting answer was provided by Asym P. Tomaniac, Ph.D., in the letter "Asymptomania?" submitted by P. Stoica to IEEE *Signal Processing Magazine* (Tomaniac, 1998):

> To summarize, a possible (brief, admittedly vague, yet deterrent) answer to the question asked … is: If "my" asymptotic results do not apply to "your" scenario, then you have a "large-error case" and, hence, you are in trouble anyway. It is true that such an answer may generate further questions:
>
> - How do I know that I am in a "large-error" case (which could also be called a "threshold regime")?
> - Is it really true that I cannot do anything about it?

This threshold effect is observed in any DOA estimation algorithm, including maximum likelihood (ML) estimation, with algorithm- and scenario-specific onset points. In the case of multiple Gaussian signals impinging on an M-element antenna array, maximum likelihood has been broadly treated as providing "benchmark" estimation accuracy if effectively implemented. But in most practical multiple-source cases, we are not able to perform a globally optimal ML search for the DOA estimates, and therefore we want to consider Stoica's concerns for other detection/estimation techniques relative to maximum likelihood performance. To conduct such an analysis in a likelihood framework, we restructure the previous "admittedly vague" problem statement into the following three specific practical questions:

1. How close to the "true" maximum likelihood is any model $R_{mod}(\hat{m})$?[1]
2. If $R_{mod}(\hat{m})$ is away from the true maximum likelihood, can I rectify the DOA estimates to come "sufficiently close" to the true ML condition?
3. If the (rectified) $R_{mod}(\hat{m})$ is sufficiently close to the true ML condition, how likely is it to still contain a severely erroneous condition (an outlier)?

Several separate important issues are raised by our questions. The first is whether there is a gap between the threshold conditions for a particular detection/estimation technique and more fundamental threshold conditions for the maximum likelihood principle itself. If such a gap exists, could the method-specific performance breakdown (i.e., when $R_{mod}(\hat{m})$ contains a severely erroneous DOA estimate) be reliably predicted and cured (Abramovich and Spencer, 2004; Hawkes et al., 2001)? What type of theoretic analysis should support such prediction and cure attempts?

The other important issue is whether anything can be done when even the truly maximum likelihood estimator produces a severe outlier. Given a particular (global) ML estimate $R_{mod}(\hat{m})$, can we predict that we are in the maximum likelihood performance breakdown region? Can we at least identify the DOA estimates that cannot be trusted?

In this chapter, we address the first issue, since our goal is to produce detection/estimation solutions that for a small-sample support and SNR are statistically as likely as the maximum likelihood solutions, without having to invoke an impractical global ML search. The second issue is not addressed here, but it must be remembered that breakdown of maximum likelihood itself cannot be addressed within the likelihood paradigm we employ. Only a change in the paradigm, and specifically an introduction of additional a priori information regarding the sources can change the maximum likelihood estimation (MLE) performance itself, often quite dramatically. For example, assumptions that the sources are all of the same power change MLE resolutions limits significantly (Abramovich et al., 2008).

7.2 DOA ESTIMATION IN THE THRESHOLD REGION

Detection/estimation techniques the performance breakdown of which we intend to address are well-known "ML-proxy" routines such as MUSIC, which asymptotically (under conditions given either in Equation 7.2 or 7.3) approach MLE performance. This asymptotic equivalence bounded by the CRB, established by Stoica and Nehorai (1989), obviously does not mean their equivalence everywhere, and especially in the threshold region. In fact, large-sample ML surrogates such as MUSIC may demonstrate a completely different threshold behavior. While it is recognized that the performance of these surrogates is "in general

---

1. $R_{mod}(\hat{m})$ can be formed by using any particular detection/estimation technique to estimate the number of sources $\hat{m}$, DOAs $\hat{\theta}_1, \ldots, \hat{\theta}_m$, and source powers $\hat{P}_1, \ldots, \hat{P}_m$, and then using these parameters to form an $M \times M$ covariance matrix model of the input data.

inferior to the ML technique, especially in the regime of low signal-to-noise ratio or small number of samples" (Hurt, 1990), such surrogates are often the only practical alternatives to the multidimensional ML global search. Thus, this difference is often not fully explored.

For example, the authors in Lee and Li (1993) argued that for very closely spaced sources, the detection SNR threshold ϵ_{ITC} below which the number of sources is underestimated by information theoretic criteria (ITC) is significantly lower than the SNR threshold ϵ_{RES} at which the sources can be resolved and estimated. Yet in order to illustrate this fact, they compared the actual simulation results on ITC-based detection with the theoretic formula on the MUSIC resolution threshold derived by Kaveh and Barabell (1986), arguing that "this expression previously has been validated via extensive simulations." The fact that MLE and MUSIC DOA estimation have completely different threshold regions was not considered because of the lack of an accurate MLE proxy in the analyzed circumstance.

This situation recently changed conceptually, after a reliable and practical approach for MUSIC-specific performance breakdown prediction and cure was suggested in Abramovich and Johnson (2008), Abramovich and Spencer (2004), and Abramovich et al. (2007c). This approach is based on two quite straightforward observations. The first is that for the (unconditional, stochastic) Gaussian model, the likelihood function (i.e., the probability density of the model parameters given the observed data) may be normalized to lie between zero and unity in a way that the resultant likelihood ratio (LR) for the true (actual) covariance matrix as an argument (i.e., $LR(R_0)$) is "scenario-invariant." By this we mean that the LR is described by a probability density function (p.d.f.) that does not depend on this matrix R_0, but rather is fully described by a function of known parameters such as M and T. The second observation is that a true ML model $R_{ML}(\hat{m})$ for $\hat{m} \geq m$ should always (deterministically) exceed the $LR(R_0)$:

$$LR[R_{ML}(\hat{m} \geq m)] > LR(R_0) \tag{7.4}$$

These two observations allow for a statistical *nonasymptotic* answer to our question regarding the quality of the model $R_{mod}(\hat{m})$. Within the approach we have called "expected likelihood" (Abramovich et al., 2004), $R_{mod}(\hat{m})$ derived by any detection/estimation technique is treated as appropriate if

$$LR[R_{mod}(\hat{m})] > \gamma_{LR}(P_{FA}) \tag{7.5}$$

where

$$\gamma_{LR}(P_{FA}) = \arg_{\gamma}\left[\int_0^{\gamma} w(x, M, T)dx = P_{FA}\right] \tag{7.6}$$

and $w(x, M, T)$ is the previously discussed p.d.f. for $LR(R_0)$. By the thresholding in Equation (7.5), we check that with probability $1 - P_{FA} \gg 0$, the tested model $R_{mod}(\hat{m})$ is at least as likely as the true covariance matrix R_0. Note that the tested model may have a likelihood that is not truly global, since for a sufficiently low

P_{FA} (10^{-3}, say), we have to set the threshold γ_{LR} well below the median value of $LR(R_0)$ and therefore, in most cases, also below the global LR maximum that remains unknown. However, even with this statistical but nonasymptotic approach, we were able to demonstrate in Abramovich and Spencer (2004) and Abramovich et al. (2007c) that for $T \to \infty$, SNR $\to \infty$, covariance models generated from MUSIC DOA estimates and source-power estimation such as in Ottersten et al. (1993) reliably exceed the threshold in Equation (7.5), in full accordance with Stoica and Nehorai's predictions (1989).

More important, we found that in cases where T and/or SNR decrease to threshold values where MUSIC loses its ability to resolve closely spaced sources and starts to produce outlier DOA estimates, the likelihood ratio for $R_{mod}(\hat{m})$ constructed from these "broken" solutions is reliably below the statistical threshold γ_{LR} in Equation (7.5). The fact that MUSIC outliers were found to be much less likely than the true covariance matrix R_0 provided a straightforward capability to specify (predict) such an outlier, as requested by our first question, and provided an approach to replace ("cure") the outliers by a solution that meets the threshold in Equation (7.5), as considered in our second question. The *nonasymptotic* and *nonclairvoyant* nature of the p.d.f. for $LR(R_0)$ played a crucial role in the success of this approach.

However, the development of this performance breakdown "predict and cure" technique (Abramovich and Spencer, 2002; Abramovich et al., 2004) was conducted for scenarios with the amount sample data T still exceeding the antenna dimension M ($T > M$). This assumption not only enabled likelihood ratio formation from quite conventional normalization of likelihood functions, but also retained the possibility for conventional asymptotic ($T \to \infty$) statistical analysis to be considered for threshold behavior description. In contrast, in this chapter we focus on applications where the sample support T, while significantly exceeding the number of sources m, is similar to or even smaller than the number of antenna elements M. That is, when

$$m < T < M \tag{7.7}$$

The number of practical applications where one has to operate in such an "undersampled" training regime is growing. Antenna systems in radars, sonar, and communications systems are either growing in array size or are increasingly relying on receiver-per-element rather than subarray architectures (Pearce, 1998). At the same time, the ability to collect the amount of sample data that exceeds this array dimension is in most cases limited by the heterogeneity of the environment and is therefore not readily expandable.

Operations in this undersampled regime require a significant shift in the statistical framework. First, it is necessary to introduce new undersampled likelihood ratios with the same invariance of p.d.f. for the true covariance matrix R_0 (see Section 7.3). We then need to demonstrate that for this undersampled training regime, "broken" MUSIC solutions may be once again reliably identified because of their significantly smaller likelihood compared with the unknown actual scenario (see Section 7.4).

Even more important than the derivation of necessary tools such as undersampled likelihood ratios is the introduction of an appropriate theoretical framework for description of MUSIC's threshold behavior that does not rely on standard asymptotic assumptions. Most of the analytical derivations for MUSIC SNR resolution thresholds (Kaveh and Barabell, 1986; Lee and Wengrovitz, 1990; Xu and Kaveh, 1994; Zhang, 1995) have been derived for conventional ($T \to \infty$) asymptotic assumptions, which for $T < M$ is obviously not valid.

Moreover, even the physical phenomena previously associated with the MUSIC breakdown phenomena have to be reconsidered. For a long time, MUSIC breakdown has been associated with "subspace swap" (Thomas et al., 1995; Tufts et al., 1991), where one or more sample eigenvector–eigenvalue pairs associated with the signal subspace "actually estimate noise eigenelements instead of signal eigenelements" (Hawkes et al., 2001). This transition behavior of the eigendecomposition of a sample covariance matrix was recently described by a statistical technique called Random Matrix theory (RMT) (Beenakker, 1997; de Monvel et al., 1995), or general statistical analysis (GSA or G-analysis) (Girko, 1987, 1998).

Unlike the traditional asymptotic ($M = \text{const}, T \to \infty$) assumption, RMT/GSA considers quite different (Kolmogorov, 1956) asymptotics, when M and T concurrently go to infinity but maintain a constant ratio ($M, T \to \infty$, $M/T \to \gamma$, $0 < \gamma < \infty$). Therefore, even as the sample support T and the antenna dimension M grow without bound, the ratio M/T tends to a fixed number that can be larger than one, meaning that the sample support as a percentage of array size remains constant, making this asymptotic framework more appropriate for signal processing applications in the small-sample support case (Mestre and Lagunas, 2008; Nadakuditi and Edelman, 2008).

To validate this new statistical tool that can embrace these undersampled training conditions, we need to conduct this alternative asymptotic analysis and compare it with Monte Carlo analysis for MUSIC and MLE breakdown conditions, allowing a statistical description of the threshold phenomena, as shown in Sections 7.5 and 7.6. This allows us to clarify the physical phenomena responsible for the threshold behavior of our detection/estimation techniques. We close the chapter with a summary of these results in Section 7.7.

7.3 EXPECTED LIKELIHOOD FORMULATIONS

The expected likelihood technique introduced in Abramovich et al. (2004) provides a number of key features desired for adaptive processing assessments: The algorithm is based on finite-sample assumptions, provides nonclairvoyant performance assessment, and operates with scenario distributions that are operationally relevant. To summarize this approach, let us first examine some fundamental results that rely on known finite-sample distributions, not for the parameters themselves but for an important function of the parameters—their likelihood ratio. Specifically, let us examine the Gaussian case, for which the covariance matrix is the complete model identifier.

A standard approach to determining how closely a particular covariance matrix model matches T samples of M-dimensional multivariate (zero-mean) complex Gaussian sample data (which is complex Wishart ($\mathcal{CW}$) distributed (Wishart, 1928) when $T > M$, denoted $X_T = [\mathbf{x}_1, \ldots, \mathbf{x}_T]$, $\mathbf{x}_t \in \mathcal{CW}(0, R_0)$ when the underlying process has covariance R_0), is to take the sample covariance matrix $\hat{R}_M$:

$$\hat{R}_M = \frac{1}{T}\sum_{j=1}^{T} \mathbf{x}_j \mathbf{x}_j^H \tag{7.8}$$

and form a likelihood function w.r.t. the covariance matrix model R_{mod}:

$$\mathcal{L}(X_T, R_{mod}) = \left[\frac{1}{\pi \det R_{mod}} \exp\{-\mathrm{Tr}[R_{mod}^{-1}\hat{R}_M]\}\right]^T \tag{7.9}$$

Normalization of this likelihood function

$$LR(R_{mod}) = \frac{\mathcal{L}(X_T, R_{mod})}{\max_{R_{mod}} \mathcal{L}(X_T, R_{mod})} \tag{7.10}$$

leads to a likelihood ratio (after the methodology in Muirhead, 1982)

$$LR(R_{mod}) = \left[\frac{\det R_{mod}^{-1}\hat{R}_M \exp M}{\exp\{\mathrm{Tr} R_{mod}^{-1}\hat{R}_M\}}\right]^T \leq 1 \tag{7.11}$$

since

$$\max_{R_{mod}} \mathcal{L}(X_T, R_{mod}) = \left[\frac{\exp(-M)}{\pi \det \hat{R}_M}\right]^T, \quad \text{for } R_{mod} = \hat{R}_M \tag{7.12}$$

Hypothesis testing in circumstances where training sample data and primary sample data have a scalar offset can also be supported. Such LR formulation is based on hypothesis testing:

$$H_0 : \mathcal{E}\{\hat{R}_M\} = cR_{mod} \quad c > 0 \tag{7.13}$$

versus

$$H_1 : \mathcal{E}\{\hat{R}_M\} \neq cR_{mod} \tag{7.14}$$

and the resultant scale-invariant "sphericity" LR (for $T \geq M$) (Anderson, 1958; Muirhead, 1982) is introduced as

$$LR_{sp}(R_{mod}) = \left[\frac{\det \hat{R}_M R_{mod}^{-1}}{\{\frac{1}{M}\mathrm{Tr}\, \hat{R}_M R_{mod}^{-1}\}^M}\right]^T \tag{7.15}$$

These likelihood ratio values are a metric on the quality of the assumed covariance matrix model relative to the observed sample covariance. The p.d.f. of the LR provides significant information about the estimation problem, and the maximal LR value is of course associated with the maximum likelihood solution, but the actual likelihood value itself is often not directly utilized. It is indirectly utilized in tests based on the ratio of different likelihood ratios, such as the generalized likelihood ratio test (GLRT), where the ratio of the maximal $LR(R_{mod})$ given a particular hypothesis ($\max_{H_1} LR(R_{mod}|H_1)$) to the maximal $LR(R_{mod})$ given the null hypothesis ($\max_{H_0} LR(R_{mod}|H_0)$) is thresholded to make a detection decision (van Trees, 1968). Expected likelihood makes much more direct use of the LR value itself. Specifically, expected likelihood compares the achieved likelihood ratio (using the form given in either Equation (7.11) or (7.15)) for any particular covariance estimate to statistical measures of the likelihood ratio associated with the (unknown) true model.

Of course, the likelihood ratio associated with the true model (denoted R_0) is in general unknown. Yet under some circumstances, we can gain a priori knowledge of the p.d.f. of the likelihood ratio of that true model, even if $LR(R_0)$ and R_0 itself remain unknown. While it has been long known (Wilks, 1938) that under mild regularity conditions on the sample data, the maximum likelihood estimator has a known asymptotic distribution, we are by definition not interested in the asymptotic case ($T = \infty$), as we consider the low training-sample volume circumstance. If instead we restrict the sample data to be complex Gaussian (denoted $\mathcal{CN}$) (i.e., $\mathbf{x}_t \sim \mathcal{CN}(0, R_0)$), then the distribution for $LR(R_0)$ is described analytically in Consul (1969) (where the real-valued case is described), without resorting to asymptotic considerations.

In fact, an exact distribution can be derived for any spherically invariant random process (SIRP) (Brehm, 1982), of which the Gaussian process is a (stationary and ergodic) subset. For the complex Gaussian case, an expression for the p.d.f. of $LR(R_0)$ has been derived (Abramovich et al., 2002a) and is described in terms of the Meijer's G-function, depending solely on the number of samples T and the data dimension M (i.e., the p.d.f. of $LR(R_0)$ is independent of any scenario parameters and is therefore scenario-invariant). Since evaluation of the G-function is difficult, the p.d.f. can be approximated (Marquesa and Coelho, 2008), or, since it does not depend on any scenario parameters, it can be generated numerically using Monte Carlo techniques. In addition, the hth moment is given by

$$\mathcal{E}[LR(R_0)]^h = \frac{T^{MT}\exp(Mh)}{(T+h)^{M(T+h)}} \prod_{j=1}^{M} \frac{\Gamma(T+1-j+h)}{\Gamma(T+1-j)} \tag{7.16}$$

where Γ represents the Gamma function. The p.d.f. can then be expressed in serial form by applying an inverse Mellin transform to this moment function (Nagarsenker and Pillai, 1973). Note that, as expected, the moments (and therefore the underlying p.d.f.) do not depend on R_0 and are fully specified by parameters M (the dimension of the data snapshots), T (the number of data snapshots), and h (the moment index).

With the p.d.f. of $LR(R_0)$ determined, any appropriate quantile can now be chosen for use in an expected likelihood estimation process. If the likelihood of any given set of estimates exceeds the chosen threshold (which statistically represents the likelihood associated with the true model), the covariance model is considered acceptable. Although there are some performance penalties asymptotically as $T \to \infty$ (Abramovich and Johnson, 2007) compared with the global ML estimate, in the low training-volume regime, which is our primary focus, the technique provides a robust means to produce a "quality assessment" of potential solutions.

This has a number of very important applications. For example, since in the low training-sample regime adaptive filters with no higher complexity than necessary should be used, a key step is to determine the appropriate free parameter space of the covariance model. Simply choosing a model that results in a maximal likelihood ratio is unsatisfactory, as it always drives the covariance matrix free parameter order to a level commensurate with the dimension of the covariance matrix (in the case of a structured model) (see Equation 7.4), or provides us with the sample covariance matrix itself (which has a value for $LR(\hat{R}_M)$ of unity) if our model is unstructured. To achieve a parsimonious model, the likelihood ratio must be driven to a high enough value but no higher. The likelihood ratio of the underlying true model is often much lower than the maximum likelihood ratio.

In Abramovich et al. (2007b), the absolute maximum $LR(R_0)$ values associated with a properly ordered model for a 12-element antenna with 24 samples (i.e., a circumstance where the sample covariance matrix is reasonably representative of the true covariance matrix R_0; Reed et al., 1974) were shown to average around 0.03 and never exceeded 0.1, compared to the LR value of unity associated with the unstructured MLE estimate ($\hat{R}_M$), which has an LR of unity. Therefore, setting the complexity of the model high enough to drive the likelihood ratio close to its ultimate maximum results in a much more intricate model than necessary. Instead, the expected likelihood approach can provide the means to estimate a sufficiently accurate, but still parsimonious, parametric model complexity.

Despite its demonstrated utility, as previously formulated, the expected likelihood technique cannot be applied to the undersampled regime (discussed in Section 7.2), where the number of i.i.d. training samples T is less than M. In that case, the sample covariance matrix is rank deficient ($\det \hat{R}_M = 0$) and the normalization used in Equation (7.11) is no longer available, even though adaptive processors can still be applied.

The lack of a classically normalized likelihood ratio in the undersampled regime means that the potential applications of expected likelihood in solution assessment or order estimation are not available or require modification in this circumstance.

For the undersampled case, a likelihood ratio $LR_u(R_{mod})$ is needed that satisfies the following conditions:

Normalization condition:

$$0 < LR_u(R_{mod}) \leq 1 \tag{7.17}$$

Transition behavior: $LR_u(R_{mod})$ should be an "analytic extension" of $LR(R_{mod})$ (7.11). That is,

$$LR_u(R_{mod}) = LR(R_{mod}) \quad \text{for } T \geq M \tag{7.18}$$

Invariance property:

$$p.d.f.[LR_u(R_0)] = f(M,T) \tag{7.19}$$

The covariance matrix $\hat{R}_M$ in Equation (7.8) is rank deficient when $T < M$ and therefore is described by the anti-Wishart distribution (Janik and Nowak, 2003). $\hat{R}_M$ cannot be used directly in the likelihood ratio (7.11) since its determinant is zero, so a substitute for it that will serve the same purpose (i.e., represent the information in the sample data), but have a nonzero determinant, needs to be found. In contrast to the oversampled case, there is no single "correct" likelihood ratio and several formulations can be considered, including

- Contraction/projection of the sample matrix to a smaller non-rank-deficient matrix
- Extension of the sample matrix in a manner that provides a nonzero determinant

The first approach selects a subset or transform of the sample matrix that is not rank deficient and uses that in the likelihood ratio itself. That option is explored in Section 7.3.1. The second approach adds to the sample matrix in some ordered way so that the rank is expanded to the point that the matrix is no longer rank deficient. This option is explored, based on a maximum entropy extension, in Section 7.3.2.

Other options certainly exist, including perturbations of existing hypothesis tests used in the statistical mathematics literature in cases where the sample covariance matrix is singular (John, 1973; Nagao, 1973). But the presented options provide all the features needed to use them as undersampled likelihood ratios in an expected likelihood framework, thereby opening up the application of expected likelihood to this important regime.

To support these derivations, let us first define the circumstance of interest. For the expected likelihood approach to function properly, the number of training samples T has to be sufficient to at least capture the salient features of the signal subspace, where the signal subspace is defined as follows: When the number m of the covariance matrix eigenvalues that exceed the minimal eigenvalue (equal to ambient white noise power) is smaller then the matrix dimension M ($m < M$), the following form for admissible covariance matrices can be introduced:

$$R_0 = \sigma_0^2 I_M + R_S; \quad R_S = \mathcal{U}_m \Lambda_0 \mathcal{U}_m^H; \quad \Lambda_0 = \Lambda_m - \sigma_0^2 I_m \tag{7.20}$$

where $\mathcal{U}_m$ is the $M \times m$ matrix of signal subspace eigenvectors and Λ_m is the m-variate diagonal matrix of (positive) eigenvalues.

We now proceed to examine ways to devise a likelihood ratio in the undersampled domain that satisfies our key requirements.

7.3.1 Subspace Projection Formulation of the Undersampled LR

As noted previously, in the undersampled regime where the number of i.i.d. training samples T is less than M, the sample covariance matrix $\hat{R}_M$ is rank deficient. This means that since the sample data spans only a limited linear subspace with dimension $T < M$, any inference on a covariance matrix model *may be provided only regarding the properties of this model within this subspace that is spanned by the sample data.*

The property of undersampled testing raises identifiability concerns, and requires that projection of any model of the covariance matrix (denoted R_{mod}) onto the subspace spanned by the sample data uniquely identify the entire model R_{mod}. Along with this, a finite subspace model as introduced in Equation (7.20) can be parameterized as a summation of m (not necessarily independent) planewave sources:

$$R_{mod}(m) = \sigma_0^2 I_M + S(\boldsymbol{\theta})_m B_m S^{\mathrm{H}}(\boldsymbol{\theta})_m; \quad B_m \in \mathcal{H}^{m \times m} \geq 0 \tag{7.21}$$

where

$$S(\boldsymbol{\theta})_m = [s(\theta_1), \ldots, s(\theta_m)] \in \mathcal{CN}^{M \times m} \tag{7.22}$$

is the $M \times m$ antenna array manifold matrix for the m DOAs $\boldsymbol{\theta}_m = [\theta_1, \ldots, \theta_m]$, B_m is the $m \times m$ Hermitian (denoted $\mathcal{H}$) p.s.d. inter-source correlation matrix, and σ_0^2 is the additive white noise power.

One can expect that for

$$T \geq r, \; r = \mathrm{rank}(B_m) \leq m \tag{7.23}$$

inferences on the covariance matrix may deliver required detection/estimation results on the $M \times m$ (or $M \times r$) signal subspace of the input covariance matrix, since the T data samples adequately span the signal subspace.

Let the eigendecomposition of the sample covariance matrix $\hat{R}_M$ in Equation (7.8) for $T < M$ be presented as

$$\hat{R}_M = \hat{\mathcal{U}}_T \hat{\Lambda}_T \hat{\mathcal{U}}_T^{\mathrm{H}} \tag{7.24}$$

where

$$\hat{\mathcal{U}}_T^{\mathrm{H}} \hat{\mathcal{U}}_T = I_T, \; \hat{\Lambda}_T = diag(\hat{\lambda}_1 \geq \hat{\lambda}_2 \cdots \hat{\lambda}_T > 0) \tag{7.25}$$

Then any inference about the covariance matrix model R_{mod} (not necessarily parameterized as $R_{mod}(m)$) may be provided only regarding its "projection" onto the linear subspace spanned by the $M \times T$ matrix of eigenvectors $\hat{\mathcal{U}}_T$ associated with the T nonzero eigenvalues.

A likelihood ratio utilizing this projection approach was derived (Abramovich and Johnson, 2008), where a properly normalized undersampled likelihood ratio, denoted $LR_u(X_T|R_{mod})$, analogous to the standard $LR(R_{mod})$ in Equation (7.11), can be presented as

$$LR_u(X_T|R_{mod}) = \left[\frac{\det(X_T^{\mathrm{H}} R_{mod}^{-1} X_T/T)\exp T}{\exp\mathrm{Tr}\,(X_T^{\mathrm{H}} R_{mod}^{-1} X_T/T)}\right]^T, \quad T \leq M \tag{7.26}$$

or

$$LR_u(X_T|R_{mod}) = \left[\frac{\prod_{j=1}^{\min(T,M)} \mathrm{eig}_j(R_{mod}^{-1}\hat{R}_M)\exp(\min(T,M))}{\exp\sum_{j=1}^{\min(T,M)} \mathrm{eig}_j(R_{mod}^{-1}\hat{R}_M)}\right]^T \tag{7.27}$$

where eig_j indicates the jth eigenvalue.

The last version clearly demonstrates that for $T \geq M$,

$$LR_u(X_T|R_{mod}) = LR(X_T|R_{mod}) \tag{7.28}$$

where $LR(X_T|R_{mod})$ is specified in Equation (7.11), thus satisfying the transition requirement in Equation (7.18).

Now the introduced LR (7.26) has a very straightforward interpretation, testing the hypothesis

$$H_0: \quad \mathcal{E}\{X_T^{\mathrm{H}} R_{mod}^{-1} X_T/T\} = MI_T \tag{7.29}$$

versus

$$H_1: \quad \mathcal{E}\{X_T^{\mathrm{H}} R_{mod}^{-1} X_T/T\} \neq MI_T \tag{7.30}$$

In fact, $LR_u(X_T|R_{mod})$ tests the quality of the prewhitened (singular) matrix $\hat{C}_{mod}$, where

$$\hat{C}_{mod} = R_{mod}^{-\frac{1}{2}} X_T X_T^{\mathrm{H}} R_{mod}^{-\frac{1}{2}} \tag{7.31}$$

Specifically, this test checks how close to unity are the *nonzero* eigenvalues of the prewhitened matrix $\hat{C}_{mod}/T$.

An undersampled scale-invariant "sphericity" test can be similarly introduced that tests how close the eigenvalues $\mu_j(R_{mod}^{-1}\hat{R}_M), j = 1, \ldots, T$ are to being equal:

$$LR_u^{(sp)}(X_T|R_{mod}) = \left[\frac{\det\left(X_T^{\mathrm{H}} R_{mod}^{-1} X_T\right)}{\left[\frac{1}{T}\mathrm{Tr}\left(X_T^{\mathrm{H}} R_{mod}^{-1} X_T\right)\right]^T}\right]^T, \quad T \leq M \tag{7.32}$$

The introduced "general" (Equations 7.26 and 7.27) and "sphericity" (7.32) undersampled LRs are properly normalized and reach an ultimate value equal to one for the (unconditional) undersampled likelihood solution, thus satisfying the normalization requirement (7.17). Both LRs stem from the undersampled likelihood function (7.9) and maximization of either LR is associated with the best "whitening" of the sample data.

For $T \geq M$, when the sample data with probability of one spans the entire $M \times M$ subspace, the likelihood ratio $LR_u(X_T|R_{mod})$ (7.26) and its associated sphericity test version $LR_u^{(sp)}(X_T|R_{mod})$ (7.32) (where for the general case, similar to Equation (7.27), T must be replaced by $\min(T,M)$) coincide with the standard LR (7.10) as well as the sphericity version (7.15).

Finally (and most important), it needs to be shown that, when the model R_{mod} in Equations (7.26) and (7.32) coincide with the actual covariance matrix of the input matrix of the input data,

$$R_{mod} = R_0 \equiv \mathcal{E}\left\{\frac{1}{T}X_T X_T^{\mathrm{H}}\right\} \tag{7.33}$$

the p.d.f. for the likelihood ratios $LR_u(X_T|R_0)$ and $LR_u^{(sp)}(X_T|R_0)$ does not depend on R_0 and is specified by M and T only. In this case, the properties of the matrix $\hat{C}_0$

$$\hat{C}_0 = R_0^{-\frac{1}{2}} X_T X_T^{\mathrm{H}} R_0^{-\frac{1}{2}} = \mathcal{E}_T \mathcal{E}_T^{\mathrm{H}} \tag{7.34}$$

are fully specified by $\mathcal{E}_T \sim \mathcal{CN}_T(0, I_M)$, while T nonzero eigenvalues of the matrix $\hat{C}_0$ are the same as the eigenvalues of the $T \times T$ matrix

$$\hat{C}_T \sim \mathcal{CW}(M \geq T, T, I_T) \tag{7.35}$$

Therefore, for $R_{mod} = R_0$,

$$LR_u(X_T|R_0) = \left[\frac{\det\left(\mathcal{E}_T^{\mathrm{H}}\mathcal{E}_T/T\right)\exp T}{\exp\mathrm{Tr}\left(\mathcal{E}_T^{\mathrm{H}}\mathcal{E}_T/T\right)}\right]^T, \quad T \leq M,\ \mathcal{E}_T^{\mathrm{H}} \sim \mathcal{CN}_M(0, I_T) \tag{7.36}$$

while the sphericity test (7.32) may be presented as

$$LR_u^{(sp)}(X_T|R_0) = \left[\frac{\det(\mathcal{E}_T^{\mathrm{H}}\mathcal{E}_T)}{\left[\frac{1}{T}\mathrm{Tr}\left(\mathcal{E}_T^{\mathrm{H}}\mathcal{E}_T\right)\right]^T}\right]^T, \quad T \leq M \tag{7.37}$$

As with the matrix extension undersampled LR, distributions for $LR_u(X_T|R_0)$ and $LR_u^{(sp)}(X_T|R_0)$ in this form can be derived similarly to Consul (1969) and Nagarsenker and Pillai (1973), where the real-valued case was considered, and Abramovich et al. (2004, 2007b), where modifications for the complex-valued case were introduced. Specifically, a serial representation of the p.d.f. for

$LR_u(X_T|R_0)$ can be found via Mellin's transform of the moment function:

$$\mathcal{E}\left\{[LR_u(X_T|R_0)]^{\frac{h}{T}}\right\} = \frac{\exp(Th)}{T^{Th}}\mathcal{E}\left[\frac{\det \mathcal{E}_T^{\mathrm{H}}\mathcal{E}_T}{\exp \mathrm{Tr}\left(\mathcal{E}_T^{\mathrm{H}}\mathcal{E}_T/T\right)}\right]^h = \frac{\exp(Th)T^{TM}}{(T+h)^{T(M+h)}}$$
$$\times \prod_{j=1}^{T}\frac{\Gamma(M+h+1-j)}{\Gamma(M+1-j)} \quad T \le M \tag{7.38}$$

where $\Gamma(x)$ represents the Gamma function (Gradshteyn and Ryzhik, 2000).

Since expected likelihood (and maximum likelihood) can use any monotonic increasing function of the LR, and since the form in Equation (7.38) often results in numerically small values, the Tth root of this LR can be equivalently used. In that case, the moments for the T^2th root of LR_u are given by

$$\mathcal{E}\left\{[LR_u(X_T|R_0)]^{\frac{h}{T^2}}\right\} = \frac{\exp(h)T^{TM}}{\left(T+\frac{h}{T}\right)^{(MT+h)}}\prod_{j=1}^{T}\frac{\Gamma\left(M+\frac{h}{T}+1-j\right)}{\Gamma(M+1-j)} \quad T \le M \tag{7.39}$$

Note that, as with the previously presented properly sampled case, these moment functions are dependent only on M and T, and therefore the p.d.f. obtained via the Mellin's transform is also only dependent on M and T, establishing the "scenario-free" requirement given in Equation (7.19).

The p.d.f. for the sphericity test (7.15) with complex Gaussian data $\mathcal{E}_T$ was introduced in Abramovich et al. (2004), denoted there as $w(\alpha)$. If we define α_0 as

$$\alpha_0 = LR_u^{(sp)}(X_T|R_0)^{\frac{1}{T}} \tag{7.40}$$

then using a similar methodology as in Abramovich et al. (2004), we get

$$w(\alpha) = C(T,M)\alpha^{\mathrm{M-T}}\, G_{T,T}^{T,0}\left(\alpha_0 \,\middle|\, \begin{matrix}\frac{T^2-1}{T}, \frac{T^2-2}{T}, \ldots, \frac{T^2-T}{T}\\ 0,1,\ldots,T-1\end{matrix}\right) \tag{7.41}$$

where $G_{c,d}^{a,b}(\cdot)$ is Meijer's Gamma function (Gradshteyn and Ryzhik, 2000), with

$$C(T,M) = (2\pi)^{\frac{1}{2}(T-1)}T^{\frac{1}{2}-TM}\frac{\Gamma(TM)}{\prod_{j=1}^{T}\Gamma(M-j+1)} \tag{7.42}$$

The moments of $\{LR_u^{(sp)}(X_T|R_0)\}^{\frac{1}{T}}$ are equal to

$$\mathcal{E}\left\{\alpha_0^h\right\} = T^{Th}\left[\prod_{j=1}^{T}\frac{\Gamma(M-j+1+h)}{\Gamma(M-j+1)}\right]\frac{\Gamma(TM)}{\Gamma(T(M+h))}, \quad T \le M \tag{7.43}$$

Note again that a p.d.f. derived from these moments is dependent only on M and T, as required in Equation (7.19).

Thus, it has been established that the undersampled LRs given in Equations (7.26) and (7.32) meet the requirements for an undersampled LR, and are therefore appropriate for use within the expected likelihood framework. The application of this undersampled LR to a DOA estimation framework is explored later in this chapter. Meanwhile, the other alternative approach presented earlier for forming an undersampled likelihood ratio using matrix extension techniques is examined.

7.3.2 A Matrix Extension Formulation of the Undersampled LR

In addition to the original sample matrix, the transformed (whitened) sample matrix

$$\hat{C}_0 = R_0^{-\frac{1}{2}} \sum_{j=1}^{T} \mathbf{x}_j \mathbf{x}_j^H R_0^{-\frac{1}{2}}; \quad \mathbf{x}_j \sim \mathcal{CN}(0, R_0) \tag{7.44}$$

is described by the anti-Wishart p.d.f. (denoted $\mathcal{ACW}(T < M, M, I_M)$) given in Equation (27) of Janik and Nowak (2003):

$$K_{T,M}\left(\det \hat{C}_{[T]}\right)^{T-M} e^{-\mathrm{Tr}\,\hat{C}_0} \prod_{l=T+1}^{M} \prod_{p=T+1}^{M} \delta\left(\frac{\det \hat{C}_{[T]lp}}{\det \hat{C}_{[T]}}\right) \tag{7.45}$$

Here δ is the Dirac delta function, $K_{T,M}$ is a normalization constant, and $\hat{C}_{[T]}$ is the upper left-hand $T \times T$ submatrix of the original matrix $\hat{C}_0$:

$$\hat{C}_0 = \left[\begin{array}{c|c} \hat{C}_{[T]} & * \\ \hline * & * \end{array}\right] \tag{7.46}$$

Furthermore, for each $l, p > T$, the $(T+1) \times (T+1)$ matrix $\hat{C}_{[T]lp}$ in Equation (7.45) is generated by adjoining the lth row and the pth column of $\hat{C}_0$ to $\hat{C}_{[T]}$:

$$\hat{C}_{[T]lm} = \left[\begin{array}{ccc|c} & & & \hat{C}_{1p} \\ & \hat{C}_{[T]} & & \vdots \\ & & & \hat{C}_{Tp} \\ \hline \hat{C}_{l1} & \cdots & \hat{C}_{lT} & \hat{C}_{lp} \end{array}\right] \tag{7.47}$$

Noting that the first T rows or columns of a rank-T $\hat{C}_0$ matrix (or any set of entries with the same number of real-valued degrees of freedom) *uniquely specify* the entire matrix $\hat{C}_0$ with rank T, one would like to involve all these independent entries in the formation of the undersampled likelihood ratio. However, the "low-rank" covariance matrix R_0 in Equation (7.20) that defines our admissible set of covariance matrices is described by a limited number of degrees of freedom (DOF) given by

$$\mathrm{DOF}(R_0) = 1 + 2Mm - m^2 \tag{7.48}$$

where $(2Mm - m^2)$ is the number of DOFs that uniquely describe the rank m signal counterpart R_S of the matrix R_0. Thus, one can expect that consistent (with SNR $\to \infty$) testing is possible when the number of degrees of freedom in the full-rank extension exceeds DOF(R_0), even if some degrees of freedom available in $\hat{C}_0$ are not utilized. In fact, this statement is just another version of the well-known requirement on a sample support ($T \geq m$) for "low-rank" covariance matrices of the form given in Equation (7.20).

For $m < T < M$, rather than the first T rows or columns, consider a $(2T-1)$-wide band of the matrix $\hat{C}_0$ (i.e., elements $\hat{c}_{ij}$, $|i-j| \leq T-1$). This band contains almost the same number of degrees of freedom as the original matrix (in fact, only $(M-T)$ fewer). Since the band no longer uniquely specifies the original matrix $\hat{C}_0$ (which is degenerate with rank T), the elements outside the band may be reconstructed in a number of different ways. Thus, by giving up a small number of degrees of freedom and no longer fully specifying $\hat{C}_0$, a series of possible extensions to the band matrix is opened up, including some nondegenerate completions. In particular, the maximum entropy completion (i.e., the band extension with the maximal determinant) for order p, denoted $\hat{C}^{(p)}$, is specified by the Dym-Gohberg band extension method (Dym and Gohberg, 1981; Woerdeman, 1989).

In Grone et al. (1984) and Woerdeman (1989) it was proven that among all band extensions, the Dym-Gohberg extension has the maximal determinant (i.e., maximal entropy), given by

$$\det[\hat{C}^{(p)}]^{-1} = \prod_{q=1}^{M} e_q^T \hat{C}_q^{-1} e_q \tag{7.49}$$

where e_q is a column vector of length $(L(q)-q+1)$ with a single unity entry at position q (where $L(q) = \min\{M, q+p\}$) and $\hat{C}_q$ is the $(L(q)-q+1) \times (L(q)-q+1)$ Hermitian central block matrix in $\hat{C}_0$ with

$$\hat{C}_q = \begin{bmatrix} \hat{c}_{q,q} & \cdots & \hat{c}_{q,L(q)} \\ \vdots & & \vdots \\ \hat{c}_{L(q),q} & \cdots & \hat{c}_{L(q),L(q)} \end{bmatrix} \tag{7.50}$$

for $q = 1, \ldots, M, p \leq T-1$.

The Dym-Gohberg band extension method applied to the rank-deficient sample matrix $\hat{C}_{mod}$ transforms this matrix into a positive definite (p.o.d) Hermitian matrix $\hat{C}^{(p)}$ with exactly the same elements as the sample matrix $\hat{C}_{mod}$ within the $(2p+1)$-wide diagonal band.

Moreover, this p.d. matrix $\hat{C}^{(p)}$ is *uniquely specified* by all different $(p+1) \times (p+1)$ central block matrices in $\hat{C}_{mod}$, and the only necessary and sufficient condition for such transformations to exist is the positive definiteness of all $(p+1) \times (p+1)$ submatrices $\hat{C}_q$ in $\hat{C}_{mod}$. Let $m < p \leq T-1$, when DOF$(R_S) <$ DOF$(\hat{C}^{(p)})$. Now the following likelihood ratio $LR_u^{(p)}(X_T|R_{mod})$ can be introduced for the undersampled scenario:

$$LR_u^{(p)}(X_T|R_{mod}) = \left[\frac{\det\left[\left(R_{mod}^{-\frac{1}{2}}\hat{R}_M R_{mod}^{-\frac{1}{2}}\right)^{(p)}\right] \exp M}{\exp\{\mathrm{Tr}\,\hat{R}_M R_{mod}^{-1}\}} \right]^{\frac{1}{M}} \leq 1 \qquad (7.51)$$

since $\mathrm{Tr}\,\hat{R}_M R_{mod}^{-1} = \mathrm{Tr}\left[\left(R_{mod}^{-\frac{1}{2}}\hat{R}_M R_{mod}^{-\frac{1}{2}}\right)^{(p)}\right]$. Here $\left(R_{mod}^{-\frac{1}{2}}\hat{R}_M R_{mod}^{-\frac{1}{2}}\right)^{(p)} = \hat{C}^{(p)}$ is the Dym-Gohberg p-band transformation of the matrix $\hat{C}_{mod}$. Note that, in fact, the $LR_u^{(p)}(X_T|R_{mod})$ calculation does not require actual reconstruction of the Dym-Gohberg extension, since its determinant is explicitly calculated via block matrix $\hat{C}_q$ in Equation (7.49). In this regard, this likelihood ratio may be treated as a test on $\mathcal{E}\{\hat{C}_q\} = I_{Lq-q+1}$, simultaneously for all q, which for $m < p$ is consistent with the original testing problem $\mathcal{E}\{\hat{C}_{mod}\} = I_M$.

For $R_{mod} = R_0$, $\hat{C}_{mod} = \hat{C}_0$, where $\hat{C}_0 \sim \mathcal{ACW}(T < M, M, I_M)$. That is, $\hat{C}_0$ is now described by the scenario-free anti-Wishart complex distribution, specified by T and M only. Therefore, the p.d.f. for the $LR_u^{(p)}(R_0)$ does not depend on scenario, and can be specified by no more than the parameters M, T, and p, establishing compliance with the desired invariance property (7.19), which is critically important in the expected likelihood methodology.

The Dym-Gohberg band extension utilized in this likelihood ratio is computationally straightforward to generate, and is described in detail in Dym and Gohberg (1981) and Johnson and Abramovich (2007). The final matrix $\hat{C}^{(p)}$ produced by applying the band extension has the property that the original band entries are preserved and the inverse has zeroes outside the $2p+1$ band, as expected for maximum-entropy completions (Lev-Ari et al., 1989).

As with Equation (7.26), the LR in Equation (7.51) effectively extends LR-based hypothesis-testing techniques into the important undersampled domain, but in a manner very different from the subspace projection approach given in Section 7.3.1. Of course, in addition to the undersampled likelihood ratios derived previously, a number of alternative hypothesis tests for the circumstance where the sample covariance matrix is singular have been proposed in the statistical mathematics literature. For example, Ledoit and Wolf (2002) recently re-examined hypothesis tests proposed by John (1973) and Nagao (1973), but under the asymptotic framework used in the Random Matrix theory. Such tests require some adjustment to be used within expected likelihood applications, as discussed in Abramovich and Johnson (2008), where they were shown to have properties quite similar to the undersampled LRs given in Sections 7.3.1 and 7.3.2.

Having now established the tools necessary to apply expected likelihood in both the properly sampled and undersampled regimes, we now turn to applications of the technique.

7.4 USE OF EXPECTED LIKELIHOOD IN THE MUSIC THRESHOLD REGION

The expected likelihood technique just outlined has been applied to the problem of assessing whether a MUSIC-generated estimate is an "outlier" when operating in a multiple-source scenario at SNRs and/or sample support levels within the threshold region for MUSIC (Abramovich et al., 2007c). That investigation (conducted primarily by Monte Carlo simulations) demonstrated a "gap" in the minimum sample support and/or SNR between the MUSIC-specific and ML-intrinsic threshold conditions. In fact, it was demonstrated that for the considered multiple-source scenarios, MLE outlier production (or "breakdown") occurs at a significantly lower SNR than for MUSIC. This gap allows LR-based techniques such as information theoretic criteria (ITC) and expected likelihood to be applied to prediction of MUSIC breakdown, since the likelihood principle continues to operate robustly at those source SNRs where MUSIC outlier production is active.

In Abramovich et al. (2007c), the sample size T was assumed to exceed the dimension M of the adaptive filter (antenna array), so that the conventional (nondegenerate) ML estimate of the covariance matrix exists and drives the LR to its ultimate limit of unity. Yet, as just discussed, for many practical applications in the low sample support regime, i.i.d. samples are not available in sufficient quantity to approach this conventional (Janik and Nowak, 2003) sample support condition, nor is the condition necessary for detection/estimation to occur. In fact, for sufficiently strong sources, it is only necessary for the number of samples to exceed the number of sources m (i.e., $T > m$ rather than $T > M$) for MUSIC to produce reasonably accurate DOA estimates. For fully correlated (coherent) sources, the minimum number of samples is even less demanding, since only an additive noise component and a complex amplitude of the compound signal snapshot changes from sample to sample. Spatial averaging, suggested in Shan et al. (1985) to resolve fully correlated sources in uniform linear antenna arrays, can in fact operate with just a single snapshot. Therefore, it is important to investigate the undersampled (anti-Wishart) sample support condition ($T < M$) for these problems.

With the use of undersampled likelihood ratios LR_u such as those introduced earlier (Equations 7.26 and 7.51), the expected likelihood methodology can be extended to these cases. This methodology relies on the fact that the p.d.f. of $LR(R_0)$ or $LR_u(R_0)$ (where R_0 represents the true (unknown) covariance matrix) depends only on the dimension of the array M and the number of snapshots T, rather than on any specific scenario parameters. A threshold LR value can therefore be precomputed and applied to any scenario that meets the underlying assumptions of i.i.d. Gaussian observations. If the likelihood of any given set of estimates exceeds the chosen threshold (set to a value statistically related to the likelihood ratio associated with the true model), then the covariance model is considered acceptable. Application of this methodology to the DOA estimation of a finite number of sources m fewer than the dimension of the array M can be reiterated as follows.

A finite subspace model $R_{mod}(m)$ can be parameterized as a summation of m (not necessarily independent) planewave sources, as shown in Equation (7.21). The problem can then be stated in this way: Given a set T of i.i.d. Gaussian data $X_T \in \mathcal{CN}_T(0, R_0)$, how does one find the optimal (in an expected likelihood sense) parametric covariance matrix model $R_{mod}(m)$, where m (the number of sources, say) is an arbitrary integer?

The expected likelihood methodology suggests that the optimum order $\hat{m}$ and parameters $\Omega_{\hat{m}}$ used in constructing the model $R(\hat{m})$ should be found as

$$h^* = \arg_h \{\text{Prob}[LR_u(X_T|R_0) < h] = P_{FA}\} \tag{7.52}$$

$$\hat{m} = \arg\min_m \left\{ \max_{\Omega_m \in \Omega} LR_u(X_T|R(\Omega_m)) \geq h^* \right\} \tag{7.53}$$

$$\Omega_{\hat{m}} = \arg\max_{\Omega_{\hat{m}}} LR_u(X_T|R(\Omega_{\hat{m}})) \geq h^* \tag{7.54}$$

Here $P_{FA} \ll 1$ is the probability that the LR produced by the true covariance matrix falls below the threshold value h^* (probability of "false alarm"). Since the p.d.f. of the undersampled $LR_u(X_T|R_0)$ does not depend on R_0, this threshold h^* can be precalculated for any given P_{FA}, M, and T.

Even for estimation only (i.e., no detection because m is known a priori), the expected likelihood methodology still suggests finding the model parameters as

$$\Omega_m = \arg \left\{ \max_{\Omega_m \in \Omega} LR_u(X_T|R_{mod}(m)) \geq h^* \right\} \tag{7.55}$$

The key element of the suggested detection/estimation technique from Equations (7.52) through (7.55) is the search for an appropriate solution, which in terms of LR is statistically as likely as the actual covariance matrix, and continues until such a solution (which exists with a probability greater than $1 - P_{FA}$) is found. Naturally, no claims regarding global optimality of such a solution can be made, but asymptotically, with $T \to \infty$, the solution found in Equation (7.54) or (7.55) is associated with the maximum likelihood estimate, as the likelihood objective function becomes "narrower" with $T \to \infty$.

This principle elucidates the important phenomenon in which the entire expected likelihood and, more important, the ML estimation approach itself "breaks down." Indeed, if the global LR_u extremum exists for a severely erroneous parameter set $\hat{\Omega}_m$ (i.e., far away from the local extremum in the vicinity of R_0), then the ML principle is no longer adequate. Obviously, this may happen only in the threshold area for MLE, and, as mentioned earlier, this remains an active area of research where a number of investigators are trying to specify the conditions (in terms of SNR, sample support, or source separation, say) under which the probability of such "ML breakdown" reaches a certain level (Athley, 2005; Richmond, 2006). Later in this chapter, some additional insight into this issue will be provided.

For now, the techniques in Equations (7.52) through (7.54) rely on the fact that some estimation method–specific threshold conditions (such as those experienced by spectral MUSIC) do not coincide with the ultimate ML breakdown threshold conditions, and therefore can be detected via the threshold derived in Equation (7.52). In this case, a set of (DOA) estimates $\boldsymbol{\theta}_{\hat{m}}$ within $\Omega_{\hat{m}}$ containing an outlier creates a covariance matrix model $R_{mod}(\hat{m})$ the LR of which does not statistically approach $LR_u(X_T|R_0)$. Determining that the DOA estimation set contains an outlier is called "MUSIC performance breakdown prediction" in Abramovich et al. (2004, 2007c).

Practical attempts to enhance the estimation set and bring the $LR_u(X_T|R_{mod}(\hat{m}))$ beyond the threshold can lead to rectification of the outlier (referred to as "performance breakdown cure"; Hawkes et al.,2001) if the algorithm is not yet operating in the ML threshold area. Particular details of this technique, applied for MUSIC threshold performance enhancement in the threshold area, are illustrated in Section 7.4.1, where the efficiency of expected likelihood for rectifying outliers caused by MUSIC threshold conditions (due to lack of SNR and/or sample support) is demonstrated. The main difference in the next section relative to previous studies is the presence of undersampled conditions and therefore the likelihood ratios used.

7.4.1 Monte Carlo Simulation Results

In this section, MUSIC and MLE performance in the threshold region (of parameters) that spans the range from "proper" MUSIC behavior (no outliers) to MLE complete "performance breakdown" is examined. To accomplish this, an undersampled scenario is considered with an $M=10$-element uniform linear array (ULA), $T=6$ samples, array element spacing of $d/\lambda=0.5$, and $m=3$ independent equal-power Gaussian sources with a per-source input SNR of 20 dB (stochastic source model) located at azimuth angles

$$\theta_m = \{\arcsin(-0.40); 0; \arcsin(0.06)\} \tag{7.56}$$

The covariance matrix R_0 for this mixture is

$$R_0 = \left[\sum_{j=1}^{m} \sigma_j^2 S(\theta_j) S^{\mathrm{H}}(\theta_j)\right] + \sigma_0^2 I_M \tag{7.57}$$

where the noise power is $\sigma_0^2=1$; source SNR is given by σ_j^2/σ_0^2; and $S(\theta)$ is the DOA θ-dependent M-variate "steering" (antenna manifold) vector, normalized to a norm of unity.

The number of sources ($m=3$) in the Monte Carlo simulations is assumed to be known a priori. This is not a particularly onerous assumption in this case. The scenario is specifically chosen at an SNR where significant MUSIC breakdown is occurring, yet the likelihood principle (which is at the heart of traditional information theoretic criteria order estimation; Akaike, 1974; Rissanen, 1978; Schwarz, 1978) is still robust and therefore a cross-check using the maximum a posteriori probability (MAP) information theoretic criteria,

$$\hat{m} = \arg\min_{m} \left[-\log LR(R_{mod}(m)) + \frac{5}{3} m \log N \right] \tag{7.58}$$

produces an order estimate that agrees with the actual number of independent sources (three in this case) for every one of the 10^3 trials conducted during the cross-check.

The MUSIC algorithm is implemented as usual by selecting the m largest maxima of the MUSIC pseudo-spectra formed over the noise subspace eigenvectors $\hat{\mathbf{e}}_j$:

$$\varphi^{-1}(\hat{R}_M, \theta) = \left[S^{\mathrm{H}}(\theta) \left(\sum_{j=m+1}^{M} \hat{\mathbf{e}}_j \hat{\mathbf{e}}_j^{\mathrm{H}} \right) S(\theta) \right]^{-1} \tag{7.59}$$

A covariance matrix model R_{mod} can then be constructed using the known order and the MUSIC DOA estimates:

$$R_{MUSIC} = \left[\sum_{j=1}^{m} \hat{\sigma}_j^2 S(\hat{\theta}_j) S^{\mathrm{H}}(\hat{\theta}_j) \right] + \sigma_0^2 I_M \tag{7.60}$$

where the noise power is σ_0^2; $\hat{\sigma}_j^2$ is the source power estimated via the methodology in Ottersten et al. (1993) (as done here) or alternatively (in the case of closely spaced sources) by the method given in Mestre et al. (2007); and $S(\hat{\theta})$ is the MUSIC DOA estimate $\hat{\theta}$-dependent M-variate steering (antenna manifold) vector.

In the Gaussian case, MLE is (theoretically) obtained by the selection of the single largest maxima of the multivariate likelihood function (LF) (Anderson, 1958):

$$\mathcal{L}[R_{mod}(\Omega)] = \left[\frac{\exp\left[-\operatorname{Tr} R_{mod}^{-1}(\Omega) \hat{R}_M \right]}{\pi^M \det R_{mod}(\Omega)} \right] \tag{7.61}$$

where Ω represents the parameter power (σ_m^2) and angle of arrival (θ_m) for the m sources. For a single source, MLE may be implemented via a one-dimensional search (similar to MUSIC) over possible directions of arrival (i.e., the traditional matched filter (beamformer) output), making comparison with the MUSIC threshold condition straightforward. For multiple sources, however, analysis of MLE threshold conditions is not as simple, primarily because the globally optimal ML solution often cannot be guaranteed in practice. Therefore, an MLE proxy algorithm is implemented utilizing the expected likelihood approach. To do so on a practical basis in the low sample support regime (and in particular the undersampled case), rather than the LF given in Equation (7.61), the LRs introduced in Equations (7.26) and (7.51) (and repeated here) with scenario-invariant p.d.f. properties are used.

$$LR_u(X_T|R_{mod}) = \left[\frac{\det\left(X_T^H R_{mod}^{-1}{}^{-1} X_T/T\right)\exp T}{\exp\mathrm{Tr}\left(X_T^H R_{mod}^{-1} X_T/T\right)}\right]^T, \quad T \leq M \tag{7.62}$$

$$LR_u^{(p)}(X_T|R_{mod}) = \left[\frac{\det\left[\left(R_{mod}^{-\frac{1}{2}}\hat{R}_M R_{mod}^{-\frac{1}{2}}\right)^{(p)}\right]\exp M}{\exp\{\mathrm{Tr}\,\hat{R}_M R_{mod}^{-1}\}}\right]^{\frac{1}{M}} \leq 1 \tag{7.63}$$

After deriving a threshold by the method given in Equation (7.52), the expected likelihood "predict and cure" procedure can be detailed as a multistep procedure as follows:

Step 1. Find those trials where the DOA estimates (produced by MUSIC) and associated source-power estimates generate a covariance matrix model R_{MUSIC} (7.60) that does not exceed the LR_u or $LR_u^{(p)}$ threshold in Equation (7.52). Otherwise, retain the trial results as the final estimates.

Step 2. Identify the particular DOA estimate within the set of m MUSIC-produced DOA estimates that is an outlier. This step relies on the fact that, statistically at least, an outlier estimate will not contribute as significantly to the LR_u or $LR_u^{(p)}$ value as a nonoutlier (as long as the scenario being dealt with is identifiable[2]). The assumption here is that, since the scenario is far from the ML breakdown condition, only close proximity to the true model R_0 may produce correspondingly high $LR_u(X_T|R_{MUSIC})$ or $LR_u^{(p)}(X_T|R_{MUSIC})$ values. Therefore, the source in the model $R_{MUSIC}(m)$ that can be deleted to create a model $R_{MUSIC}(m-1)$ with the highest residual LR_u or $LR_u^{(p)}$ is treated as an outlier. This step requires only relatively minor computations to construct m separate covariance matrices, based on the remaining $(m-1)$ DOA estimates.

Step 3. Provide outlier "replacement." Here, instead of the outlier identified at step 2, the source with a DOA estimate that maximally contributes to the $LR_u(X_T|R_{mod})$ or $LR_u^{(p)}(X_T|R_{mod})$ is sought. A one-dimensional search (similar to MUSIC) across possible DOAs is suggested to find this maximum. While the increase in computation relative to classical MUSIC is large, the process of calculating the LR_u or $LR_u^{(p)}$ values along a single variable is still much more tractable computationally compared to a $(2m+1)$-variate search, as demanded by an ML global search routine.

2. Identifiable scenarios are ones in which a particular source wavefront can be unambiguously and uniquely associated with a single DOA. See Stuart and Ord (1991) for a definition of nonidentifiability and Abramovich et al. (1999) for a discussion of identifiability across the combination of scenario and array.

Step 4. This step, in most cases, concludes the routine and considers "final refinement" of the estimate set. Here any gradient-type local optimization routine (Gauss-Newton, for example) algorithm can be used to deliver a solution. The specific method used in this case is the Neder-Mead simplex method (Lagarias et al., 1998) provided via the MATLAB routine `fminsearch`. This optimization can be conducted both on the rectified trials from step 3 and on trials that exceeded the LR threshold imposed in step 1 (if improvement in nonthreshold region accuracy is desired). This final step implements in full the expected likelihood estimation approach (Equation 7.55) and asymptotically ($T \to \infty$) allows us to treat these estimates as the ML estimates. For finite-sample support, such a solution can only be defined as appropriate (in terms of Equation 7.55) rather than globally optimal. Practically, though, this procedure may be terminated even after step 3, if the LR_u or $LR_u^{(p)}$ threshold is exceeded.

Step 5. If the original set includes more than a single outlier and, as a result, the threshold is not exceeded after step 4, the procedure can be repeated until the threshold is exceeded. One of the obvious drawbacks of this step is that with probability P_{FA} (at most), an already truly ML solution may be involved, and therefore this procedure can never reach the threshold and must be terminated at some stage. Because our interest is in robust comparison of MLE and MUSIC in the threshold region rather than in an algorithm that must deliver an estimate at every trial, in cases where the threshold is not reached on the first iteration, we treat the result as a failure to find a global maxima and discard the result rather than iterate the procedure until the threshold is exceeded.

First, the performance of the expected likelihood approach in "predicting" MUSIC performance breakdown in step 1 is examined. The input source SNR (20 dB) for Equation (7.57) has been chosen to result in approximately 50% of the MUSIC estimates having an outlier (specifically 42.6% in 10^6 Monte Carlo trials). The distribution of the MUSIC-generated DOA errors (for the two closely spaced sources) illustrated in Figure 7.1 clearly demonstrates the existence of outliers with errors that exceed 20°, while virtually all the remaining "proper" MUSIC trials exhibit DOA errors under 2°. So in the tabled results that follow (Table 7.1), outliers are defined as those estimates with clairvoyantly determined angular errors that exceed 2°. The percentage of trials that do not contain outliers ("good" trials) or contain at least one outlier ("bad" trials) are shown in the Truth column, along with the mean LR_u or $LR_u^{(p)}$ for each good and bad population. Although a clairvoyant analysis is employed to verify the results, the outlier detection and cure procedure that generates the results in the final columns does not rely on any clairvoyant knowledge.

In Figure 7.2, with the $M = 10$ scenario (7.56), the p.d.f. of the maximum (over the three DOA estimates) errors for the 5742 good and 4258 bad trials is introduced. These distributions demonstrate that in Monte Carlo simulations with this scenario, problems in reliably identifying trials that contain an outlier are not experienced. In Figures 7.2(a) and 7.2(b) for the identified subsets of good and

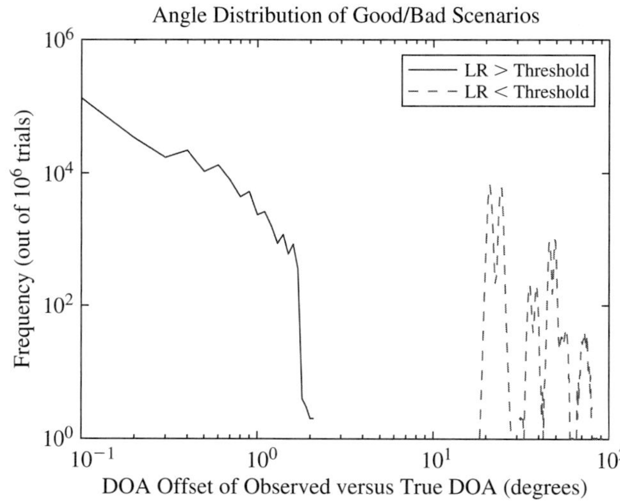

FIGURE 7.1 Histogram of angle errors for sources ($\theta_2 = 0$, $\theta_3 = \arcsin(0.06)$) with input SNR = 20 dB. *Source:* © 2008 IEEE, with permission.

TABLE 7.1 Outlier Detection Statistics with Practical Threshold

Step	Outlier Detected	Truth	Mean LR_u	$P_{FA} = 10^{-3}$ $\alpha = 0.363$	Mean $LR_u^{(p)}$	$P_{FA} = 10^{-3}$ $\alpha = 0.065$
Breakdown	No	57.4%	0.6288	57.2%	0.2451	57.4%
Prediction	Yes	42.6%	0.0008	42.8%	0.0015	42.6%

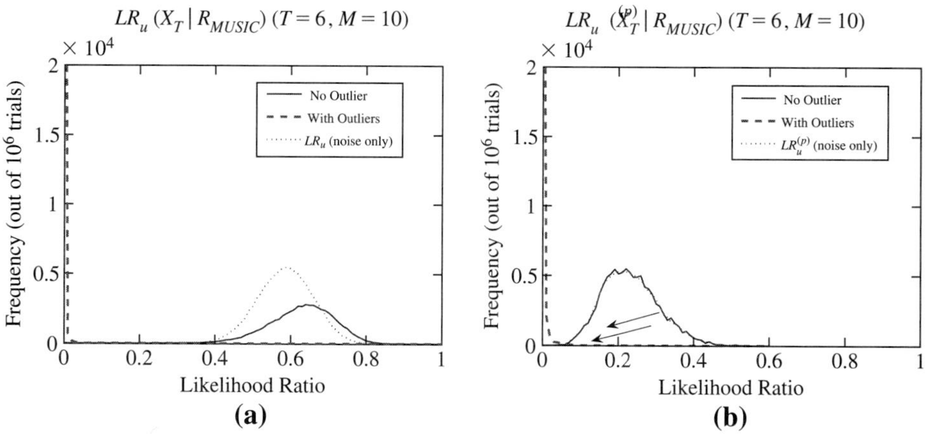

FIGURE 7.2 Likelihood ratio histograms with and without outliers: (a) $LR_u(X_T|R_{mod})$; (b) $LR_u^{(p)}(X_T|R_{mod})$. *Source:* © 2008 IEEE, with permission.

bad trials, sample p.d.f.s for the derived $LR_u(X_T|R_{MUSIC})$ and $LR_u^{(sp)}(X_T|R_{MUSIC})$ are illustrated.

While the exact form of each of the two types of undersampled likelihood ratios differ, they both are very sensitive to the covariance matrix model mismatch caused by DOA estimation outliers. As a result, the likelihood values associated with the good and bad trials are well separated, making identification of bad trials containing outliers quite reliable by all considered tests. Efficiency of such outlier prediction in the absence of clairvoyant knowledge is illustrated below for the test $LR_u(X_T|R_{MUSIC})$ (7.62) and $LR_u^{(sp)}(X_T|R_{MUSIC})$ (7.63).

For the expected likelihood methodology to be used, an LR threshold must first be derived based on the p.d.f. of the LR for the true model. For expediency, rather than compute the p.d.f. analytically, the scenario-invariant p.d.f. of the tests $LR_u(X_T|R_0)$ and $LR_u^{(sp)}(X_T|R_0)$ were generated using 10^6 Monte Carlo runs for the a priori known values $M = 10$, $T = 6$, $m = 3$ associated with the scenario in Equation (7.56). Examination of the resultant distribution indicated that for a statistical approach using a false alarm rate of $P_{FA} = 10^{-3}$, the threshold h^* in Equation (7.52) needs to be set to 0.363 and 0.065 for the two likelihood ratios, respectively. Table 7.1 contains the results of the outlier prediction and shows that of the 57.4% of the 10^6 trials that were good, the expected likelihood test was able to identify 57.2% in the case of LR_u and 57.4% in the case of $LR_u^{(p)}$, meaning that $< 0.2\%$ of the trials were erroneously identified as bad, a value consistent with the considered false alarm rate of 10^{-3}.

More important, all 42.6% of the trials clairvoyantly identified as bad using a 2° DOA error threshold were also properly identified by the statistical nonclairvoyant thresholds. Thus, the MUSIC-specific threshold phenomenon, even at a source separation resulting in a significant failure rate ($> 40\%$), is observed to be sufficiently removed from the ultimate ML breakdown conditions to be reliably identified (predicted) by expected likelihood in the considered undersampled ($T < M$) scenarios.

Let us now examine the ability of the expected likelihood methodology to support outlier "cure" in steps 2 through 4. The "predict and cure" routine does provide (as illustrated next) significant performance improvement in considered scenarios. However, special circumstances may be found where this approach fails, as with any ML "surrogate," in scenarios where an accurate global ML maxima in the region of the true model still exists. To account for this, the algorithm has been designed to discard solutions in those circumstances (as detailed in step 5), so that a valid comparison between MUSIC and MLE in the threshold region can be made. This emphasizes the fact that the algorithm is being used to illustrate the value of solution performance assessment using expected likelihood, rather than being optimized to perform as an operational algorithm. While the algorithm has clear advantages in such an operational regime (including MUSIC-type computational costs in "easy" scenarios and a mechanism to decouple the multidimensional search for MLE to a series of one-dimensional searches), further necessary analysis of computational costs and appropriate P_{FA} selection criteria to make the algorithm operationally relevant is not included here.

The efficiency of this rectification technique can be demonstrated using the scenario given in Equation (7.56). The results of the routine are presented in Table 7.2. It shows that in 99.2% to 99.3% of all trials, the algorithm succeeded in generating a scenario with an LR above the threshold $h^* = 0.363$ or 0.065 calculated for the two different undersampled likelihood ratios with a P_{FA} fixed at 10^{-3}. At the same time, 99.2% of all trials resulting from the rectification routine did not contain an outlier, versus the original 57.4% produced by MUSIC. Therefore, the introduced technique produced solutions above the imposed threshold but still containing an outlier in 0.1% to 0.2% of the cases.

In fact, the 0.1% to 0.2% of misidentified trials illustrates instances of the statistical "ML breakdown" phenomenon where erroneous solutions (derived via the surrogate ML rectification routine) still generated a sufficiently high (above-the-threshold) LR. According to Table 7.2, in 0.7% to 0.8% of all trials, neither LR could be driven above its respective threshold, but unlike the prediction step, all of these nonrectified trials remaining below the threshold still contained an outlier.

Thus, in 99.2% of all trials, solutions without an outlier were obtained and, while in 0.7% to 0.8% the rectification procedure failed (for the reasons described in step 5 of the procedure), all of them still containing an outlier after the rectification process were properly identified as bad whereas only in 0.1% to 0.2% did the outlier remain undetected and therefore unrectified. Obviously, 0.1% to 0.2% versus 42.6% of "broken-down" solutions means considerable performance improvement.

One can note some differences between the two undersampled likelihood ratios given in Equations (7.62) and (7.63), both in computation and in prediction/cure performance. In this scenario, the differences are essentially immaterial, but some qualitative differences were observed while using the like-

TABLE 7.2 Outlier Predict and Cure Statistics Using $LR_u(X_T|R_{MUSIC})$ and Practical Threshold

Step	Outlier Detected	Truth	Mean LR_u	$P_{FA}=10^{-3}$ $\alpha=0.363$	Mean $LR_u^{(p)}$	$P_{FA}=10^{-3}$ $\alpha=0.065$
1. Breakdown Prediction	No	57.4%	0.6288	57.2%	0.2451	57.4%
	Yes	42.6%	0.0008	42.8%	0.0015	42.6%
2/3. Outlier Prediction/Cure	No	99.2%	0.6251	99.2%	0.2295	99.0%
	Yes	0.8%	0.5111	0.8%	0.1175	1.0%
4. Final Refinement	No	99.2%	0.6255	99.3%	0.2297	99.2%
	Yes	0.8%	0.5111	0.7%	0.1175	0.8%

lihood ratios in other scenarios. For example, it was observed that the ability of the subspace projection–likelihood ratio given in Equation (7.62) to discriminate between scenarios with and without outliers exceeded the performance of the Dym-Gohberg matrix extension–likelihood ratio given in Equation (7.63) at extremely low P_{FA} values. Also, the matrix extension–likelihood ratio appeared to be less sensitive to errors in estimation of the noise floor (a problem that occurs when the number of snapshots T is only slightly larger than the rank of the signal subspace R_S). But in most cases, performance was similar.

Graphically, the performance of the expected likelihood approach can be further examined across a broader range of input SNRs. To do so, a slightly different (but still undersampled scenario) used by a range of investigators (Abramovich et al., 2007a; Mestre, 2006a; Steyerl and Vaccaro, 2007) is employed. The scenario uses an $M=20$-element ULA, $T=15$ samples, array element spacing of $d/\lambda=0.5$, and $m=4$ independent equal-power Gaussian sources (stochastic source model) located at azimuth angles

$$\theta_m=\{-20°, -10°, 35°, 37°\} \tag{7.64}$$

immersed in white noise, with various per-element source SNRs (ranging from -15 to $+25$ dB, or set to specific SNRs for more detailed investigation).

Figure 7.3(a) shows the mean-square error (MSE), averaged over 300 trials for each 1-dB SNR increment, for DOA estimates of the two closely spaced sources (at 35° and 37°). It demonstrates the familiar threshold effect in MSE for the DOA estimation process, with the sudden degradation in DOA accuracy (due to outliers) as the SNR is decreased. The threshold point predicted for MUSIC of 20 dB is consistent with similar results shown in Mestre (2006a) and

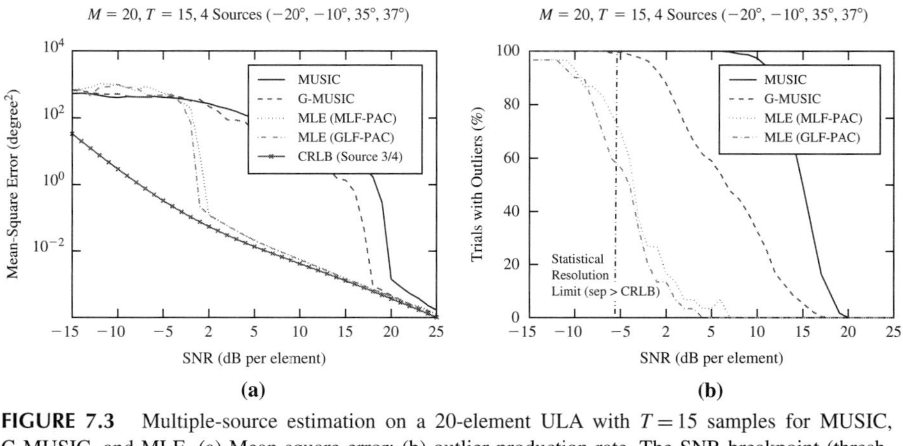

FIGURE 7.3 Multiple-source estimation on a 20-element ULA with $T=15$ samples for MUSIC, G-MUSIC, and MLE. (a) Mean-square error; (b) outlier production rate. The SNR breakpoint (threshold) decreases from around 20 dB for MUSIC to 17 dB for G-MUSIC, but is still dramatically greater than the MLE proxy (LR-PAC) threshold observed at around 0 dB. *Source:* © 2008 IEEE, with permission.

Steyerl and Vaccaro (2007). The MLE breakdown is demonstrated with the MLE proxy algorithm discussed earlier, using the undersampled likelihood ratio given in Equation (7.62) and "seeding" solutions produced by MUSIC (and therefore labeled here as "MLR-PAC" for MUSIC-seeded likelihood ratio "Predict and Cure"). Also shown is the stochastic Cramer-Rao bound (CRB) for the two sources at 35° and 37° (averaged together).

Figure 7.3(a) also shows the threshold region performance of an alternative DOA estimation algorithm referred to as G-MUSIC. In Mestre (2006a,b), G-MUSIC was derived by Mestre, seeking an improvement in MUSIC's threshold performance in undersampled scenarios by deriving a DOA estimator that is consistent in the Random Matrix theory asymptotic conditions, where both the array dimension M and the number of snapshots T grow without bound but at the same rate:

$$M, T \to \infty, \; M/T \to \gamma < \infty \tag{7.65}$$

One can observe from Figure 7.3(a) two important items for the G-MUSIC algorithm. First, the MLE proxy algorithm provides essentially the same performance independent of the seeding mechanism (i.e., the MLR-PAC and GLR-PAC curves use MUSIC and G-MUSIC DOA estimates as seeds, but still provide the same final MSE profile versus SNR), indicating reliable association of the results with true MLE performance in the threshold area, rather than performance specific to the method used to generate the initial DOA estimates. Second, while there is a performance difference in the threshold region between MUSIC and G-MUSIC (consistent with similar figures in Mestre, 2006a,b) indicating some improvement in threshold conditions for G-MUSIC with respect to conventional MUSIC, Mestre and Lagunas (2008) noted, "It was rather disappointing to observe that the use of M, T-consistent estimates does not cure the breakdown effect of subspace-based techniques (in MUSIC) and it merely moves to a lower SNR." This indicates that G-MUSIC is not able to avoid the fundamental phenomenon that separates MUSIC from MLE breakdown conditions, which is surprising given that the G-MUSIC derivations and actual Monte Carlo simulations were conducted under conditions that "guarantee separation of the noise and first signal eigenvalue cluster of the asymptotic eigenvalue distribution of $\hat{R}$" (Mestre, 2006b). Therefore, asymptotically the "subspace swap" phenomenon to which the threshold effect in MUSIC estimation is attributed (Hawkes et al., 2001; Thomas et al., 1995; Tufts et al., 1991) should have been precluded by these conditions, and yet G-MUSIC's breakdown was regularly observed in the conducted Monte Carlo trials.

Clearly the connections between "subspace swap" in MUSIC, G-MUSIC, and MLE performance breakdown, as well as the relevance of the GSA/RMT methodology for practically limited M and T values, needs to be clarified. These connections are explored in the next section.

7.5 SUBSPACE SWAP AND MUSIC PERFORMANCE BREAKDOWN

The subspace swap phenomenon has often been treated as the sole apparent mechanism responsible for performance breakdown in subspace-based techniques. This phenomenon is specified (Thomas et al., 1995) as a case where the estimates of the noise subspace eigenvalues $\lambda_{m+1}=\lambda_{m+2}=\ldots=\lambda_M=\sigma_0^2$ with increasing probability become larger than the estimates of the signal subspace eigenvalues $\lambda_m \leq \lambda_{m-1} \leq \ldots \leq \lambda_1$. "More precisely, in such a case one or more pairs in the set $\{\hat{\lambda}_j, \hat{\mathbf{e}}_j\}, j=1,\ldots,m$ actually estimate noise (subspace) eigenelements instead of signal elements" (Hawkes et al., 2001).

The fact that subspace swap is associated with MUSIC breakdown has been well demonstrated in the literature, with several (only partially) successful attempts undertaken to analytically specify the threshold conditions for a given scenario. In Hawkes et al. (2001), for example, it was admitted that analytical predictions "grossly under-estimate probability of subspace swap in and below the threshold region." This lack of complete success may be attributed to the traditional (M = constant, $T \rightarrow \infty$) asymptotic perturbation eigendecomposition analysis adopted for these derivations. Actual breakdown is of course observed in finite M/T conditions, and the scenario in Equation (7.56) with $T<M$ is quite far from classical asymptotic assumptions.

To explore the conditions under which subspace algorithms break down and establish the potential relationship with subspace swaps, it is convenient to introduce some basics notions of large random matrix analysis related to sample covariance matrices. In particular, it is important to establish the behavior of the sample eigenvalues and eigenvectors—namely, $\{\hat{\lambda}_j, \hat{\mathbf{e}}_j\}, j=1,\ldots,M$—as both the sample size T and the observation dimension M increase without bound at the same rate. This will be the objective of the following subsection.

7.5.1 Fundamentals of Large-Sample Covariance Matrices

The theory of large random matrices studies the behavior of the eigenvalues and eigenvectors as the dimensions of the matrices tend to infinity. It turns out that, for several random matrix models, the histogram of the eigenvalues converges to a fixed density function as the dimensions of the matrix grow without bound. To further understand this property, consider once more the $M \times M$ sample covariance matrix constructed from T snapshots:

$$\hat{R}_M = \frac{1}{T}\sum_{j=1}^{T} \mathbf{x}_j \mathbf{x}_j^H$$

Figure 7.4 represents (solid line) the histogram of the eigenvalues of a particular realization of $\hat{R}_M$ for different values of M in a situation where the number of samples per antenna was fixed to be equal to 5 (so that $M/T=0.2$). In this

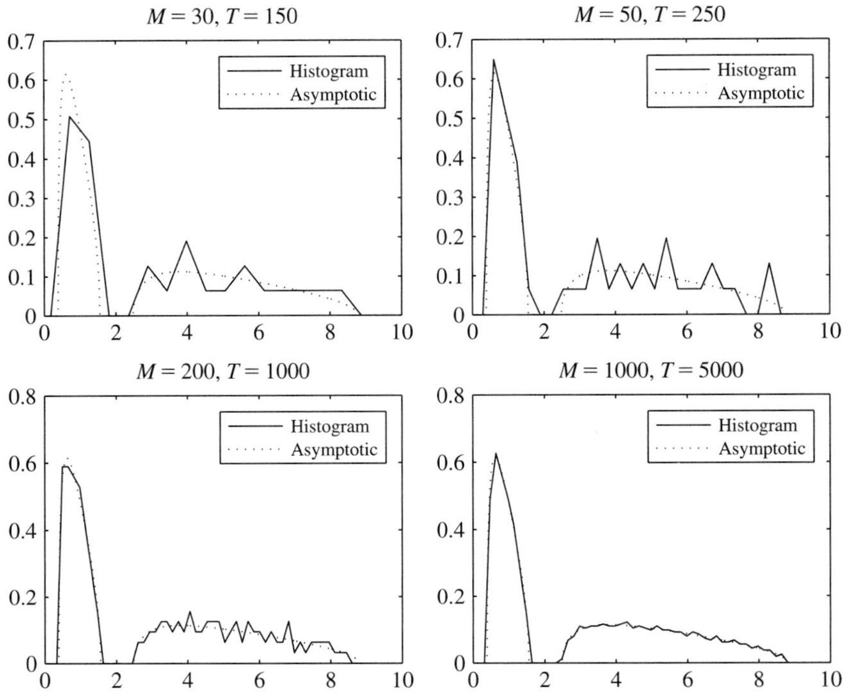

FIGURE 7.4 Histogram of the eigenvalues of a particular realization of $\hat{R}_M$ for different values of M when five samples per antenna are available ($M/T = 0.2$). In this example, the true covariance matrix R_0 always presents two eigenvalues $\{1, 5\}$ with equal multiplicity.

particular example, the true covariance matrix R_0 presented two eigenvalues $\{1, 5\}$ with equal multiplicity for all M. This means that the distribution of eigenvalues of the true covariance matrix consisted of two Dirac delta functions with equal mass located at 1 and 5, and this distribution did not vary with M. We can observe that, as M, T grow large with a constant ratio, the empirical distribution of the eigenvalues of any realization of $\hat{R}_M$ converges to a nonrandom density (dotted line). It is also quite visible that each of the two true eigenvalues $\{1, 5\}$ shows up in the distribution of eigenvalues of $\hat{R}_M$ as separated eigenvalue clusters. For instance, in Figure 7.4 we can clearly distinguish two sample eigenvalue clusters, each one being located around the two true eigenvalues $\{1, 5\}$.

Before going into a more detailed characterization of this behavior, it seems quite reasonable to ask whether something can be said in a situation where the eigenvalue distribution of R_0 is not kept constant for all M, and thus is allowed to vary with the matrix dimension. Figure 7.5 represents once again the evolution of a particular realization of the histogram of the eigenvalues of $\hat{R}_M$ as the matrix dimension M and the number of samples T increase without bound, with a constant ratio in the scenario considered in Figure 7.4. In this situation, however, the multiplicity of the highest true eigenvalue (equal to 5) was set to

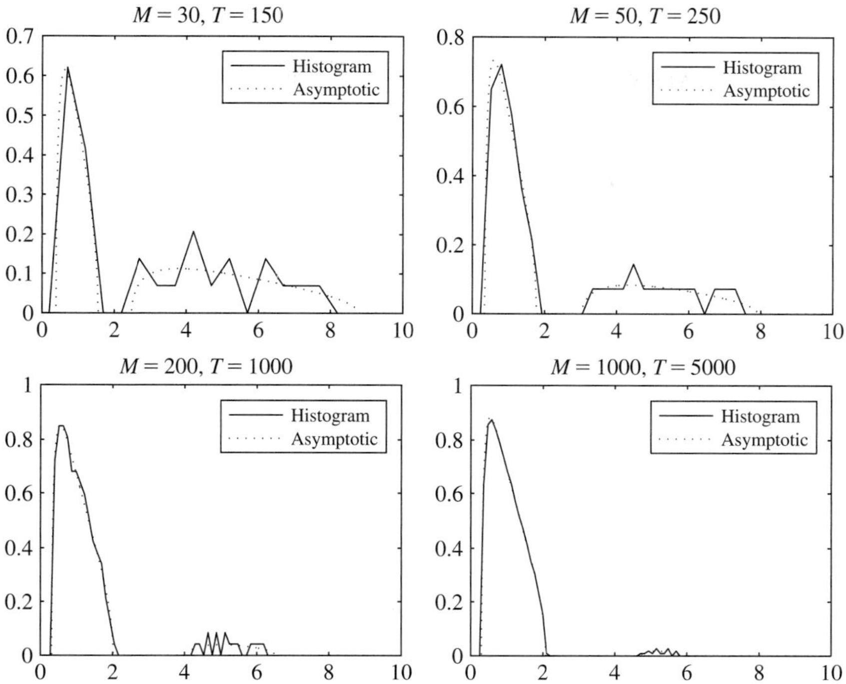

FIGURE 7.5 Histogram of the eigenvalues of a particular realization of $\hat{R}_M$ for different values of M when five samples per antenna are available ($M/T = 0.2$). In this example, the true covariance matrix R_M presents two eigenvalues $\{1, 5\}$. Eigenvalue 5 has fixed multiplicity equal to 15 for all M, whereas eigenvalue 1 presents a multiplicity equal to $M - 15$.

15 regardless of the matrix dimension, whereas the multiplicity of the smallest one (equal to 1) was fixed at $M - 15$, and therefore increased with the matrix dimension. This implies that the contribution of the highest eigenvalue to the eigenvalue distribution of $\hat{R}_M$ becomes negligible as the matrix dimension M grows to infinity, and therefore the density of eigenvalues of $\hat{R}_M$ converges to a single Dirac delta function centered at 1.

Observe that this phenomenon is reflected in the evolution of the density of eigenvalues of the sample covariance matrix $\hat{R}_M$ (solid line), in the sense that the sample eigenvalues linked to the highest true eigenvalue contribute with a diminishing cluster in the asymptotic eigenvalue distribution in comparison with the cluster of sample eigenvalues associated with the smallest true eigenvalue. On the other hand, it is worth pointing out that, even in this situation, we can approximate the nonasymptotic sample eigenvalue distribution by an equivalent deterministic counterpart (dotted line), which describes quite closely the nonasymptotic sample eigenvalue distribution. In fact, it will be shown that the behavior of the eigenvalues of R_0 as $M \to \infty$ is not important as long as they remain bounded for all M. Thus, we will be able to provide a nonasymptotic description of the sample eigenvalue distribution, which becomes a good approximation as $M, T \to \infty$

regardless of the behavior of R_0. Let us now provide a more accurate description of this behavior in analytical terms.

The study of the asymptotic behavior of the eigenvalues and eigenvectors of $\hat{R}_M$ is usually carried out by examining the so-called resolvent of $\hat{R}_M$, which is a matrix-valued complex function defined as

$$Q_M(z) = \left[\hat{R}_M - zI_M\right]^{-1}$$

where z can take any value outside the real axis—namely, $z \in \mathcal{C} \backslash \Re$. The resolvent is a very important function, because many interesting quantities related to the asymptotic behavior of $\hat{R}_M$ can be retrieved from it. For example, one may consider the normalized trace of the resolvent:

$$m_M(z) = \frac{1}{M}\operatorname{tr}[Q_M(z)] = \frac{1}{M}\operatorname{tr}\left[\left(\hat{R}_M - zI_M\right)^{-1}\right]$$

which is usually referred to as the Stieltjes transform of the sample eigenvalues. This quantity is very strongly related to the eigenvalues of $\hat{R}_M$, denoted $\hat{\lambda}_1 \geq \hat{\lambda}_2 \geq \ldots \geq \hat{\lambda}_M$, in the sense that we can express

$$m_M(z) = \frac{1}{M}\sum_{k=1}^{M}\frac{1}{\hat{\lambda}_k - z} \tag{7.66}$$

and therefore we see that $m_M(z)$ is completely determined by the eigenvalues of $\hat{R}_M$. But the true importance of $m_M(z)$ stems from its close relationship with the empirical distribution of the eigenvalues of $\hat{R}_M$, denoted $F_{\hat{R}_M}(\lambda)$ and defined as

$$F_{\hat{R}_M}(\lambda) = \frac{1}{M}\#\left\{\hat{\lambda}_k \leq \lambda, \quad k = 1\ldots M\right\}$$

where $\#\{\cdot\}$ denotes the cardinality of a set. The function $F_{\hat{R}_M}(\lambda)$ returns, for each λ, the percentage of sample eigenvalues that is lower than or equal to λ. This function can be expressed in terms of $m_M(z)$ using the Stieltjes inversion formula:

$$F_{\hat{R}_M}(\lambda) = \lim_{y \to 0^+}\frac{1}{\pi}\int_{-\infty}^{\lambda}\operatorname{Im}[m_M(x + \mathrm{i}y)]\,dx \tag{7.67}$$

Thanks to this identity, we can study the properties of the empirical distribution of eigenvalues of $\hat{R}_M$ by studying the properties $m_M(z)$ or, equivalently, the normalized trace of the resolvent $Q_M(z)$.

Another interesting quantity that is strongly related to the resolvent is the MUSIC cost function:

$$\varphi\left(\hat{R}_M, \theta\right) = S^{\mathrm{H}}\hat{E}_N\hat{E}_N^{\mathrm{H}}S \tag{7.68}$$

where $\hat{E}_N$ contains the noise sample eigenvectors and $S = S(\theta)$ is the steering vector evaluated at a particular angle of arrival. It turns out that, making use of the Cauchy integral formula, one can express

$$\varphi\left(\hat{R}_M, \theta\right) = \frac{1}{2\pi \mathrm{i}} \oint_{C^-} s_M(z) dz$$

where

$$s_M(z) = S^{\mathrm{H}} Q_M(z) S \tag{7.69}$$

and where C^- is a negatively oriented contour that encloses (with index 1) the noise sample eigenvalues on the complex plane. To see this, observe that we can express $\varphi\left(\hat{R}_M, \theta\right)$ as

$$\varphi\left(\hat{R}_M, \theta\right) = \sum_{k=1}^{M} \frac{1}{2\pi \mathrm{i}} \oint_{C^-} \frac{S^{\mathrm{H}} \hat{\mathbf{e}}_k \hat{\mathbf{e}}_k^{\mathrm{H}} S}{\hat{\lambda}_k - z} dz \tag{7.70}$$

and the function inside the complex integral is analytic over all the complex plane except for a set of simple poles at the eigenvalues $\{\hat{\lambda}_k, k = 1 \ldots M\}$. Assuming that all sample eigenvalues are different, and since the pole located at $z = \hat{\lambda}_k$ has a residue equal to $-S^{\mathrm{H}} \hat{\mathbf{e}}_k \hat{\mathbf{e}}_k^{\mathrm{H}} S$, we can obtain Equation (7.68) as a simple application of the Cauchy integral formula (Marsden and Hoffman, 1987).

It is therefore clear that we can study the behavior of quantities associated with the eigenvalues and eigenvectors of the sample covariance matrix by studying the behavior of the resolvent $Q_M(z)$. In particular, the asymptotic behavior of the normalized trace of $Q_M(z)$ will describe the asymptotic correspondence between the true eigenvalues $\{\lambda_k\}$ and the sample eigenvalues $\{\hat{\lambda}_k\}$. On the other hand, the asymptotic form of a generic quadratic form of the resolvent (i.e., $S^{\mathrm{H}} Q_M(z) S$ for a generic column vector S) will be very useful in studying the behavior of the MUSIC DOA estimation algorithm.

Next, we provide a fundamental result that will be the basis for the asymptotic characterization of these quantities. Before presenting the result, it is worth pointing out that, since $Q_M(z)$ depends on the sample covariance matrix $\hat{R}_M$, it is a random quantity and so therefore are both its normalized trace $m_M(z)$ and its quadratic forms $s_M(z)$. Theorem 7.1 establishes the fact that, when we let both the sample size T and the matrix dimension M increase without bound at the same rate, these two random quantities behave as two equivalent deterministic counterparts, which will be denoted as $\bar{m}_M(z)$ and $\bar{s}_M(z)$, respectively.

Theorem 7.1. Assume that $\|S\| = 1$ and that the true covariance matrix has a uniformly bounded spectral norm for all M. Assume that we can express $\hat{R}_M = R_M^{1/2} U U^{\mathrm{H}} R_M^{1/2}$, where $R_M^{1/2}$ is the Hermitian positive definite square root of R_M and U has independent and identically distributed complex entries with independent real and imaginary parts having zero mean, variance $1/2$, and bounded moments of higher order. Then, as $M, T \to \infty$ and $M/T \to \gamma$, $0 < \gamma < \infty$, the quantities $m_M(z)$ and $s_M(z)$ are almost surely close to two deterministic counterparts, $\bar{m}_M(z)$ and $\bar{s}_M(z)$, respectively, for all $z \in \mathcal{C}^+ \equiv \{z \in \mathcal{C}: \text{Im}(z) > 0\}$, in the sense that

$$|m_M(z) - \bar{m}_M(z)| \to 0, \quad |s_M(z) - \bar{s}_M(z)| \to 0$$

where

$$\bar{s}_M(z) = \sum_{k=1}^{M} \frac{s^H \mathbf{e}_k \mathbf{e}_k^H s}{\lambda_k (1 - \gamma - \gamma z \bar{m}_M(z)) - z} \tag{7.71}$$

and where $\bar{m}_M(z)$ is the unique solution to the following equation:

$$\bar{m}_M(z) = \frac{1}{M} \sum_{k=1}^{M} \frac{1}{\lambda_k (1 - \gamma - \gamma z \bar{m}_M(z)) - z} \tag{7.72}$$

located on the set $\{\bar{m} \in \mathcal{C} : -(1-\gamma)/z + \gamma \bar{m} \in \mathcal{C}^+\}$.

The proof of this theorem has appeared in different papers under different assumptions. For instance, in Girko (1998) this theorem appears under slightly weaker statistical assumptions, where the existence of only fourth-order moments of the entries of U is required. In Silverstein (1995) and Bai et al. (2007) it is proven without the need of assuming that the entries of U have moments of order higher than 2. In Mestre (2006c) a simpler alternative proof is given under the assumptions stated here.

An important observation concerning the theorem is the fact that no asymptotic expression is given for either $m_M(z)$ or $s_M(z)$. Instead, these two random quantities are found to be equivalent to two deterministic counterparts, $\bar{m}_M(z)$ and $\bar{s}_M(z)$, which depend on the matrix dimension M. Thus, the result is valid regardless of how these quantities $\bar{m}_M(z)$ and $\bar{s}_M(z)$ behave as M increases (in fact, these quantities do not need to converge at all as $M \to \infty$). Thus, the theorem establishes a nonasymptotic description that is a valid approximation in the asymptotic regime when M and T increase without bound at the same rate. This in turn implies that our characterization of the eigenvalues and eigenvectors of $\hat{R}_M$ does not depend on how the eigenvalues and eigenvectors of R_M behave as M increases to infinity.

Let us first investigate how the above result may be helpful in order to characterize the behavior of the eigenvalues of $\hat{R}_M$. We recall first that in using the inverse Stieltjes transform in Equation (7.67) one can easily recover the empirical distribution of eigenvalues of $\hat{R}_M$ from the normalized trace of the resolvent, $m_M(z)$. On the other hand, since $m_M(z)$ is asymptotically close to the deterministic counterpart $\bar{m}_M(z)$, one can prove that the empirical distribution of eigenvalues of $\hat{R}_M$ will be close to the equivalent distribution associated with

$\bar{m}_M(z)$ via the inverse Stieltjes transform:

$$G_M(\lambda) = \lim_{y \to 0^+} \frac{1}{\pi} \int_{-\infty}^{\lambda} \mathrm{Im}[\bar{m}_M(x + \mathrm{i}y)]\, dx$$

Now one can easily see that the associated measure is absolutely continuous (except at zero, in the undersampled case), and therefore we can take derivatives in order to investigate the associated density, which can be expressed as

$$g_M(\lambda) = \lim_{y \to 0^+} \frac{1}{\pi} \mathrm{Im}[\bar{m}_M(\lambda + \mathrm{i}y)]$$

This means that, by examining the form of the imaginary part of $\bar{m}_M(z)$ when z tends to the real axis, we are able to establish the form of the density $g_M(\lambda)$, which is an asymptotically good approximation of the "random density" (i.e., histogram) of the sample eigenvalues. Once again, it is worth pointing out that $g_M(\lambda)$ is a deterministic approximation of the random eigenvalue density regardless of how the true eigenvalues $\{\lambda_m\}$ behave as $M \to \infty$.

In Figure 7.6 we represent the form of $g_M(\lambda)$ for different values of γ in a scenario where the true covariance matrix presented three different eigenvalues, $\{15, 5, 1\}$, with equal multiplicities. It can readily be observed that, as $\gamma \to 0$ (meaning that the number of samples per antenna increases to infinity), the sample eigenvalues tend to group around the true eigenvalues, forming clusters. If γ is sufficiently high, the density $g_M(\lambda)$ consists of a single cluster, whereas if γ goes to zero, the density $g_M(\lambda)$ splits up into smaller clusters. If the number of samples per antenna is sufficiently high (i.e., γ is sufficiently low), each true eigenvalue generates a single cluster in $g_M(\lambda)$.

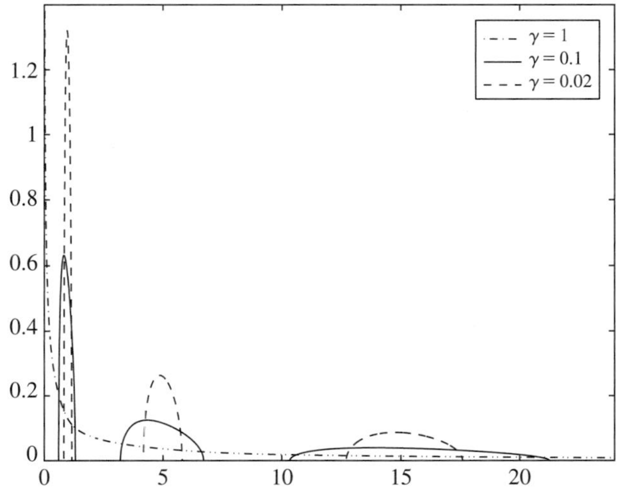

FIGURE 7.6 Representation of the density $g_M(\lambda)$ for different values of γ where the true covariance matrix presents three equal-multiplicity eigenvalues $\{15, 5, 1\}$.

The number and position of the different clusters in $g_M(\lambda)$ can be determined as follows. Let us consider the following equation in an unknown f:

$$\frac{1}{M}\sum_{k=1}^{M}\left(\frac{\lambda_k}{\lambda_k - f}\right)^2 = \frac{1}{\gamma} \tag{7.73}$$

In Figure 7.7 we provide a graphical representation of the above equation. The solutions can be characterized by the intersection points between the function on the right side of Equation (7.73), which does not depend on γ, and a horizontal line at γ^{-1}. It can be observed from Figure 7.7 that the number of solutions to the Equation (7.73) is always even (counting multiplicities). Let us denote the number of such solutions as $2Q$, and the solutions themselves as

$$f_Q^- < f_Q^+ \leq \ldots \leq f_2^- < f_2^+ \leq f_1^- < f_1^+$$

where equalities hold only when there exist solutions with multiplicity 2. Now, it turns out that Q is the number of clusters in $g_M(\lambda)$ and that the support of each of these Q clusters is $\left[x_q^-, x_q^+\right]$, $q = 1 \ldots Q$, where $x_q^- = \Phi\left(f_q^-\right)$, $x_q^+ = \Phi\left(f_q^+\right)$, and where we have defined

$$\Phi(f) = f\left(1 - \frac{\gamma}{M}\sum_{k=1}^{M}\frac{\lambda_k}{\lambda_k - f}\right)$$

Figure 7.8(a) provides a typical representation of the function $\Phi(f)$ above. It turns out that, as already observed in the figure, the function $\Phi(f)$ preserves the ordering in the values of $f_Q^- \ldots f_1^+$. This means that $x_Q^- < x_Q^+ \leq \ldots \leq x_2^- < x_2^+ \leq x_1^- < x_1^+$, and therefore we can express the support of $g_M(\lambda)$ (excluding Dirac

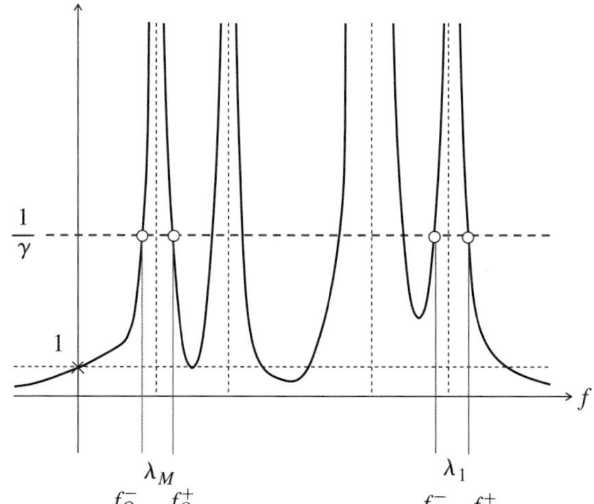

FIGURE 7.7 Graphical representation of the solutions to Equation (7.73).

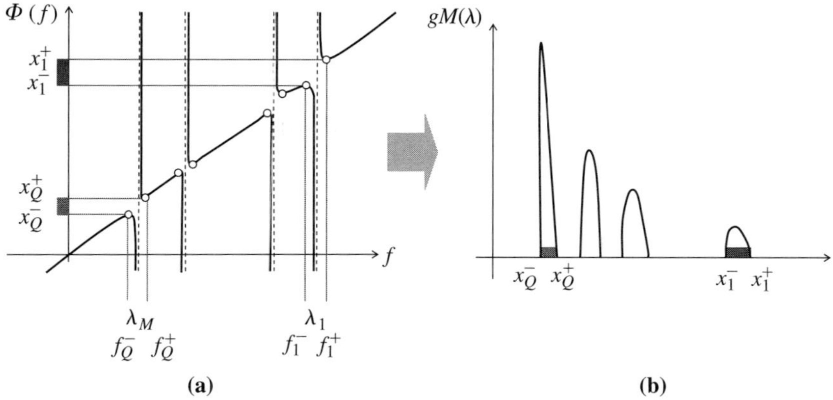

FIGURE 7.8 Graphical representation of (a) transformation $\Phi(f)$ and (b) corresponding support of the clusters in $g_M(\lambda)$.

deltas at zero when $\gamma > 1$) as the union of Q closed intervals

$$\mathcal{S}_M = \left[x_Q^-, x_Q^+\right] \cup \ldots \cup \left[x_1^-, x_1^+\right]$$

where each interval $[x_q^-, x_q^+]$ corresponds to the transformation of an interval $[f_q^-, f_q^+]$ obtained from the solutions to Equation (7.73) (see the right side of Figure 7.8(b)).

By observing the function on the left side of Equation (7.73) in Figure 7.7, one can conclude that each eigenvalue of the true covariance matrix λ_k can be associated with a particular cluster supported by $g_M(\lambda)$, in the sense that there exists one interval $[f_q^-, f_q^+]$ for which $\lambda_m \in [f_q^-, f_q^+]$. This way, one can associate λ_m with the cluster having support $[x_q^-, x_q^+]$. Now it is also clear from Figure 7.7 that when $\gamma \to 0$ (and thus when the horizontal line at γ^{-1} moves upward), the original clusters split into smaller ones, up to the point where each eigenvalue of the true covariance matrix will be associated with a unique cluster of $g_M(\lambda)$. In other words, if the number of samples per antenna is sufficiently high (or, equivalently, γ is sufficiently low), each distinct eigenvalue of the true covariance matrix will be associated with only one cluster in $g_M(\lambda)$.

From this description, one can readily derive the minimum number of samples per antenna required for a cluster associated with a particular eigenvalue to split from the rest of the asymptotic eigenvalue distribution. To do that, one only needs to find the values of the function on the right side of Equation (7.73) at the corresponding inflexion points. It is particularly interesting to characterize the minimum number of samples per antenna that guarantees that the cluster associated with the noise eigenvalue (i.e., the lowest eigenvalue) is separated from the clusters associated with the other (i.e., signal) eigenvalues. If we denote by $f_{\min}^*$ the minimum inflexion point of the function on the right side of

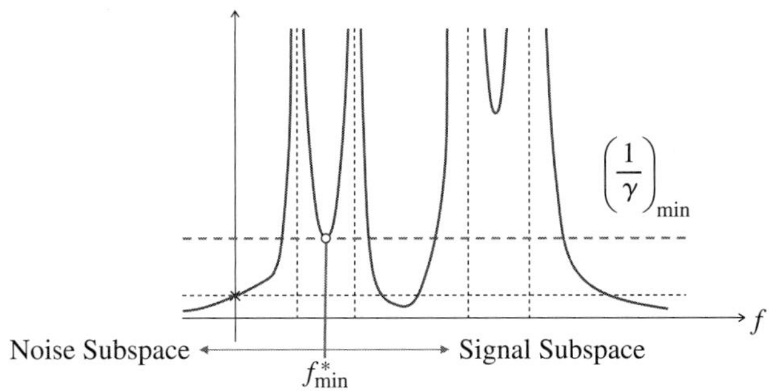

FIGURE 7.9 Graphical representation of the minimum number of samples per antenna needed to guarantee asymptotic separability of signal and noise subspaces.

Equation (7.73) (see Figure 7.9)—namely, the lowest solution to the following equation in f^*:

$$\frac{1}{M}\sum_{k=1}^{M}\frac{\lambda_k^2}{(\lambda_k - f^*)^3} = 0 \tag{7.74}$$

then we can establish that the noise cluster of $g_M(\lambda)$ is asymptotically separated from the rest of the eigenvalue distribution whenever

$$T/M > \left(\frac{1}{\gamma}\right)_{\min} = \frac{1}{M}\sum_{r=1}^{M}\left(\frac{\lambda_r}{\lambda_r - f^*_{\min}}\right)^2 \tag{7.75}$$

Another crucial aspect that must be addressed refers to the concentration and separation properties of sample eigenvalues with respect to the deterministic density $g_M(\lambda)$. It should be clear that the histogram of the eigenvalues of the sample covariance matrix is somehow close to the deterministic density $g_M(\lambda)$ as M and T grow without bound at the same rate. However, this property is not useful in characterizing the behavior of a particular eigenvalue because the contribution of a single eigenvalue is lost in the asymptotic density. Thus, some stronger results are needed to characterize where a particular eigenvalue converges as M and T grow to infinity. Part of the answer to this was given in Bai and Silverstein (1998), where it was established that, for sufficiently high M, T, all the eigenvalues of the sample covariance matrix are located inside the support of $g_M(\lambda)$—that is, $\mathcal{S}_M$. This was later refined in Bai and Silverstein (1999), where it was shown that separation is exact in the sense that, with probability one for all large M, T, the number of sample eigenvalues inside a particular cluster of $g_M(\lambda)$ is equal to the number of true eigenvalues associated with that particular cluster (counting multiplicities). The only exception to this is $\gamma > 1$, where $T - M$ sample eigenvalues are equal to zero and thus do not contribute to the lowest cluster.

An interesting consequence of the above result is the fact that, if we assume that the noise (lowest) eigenvalue cluster is separated from the signal cluster, then the probability that a noise sample eigenvalue moves to the signal part of the spectrum is equal to zero for all sufficiently large M, T. In other words, if the noise subspace has dimension m, one can ensure that the m lowest sample eigenvalues are located in the cluster with support $\left[x_1^-, x_1^+\right]$ for all large M, T, provided that the number of samples T per antenna element M is greater than the right side of Equation (7.75). In this situation, one can ensure that signal and noise sample eigenvalues will be separated with probability one for all M, T that are sufficiently large.

7.5.2 Using Large Random Matrix Theory to Describe MUSIC Behavior

It turns out that whenever there is asymptotic separation between signal and noise subspaces, one can effectively describe the behavior of the sample eigenvalues and eigenvectors using the technique presented in Mestre (2006b): Let R_M and $\hat{R}_M$ have the following eigendecomposition:

$$R_M = \sum_{k=1}^{M} \lambda_k \mathbf{e}_k \mathbf{e}_k^{\mathrm{H}}, \quad \hat{R}_M = \sum_{k=1}^{M} \hat{\lambda}_k \hat{\mathbf{e}}_k \hat{\mathbf{e}}_k^{\mathrm{H}}$$

where

$$\lambda_{m+1} = \ldots = \lambda_M = \sigma_0^2$$

is the true noise eigenvalue, which has multiplicity $M - m$ (the true noise eigenvectors are taken to form a basis for the corresponding noise subspace). We also denote as E_N (resp. $\hat{E}_N$) a matrix containing the true (resp. sample) noise eigenvectors, and as E_S (resp. $\hat{E}_S$) the matrix that contains the true (resp. sample) signal eigenvectors. Now, in order to investigate the asymptotic behavior of the MUSIC algorithm it is sufficient to investigate the following quadratic form:

$$\varphi\left(\hat{R}_M, \theta\right) = S^{\mathrm{H}} \hat{E}_N \hat{E}_N^{\mathrm{H}} S$$

which, as indicated in Equation (7.70) can be expressed as

$$\varphi\left(\hat{R}_M, \theta\right) = \frac{1}{2\pi \mathrm{i}} \oint_{C^-} s_M(z) dz$$

where $s_M(z)$ is defined in Equation (7.69) and where, as before, C^- is a negatively oriented simple contour enclosing only the noise sample eigenvalues. Using Theorem 7.1 we can see that the integrand is almost surely close to $\bar{s}_M(z)$ as defined in Equation (7.71), in the sense that $|s_M(z) - \bar{s}_M(z)| \to 0$ with probability one as $M, T \to \infty$ at the same rate. Assuming that noise and signal eigenvalue clusters in $g_M(z)$ are separated, and obviating some technicalities (see Mestre, 2008b, for further details), we can state that

$$\left| \varphi\left(\hat{R}_M, \theta\right) - \frac{1}{2\pi \mathrm{i}} \oint_{C^-} \bar{s}_M(z) dz \right| \to 0$$

almost surely, where we have basically replaced $s_M(z)$ by its asymptotic deterministic equivalent, $\bar{s}_M(z)$. As a consequence, we only need to investigate the form of the above integral to find the form of the asymptotic MUSIC cost function.

In order to do this, we insert the form of $\bar{s}_M(z)$ given in Equation (7.71), so that

$$\frac{1}{2\pi \mathrm{i}} \oint_{C^-} \bar{s}_M(z) dz = \sum_{k=1}^{M} w_{MUSIC}(k) S^{\mathrm{H}} \mathbf{e}_k \mathbf{e}_k^{\mathrm{H}} S \tag{7.76}$$

where

$$w_{MUSIC}(k) = \frac{1}{2\pi \mathrm{i}} \oint_{C^-} \frac{1}{\lambda_k (1 - \gamma - \gamma z \bar{m}_M(z)) - z} dz$$

We can already observe in Equation (7.76) that the MUSIC cost function is asymptotically equivalent to a linear combination of the true signal and noise subspaces. The weights $w_{MUSIC}(k)$ corresponding to the signal subspace, $k \leq m$, indicate that part of the content in the noise sample subspace asymptotically leaks into the true signal subspace. To determine the degree of leakage between the sample and the true subspaces, one needs to characterize the weights $w_{MUSIC}(k)$ in closed form. To do this, we define the complex function

$$f_M(z) = \frac{z}{1 - \gamma - \gamma z \bar{m}_M(z)} \tag{7.77}$$

so that we can express $w_{MUSIC}(k)$ as

$$w_{MUSIC}(k) = \frac{1}{2\pi \mathrm{i}} \oint_{C^-} \frac{1}{z} \frac{f_M(z)}{\lambda_k - f_M(z)} dz \tag{7.78}$$

Inserting Equation (7.77) into Equation (7.72) one can readily see that $f_M(z)$ is the solution to

$$z = f_M(z) \left(1 - \frac{\gamma}{M} \sum_{k=1}^{M} \frac{\lambda_k}{\lambda_k - f_M(z)} \right) \tag{7.79}$$

On the other hand, taking derivatives of Equation (7.79) with respect to z on both sides, we obtain

$$1 = f_M'(z) \left(1 - \frac{\gamma}{M} \sum_{k=1}^{M} \left(\frac{\lambda_k}{\lambda_k - f_M(z)} \right)^2 \right) \tag{7.80}$$

Now, inserting Equations (7.79) and (7.80) into Equation (7.78) we get to the key expression:

$$w_{MUSIC}(k) = \frac{1}{2\pi \mathrm{i}} \oint_{C^-} \frac{1 - \frac{\gamma}{M}\sum_{r=1}^{M}\left(\frac{\lambda_r}{\lambda_r - f_M(z)}\right)^2}{1 - \frac{\gamma}{M}\sum_{r=1}^{M}\frac{\lambda_r}{\lambda_r - f_M(z)}} \frac{1}{\lambda_k - f_M(z)} f_M'(z)dz$$

At this point, we make the nontrivial observation that, when z moves along the real axis, $f_M(z)$ describes a contour as illustrated in Figure 7.10, where we also show the corresponding complex conjugate curve. Observe from the figure that $f_M(z)$ together with its conjugate $f_M^*(z)$ can be concatenated to form a contour that encloses the true noise eigenvalue only. Thus, it is possible (see Mestre, 2008b, for further details) to use f as the complex integration variable in the above expression:

$$w_{MUSIC}(k) = \frac{1}{2\pi \mathrm{i}} \oint_{C^-} \frac{1 - \frac{\gamma}{M}\sum_{r=1}^{M}\left(\frac{\lambda_r}{\lambda_r - f}\right)^2}{1 - \frac{\gamma}{M}\sum_{r=1}^{M}\frac{\lambda_r}{\lambda_r - f}} \frac{1}{\lambda_k - f} df$$

where here again C^- is a negatively oriented contour enclosing only the true noise eigenvalue and having index 1 with respect to it. The resulting integral is easily solved using the residue theorem (Marsden and Hoffman, 1987), noting that the integrand has only two poles inside C^-: one located at the true noise eigenvalue $\lambda_{m+1} = \ldots = \lambda_M = \sigma_0^2$ and one located at the value that nulls out the denominator, which corresponds to the minimum solution to the following equation (denoted $\mu_{\min}$):

$$\frac{1}{M}\sum_{r=1}^{M}\frac{\lambda_r}{\lambda_r - \mu_{\min}} = \frac{1}{\gamma} \tag{7.81}$$

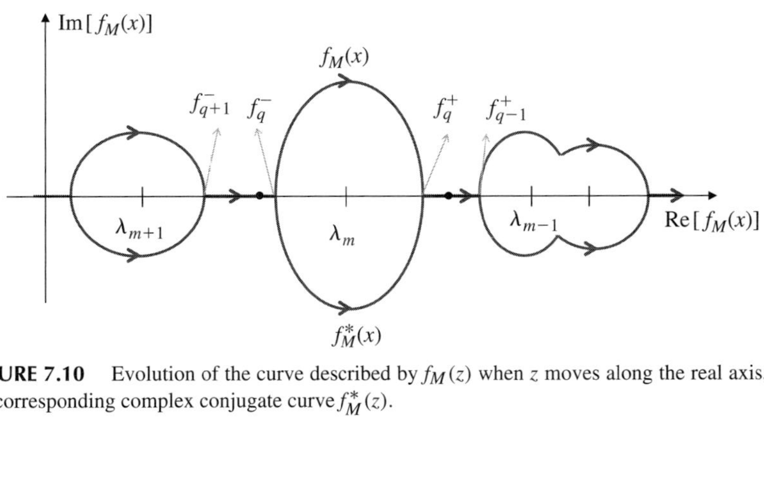

FIGURE 7.10 Evolution of the curve described by $f_M(z)$ when z moves along the real axis, and the corresponding complex conjugate curve $f_M^*(z)$.

After some calculation, one obtains

$$w_{MUSIC}(k) = \begin{cases} 1 - \frac{1}{M-m}\sum_{r=1}^{m}\left(\frac{\sigma_0^2}{\lambda_r - \sigma_0^2} - \frac{\mu_{\min}}{\lambda_r - \mu_{\min}}\right) & k > m \\ \frac{\sigma_0^2}{\lambda_k - \sigma_0^2} - \frac{\mu_{\min}}{\lambda_k - \mu_{\min}} & k \leq m \end{cases} \tag{7.82}$$

In conclusion, we can state that, assuming that the number of samples per antenna is sufficiently high so that the eigenvalue-splitting condition in Equation (7.75) is fulfilled, as $M, T \to \infty$ at the same rate, the MUSIC cost function $\varphi(\hat{R}_M, \theta) = S^{\mathrm{H}}\hat{E}_N\hat{E}_N^{\mathrm{H}}S$ is asymptotically close to a deterministic counterpart $\bar{\varphi}(R_M, \theta)$ in the sense that $\left|\varphi(\hat{R}_M, \theta) - \bar{\varphi}(R_M, \theta)\right| \to 0$ with probability one, where

$$\bar{\varphi}(R_M, \theta) = w_{MUSIC}(M) \underbrace{S^{\mathrm{H}}E_N E_N^{\mathrm{H}} S}_{\text{Desired cost function}} + \underbrace{\sum_{k=1}^{m} w_{MUSIC}(k) S^{\mathrm{H}}\mathbf{e}_k\mathbf{e}_k^{\mathrm{H}} S}_{\text{Leakage into the true signal subspace}} \tag{7.83}$$

Thus, we see that the sample noise subspace can be asymptotically described as having a component on the true noise subspace plus an additional undesired component on the true signal subspace. When the number of samples per antenna increases (or, equivalently, when $\gamma \to 0$), we recover the desired cost function $S^{\mathrm{H}}E_N E_N^{\mathrm{H}}S$. Indeed, observe from Equation (7.81) that $\mu_{\min} \to \sigma_0^2$ as $\gamma \to 0$, and consequently

$$w_{MUSIC}(k) \to \begin{cases} 1 & k > m \\ 0 & k = 1 \ldots m \end{cases}$$

so that $\bar{\varphi}(R_M, \theta) \to S^{\mathrm{H}}E_N E_N^{\mathrm{H}}S$. For the general case where $\gamma > 0$, we see that $w_{MUSIC}(k) > 0$ and therefore $\varphi(\hat{R}_M, \theta)$ is no longer a consistent estimator of $S^{\mathrm{H}}E_N E_N^{\mathrm{H}}S$ when $M, T \to \infty$.

It is worth pointing out that an equivalent result can be obtained for the signal eigenvalues using exactly the same procedure. Assume that we need to characterize the asymptotic behavior of the quantity $S^{\mathrm{H}}\hat{\mathbf{E}}_j\hat{\mathbf{E}}_j^{\mathrm{H}}S$, where $\hat{\mathbf{E}}_j$ contains the eigenvectors associated with an eigenvalue λ_j (if λ_j has multiplicity one, $\hat{\mathbf{E}}_j = \hat{\mathbf{e}}_j$).

Theorem 7.2. Under the statistical assumptions just specified, and assuming that the eigenvalue cluster associated with λ_j is separated from the rest of the eigenvalue distribution $g_M(\lambda)$, we can establish that (as $M, T \to \infty$ at the same rate)

$$\left|\left|S^{\mathrm{H}}\hat{\mathbf{E}}_j\right|^2 - \sum_{k=1}^{M} w_j(k)\left|S^{\mathrm{H}}\mathbf{e}_k\right|^2\right| \to 0 \tag{7.84}$$

where we have defined

$$w_j(k) = \begin{cases} 1 - \frac{1}{K_j} \sum_{\substack{r=1 \\ \lambda_r \neq \lambda_j}}^{M} \left(\frac{\lambda_j}{\lambda_r - \lambda_j} - \frac{\mu_j}{\lambda_r - \mu_j} \right) & k : \lambda_k = \lambda_j \\ \frac{\lambda_j}{\lambda_k - \lambda_j} - \frac{\mu_j}{\lambda_k - \mu_j} & k : \lambda_k \neq \lambda_j \end{cases}$$

where K_j is the multiplicity of λ_j and where now $\mu_j, j = 1 \ldots M$ are the solutions to Equation (7.81) repeated according to the multiplicity of each λ_j.

The proof of this general result can be found in Mestre (2008b).

7.5.3 A Simplified Scenario: The Spiked Population Model

The results just presented are quite powerful, but allow for little interpretation. The reason for that stems in part from the fact that the derived expressions, although asymptotically valid in the large M, T regime, are nonasymptotic, meaning that they depend on fixed values of M, T. Nothing has been assumed about how the eigenvalues and eigenvectors of R_M behave as $M \to \infty$, and thus the provided expressions are quite general but difficult to study. A potential simplification of the above expressions comes from assuming a particular asymptotic behavior of the eigenvalues and eigenvectors of R_M. More specifically, the so-called "spiked population covariance matrix model" (Johnstone, 2001) describes the asymptotic behavior corresponding to the case where the signal subspace dimension becomes negligible compared to the observation dimension. This means that m remains fixed, whereas $M \to \infty$.

Therefore, this model is essentially an asymptotic particularization of the situation analyzed up to now, based on the simplifying assumption that the contribution of the signal subspace is negligible in the asymptotic regime, in the sense that only the dimension of the noise subspace scales up with the number of antenna elements, whereas the dimension of the signal subspace m remains fixed. The immediate consequence of this specific convergence behavior of the spectrum of R_M is the fact that, in the asymptotic regime (when $M \to \infty$), the true covariance matrix consists of infinitely more noise eigenvalues than signal eigenvalues. In other words, the signal subspace does not contribute to the asymptotic density of eigenvalues of R_M, which consists of a Dirac delta function centered at σ_0^2.

Let us analyze the implications of the spiked population model on the eigenvalue separability conditions. To illustrate this, we consider a scenario in which the true covariance matrix presented three eigenvalues at $\{4, 3, 1\}$, where the first two eigenvalues (signal eigenvalues, 4 and 3) had multiplicity 1 regardless of the matrix dimension, whereas the multiplicity of the smallest eigenvalue (noise eigenvalue) was fixed at $M-2$. In Figure 7.11 we represent the evolution of the support of $g_M(\lambda)$ as a function of M. The support is described as $\mathcal{S}_M = [x_3^-, x_3^+] \cup [x_2^-, x_2^+] \cup [x_1^-, x_1^+]$, where each interval is associated with a different true eigenvalue. Now we can observe from Figure 7.11 that the support of the clusters associated with the signal eigenvalues shrink as $M \to \infty$. This is

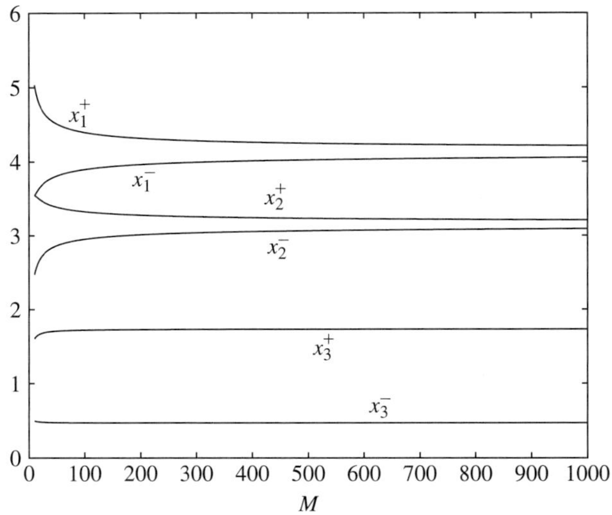

FIGURE 7.11 Evolution of asymptotic support as a function of M in a spiked population covariance matrix model ($c=0.1$) with three eigenvalues $\{4,3,1\}$ with multiplicities $\{1,1,M-2\}$. The number of samples per antenna was fixed equal to $10\,(\gamma=0.1)$.

quite reasonable since the contribution of these eigenvalues to the global density becomes negligible as $M\to\infty$ because their multiplicity does not scale up with the matrix dimension. Thus, we can readily conclude that, in a spiked population model, there is always separation between the signal eigenvalue clusters. This is not necessarily the case for the separability between signal and noise subspaces, because the noise sample eigenvalue cluster does not disappear as $M\to\infty$. Let us derive the actual condition that guarantees separability of the signal and noise subspaces.

Under the spiked population simplification of the original model (which implies letting $M\to\infty$ for m fixed in the original formulas), the asymptotic subspace-splitting condition in Equation (7.75) becomes

$$T/M > \left(\frac{\lambda_M}{\lambda_M-\lambda_m}\right)^2 = \left(\frac{\sigma_0^2}{\sigma_0^2-\lambda_m}\right)^2 \tag{7.85}$$

which can also be expressed as

$$\lambda_m > \sigma_0^2\left(1+\sqrt{\gamma}\right) \tag{7.86}$$

where we recall that γ is the asymptotic inverse of the number of samples per antenna, $(T/M)^{-1}$. If the above condition holds, all the clusters are separated in the asymptotic eigenvalue distribution. Indeed, as previously shown we can ensure that the cluster associated with a particular signal eigenvalue is always separated from the cluster associated with another signal eigenvalue for sufficiently high M. On the other hand, the cluster associated with the noise subspace does not disappear (because the multiplicity of the noise eigenvalue scales up with M), and thus we can guarantee that the noise eigenvalue is separated from the minimum signal eigenvalue λ_m as long as Equation (7.85) holds.

Next let us investigate the behavior of the weights $w_{MUSIC}(k)$ under the spiked population covariance model. Note first that Equation (7.81) can equivalently be written as

$$\frac{1}{M}\left[\sum_{j=1}^{m}\frac{\lambda_j}{\lambda_j-\mu_{\min}}+(M-m)\frac{\sigma_0^2}{\sigma_0^2-\mu_{\min}}\right]=\frac{1}{\gamma} \tag{7.87}$$

By definition, $\mu_{\min}<\lambda_M=\ldots=\lambda_{m+1}<\lambda_m\leq\ldots\leq\lambda_1$. Thus, $\lambda_j-\mu_{\min}$ can never go to zero for any $1\leq j\leq m$. Consequently, the first term of Equation (7.87) will go to zero as $M\to\infty$ for a fixed m, and $\mu_{\min}$ will converge to the solution of

$$\frac{\sigma_0^2}{\sigma_0^2-\mu_{\min}}=\frac{1}{\gamma} \tag{7.88}$$

namely,

$$\mu_{\min}\to\sigma_0^2(1-\gamma) \tag{7.89}$$

Consequently, for the spiked population model, we can express

$$w_{MUSIC}(k)=\begin{cases}1 & k>m\\[2ex] \gamma\sigma_0^2\frac{\lambda_k}{(\lambda_k-\sigma_0^2)^2}\Big/\left(1+\gamma\frac{\sigma_0^2}{\lambda_k-\sigma_0^2}\right) & k\leq m\end{cases}$$

Under the spiked population model, the MUSIC cost function is thus asymptotically equivalent to

$$\bar{\varphi}(R_M,\theta)=S^{\mathrm{H}}E_NE_N^{\mathrm{H}}S+\gamma\sigma_0^2\sum_{k=1}^{m}\frac{\frac{\lambda_k}{(\lambda_k-\sigma_0^2)^2}}{1+\gamma\frac{\sigma_0^2}{\lambda_k-\sigma_0^2}}S^{\mathrm{H}}\mathbf{e}_k\mathbf{e}_k^{\mathrm{H}}S \tag{7.90}$$

A similar simplification can be carried out for the expressions associated with the signal sample eigenvectors in Equation (7.84). Assume for simplicity that $K_j=1$, so that we are dealing with single-multiplicity signal eigenvectors. Expressing the weights $w_j(k)$, $1\leq j\leq m$ as

$$w_j(k)=\begin{cases}1-\sum_{\substack{r=1\\r\neq j}}^{m}\left(\frac{\lambda_j}{\lambda_r-\lambda_j}-\frac{\mu_j}{\lambda_r-\mu_j}\right)-(M-m)\,\sigma_0^2\frac{\lambda_j-\mu_j}{(\sigma_0^2-\lambda_j)(\sigma_0^2-\mu_j)} & k=j\\[2ex] \frac{\lambda_j}{\lambda_k-\lambda_j}-\frac{\mu_j}{\lambda_k-\mu_j} & k\neq j\end{cases}$$

and noting that for the spiked population model (see Johnson et al., 2008) $\mu_j \to \lambda_j$ and

$$M(\lambda_j - \mu_j) \to \frac{\gamma\lambda_j}{1-\gamma\frac{\sigma_0^2}{\sigma_0^2-\lambda_j}}, \quad 1 \le j \le m$$

we obtain

$$w_j(k) \to \begin{cases} \left(1-\gamma\left(\frac{\sigma_0^2}{\sigma_0^2-\lambda_j}\right)^2\right) \Big/ \left(1-\gamma\frac{\sigma_0^2}{\sigma_0^2-\lambda_j}\right) & k=j \\ 0 & k \neq j \end{cases}$$

Thus, one can ensure that, under the spiked population covariance simplification, and assuming that the separability condition (7.86) holds, one has

$$\left|\langle S^{\mathrm{H}}\hat{\mathbf{e}}_j\rangle\right|^2 \to \frac{1-\gamma\left(\frac{\sigma_0^2}{\sigma_0^2-\lambda_j}\right)^2}{1-\gamma\frac{\sigma_0^2}{\sigma_0^2-\lambda_j}}\left|\langle S^{\mathrm{H}}\mathbf{e}_j\rangle\right|^2 \tag{7.91}$$

where S can be any deterministic column vector with unit norm. This is precisely the result introduced by Paul (2004) for the specific class of spiked population covariance matrices, where again a fixed limited number of eigenvalues is greater than the smallest eigenvalue, the multiplicity of which grows with the number of antennas. If S is replaced with an eigenvector of the true covariance matrix $\mathbf{e}_k$, then (under the spiked population covariance matrix model, and assuming asymptotic subspace separation) the projection of a sample eigenvector onto the linear space spanned by an eigenvector associated with a different eigenvalue converges to zero:

$$\left|\langle \mathbf{e}_k^{\mathrm{H}}\hat{\mathbf{e}}_j\rangle\right|^2 \to \frac{1-\gamma\left(\frac{\sigma_0^2}{\sigma_0^2-\lambda_j}\right)^2}{1-\gamma\frac{\sigma_0^2}{\sigma_0^2-\lambda_j}}\delta_{j-k} \tag{7.92}$$

Additionally, Paul (2004) studied the convergence of the sample eigenvector $\hat{\mathbf{e}}_j$ when there is no asymptotic separation between signal and noise subspaces: $\lambda_m < \sigma_0^2(1+\sqrt{\gamma})$. In particular, he established that, under the spiked population covariance matrix model,

$$\left|\langle \mathbf{e}_j^{\mathrm{H}}\hat{\mathbf{e}}_j\rangle\right|^2 \to 0 \tag{7.93}$$

almost surely as $M, T \to \infty$ at the same rate. (In fact, Paul proved this for real-valued Gaussian observations with a diagonal covariance matrix; we can conjecture that the result is also valid for the observation model considered here.) Paul (2004) admitted that a "crucial aspect of the work (Baik and Silverstein, 2006; Baik et al., 2005) is the discovery of a phase transition phenomenon," which is clearly analogous to the subspace swap phenomenon known in the signal processing literature for 20 years (Tufts et al., 1988). Here it has been shown that the condition (7.75), or the simplified condition for the spiked population matrix model (7.86), which asymptotically prevents the phase transition phenomenon from occurring, is in fact the condition that guarantees the separability of the signal and noise subspaces in the asymptotic sample eigenvalue distribution.

Tufts et al. (1991) stated that the threshold effect in subspace techniques (such as MUSIC) is associated with the probability that the measured data is better approximated by some components of the orthogonal subspace than by some components of the signal subspace. A narrow investigation of the relationship between these asymptotic phase transitions and the subspace swap would therefore pinpoint the conditions under which the norm of the scalar product between the true and estimated eigenvectors in the signal subspace falls below 0.5. In this case, Tufts's description of the subspace swap becomes clearly equivalent. Therefore, the "signal processing" subspace swap definition implies that, for the last signal eigenvector of the true covariance matrix $\mathbf{e}_m$,

$$\mathbf{e}_m^H \hat{E}_N \hat{E}_N^H \mathbf{e}_m > \mathbf{e}_m^H \hat{E}_S \hat{E}_S^H \mathbf{e}_m \tag{7.94}$$

or, equivalently,

$$\mathbf{e}_m^H \hat{E}_S \hat{E}_S^H \mathbf{e}_m < 0.5 \tag{7.95}$$

That is, the signal eigenvector associated with the minimum signal eigenvalue (λ_m) is better represented by the noise sample subspace than the signal sample subspace. Accordingly, the behavior of the projection $\mathbf{e}_j^H \hat{E}_S \hat{E}_S^H \mathbf{e}_j$ is of prime importance for this analysis.

7.5.4 Simulation Analysis of MUSIC Breakdown

There is an immediate temptation to associate the MUSIC-specific performance breakdown with the phase transition described by Paul (2004) and the subspace swap threshold (7.95) associated with $\mathbf{e}_m^H \hat{E}_S \hat{E}_S^H \mathbf{e}_m$. However, detailed analysis of the previously introduced MUSIC breakdown scenario demonstrates that the relationship is more complicated. Examining the eigendecomposition of the true covariance matrix associated with the scenario given in Equation (7.56):

$$T = 6, M = 10, m = 3$$

$$(\sigma_1^2 = \sigma_2^2 = \sigma_3^2)/\sigma_0^2 = 20\text{-dB input SNR} \tag{7.96}$$

$$\theta_m = \{\arcsin(-0.40); 0; \arcsin(0.06)\} \tag{7.97}$$

we see that the first three eigenvectors of the true covariance matrix R_0 are equal to

$$\lambda_1 = 1869,\ \lambda_2 = 1002,\ \lambda_3 = 132,\ \left(\sum_{j=1}^{3} \lambda_j = 3003\right) \tag{7.98}$$

For $\lambda_3 = 132$, Paul's condition (7.86) evaluates as

$$\lambda_3 > 1 + \sqrt{(M/T)}\ (= 1 + \sqrt{10/6} = 2.29 \tag{7.99}$$

and is strongly satisfied. Therefore, according to Equation (7.92),

$$|\hat{\mathbf{e}}_m^{\mathrm{H}} \mathbf{e}_m| \xrightarrow{a.s.} \sqrt{(1 - 1.66/(131)^2)/(1 + 1.66/131)} = 0.9937 \tag{7.100}$$

which still may be treated as "extraordinarily good" estimation accuracy for the sample eigenvector $\hat{\mathbf{e}}_m$. A threshold phenomenon in the sense described in Paul (2004) is therefore not expected. Yet, as shown earlier in Table 7.1, almost half the MUSIC-derived estimates in this scenario contain outliers.

Therefore, quite subtle differences in the eigensubspace of the sample matrix $\hat{R}_M$ transform into catastrophic threshold behavior during the MUSIC transformation ($\hat{R}_M \rightarrow R_{MUSIC}$) for this scenario. One has to conclude that breakdown in the MUSIC-produced covariance matrix model R_{MUSIC} for such multiple-source scenarios may occur much more readily than subspace swaps or phase transitions (Paul, 2004) in the sample covariance matrices rather than matrix $\hat{R}_M$ itself.

The transition between the MUSIC-specific threshold phenomenon and ML performance breakdown can be illustrated by modifying the closely spaced sources' separation to ever-smaller values. When the separation for the $M = 10$, $T = 6, m = 3$ scenario drops to arcsin(0.003) (close to the statistical resolution limit (Smith, 2005), because the Cramer-Rao accuracy for isolated 20-dB input SNR targets in this scenario is 0.0018 radians (Swingler, 1993)), MUSIC fails completely, with all trials producing outliers, while the statistical outlier prediction also fails regularly, with 57.7% of the trials still identified as good by comparison of the LR_u with the threshold 0.363 calculated for a false alarm rate of 10^{-3}. Interestingly enough, within these misidentified trials, 15% also exceed a "strict" trial-dependent threshold based on the LR associated with the clairvoyant model. That is,

$$LR_u(X_T, R_{MUSIC}) > LR(X_T, R_0) \tag{7.101}$$

which means the accurate (not statistical) ML "breakdown" phenomenon has been reached, which is associated with the generation of a covariance matrix model R_{MUSIC} that contains a severely erroneous DOA estimate (an outlier) but is still considered more likely than the true covariance matrix by the LR test. Note that in this scenario, only a very small portion of the signal energy resides in the third eigenvector:

$$\lambda_1 = 1996.5,\ \lambda_2 = 1001,\ \lambda_3 = 5.5,\ \left(\sum_{j=1}^{3} \lambda_j = 3003\right) \tag{7.102}$$

which corresponds to a relatively small projection along the source directions of arrival

$$\left| \frac{\mathbf{e}_3^{\mathrm{H}} S(\theta_j)}{S_j^{\mathrm{H}}(\theta_j)\, S_j(\theta_j)} \right|, \quad j=1,2,3 \tag{7.103}$$

equal to 0.0014, 0.0475, and 0.0475, correspondingly.

Therefore, the "threshold" condition (7.99) (i.e., $\lambda_3 \gtrless 1+\sqrt{\gamma}$) of the subspace swap in the sample covariance matrix $\hat{R}_M$ (which is the unconstrained ML covariance matrix estimate and a sufficient statistic for any inference regarding this matrix in the complex Gaussian case) is associated with the condition of *ultimate ML performance breakdown.*

Indeed, for the scenario with $\theta_3 = \arcsin(0.003)$, which experiences 100% MUSIC breakdown, 57.7% statistical (7.55) ML breakdown, and 15% strict (7.101) ML breakdown, the third true eigenvalue of 5.5 is close to the Paul condition (7.99) of 2.29. Compare this condition with the 42.7% MUSIC breakdown and 0.1% ML breakdown experienced at $\theta_3 = \arcsin(0.06)$ and $\lambda_3 = 132$ (see Equation 7.98).

When the difference between the threshold conditions for the MUSIC-transformed model R_{MUSIC} and for the sample covariance matrix $\hat{R}$ itself is clearly demonstrated, the nature of the "gap" between MUSIC and ML breakdown threshold conditions observed in previous studies (Abramovich et al., 2002b, 2004, 2007a; Johnson and Abramovich, 2007; Spencer and Abramovich, 2004) has another justification. Because of this "gap" between ML and MUSIC breakdown conditions, it comes as no surprise that erroneous models R_{MUSIC} involving closely spaced sources may be identified (predicted) and rectified (cured) as long as the threshold condition (7.99) is strongly satisfied.

To further examine the large M, T asymptotic analytic predictions with respect to MUSIC breakdown, let us return to the four-source scenario (7.64) for the following four source SNR values:

- Input SNR = 25 dB → no MUSIC outliers
- Input SNR = 14 dB → ~50% MUSIC outliers
- Input SNR = 9 dB → almost 100% MUSIC outliers
- Input SNR = 0 dB → onset of ML breakdown

First, in Figure 7.12, for each of these four SNR values, the sample distributions of the four eigenvalues $\hat{\lambda}_1, \ldots, \hat{\lambda}_4$ in the signal subspace are separately shown, along with the distribution of all nonzero eigenvalues in the noise subspace. As expected, the separation between the noise and signal subspace eigenvalues decrease as the source SNR decreases, but one can see that even for the SNR of 9 dB with practically 100% MUSIC breakdown, the nonzero noise subspace eigenvalues are still well separated from the minimal signal subspace eigenvalue $\hat{\lambda}_4$. It is only at the lowest plotted SNR of 0 dB that significant overlap between the noise and signal subspace eigenvalues is observed. The true

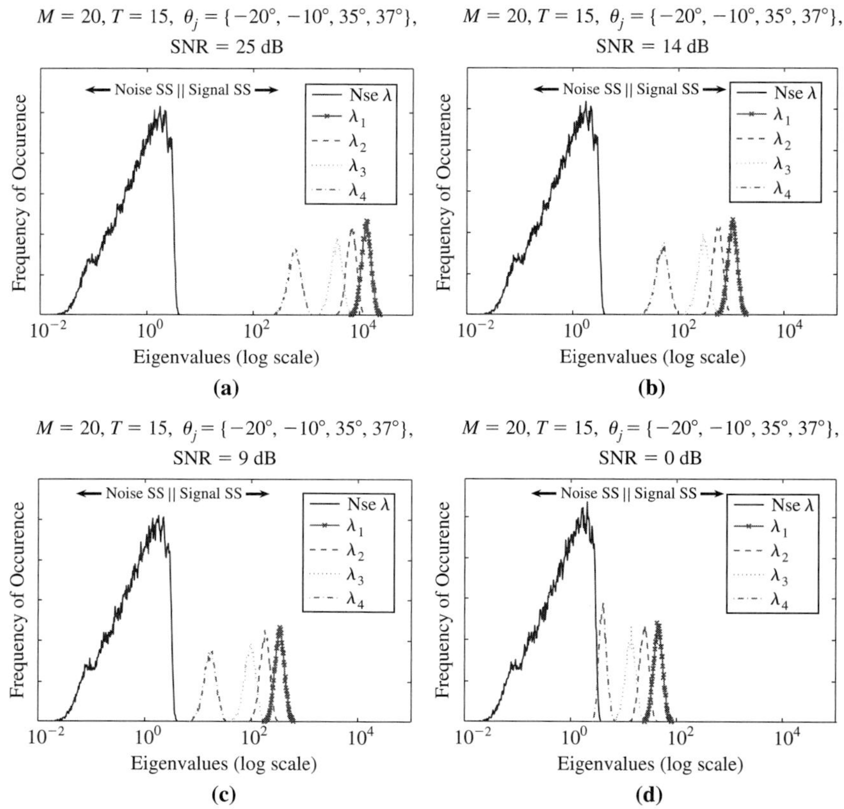

FIGURE 7.12 Eigenvalue distributions for the scenario in Equation (7.64): (a) 25-dB SNR—no MUSIC breakdown; (b) 14-dB SNR—~50% MUSIC outliers; (c) 9-dB SNR—~100% MUSIC outliers; (d) 0-dB SNR—start of ML breakdown. Even in the presence of significant MUSIC breakdown for the scenario SNR of 9 and 14 dB, the signal subspace eigenvalues remain well separated from the noise subspace eigenvalues. *Source:* © 2008 IEEE, with permission.

eigenvalues of the underlying true covariance matrices (denoted $Eig(M, \text{SNR})$) are

$$Eig(20, 25) = [11896, 7341, 5271, 794, 1, \ldots, 1] \tag{7.104}$$

$$Eig(20, 14) = [945, 584, 419, 64, 1, \ldots, 1] \tag{7.105}$$

$$Eig(20, 9) = [300, 185, 133, 21, 1, \ldots, 1] \tag{7.106}$$

$$Eig(20, 0) = [38.6, 24.2, 17.7, 3.5, 1, \ldots, 1] \tag{7.107}$$

which means that the subspace-splitting condition $\lambda_j > \lambda_M(1+\sqrt{\gamma})$ from Equation (7.86),

$$\lambda_4 > 1 \times (1+\sqrt{1.33}) = 2.15 \tag{7.108}$$

is satisfied for all four SNR values (although the 0-dB SNR case is marginal). The subspace-splitting condition given in Equation (7.75) can be computed for the transition from the signal subspace to the noise subspace, which is the splitting condition between the fourth and fifth eigenvalues. This gives values of 0.21, 0.24, 0.29, and 1.08 for 25 dB, 14 dB, 9 dB, and 0 dB, respectively. This value is clearly less than $T/M = 0.75$ for all but the last SNR value.

Based on these asymptotic splitting metrics only, one would conclude that the noise and signal subspace eigenvalues are distinct for all but the last case at 0-dB SNR, and therefore subspace techniques should operate robustly at the higher SNRs. Yet significant MUSIC breakdown occurs in the 9- and 14-dB SNR case. It is also important to establish that the finite M, T conditions examined here do not change dramatically asymptotically (in the sense of Equation 7.65), so in addition to the $M = 20$, $T = 15$ scenario considered above, let us examine the following three scenarios with an increased M and T dimension but with the ratio M/T held constant.

- The original scenario where $M = 20$:

$$\begin{aligned} &T = 15;\ \theta_m = \{-20^\circ, -10^\circ, 35^\circ, 37^\circ\}\ \text{SNR} = 14\,\text{dB} \\ &Eig(20, 14) = [945, 584, 419, 64, 1, \ldots, 1] \end{aligned} \tag{7.109}$$

- A 200-element array:

$$\begin{aligned} &T = 150;\ \theta_m = \{-20^\circ, -10^\circ, 35^\circ, 35.2^\circ\}\ \text{SNR} = 4\,\text{dB} \\ &Eig(200, 4) = [941, 508, 498, 65, 1, \ldots, 1] \end{aligned} \tag{7.110}$$

- A 400-element array with

$$\begin{aligned} &T = 300;\ \theta_m = \{-20^\circ, -10^\circ, 35^\circ, 35.1^\circ\}\ \text{SNR} = 1\,\text{dB} \\ &Eig(400, 1) = [943, 508, 500, 66, 1, \ldots, 1] \end{aligned} \tag{7.111}$$

Inter-source separations and SNR in the increased array size scenarios in Equations (7.110) and (7.111) have been chosen to produce essentially the same signal eigenvalues as per the original scenario in Equation (7.109) with $M = 20$. All three eigenspectra have minimal signal subspace eigenvalues in the range of 64 to 66, which allows us to expect the same asymptotic behavior under the spiked population covariance model. In Figure 7.13, sample distributions of the projection $\mathbf{e}_4^{\mathrm{H}} \hat{E}_S \hat{E}_S^{\mathrm{H}} \mathbf{e}_4$, calculated for all three scenarios in Equations (7.109), (7.110), and (7.111) (with $M/T = 1.33$ in all cases) are introduced. First, observe that, in full agreement with the results derived from large Random Matrix theory, the projection is converging as $M, T \to \infty$ ($M/T \to$ constant) to a nonstatistical deterministic value. Indeed, one can observe quite a consistent convergence of the sample distributions for $\mathbf{e}_4^{\mathrm{H}} \hat{E}_S \hat{E}_S^{\mathrm{H}} \mathbf{e}_4$ to a delta function, whether the scenario contains a MUSIC outlier or not. Furthermore, while the results are converging asymptotically, the mean values observed with a modest array dimension of 20 elements are already quite accurate (to within 0.5% of the mean values observed with 400 elements).

To validate these asymptotic deterministic values, we point out that

$$\mathbf{e}_4^{\mathrm{H}}\hat{E}_S\hat{E}_S^{\mathrm{H}}\mathbf{e}_4 = 1 - \mathbf{e}_4^{\mathrm{H}}\hat{E}_N\hat{E}_N^{\mathrm{H}}\mathbf{e}_4$$

where the convergence of the term $\mathbf{e}_4^{\mathrm{H}}\hat{E}_N\hat{E}_N^{\mathrm{H}}\mathbf{e}_4$ has been characterized in Equation (7.83) for the general model and in Equation (7.90) for the specific model of the spiked covariance matrix limit. More specifically, if we focus on the spiked model simplification and replace S with $\mathbf{e}_4$ in these expressions, one can state that

$$\mathbf{e}_4^{\mathrm{H}}\hat{E}_S\hat{E}_S^{\mathrm{H}}\mathbf{e}_4 \to \frac{1-\gamma\left(\frac{\sigma_0^2}{\lambda_4-\sigma_0^2}\right)^2}{1+\gamma\frac{\sigma_0^2}{\lambda_4-\sigma_0^2}} \tag{7.112}$$

with probability one. This asymptotic expression is in fact the same as that presented in Equation (7.92), but now the projection is onto the entire sample subspace (rather than a single eigenvector). This is because, as pointed out in Figure 7.11, in the spiked population model separability between signal and noise subspaces precludes separability within the signal subspace.

As can be observed in Figure 7.13, the discrepancy between the estimated mean values for $\mathbf{e}_4^{\mathrm{H}}\hat{E}_S\hat{E}_S^{\mathrm{H}}\mathbf{e}_4$ and the prediction in Equation (7.112) is within the fourth decimal point for a set of 10^2 Monte Carlo trials and an array of $M=400$ elements. While the match for Equation (7.112) is quite good even for small arrays, the projections of the sample eigenvectors onto the individual true signal subspace eigenvectors, $|\mathbf{e}_j^T\hat{\mathbf{e}}_j|$, are separately observed to deviate significantly

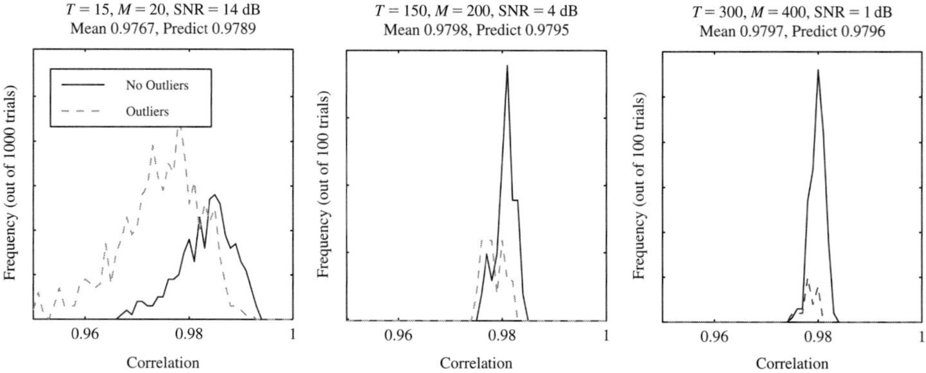

FIGURE 7.13 Projection of the fourth true eigenvector onto the sample signal subspace (scenarios 7.109 through 7.111). All scenarios show significant MUSIC breakdown (~60%, ~30%, and ~15%, respectively), but the mean projection is converging to the predicted value in Equation (7.112), with a high value that indicates little subspace leakage. *Source:* © 2008 IEEE, with permission.

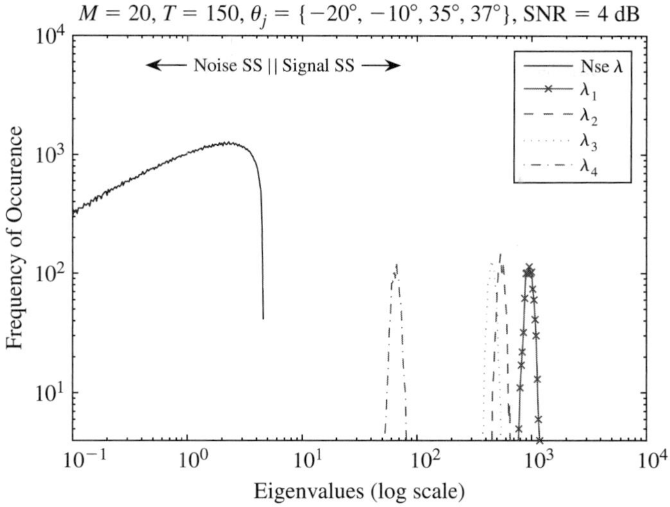

FIGURE 7.14 Eigenvalue distributions for the first four eigenvalues and noise eigenvalues for the scenario in Equation (7.110), SNR = 4 dB. Note the significant overlap between the second and third eigenvalue distributions. *Source:* © 2008 IEEE, with permission.

from Equation (7.92) even for large arrays under some circumstances. The problem occurs in scenarios where the spiked population model is too restrictive and the eigenvalue-splitting condition is not met for all signal subspace eigenvalues, contradicting the general assumption of spiked population models. For example, in the scenario in Equation (7.110) with $M = 200$ elements, $T = 150$ samples, we observe a very small difference between the second ($\lambda_2 = 508$) and third ($\lambda_3 = 498$) eigenvalue (see Figure 7.14), which indicates that the associated clusters in $g_M(\lambda)$ may not be separated. Indeed, a rapid calculation according to the procedure in Figure 7.7 indicates that the support of $g_M(\lambda)$ in this scenario is given by

$$\mathcal{S}_M = \underbrace{[0.025, 4.53]}_{\lambda_5} \cup \underbrace{[54.82, 75.78]}_{\lambda_4} \cup \underbrace{[389.42, 618.88]}_{\lambda_2, \lambda_3} \cup \underbrace{[811.45, 1116]}_{\lambda_1}$$

so that clearly the cluster associated with λ_2, λ_3 is not separated, as the spiked model implicitly assumes. This is an intrinsic limitation of the spiked covariance matrix model, which does not accurately model the finite M situation for the second and third eigenvalues. Of course, this limitation would be mitigated by allowing M to increase further while keeping the multiplicities of the signal eigenvalues fixed. This procedure would shrink the support of the signal eigenvalue clusters to zero, guaranteeing that separability holds for all eigenvalues (as is the case in the spiked population model).

As a consequence of this limitation of the spiked population model, intra-subspace swap for eigenvectors 2 and 3 was frequently observed, and the sample

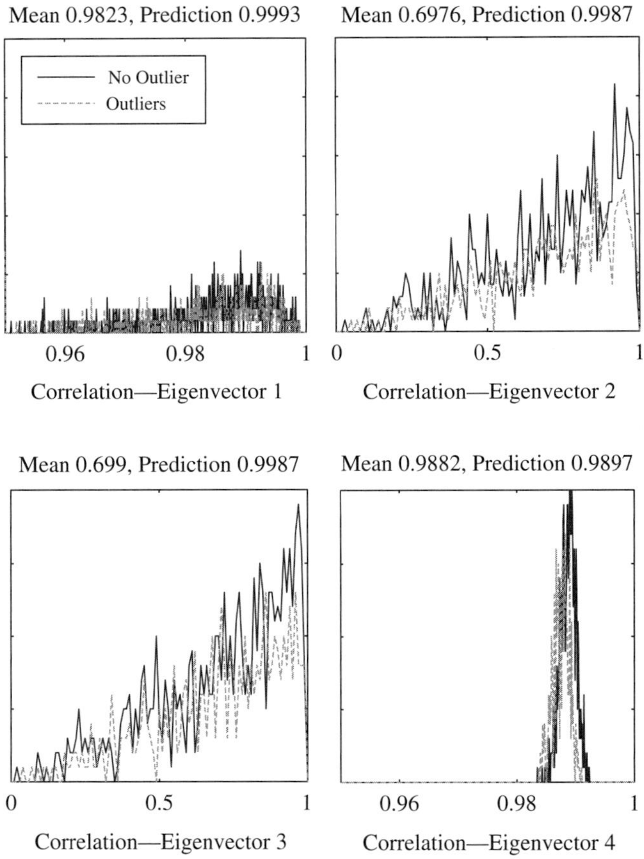

FIGURE 7.15 Inner product of sample and true eigenvectors for an M = 200-element array and the scenario in Equation (7.110). Correlation of the second and third sample eigenvectors with their associated true eigenvectors $|\langle^{H}_{k\ j}\rangle|$ is poor because of frequent eigenvector swap, as is the match between the observed mean and the prediction from Equation (7.92). Such an intra-subspace swap should not affect MUSIC or any other subspace-based technique. *Source:* © 2008 IEEE, with permission.

distribution of $|\langle \mathbf{e}_2^T \hat{\mathbf{e}}_2 \rangle|$ is spread widely over the [0 1] interval, as seen in Figure 7.15. While this intra-subspace swap phenomenon does not persist asymptotically ($M, T \rightarrow \infty, M/T \rightarrow \gamma$) as the relative dimension of the signal subspace vanishes, it still indicates that, for breakdown analysis on finite arrays using the spiked covariance model, the projection onto the entire subspace rather than individual eigenvectors is the appropriate metric.

Finally, the most important observation from the MUSIC breakdown standpoint is that for both "proper" trials with no outliers and "improper" trials with at least one outlier, the minimal signal subspace eigenvector still resides in the sample signal subspace with more than 95% of its power, converging asymptot-

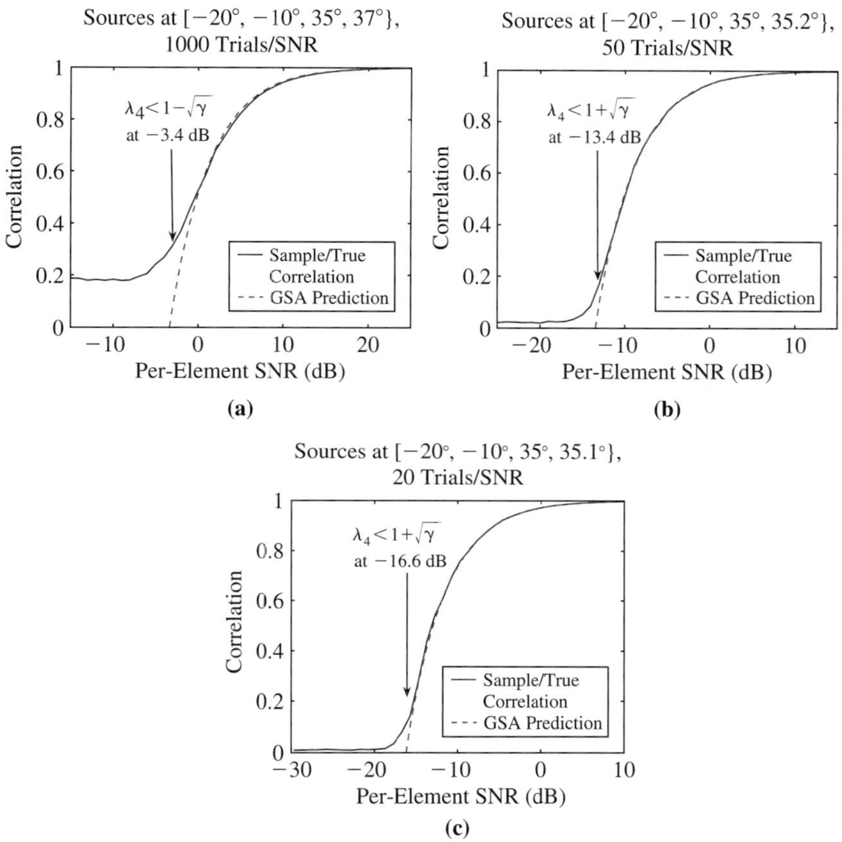

FIGURE 7.16 Comparison of predicted and observed projection of the fourth sample eigenvector onto the true signal subspace. (a) $M = 20, T = 15$; (b) $M = 200, T = 150$; (c) $M = 400, T = 300$. The correspondence between the observations and the predictions above $\lambda_4 < 1 + \sqrt{\gamma}$ is accurate even at small array sizes such as $M = 20$. *Source:* © 2008 IEEE, with permission.

ically to 98%. This convergence is accurately predicted by Equation (7.112) for multiple SNR values, as indicated in Figure 7.16.

The main conclusion that is now supported both by large Random Matrix theory and direct Monte Carlo simulations is that, for the considered scenario, subspace swap *is not responsible* for the MUSIC breakdown phenomenon observed, and the underlying mechanism requires further exploration.

7.5.5 Source Resolution and MUSIC Performance Breakdown

Careful examination of the pseudo-spectrum produced during trials with MUSIC outliers, such as in the scenario in Equation (7.64) with SNR = 14 dB, shows that in many trials the MUSIC algorithm selected an erroneous peak at θ_{er} despite the fact that the pseudo-spectrum value at that peak was significantly smaller than the pseudo-spectrum values at any true source direction:

$$S^{\mathrm{H}}(\theta_{er})\hat{E}_S\hat{E}_S^{\mathrm{H}}S(\theta_{er}) \ll S^{\mathrm{H}}(\theta_j)\hat{E}_S\hat{E}_S^{\mathrm{H}}S(\theta_j), \quad j=1,\ldots,4 \tag{7.113}$$

This happened only because MUSIC was unable to resolve the third and fourth closely located sources ($\theta_3 = 35°$, $\theta_4 = 37°$) and instead found a single maxima in their vicinity. This well-known phenomenon of *loss of MUSIC resolution capability* (Kaveh and Barabell, 1986) is not directly associated with the "subspace swap" phenomenon, and in fact is associated with a significantly smaller portion of sample signal subspace energy residing in the noise subspace than is required for subspace swap as defined in Equation (7.94). This fact has been demonstrated by the experimental data in Figures 7.13 through 7.16 as well as by the large Random Matrix theory prediction (7.112).

To examine the effect of this loss of resolution further, a "resolution event" needs to be defined. In Cox (1973), Cox defines a resolution event for closely spaced sources, such as the DOAs θ_3 and θ_4 in the scenario in Equation (7.64), as

$$\begin{aligned}\hat{\alpha}(\theta_3,\theta_4) = &-2S^{\mathrm{H}}(\theta_*)\hat{E}_S\hat{E}_S^{\mathrm{H}}S(\theta_*)\\ &+S^{\mathrm{H}}(\theta_3)\hat{E}_S\hat{E}_S^{\mathrm{H}}S(\theta_3)+S^{\mathrm{H}}(\theta_4)\hat{E}_S\hat{E}_S^{\mathrm{H}}S(\theta_4) > 0\end{aligned} \tag{7.114}$$

where $\theta_* = (\theta_3 + \theta_4)/2$. This condition means that the sample MUSIC pseudo-spectrum in the mid-point θ_* between the true DOAs (θ_3 and θ_4) lies *below* the line that connects the pseudo-spectrum values at the true DOAs.

Although this definition is commonly used, it should be noted that this metric ($\hat{\alpha} \gtrless 0$) may only *approximate* the actual resolution event. Indeed, in Figure 7.17, two actual examples from the set of Monte Carlo trials in the scenario in

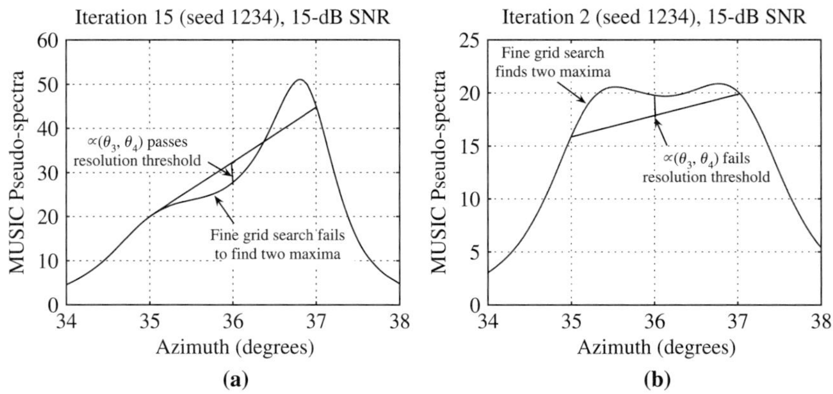

FIGURE 7.17 Resolution of two closely spaced sources from the scenario in Equation (7.64), SNR = 15 dB. (a) $\hat{\alpha} > 0$, but only one maxima; (b) $\hat{\alpha} < 0$, but two maxima. The two example trials illustrate the difference between the resolution event criteria (7.114) and the local maxima criteria used in the simulation. *Source:* © 2008 IEEE, with permission.

Equation (7.64) with an SNR of 15 dB are illustrated. The resolution definition is compared to a local maxima criterion, with Figure 7.17(a) showing success of the resolution event when local maxima tests fail, and Figure 7.17(b) conversely showing failure of the resolution event when local maxima tests succeed. These variations are statistical, however, and on average the probability that Equation (7.114) is positive should correspond well to the probability of actual resolution. In Lee and Wengrovitz (1990), Xu and Kaveh (1994), and Zhang (1995), analytic approximations of this resolution probability are found for the standard asymptotic conditions ($M=\text{constant}, T\to\infty$) and proven to be quite accurate for $T\gg M$, as in the example with $M=8, T=100$, considered in Zhang (1995). However, for $M/T>1$, it is once again more appropriate to apply large Random Matrix theory asymptotic analysis. In accordance with the introduced methodology, one can find a nonrandom (deterministic) function $g(\theta_3,\theta_4)$, such that

$$|\hat{\alpha}(\theta_3,\theta_4)-g(\theta_3,\theta_4)|\longrightarrow 0 \tag{7.115}$$

with probability one as $M,T\to\infty$, $M/T\to\gamma<\infty$. The condition $g(\theta_3,\theta_4)=0$ is then associated with roughly 50% probability of resolution, and the decision threshold

$$g(\theta_3,\theta_4) \underset{\text{no res}}{\overset{\text{res}}{\gtrless}} 0 \tag{7.116}$$

with high (and low) resolution probability correspondingly.

First of all, in Figure 7.18 sample distributions for the $\hat{\alpha}(\theta_3,\theta_4)$ resolution metric are shown, calculated for SNR = 9, 14, and 25 dB, correspondingly. As usual, sample distributions (for SNR = 14 dB) averaged over the subsets of improper and proper MUSIC trials are introduced (recalling that for SNR = 9 dB, practically all MUSIC trials are improper while for SNR = 25 dB, all trials are proper). Although trials with outliers in general have a negative resolution metric (regardless of SNR) and those without outliers have a positive metric, there is some overlap at SNR = 14 dB. The overlap between these two distributions is explained by the circumstance instead of the phenomenon illustrated in Figure 7.17, since the Monte Carlo simulation DOA estimation algorithm searches for strict maxima in the pseudo-spectrum rather than applying the subject resolution metric.

To get an analytic expression for the asymptotic function $g(\theta_3,\theta_4)$ in Equation (7.115), let us once again consider Theorem 7.2, or the corresponding version for the noise eigenvectors given in Equation (7.82). According to this result, if the signal/noise subspace-splitting condition (7.75) is satisfied, then the MUSIC random pseudo-spectrum $\varphi(\hat{R}_M,\theta)$ in Equation (7.68) asymptotically ($M,T\to\infty, M/T\to\text{const}$) tends to the nonrandom function $\overline{\varphi}(R_M,\theta)$:

$$|\varphi(\hat{R}_M,\theta)-\overline{\varphi}(R_M,\theta)|\to 0 \text{ as } M,T\to\infty,\ M/T\to\gamma,\ 0<\gamma<\infty \tag{7.117}$$

where

$$\bar{\varphi}(R_M,\theta)=\sum_{k=1}^{M} w_{MUSIC}(k)S^{\mathrm{H}}\mathbf{e}_k\mathbf{e}_k^{\mathrm{H}}S$$

and where $w_{MUSIC}(k)$ are specified in Equation (7.82).

According to this result,

$$S^{\mathrm{H}}(\theta)\hat{E}_S\hat{E}_S^{\mathrm{H}}S(\theta)\Rightarrow 1-\overline{\varphi}(R_M,\theta) \tag{7.118}$$

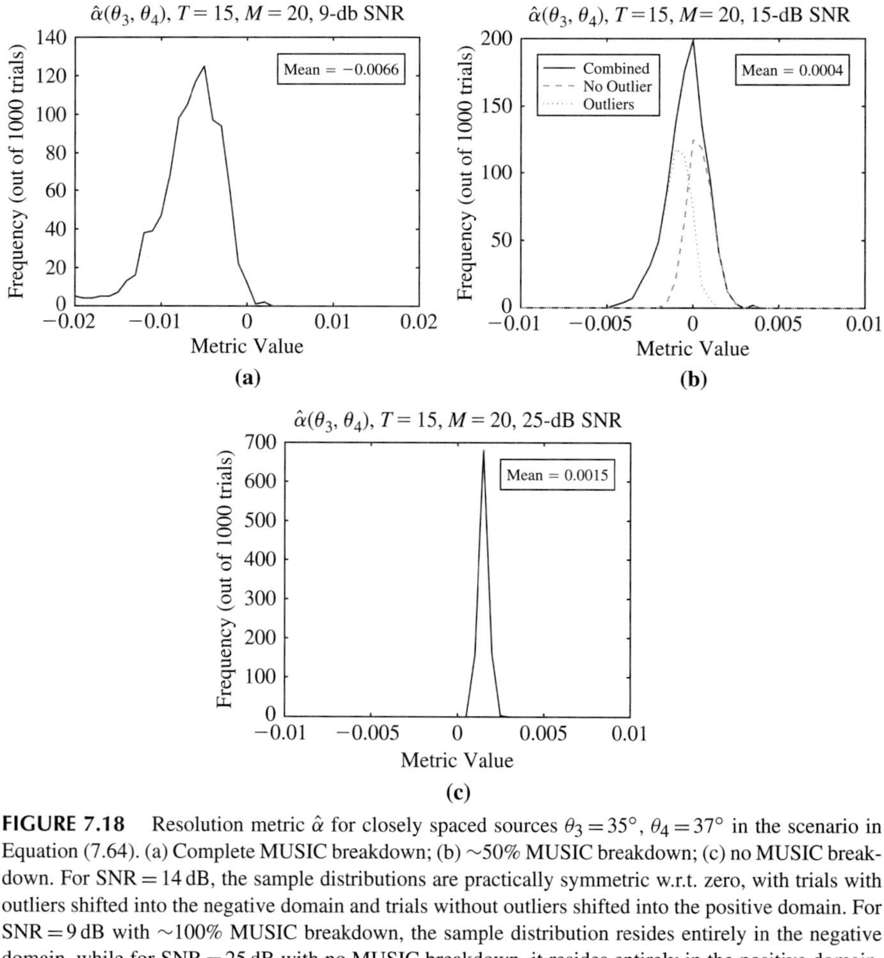

FIGURE 7.18 Resolution metric $\hat{\alpha}$ for closely spaced sources $\theta_3=35°$, $\theta_4=37°$ in the scenario in Equation (7.64). (a) Complete MUSIC breakdown; (b) ~50% MUSIC breakdown; (c) no MUSIC breakdown. For SNR = 14 dB, the sample distributions are practically symmetric w.r.t. zero, with trials with outliers shifted into the negative domain and trials without outliers shifted into the positive domain. For SNR = 9 dB with ~100% MUSIC breakdown, the sample distribution resides entirely in the negative domain, while for SNR = 25 dB with no MUSIC breakdown, it resides entirely in the positive domain. *Source:* © 2008 IEEE, with permission.

since we assume $\|S(\theta)\| = 1$, and therefore the function $g(\theta_3, \theta_4)$ in Equation (7.115) is expressed as

$$g(\theta_3,\theta_4) = -\sum_{k=1}^{M} w_{MUSIC}(k)\left[2\left|S^{\mathrm{H}}(\theta_*)\mathbf{e}_k\right|^2 - \left|S^{\mathrm{H}}(\theta_3)\mathbf{e}_k\right|^2 - \left|S^{\mathrm{H}}(\theta_4)\mathbf{e}_k\right|^2\right] \tag{7.119}$$

According to this expression, one gets the asymptotic values of $g(\theta_3, \theta_4) = -0.0066, -0.0004$, and $+0.0015$ for input SNRs of 25, 14, and 9 dB, respectively, which agree precisely with the mean values in Figure 7.18.

Consistent with the earlier conclusion that subspace swap was not the sole reason for MUSIC outlier production, one can now see that MUSIC's performance breakdown can occur when it is unable to resolve some closely spaced sources; for this to happen, quite negligible power of the actual signal subspace has to reside in the sample noise subspace ("subspace leakage"). Another important conclusion is that the large Random Matrix theory methodology has been proven to be sufficiently accurate for scenarios with surprisingly small M $(=20)$ and T $(=15)$, compared with the asymptotic requirements $(M, T \to \infty,\ M/T \to \text{constant})$, and can provide accurate predictions of MUSIC resolution performance.

The introduced analysis sheds some light on the reasons for the disappointing threshold performance improvement delivered by the G-MUSIC algorithm (Mestre, 2006a,b; see also Mestre, 2008a, and Mestre and Lagunas, 2008), as observed in Section 7.4. While G-MUSIC is indeed able to counter the bias in noise subspace estimation caused by the low sample support, the almost sure convergence to the *accurate* zeros of the cost function still may possess sufficient variance such that the value $\varphi(R_M, \theta_*)$ at a midpoint between the two closely spaced sources is even smaller, leading to a single minima (or maxima for the inverse pseudo-spectrum) and corresponding loss of resolution.

7.6 SUBSPACE SWAP AND MLE PERFORMANCE BREAKDOWN

In a way, the improved performance of MLE in the threshold region relative to MUSIC (and G-MUSIC) is reflected by the fact that MUSIC breaks down significantly earlier than estimation of the number of sources m by ITC (Lee and Li, 1993). Both the MLE criterion and the ITC approach test the *entire covariance matrix model* to fit the sample data, while MUSIC (and G-MUSIC) selects each DOA estimate independently, with no regard to how the entire set of produced estimates fits the input data. Of course, there is still the possibility of different threshold conditions for MLE and ITC as well, but with a significantly smaller gap. The significant difference between MUSIC-specific and ML-intrinsic threshold conditions is the penalty one has to pay for replacing the multivariate ML optimization problem that finds the set of estimates that jointly best fit the input data, by univariate search of the function that only has the same solution as $T \to \infty$.

MLE breakdown is observed when a set of DOA estimates that contains a severely erroneous estimate (an outlier) generates an LF value that exceeds the local extremum in the vicinity of the true model. In other words, there are maxima of the LF (including the global maximum) that exceed the local maximum of the LF considered by traditional asymptotic ML analysis ($T \to \infty$) as the ML solution. Generation of such an outlier is dependent on proper detection of the number of sources present, which may also be questionable in this ML-intrinsic threshold condition (an issue not explored in detail here).

For a solution that contains an outlier to be "more likely" than the actual covariance matrix, the sample data should indeed generate a sample signal subspace with some of its elements better represented by the true noise subspace. Therefore, the subspace swap phenomenon is more likely to be associated with MLE breakdown than with the breakdown in subspace techniques. To demonstrate this, let us analyze MLE performance in the scenario in Equation (7.64) at three SNR values: 2 dB with $\lambda_4 = 4.97$, 0 dB with $\lambda_4 = 3.51$ (again as in Equation 7.107), and -4 dB with $\lambda_4 = 2.00$.

For SNR = 2 dB and $\lambda_4 = 4.97$, the Mestre eigenvalue-splitting condition (7.75) for the fourth eigenvalue is just satisfied ($0.68 < T/M = 0.75$), indicating (almost sure) separation of the signal and noise subspaces asymptotically (in the GSA sense), while according to Equation (7.112) for this SNR,

$$\mathbf{e}_4^{\mathrm{H}} \hat{E}_S \hat{E}_S^{\mathrm{H}} \mathbf{e}_4 \xrightarrow{a.s.} 0.69 \tag{7.120}$$

Monte Carlo simulations show a mean for $\mathbf{e}_4^{\mathrm{H}} \hat{E}_S \hat{E}_S^{\mathrm{H}} \mathbf{e}_4$ of 0.6815, agreeing well with the prediction and indicating that the fourth eigenvector projects more onto its proper signal subspace than the noise subspace.

For SNR = 0 dB and $\lambda_4 = 3.51$, the Mestre eigenvalue-splitting condition (7.75) is marginally violated ($1.07 > T/M = 0.75$), while the projection of signal eigenvectors onto the signal subspace is forecast via Equation (7.112) as

$$\mathbf{e}_4^{\mathrm{H}} \hat{E}_S \hat{E}_S^{\mathrm{H}} \mathbf{e}_4 \xrightarrow{a.s} 0.51 \tag{7.121}$$

Monte Carlo simulations show a mean for $\mathbf{e}_4^{\mathrm{H}} \hat{E}_S \hat{E}_S^{\mathrm{H}} \mathbf{e}_4$ of 0.5098. While the forecast in Equation (7.112) is not technically valid when the splitting condition is violated, at this marginal condition it still provides a projection value prediction that is consistent with simulation results. The projection value of around 50% indicates that the subspace swap condition given in Equation (7.95) is essentially satisfied and a subspace swap is statistically likely, consistent with the observation in Figure 7.3 that MLE breakdown starts to occur at 0-dB input SNR.

For SNR = -4 dB, $\lambda_4 = 2.0$, and the condition for asymptotic convergence ($\lambda_4 > 1 + \sqrt{\gamma} = 2.15$) given in Equation (7.86) is violated, so the projection should converge to zero asymptotically and Equation (7.112) is clearly no longer valid. Monte Carlo simulations show a mean for $\mathbf{e}_4^H \hat{E}_S \hat{E}_S^H \mathbf{e}_4$ of 0.2502. The scenario is obviously experiencing significant subspace swap, once again consistent with results given in Figure 7.3.

Having established that MLE performance breakdown in the examined multiple-source scenario is reliably associated with *subspace swap* (whereas MUSIC breakdown in the same scenario is associated with *subspace leakage*, as shown in Section 7.5, and resultant loss of resolution, as shown in Section 7.5.5), let us next examine a single-source scenario, such as that studied in Athley (2002, 2005), where one expects that subspace swap is indeed the sole mechanism responsible for both MLE and MUSIC DOA estimation performance breakdown.

To this end, an additional scenario with a single target is introduced that is based, as in Athley (2002, 2005), on a sparse minimum redundancy array (MRA) (Moffet, 1968), where the generation of outliers is more likely as a result of poor sidelobe performance. The following specific $M = 18$ configuration ($d = [0, 2, 10, 22, 53, 56, 82, 83, 89, 98, 130, 148, 153, 167, 188, 192, 205, 216]$) is used, as suggested for the MRA context in Sverdlik (1975) and confirmed to be minimally redundant (Dollas et al., 1998).

The threshold effect of MLE estimation in this scenario (provided by the Barlett spectrum or conventional beamforming, CBF) can be observed in Figure 7.19 to occur around −5 dB for $T = 14$.

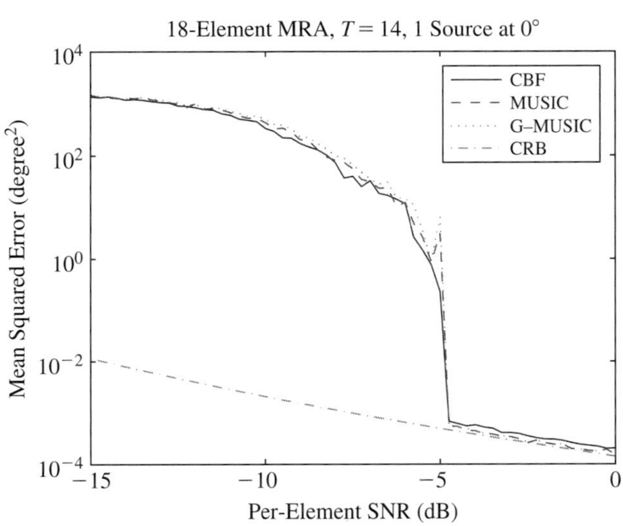

FIGURE 7.19 MSE for MUSIC, G-MUSIC, and MLE (CBF) DOA estimation on an 18-element minimum redundant array with 1000 trials/SNR step. Note that the MUSIC and G-MUSIC estimators deliver essentially the same performance in this circumstance, as expected for single sources. *Source:* © 2008 IEEE, with permission.

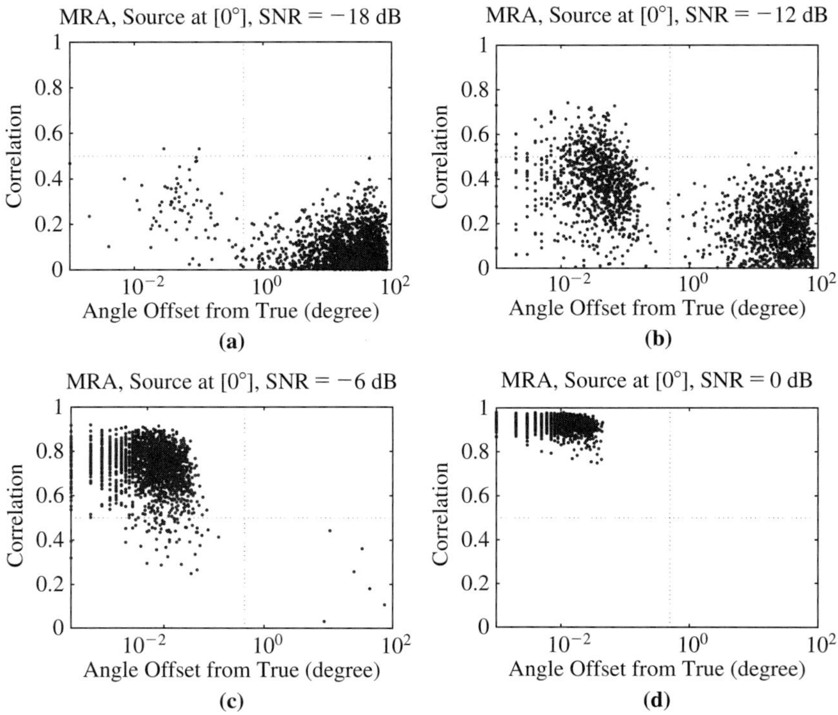

FIGURE 7.20 Distribution of DOA estimation errors versus projection $\mathbf{e}_j^H \hat{E}_S \hat{E}_S^H \mathbf{e}_j$ for a single target scenario on an 18-element minimum redundant array with $T = 14$ samples. (a) ~100% MLE breakdown; (b) ~50% MLE breakdown; (c) rare MLE breakdown; (d) no MLE breakdown. As SNR is increased, the projection approaches unity and the estimation accuracy improves. Note, however, that trials with low projection values still frequently have low estimation errors. *Source:* © 2008 IEEE, with permission.

In Equation (7.94) subspace swap was defined as occurring when the projection of the last true eigenvector into the underlying sample noise subspace was higher than that into the sample signal subspace. To examine whether this subspace swap is the sole mechanism for MLE breakdown, the DOA error of a single source estimated with the MRA, versus the correlation between the "maximal" sample and the true eigenvector, can be plotted for each of 1000 Monte Carlo trials and a sample size of $T = 14$. These plots are shown in Figure 7.20 for source SNRs ranging from very low values that result in complete MLE breakdown (input SNR of −18 dB, as shown in Figure 7.20(a)) to values where there is no MLE breakdown (input SNR of 0 dB, as shown in Figure 7.20(d)).

Figure 7.20 clearly demonstrates that when projection of the signal true eigenvector onto the sample signal subspace is high, there is no MLE breakdown (i.e., the upper right quadrant of Figures 7.20(a)–(d) are all free of any Monte Carlo trials). Interestingly, however, the converse is not true. When the projection of the signal true eigenvector onto the sample signal subspace is low, a DOA outlier estimate may or may not be produced. Thus, subspace swap is a necessary but not sufficient condition for MLE breakdown to occur.

7.7 CONCLUSION

In this chapter, the expected likelihood approach was introduced as a mechanism that allows assessment of the quality of estimates without resort to asymptotic or clairvoyant analysis. To address an important low-sample support regime, the expected likelihood approach was expanded into the sample data circumstances where the number of samples T does not exceed the dimension M of each snapshot. Three basic requirements for likelihood ratios to be used in an undersampled expected likelihood framework were detailed, the most important being the property that the p.d.f. of the likelihood ratio for the unknown true model R_0 must be independent of any scenario. Undersampled likelihood ratios that satisfy these requirements were then derived via two different mechanisms: projection of the covariance matrix model onto the subspace spanned by the sample covariance matrix, and a maximum-entropy extension of the central band of the rank-deficient sample covariance matrix.

The expected likelihood detection/estimation methodology was applied over an important class of "undersampled" scenarios, where the number of independent Gaussian samples T does not exceed the number of antenna array elements M, by applying likelihood ratios formulated for operation with singular sample covariance matrices. Specific application of the introduced methodology dealt with the well-known "threshold effect" in MUSIC, where the lack of SNR and/or sample support T below certain threshold values causes MUSIC to generate severely erroneous DOA estimates (outliers) with high probability.

It was shown that a gap between MUSIC-specific breakdown and ML-intrinsic breakdown in the sample covariance matrix $\hat{R}_M$ itself can be exploited to reliably identify outliers, since they generate $LR_u(X_T|R_{MUSIC})$ values that are statistically much smaller than $LR_u(X_T|R_0)$. Analytically or experimentally calculated "scenario-free" threshold values h^* with

$$Prob[LR_u(X_T|R_0) < h^*] = P_{FA} \ll 1 \tag{7.122}$$

were then used, and demonstrated quite remarkable efficiency of the MUSIC-specific breakdown prediction.

Examination of the performance of another DOA estimation algorithm (G-MUSIC) derived from consistency requirements in a doubly asymptotic framework rather than via the standard large-sample asymptotic approach showed some improvement in outlier production, allowing reliable operation at lower SNRs. However, a gap between the G-MUSIC breakdown and the ML-intrinsic breakdown was still shown to exist.

To explore this further, a number of results from Random Matrix theory based on the asymptotic environment, where both the sample support T and the array dimension M go to infinity at the same rate, were studied. Interestingly, it was demonstrated that the breakdown in R_{MUSIC} takes place at much higher SNR and/or sample support values T than those predicted by the Random Matrix theory subspace swap condition for the sample covariance matrix $\hat{R}_M$. Specifically, it was demonstrated that the "phase transition" predicted by RMT for the scenario with $M \times m$ signal subspace of the covariance matrix ($\lambda_m < 1 + \sqrt{(M/T)}$) is associated in multiple-source scenarios not with MUSIC-specific or G-MUSIC specific breakdown but rather with ultimate ML-intrinsic breakdown that cannot be recovered. This was shown by verifying that MUSIC-generated covariance matrix models R_{MUSIC} contain outliers in scenarios with $\lambda_m(R_0) \gg 1 + \sqrt{(M/T)}$. It appears that MUSIC-specific threshold conditions in multiple-source scenarios are triggered by very subtle changes in the eigendecomposition of the sufficient statistic (sample covariance matrix) $\hat{R}_M$ ("subspace leakage") rather than the full subspace swap traditionally associated with MUSIC breakdown in the literature.

Customarily, subspace swap is described as an event where a particular (minimal) signal subspace eigenvector is better represented (expanded) by the noise subspace of the sample covariance matrix than by the sample signal subspace. The analysis here demonstrated that the RMT/GSA methodology very accurately predicts the subspace swap conditions, even for antenna dimensions and sample volume that are far from the M, T asymptotic regime. The same analysis showed that MUSIC (and G-MUSIC) performance breakdown can take place when less than 5% of the minimal signal subspace eigenvector's power resides in the sample noise subspace, leaving more than 95% of this power residing in the sample signal subspace. Focusing on the multiple-source case, GSA methodology was found to be able to predict this breakdown condition using a standard resolution test, but with asymptotically justified weighting factors, allowing the definition of the threshold SNR and/or sample volume required for reliable resolution for a given array configuration and scenario.

Returning to consideration of the full subspace swap condition, it was established that MLE breakdown is reliably associated with the subspace swap phenomenon and well predicted by the GSA methodology. It is obvious that scenarios where the MUSIC (and G-MUSIC) pseudo-spectrum does not differ significantly from the conventional Barlett spectrum (single or well-separated sources, very low SNR, $T \to \infty$), MUSIC and MLE techniques demonstrate similar threshold performance, with full subspace swap becoming the common reason for breakdown in both techniques. For single-source cases, the similar breakdown point for MLE and MUSIC is well correlated with GSA-derived eigenvalue-splitting and subspace swap predictions. It was noted, however, that while high projection values of the minimal signal eigenvector onto the sample signal subspace precludes the formation of outliers leading to performance breakdown, low projection values (indicating in some cases almost complete subspace swap) did not always lead to performance breakdown. Therefore, subspace swap is a necessary, but not sufficient, condition for DOA estimation breakdown, with other factors also influencing outlier production, such as statistical variations of the source power and manner of distribution of the signal subspace power across the sample noise subspace.

These investigations once again point out the need for performance assessment tools other than the standard large-sample asymptotics when dealing with finite-sample threshold region behavior. The power of the RMT/GSA asymptotic approach is particularly evident here, as is the capability of expected likelihood to provide nonclairvoyant solution assessment in scenarios outside the ML-breakdown region but within SNR and/or sample support regimes whether or not other estimation methods fail.

In summary, the finite-sample performance assessment objectives laid out at the beginning of the chapter can be achieved using new applications of likelihood ratio formulations as well as Random Matrix theory, providing new tools for the assessment of DOA estimation algorithms in the low training-sample environment, which is increasingly dominating the field of adaptive processing.

REFERENCES

Abramovich, Y.I., Johnson, B.A., 2007. Expected likelihood estimation: asymptotic properties for "stochastic" complex Gaussian models. In: Proc. 2nd International Workshop Computational Advances in Multi-Sensor Adaptive Processing, pp. 33–36.

Abramovich, Y.I., Johnson, B.A., 2008. GLRT-based detection-estimation for under-sampled training conditions. IEEE Trans. Signal Process. 56 (8), 3600–3612.

Abramovich, Y.I., Johnson, B.A., Mestre, X., 2007a. Performance breakdown in MUSIC, G-MUSIC and maximum likelihood estimation. In: Proc. ICASSP, vol. 2, pp. 1049–1052.

Abramovich, Y.I., Johnson, B.A., Spencer, N.K., 2008. Statistical non-identifiability of close emitters: part I—maximum likelihood estimation breakdown. In: Proc. ICASSP, IEEE. In press.

Abramovich, Y.I., Spencer, N.K., 2002. DOA estimation performance breakdown: a new approach to prediction and cure. In: Proc. European Signal PROCESSING Conference, pp. 89–92.

Abramovich, Y.I., Spencer, N.K., 2004. Performance breakdown of subspace-based methods in arbitrary antenna arrays: GLRT-based prediction and cure. In: Proc. ICASSP, vol. 2, pp. 117–120.

Abramovich, Y.I., Spencer, N.K., Gorokhov, A.Y., 1999. Resolving manifold ambiguities in direction-of-arrival estimation for nonuniform linear antenna arrays. IEEE Trans. Signal Process. 47 (10), 2629–2643.

Abramovich, Y.I., Spencer, N.K., Gorokhov, A., 2002a. A lower bound for the maximum likelihood ratio: Sparse antenna array applications. In: Proc. ICASSP, vol. 2, pp. 1145–1148.

Abramovich, Y.I., Spencer, N.K., Gorokhov, A.Y., 2002b. Detection-estimation of distributed Gaussian sources. In: Proc. Sensor Array and Multichannel Signal Processing Workshop, pp. 513–517.

Abramovich, Y.I., Spencer, N.K., Gorokhov, A.Y., 2004. Bounds on maximum likelihood ratio—part I: application to antenna array detection-estimation with perfect wavefront coherence. IEEE Trans. Signal Process. 52 (6), 1524–1536.

Abramovich, Y.I., Spencer, N.K., Gorokhov, A.Y., 2007b. A modified GLRT and AMF framework for adaptive detectors. IEEE Trans. Aerosp. Electron. Syst. 43 (3), 1017–1051.

Abramovich, Y.I., Spencer, N.K., Gorokhov, A.Y., 2007c. GLRT-based threshold detection-estimation performance improvement and application to uniform circular antenna arrays. IEEE Trans. Signal Process. 55 (1), 20–31.

Akaike, H., 1974. A new look at the statistical model identification. IEEE Trans. Autom. Control 19 (6), 716–723.

Anderson, T., 1958. An Introduction to Multivariate Statistical Analysis. John Wiley & Sons.

Athley, F., 2002. Performance analysis of DOA estimation in the threshold region. In: Proc. ICASSP, vol. 3, pp. 3017–3020.

Athley, F., 2005. Threshold region performance of maximum likelihood direction of arrival estimators. IEEE Trans. Signal Process. 53 (4), 1359–1373.

Bai, Z., Miao, B., Pan, G., 2007. On asymptotics of eigenvectors of large sample covariance matrix. Ann. Probab. 35 (4), 1532–1572.

Bai, Z., Silverstein, J., 1998. No eigenvalues outside the support of the limiting spectral distribution of large dimensional sample covariance matrices. Ann. probab. 26, 316–345.

Bai, Z., Silverstein, J., 1999. Exact separation of eigenvalues of large dimensional sample covariance matrices. Ann. Probab. 27 (3), 1536–1555.

Baik, J., Ben Arous, G., Péché, S., 2005. Phase transition of the largest eigenvalue for non-null complex covariance matrices. Ann. Probab. 33 (5), 1643–1697.

Baik, J., Silverstein, J.W., 2006. Eigenvalues of large sample covariance matrices of spiked population models. J. Multivar Anal. 97 (6), 1382–1408.

Beenakker, C.W.J., 1997. Random-matrix theory of quantum transport. Rev. Mod. Phys. 69 (3), 731–808.

Brehm, H., 1982. Combinatorial Theory, Lecture Notes in Mathematics. Springer.

Consul, P., 1969. The exact distributions of likelihood criteria of different hypotheses. In: Krishnaian, P. (Ed.), Multivariate Analysis II, pp. 171–181. Academic Press.

Cox, H., 1973. Resolving power and sensitivity to mismatch of optimum array processors. J. Acoust. Soc. Amer. 54, 771–785.

de Monvel, A.B., Pastur, L., Shcherbina, M., 1995. On the statistical mechanics approach in the random matrix theory: integrated density of states. J. Stat. Phys. 79 (3–4), 585–611.

Dollas, A., Rankin, W., McCracken, D., 1998. An new algorithm for Golomb ruler derivation and proof of the 19 mark ruler. IEEE Trans. Inf. Theory 44 (1), 379–382.

Dym, H., Gohberg, I., 1981. Extensions of band matrices with band inverses. Linear Algebra Appl. 36, 1–24.

Girko, V.L., 1987. An introduction to general statistical analysis. Theory Probab. Applic. 32 (2), 229–242.

Girko, V.L., 1998. An Introduction to Statistical Analysis of Random Arrays. VSP.

Gradshteyn, I., Ryzhik, I., 2000. Tables of Integrals, Series, and Products. Academic Press.

Grone, R., Johnson, C., de Sa, E., Wolkowicz, H., 1984. Positive definite completions of partial Hermitian matrices. Linear Algebra Appl. 58, 109–124.

Hawkes, M., Nehorai, A., Stoica, P., 2001. Performance breakdown of subspace-based methods: prediction and cure. In: Proc. ICASSP, vol. 6, pp. 4005–4008.

Hurt, N.E., 1990. Maximum likelihood estimation and MUSIC in array localization signal process. A review. Multidimension. Syst. Signal Process. 1 (3), 279–325.

Janik, R.A., Nowak, M.A., 2003. Wishart and anti-Wishart random matrices. J. Phys. A Math. Gen. 36 (12), 3629–3637.

John, S., 1973. The distribution of a statistic used for testing sphericity of normal distributions. Biometrika 59, 169–173.

Johnson, B.A., Abramovich, Y.I., 2007. A matrix extension built under-sampled likelihood ratio test with application to MUSIC breakdown prediction and cure. J. Commun. (Academy Publishing) 2 (3), 64–72. Accessed May 23, 2009, at *http://www.academypublisher.com/jcm/vol02/no03/jcm02036472.pdf.*

Johnson, B.A., Abramovich, Y.I., Mestre, X., 2008. MUSIC, G-MUSIC and maximum likelihood performance breakdown. IEEE Trans. Signal Process. 56 (8), 3944–3958.

Johnstone, I.M., 2001. On the distribution of the largest eigenvalue in principal component analysis. Ann. Stat. 29 (2), 295–327.

Kaveh, M., Barabell, A., 1986. The statistical performance of the MUSIC and the minimum-norm algorithms in resolving plane waves in noise. IEEE Trans. Signal Process. 34 (2), 331–341.

Kolmogorov, A.N., 1956. Asymptotic characteristics of some completely bounded metric spaces. Dokl. Akad. Nauk SSSR. 108, 585–589.

Lagarias, J., Reeds, J., Wright, M., Wright, P.E., 1998. Convergence properties of the Nelder-Mead simplex method in low dimensions. SIAM J. Optim. 9 (1), 112–147.

Ledoit, O., Wolf, M., 2002. Some hypothesis tests for the covariance matrix when the dimension is large compared to the sample size. Ann. Stat. 30 (4), 1081–1102.

Lee, H., Li, F., 1993. Quantification of the difference between detection and resolution thresholds for multiple closely spaced emitters. IEEE Trans. Signal Process. 41 (6), 2274–2277.

Lee, H.B., Wengrovitz, M.S., 1990. Resolution threshold of beamspace MUSIC for two closely spaced emitters. IEEE Trans. Signal Process. 38 (9), 1545–1559.

Lev-Ari, H., Parker, S.R., Kailath, T., 1989. Multidimensional maximum-entropy covariance extension. IEEE Trans. Inf. Theory. 35 (3), 497–508.

Marquesa, F.J., Coelho, C.A., 2008. Near-exact distributions for the sphericity likelihood ratio test statistic. J. Stat. Planning Inference 138 (3), 726–741.

Marsden, J., Hoffman, M., 1987. Basic Complex Analysis, 3rd ed. W. H. Freeman and Company.

Mestre, X., 2006a. An improved subspace-based algorithm in the small sample size regime. In: Proc. ICASSP.

Mestre, X., 2006b. An improved weighted MUSIC algorithm for small sample size scenarios. In: Proc. 14th Workshop Adaptive Sensor and Array Processing. Accessed May 23, 2009, at *www.ll.mit.edu/asap/asap_06/pdf/Papers/19_mestre_pa.pdf.*

Mestre, X., 2006c. On the asymptotic behavior of quadratic forms of the resolvent covariance-type matrices. Centre Tecnològic de Telecomunicacions de Catalunya (CTTC), Technical Report CTTC/RC/2006-001. Accessed May 23, 2009 at *http://www.cttc.cat/en/person/xmestre/publications.jsp.*

Mestre, X., 2008a. Improved estimation of eigenvalues and eigenvectors of covariance matrices using their sample estimates. IEEE Trans. Inf. Theory 54 (11), 5113–5129.

Mestre, X., 2008b. On the asymptotic behavior of the sample estimates of eigenvalues and eigenvectors of covariance matrices. IEEE Trans. Signal Process. 56 (11), 5353–5368.

Mestre, X., Johnson, B.A., Abramovich, Y.I., 2007. Source power estimation for array processing applications under low sample size constraints. In: Proc. ICASSP, vol. 2, IEEE, pp. 897–900.

Mestre, X., Lagunas, M.A., 2008. Modified subspace algorithms for DOA estimation with large arrays. IEEE Trans. Signal Process. 56 (2), 598–614.

Moffet, A., 1968. Minimum-redundancy linear arrays. IEEE Trans. Antennas Propag. 16 (2), 172–175.

Muirhead, R., 1982. Aspects of Multivariate Statistical Theory. John Wiley & Sons.

Nadakuditi, R.R., Edelman, A., 2008. Sample eigenvalue based detection of high-dimensional signals in white noise using relatively few samples. IEEE Trans. Signal Process. 56, 2625–2638.

Nagao, H., 1973. On some test criteria for covariance matrix. Ann. Stat. 1 (4), 700–709.

Nagarsenker, B., Pillai, K., 1973. Distribution of the likelihood ratio criterion for testing a hypothesis specifying a covariance matrix. Biometrika 60 (2), 359–361.

Ottersten, B., Viberg, M., Stoica, P., Nehorai, A., 1993. Exact and large sample maximum likelihood techniques for parameter estimation and detection in array processing. In: Haykin, S.S., Litva, J., Shepherd, T.J. (Eds.), Radar Array Processing, pp. 99–151. Springer-Verlag.

Paul, D., 2004. Asymptotics of the leading sample eigenvalues for a spiked covariance model. Stanford University, Technical Report. Accessed May 23, 2009, at *http://anson.ucdavis.edu/ {~}debashis/techrep/eigenlimit.pdf.*

Pearce, T., 1998. Receiving array design for over-the-horizon radar. GEC J. Technol. 15 (1), 47–55.

Reed, I.S., Mallett, J.D., Brennan, L.E., 1974. Rapid convergence rate in adaptive arrays. IEEE Trans. Aerosp. Electron. Syst. 10 (6), 853–863.

Renaux, A., Forster, P., Boyer, E., Larzabal, P., 2004. Non-efficiency and non-Gaussianity of a maximum likelihood estimator at high signal-to-noise ratio and finite number of samples. In: Proc. ICASSP, vol. 2, IEEE, pp. 121–124.

Richmond, C.D., 2006. On the threshold region mean-squared error performance of maximum-likelihood direction-of arrival estimation in the presence of signal model mismatch. In: Proc. Sensor Array and Multichannel Signal Processing Workshop, pp. 268–272.

Rissanen, J., 1978. Modeling by shortest data description. Automatica 14, 465–471.

Schwarz, G., 1978. Estimating the dimension of a model. Ann. Stat. 6 (2), 461–464.

Shan, T., Wax, M., Kailath, T., 1985. On spatial smoothing for direction-of-arrival estimation of coherent signals. IEEE Trans. Acoust. Speech Signal Process. 33 (4), 806–811.

Silverstein, J., 1995. Strong convergence of the empirical distribution of eigenvalues of large dimensional random matrices. J. Multivar. Anal. 5, 331–339.

Smith, S., 2005. Statistical resolution limits and the complexified Cramér-Rao bound. IEEE Trans. Signal Process. 53 (5), 1597–1609.

Spencer, N.K., Abramovich, Y.I., 2004. Performance analysis of DOA estimation using uniform circular antenna arrays in the threshold region. In: Proc. ICASSP, vol. 2, IEEE, pp. 233–236.

Steyerl, A., Vaccaro, R., 2007. Performance of direction-of-arrival estimators at low signal-to-noise ratios. In: Proc. Adaptive Sensor and Array Processing Workshop.

Stoica, P., Nehorai, A., 1989. MUSIC, maximum likelihood and Cramer-Rao bound. IEEE Trans. Signal Process. 37 (5), 720–741.

Stuart, A., Ord, J., 1991. Kendall's Advanced Theory of Statistics. Halsted Press.

Sverdlik, M., 1975. Optimal Discrete Signals. Sovietsckoe Radio.

Swingler, D.N. 1993. Simple approximations to the Cramer-Rao lower bound on direction of arrival for closely spaced sources. IEEE Trans. Signal Process. 41 (4), 1668–1672.

Thomas, J., Scharf, L., Tufts, D., 1995. The probability of a subspace swap in the SVD. IEEE Trans. Signal Process. 43 (3), 730–736.

Tomaniac, A.P., 1998. Asymptomania?—Signal processing forum. IEEE Signal Process. Mag. 15 (1), 16.

Tufts, D.W., Kot, A.C., Vaccaro, R.J., 1988. The threshold analysis of SVD-based algorithms. In: Proc. ICASSP, pp. 2416–2419.

Tufts, D., Kot, A., Vaccaro, R., 1991. The threshold effect in signal processing algorithms which use an estimated subspace. In: Vaccaro, R. (Ed.), SVD and Signal Processing II: Algorithms, Analysis and Applications, pp. 301–320. Elsevier.

van Trees, H.L., 1968. Detection, Estimation, and Modulation Theory: Part I. John Wiley & Sons.

Wilks, S.S., 1938. The large-sample distribution of the likelihood ratio for testing composite hypotheses. Ann. Math. Stat. 9 (1), 60–63.

Wishart, J., 1928. The generalised product moment distribution in samples from a normal multivariate population. Biometrika 20, 32–52.

Woerdeman, H., 1989. Matrix and Operator Extensions. Ph.D. Dissertation. Vrije Universiteit, Amsterdam.

Xu, W., Kaveh, M., 1994. Alternatives for the definition and evaluation of resolution thresholds of signal-subspace parameter estimators. In: Proc. ICASSP.

Zhang, Q., 1995. Probability of resolution of the MUSIC algorithm. IEEE Trans. Signal Process. 43 (4), 978–987.

Chapter 8

High-Resolution DOA Estimation with Higher-Order Statistics

Pascal Chevalier, Anne Ferréol, Laurent Albera

8.1 INTRODUCTION

Direction-of-arrival (DOA) estimation is one of the main components of spectrum-monitoring and radio surveillance systems, but it also finds its place in radar, sonar, and human electrophysiology for localization of electromagnetic activities. From the beginning of the 1980s, many second-order (SO) high-resolution (HR) DOA estimation methods have been developed, mainly to overcome the limitations of classical methods such as Watson-Watt or interferometry in multiple-source contexts or for multipath propagation channels. Among these, subspace-based methods such as MUSIC (or 2-MUSIC) (Bienvenu and Kopp, 1983; Schmidt, 1986) and ESPRIT (Paulraj et al., 1986; Roy and Kailath, 1989), very powerful in multiple-source environments, are the most popular. Indeed, in the absence of modeling errors and for a background noise with a spatial coherence that is known (MUSIC) or equal to the Identity matrix (ESPRIT), they generate infinite asymptotic precision and resolution whatever the source signal-to-noise ratio (SNR) (Germain et al., 1989; Kaveh and Barabell, 1986; Porat and Friedlander, 1988; Stoica and Nehorai, 1989, 1990).

However, these methods suffer from serious drawbacks. For one, they are not able to process more than $N-1$ noncoherent sources from an array of N sensors. Moreover, they are weakly robust both to modeling errors (Ferreol et al., 2006; Friedlander, 1990; Li and Vaccaro, 1992; Swindlehurst and Kailath, 1992) (always present in operational contexts) and to the presence of strong spatially correlated background noise with a spatial coherence that is unknown (Paulraj and Kailath, 1986), such as that in the HF band (Demeure and Chevalier, 1998) or in human electrophysiology. Finally, their performance may be strongly affected when several sources with a small angular separation have to be separated from

Copyright © 2009 by Academic Press, Inc. All rights of reproduction in any form reserved.

a limited number of snapshots (Kaveh and Barabell, 1986; Stoica and Nehorai, 1989, 1990).

Since the end of the 1980s, mainly to overcome most of the previously mentioned limitations, higher-order (HO) HR DOA estimation methods (mainly fourth-order, FO), have been developed for non-Gaussian sources, omnipresent in radio communications contexts, for which HO statistics convey information. The first works on HO HR DOA estimation (Cardoso, 1990; Chiang and Nikias, 1989; Forster and Nikias, 1991; Porat and Friedlander, 1991; Swami and Mendel, 1991) were mainly devoted to the introduction of HO methods that are asymptotically insensitive to the correlation of the noise between the sensors when such noise is Gaussian. Methods proposed in Cardoso (1990), Chiang and Nikias (1989), Porat and Friedlander (1991), and Swami and Mendel (1991) exploit the FO cumulants (Amblard et al., 1996; Mendel, 1991) of the data. Among these, Chiang and Nikias (1989) and Porat and Friedlander (1991) proposed FO extensions to the ESPRIT and MUSIC methods, respectively (called 4-ESPRIT and 4-MUSIC in this chapter). In contrast, the method proposed in Forster and Nikias (1991) works in the bispectrum (Mendel, 1991) domain.

At the same time, it was proved that both FO (Cardoso, 1990) and HO (Shamsunder and Giannakis, 1993) cumulant-based methods allow for an increase in both resolution and the number of sources to be processed, beyond the number of sensors. These properties are shown in Chevalier and Ferreol (1999) and Dogan and Mendel (1995a) for FO methods, and recently in Chevalier et al. (2005), for HO methods at an even order, to be directly related to a virtual increase in both the effective aperture and the number of sensors in the array, induced by the exploitation of the FO or HO data cumulants. Thus, both the FO and HO virtual array (VA) concepts, introduced in Dogan and Mendel (1995a), Chevalier and Ferreol (1999), and Chevalier et al. (2005), respectively, which give a geometrical interpretation of cumulants and explain most of the advantages of HO cumulant-based DOA estimation methods. In Dogan and Mendel (1995a) even the symmetries of the FO VA were exploited to propose a kind of FO ESPRIT method, called FO Virtual ESPRIT, valid whatever the geometry of the array provided that at least two sensors share the same response.

Following these first works, other FO cumulant-based methods were developed with specific properties. Such properties concern in particular robustness to non-Gaussian noise (Chen and Lin, 1994a; Dogan and Mendel, 1995b); the ability to process coherent multipaths (Chen and Lin, 1994b; Gönen et al., 1997; Yuen and Friedlander, 1997), near-field sources (Challa and Shamsunder, 1998), nonstationary sources (Liu and Mendel, 1999a), and, in a selective manner, cyclo-stationary sources (Shamsunder and Giannakis, 1994); and the capability to estimate both DOA and polarization parameters from an array with diversely polarized sensors (Gönen and Mendel, 1999). At the same time, asymptotic sensitivities of FO MUSIC–like methods (Cardoso and Moulines, 1995; Fan and Younan, 1995), 4-ESPRIT (Yuen and Friedlander, 1996, 1998), and FO Virtual ESPRIT (Liu and Mendel, 1999b; Yuen and Friedlander, 1996, 1998) with respect to both finite-sample effect (Cardoso and Moulines, 1995; Fan and Younan, 1995; Yuen and Friedlander, 1996, 1998) and model errors (Fan and Younan, 1995; Liu and Mendel, 1999b) were analyzed. It has been shown that FO methods are more robust to modeling errors and more sensitive to finite-sample effect than SO methods, especially for weak sources that are properly angularly separated (Cardoso and Moulines, 1995).

This latter point, jointly with a higher numerical complexity, is the main weakness of FO and HO cumulant-based methods. One way to decrease the numerical complexity is to implement HO-contracted methods (Cardoso and Moulines, 1995). These methods exploit only a part of the HO information and are intermediate between SO and HO methods, but at the expense of the number of sources to be processed, which cannot be greater than the number of sensors minus one. One way to decrease the sensitivity of HO methods to finite-sample effect is to exploit both the SO and HO data statistics (Cardoso, 1994) in the DOA estimation process. For example, one may blindly identify the steering vectors of the sources with a well-known blind identification method that uses both SO and FO data statistics (Cardoso and Souloumiac, 1993; Comon, 1994), and then estimate the DOA of the sources from the estimated steering vectors (Chevalier et al., 1996). Such methods are robust to both finite-sample effects and model errors but at the expense of both robustness with respect to an unknown colored Gaussian noise and the number of sources to be processed, which cannot be greater than the number of sensors. Another way to overcome the drawbacks of HO methods in the presence of a high dynamic range between sources is to add some deflation schemes, as was proposed recently in Albera et al. (2008) and Birot et al. (2007), especially for brain source localization.

Although most of the developed HR DOA estimation methods exploit the FO cumulants of the data, some use HO statistics such as hybrid nonlinear moments (Jacovitti and Scarano, 1994) or generalized HO cumulants (Scarano and Jacovitti, 1996). Moreover, recent (Chevalier et al., 2005) important insights into the mechanisms of array-processing methods exploiting the cumulants of the data at an arbitrary even order, $m=2q (q \geq 1)$, were achieved. This was accomplished through the introduction of the $2q$th-order VA concept, for several arrangements of the $2q$th-order data statistics and for arrays with either identical or diversely polarized sensors. This concept allows for showing the existence of an optimal arrangement of the $2q$th-order data statistics and the increasing resolution and increasing processing capacity of $2q$th-order array-processing methods as q increases, while keeping their robustness to a colored Gaussian noise.

To exploit these properties for DOA estimation, an extension of MUSIC to $2q$th order was introduced very recently in Chevalier et al. (2006, 2007) for arrays with identical and diversely polarized sensors, respectively. This gave rise to $2q$-MUSIC and several PD-$2q$-MUSIC (Polarization Diversity $2q$-MUSIC) methods. One PD-$2q$-MUSIC method is nothing more than the $2q$th-order extension of Ferrara and Parks's method (Ferrara and Parks, 1983).

The increased processing capacity offered by the $2q$-MUSIC algorithm as q increases allows for minimizing the number of sensors and reception chains for a given number of sources. Also, despite its higher variance, it was shown in Chevalier et al. (2006) that, in the presence of multiple sources, $2q$-MUSIC also has a robustness against modeling errors that increases with q. This may allow in particular some constraints about antenna calibration or receiver chain equalization to be relaxed. Finally, a noncircular extension to the $2q$-MUSIC method was proposed in Liu et al. (2008) to further increase the performance of the latter for BPSK sources.

The aim of this chapter is to present the philosophy, properties, implementations (both parallel and sequential), and performance of HO HR DOA estimation methods through description of the families of the $2q$-MUSIC and PD-$2q$-MUSIC algorithms, which, in contrast to HO extensions of the ESPRIT algorithm, are well suited for any kind of array geometries. HO cumulants, along with the observation model and its HO statistics, are presented in Section 8.2. The $2q$-MUSIC method for arrays with identical sensors, along with its possible parallel and sequential implementations, is presented in Section 8.3. Section 8.4 introduces the HO VA concept and the identifiability issue of the $2q$-MUSIC method. Asymptotic performance, in terms of precision and resolution, of $2q$-MUSIC in the presence of modeling errors is computed analytically and illustrated in Section 8.5. Some computer simulations illustrating the performance of $2q$-MUSIC methods from a finite number of snapshots are presented in Section 8.6. Section 8.7 introduces the philosophy and performance of PD-$2q$-MUSIC methods, which correspond to extensions of $2q$-MUSIC to arrays with diversely polarized antennas. Finally, Section 8.8 concludes the chapter.

8.2 OBSERVATION MODEL AND DATA STATISTICS

The observation model is presented in Section 8.2.1 while SO and HO statistics of the data are described in Section 8.2.2.

8.2.1 Observation Model

We consider an array of N narrowband (NB) identical sensors, and we call $\boldsymbol{x}(t)$ the vector of complex amplitudes of the signals at the sensors' output. Each sensor is assumed to receive the contribution of P (P is potentially greater than or equal to N) zero-mean stationary (to simplify the presentation but not required in practice) NB sources, which may be statistically independent or not, corrupted by a noise. We assume that the P sources can be divided into G groups, with P_g sources in group g, such that the sources in each group are assumed to be statistically dependent but not coherent (i.e., not fully correlated), while sources belonging to different groups are assumed to be statistically independent. In particular, $G = P$ corresponds to P statistically independent sources, whereas $G = 1$ corresponds to the case where all the sources are dependent. Of course, the P_g parameters are such that

$$P = \sum_{g=1}^{G} P_g \tag{8.1}$$

Under these assumptions, the observation vector can approximately be written as follows:

$$\begin{aligned}\boldsymbol{x}(t) &= \sum_{i=1}^{P} m_i(t)\boldsymbol{a}(\boldsymbol{\theta}_i, \boldsymbol{\beta}_i) + \boldsymbol{v}(t) \triangleq \boldsymbol{A}\,\boldsymbol{m}(t) + \boldsymbol{v}(t) \\ &= \sum_{g=1}^{G} \boldsymbol{A}_g \boldsymbol{m}_g(t) + \boldsymbol{v}(t) = \sum_{g=1}^{G} \boldsymbol{x}_g(t) + \boldsymbol{v}(t)\end{aligned} \tag{8.2}$$

In Equation (8.2), $\boldsymbol{v}(t)$ is the noise vector, assumed zero mean and Gaussian; $\boldsymbol{m}(t)$, independent of $\boldsymbol{v}(t)$, is the vector with components $m_i(t)$ that are the complex amplitudes of the sources, $\boldsymbol{\theta}_i \triangleq (\theta_i, \varphi_i)$, where θ_i and φ_i are the azimuth and the elevation angles of source i (Figure 8.1); $\boldsymbol{\beta}_i$ (characterized by two parameters in the wave plane) (Compton, 1988) characterizes the state of polarization of source i; $\boldsymbol{A}$ is the $(N \times P)$ matrix of the source steering vectors $\boldsymbol{a}(\boldsymbol{\theta}_i, \boldsymbol{\beta}_i)$ $(1 \leq i \leq P)$, which contains in particular information about the DOA of the sources; $\boldsymbol{A}_g$ is the $(N \times P_g)$ submatrix of $\boldsymbol{A}$ corresponding to the gth group of sources; $\boldsymbol{m}_g(t)$ is the corresponding $(P_g \times 1)$ subvector of $\boldsymbol{m}(t)$ and $\boldsymbol{x}_g(t) \triangleq \boldsymbol{A}_g\boldsymbol{m}_g(t)$. In the absence of modeling errors, and of mutual coupling between the sensors in particular, assuming a planewave propagation, component n of vector $\boldsymbol{a}(\boldsymbol{\theta}_i, \boldsymbol{\beta}_i)$, denoted $a_n(\boldsymbol{\theta}_i, \boldsymbol{\beta}_i)$, can be written as (Compton, 1988)

$$\begin{aligned}a_n(\boldsymbol{\theta}_i, \boldsymbol{\beta}_i) = &f_n(\boldsymbol{\theta}_i, \boldsymbol{\beta}_i)\exp\{j2\pi[x_n \cos(\theta_i)\cos(\varphi_i) \\ &+ y_n \sin(\theta_i)\cos(\varphi_i) + z_n \sin(\varphi_i)]/\lambda\}\end{aligned} \tag{8.3}$$

In Equation (8.3) λ is the wavelength; (x_n, y_n, z_n) are the coordinates of sensor n of the array; and $f_n(\boldsymbol{\theta}_i, \boldsymbol{\beta}_i)$ is a complex number corresponding to the response of sensor n to a unit electric field coming from direction $\boldsymbol{\theta}_i$ and having the state of polarization $\boldsymbol{\beta}_i$ (Compton, 1988). For an array with identical sensors, $f_n(\boldsymbol{\theta}_i, \boldsymbol{\beta}_i) \triangleq$

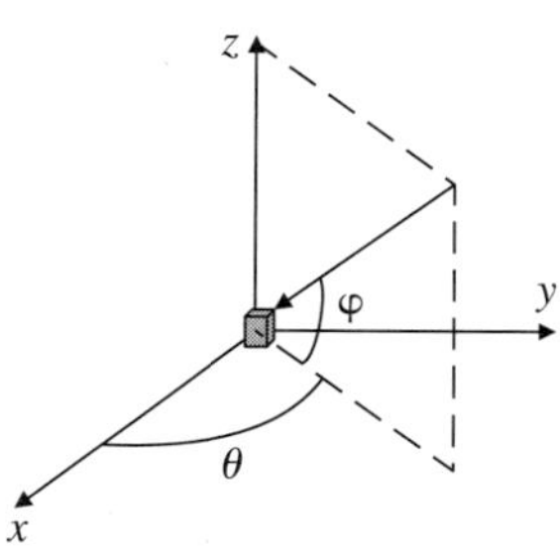

FIGURE 8.1 An incoming signal in three dimensions (*Source:* Reprinted from Chevalier et al., 2005, © IEEE, with permission.)

$f(\boldsymbol{\theta}_i, \boldsymbol{\beta}_i)$ is the same for all the sensors n, and $a_n(\boldsymbol{\theta}_i, \boldsymbol{\beta}_i)$ and $\boldsymbol{a}(\boldsymbol{\theta}_i, \boldsymbol{\beta}_i)$ can then be written as $a_n(\boldsymbol{\theta}_i, \boldsymbol{\beta}_i) = f(\boldsymbol{\theta}_i, \boldsymbol{\beta}_i)\, a_n(\boldsymbol{\theta}_i)$ and $\boldsymbol{a}(\boldsymbol{\theta}_i, \boldsymbol{\beta}_i) = f(\boldsymbol{\theta}_i, \boldsymbol{\beta}_i)\, \boldsymbol{a}(\boldsymbol{\theta}_i)$, respectively, where $a_n(\boldsymbol{\theta}_i)$ is the exponential term of the right side of Equation (8.3) and where $\boldsymbol{a}(\boldsymbol{\theta}_i)$ is the vector with components $a_n(\boldsymbol{\theta}_i)$. In such situations, the term $f(\boldsymbol{\theta}_i, \boldsymbol{\beta}_i)$ may be inserted in the complex envelope, $m_i(t)$, of source i and the steering vector of source i may be reduced to $\boldsymbol{a}(\boldsymbol{\theta}_i)$, which is done in the following up to Section 8.6. In other words, for arrays with identical sensors, the observation model is assumed to be given by Equation (8.2) with $\boldsymbol{a}(\boldsymbol{\theta}_i)$ instead of $\boldsymbol{a}(\boldsymbol{\theta}_i, \boldsymbol{\beta}_i)$.

8.2.2 Data Statistics

SO and HO statistics of the data are described in the following subsections.

Interest of Cumulants

Let $\boldsymbol{u}$ and $\boldsymbol{x}$ be two $(N \times 1)$ real-valued vectors such that $\boldsymbol{x}$ is random. The qth-order cumulants of $\boldsymbol{x}$, with components $x_i (1 \leq i \leq N)$, are defined (Priestley, 1981) as the coefficients of the terms of order q in the Taylor series expansion of the second characteristic function, $\Psi(\boldsymbol{u}) = \ln \mathrm{E}\{\exp(j\boldsymbol{u}^{\mathrm{T}}\boldsymbol{x})\}$, of $\boldsymbol{x}$. The qth-order cumulants of $\boldsymbol{x}$ are linked to its pth-order moments $(1 \leq p \leq q)$ by the so-called Leonov-Shiryaev formula (McCullagh, 1987), given by

$$\mathrm{Cum}[x_{i_1}, x_{i_2}, \ldots, x_{i_q}] = \sum_{p=1}^{q} (-1)^{p-1}(p-1)!\mathrm{E}\left\{\prod_{j\in S1} x_{i_j}\right\}\mathrm{E}\left\{\prod_{j\in S2} x_{i_j}\right\}..\mathrm{E}\left\{\prod_{j\in Sp} x_{i_j}\right\} \tag{8.4}$$

where $1 \leq i_j \leq N$ for $1 \leq j \leq q$, and where $(S1, S2, \ldots, Sp)$ describes all the partitions in p sets of $(1, 2, \ldots, q)$. In particular, for a zero-mean vector $\boldsymbol{x}$, the second-order, third-order, and FO cumulants of $\boldsymbol{x}$ are given by

$$\mathrm{Cum}[x_i, x_j] = \mathrm{E}\{x_i x_j\} \tag{8.5a}$$

$$\mathrm{Cum}[x_i, x_j, x_k] = \mathrm{E}\{x_i x_j x_k\} \tag{8.5b}$$

$$\begin{aligned}\mathrm{Cum}[x_i, x_j, x_k, x_l] = {} & \mathrm{E}\{x_i x_j x_k x_l\} - \mathrm{E}\{x_i x_j\}\mathrm{E}\{x_k x_l\} \\ & - \mathrm{E}\{x_i x_k\}\mathrm{E}\{x_j x_l\} - \mathrm{E}\{x_i x_l\}\mathrm{E}\{x_j x_k\}\end{aligned} \tag{8.5c}$$

for $1 \leq i, j, k, l \leq N$. In the case of nonzero-mean vector $\boldsymbol{x}$, one has to replace x_i with $x_i - \mathrm{E}\{x_i\}$ in these formulas. In the case of a complex vector $\boldsymbol{x}$, one has to replace x_i by $x_i^{\varepsilon i}$ in these formulas and in Equation (8.4), where $\varepsilon_i = \pm 1$ with the convention $x^1 = x$ and $x^{-1} = x^*$, the complex conjugate of x. In this case, the qth-order cumulants of $\boldsymbol{x}$ correspond to the set of quantities $\mathrm{Cum}\left[x_{i_1}^{\varepsilon i_1}, x_{i_2}^{\varepsilon i_2}, \ldots, x_{i_q}^{\varepsilon i_q}\right]$ such that $1 \leq i_j \leq N$ and $\varepsilon_{i_j} = \pm 1$ for $1 \leq j \leq q$.

Cumulants are frequently used in signal processing, and in array processing in particular, since they have many important properties (Mendel, 1991):

[P1]. Cumulants are symmetric in their arguments:

$$\mathrm{Cum}[x_1, \ldots, x_q] = \mathrm{Cum}[x_{i_1}, \ldots, x_{i_q}] \tag{8.6a}$$

where $(i_1, \ldots, i_q)$ is a permutation of $(1, \ldots, q)$

[P2]. Cumulants are multilinear—that is, linear with respect to each of their arguments:

$$\mathrm{Cum}[x_1 + y_1, \ldots, x_q] = \mathrm{Cum}[x_1, \ldots, x_q] + \mathrm{Cum}[y_1, \ldots, x_q] \tag{8.6b}$$

$$\mathrm{Cum}[\lambda x_1, \ldots, x_q] = \lambda \mathrm{Cum}[x_1, \ldots, x_q] \tag{8.6c}$$

where y_1 is a random variable and λ is a constant.

[P3]. Cumulants are invariant by a constant translation:

$$\mathrm{Cum}[x_1 + \lambda, \ldots, x_q] = \mathrm{Cum}[x_1, \ldots, x_q] \tag{8.6d}$$

where λ is a constant.

[P4]. If collections $(x_1, \ldots, x_q)$ and $(y_1, \ldots, y_q)$ are statistically independent, then

$$\mathrm{Cum}[x_1 + y_1, \ldots, x_q + y_q] = \mathrm{Cum}[x_1, \ldots, x_q] + \mathrm{Cum}[y_1, \ldots, y_q] \tag{8.6e}$$

[P5]. If a subset of the collection $(x_1, \ldots, x_q)$ is independent of the rest, then

$$\mathrm{Cum}[x_1, \ldots, x_q] = 0 \tag{8.6f}$$

[P6]. If $(x_1, \ldots, x_N)$ are jointly Gaussian, then

$$\mathrm{Cum}[x_1, \ldots, x_q] = 0 \quad \text{for } q \geq 3 \tag{8.6g}$$

Second- and Higher-Order Statistics

The $2q$th-order cumulants of the data, $\mathrm{Cum}[x_{i_1}(t), \ldots, x_{i_r}(t), x_{i_{r+1}}(t)^*, \ldots, x_{i_{2q}}(t)^*](1 \leq i_j \leq N)(1 \leq j \leq 2q)$, such that $r \neq q$, are equal to zero for $2q$th-order circular observations (Amblard et al., 1996; Picinbono, 1994). This is particularly the case in the presence of both a Gaussian noise and $2q$th-order circular sources such as M-PSK sources (Proakis, 1995) for $M > 2q$. For this reason, the $2q$-MUSIC $(q \geq 1)$ method (Chevalier et al., 2006) exploits the information contained in the $(N^q \times N^q)$ $2q$th-order circular covariance matrix, $\boldsymbol{C}_{2q,x}$, with entries that are the $2q$th-order circular cumulants of the data, $\mathrm{Cum}[x_{i_1}(t), \ldots, x_{i_q}(t), x_{i_{q+1}}(t)^*, \ldots, x_{i_{2q}}(t)^*](1 \leq i_j \leq N)(1 \leq j \leq 2q)$. Note that the previous statistics are called *circular* since they are invariant by a phase rotation of the components $x_{i_j}(t)$ (Amblard et al., 1996; Picinbono, 1994). However, the previous entries—$\mathrm{Cum}[x_{i_1}(t), \ldots, x_{i_q}(t), x_{i_{q+1}}(t)^*, \ldots, x_{i_{2q}}(t)^*]$ $(1 \leq i_j \leq N)(1 \leq j \leq 2q)$—can be arranged in the $\boldsymbol{C}_{2q,x}$ matrix in different ways that determine in particular both the resolution and the maximal processing power of the $2q$-MUSIC method (Chevalier et al., 2005, 2006).

To show this result, let us introduce an arbitrary integer l such that $(0 \leq l \leq q)$ and let us arrange the $2q$-uplet, $(i_1, \ldots, i_q, i_{q+1}, \ldots, i_{2q})$, of indices $i_j(1 \leq j \leq 2q)$ into two q-uplets indexed by l and defined by $(i_1, i_2, \ldots, i_l, i_{q+1}, \ldots, i_{2q-l})$ and

$(i_{2q-l+1},\ldots,i_{2q},i_{l+1},\ldots,i_q)$, respectively. As the indices $i_j(1\leq j\leq 2q)$ vary from 1 to N, the two latter q-uplets take N^q values.

Numbering, in a natural way, the N^q values of each of the two latter q-uplets by the integers I_l and J_l, respectively, such that $1\leq I_l, J_l\leq N^q$, we obtain

$$I_l \triangleq \sum_{j=1}^{l} N^{q-j}(i_j-1)+\sum_{j=1}^{q-l} N^{q-l-j}(i_{q+j}-1)+1 \tag{8.7a}$$

$$J_l \triangleq \sum_{j=1}^{l} N^{q-j}(i_{2q-l+j}-1)+\sum_{j=1}^{q-l} N^{q-l-j}(i_{l+j}-1)+1 \tag{8.7b}$$

In particular, for $(q,l)=(2,1)$, the integers I_l and J_l are given by $I_1=N(i_1-1)+i_3$ and $J_1=N(i_4-1)+i_2$, respectively, whereas for $(q,l)=(2,2)$ we obtain $I_2=N(i_1-1)+i_2$ and $J_2=N(i_3-1)+i_4$. Using the permutation invariance property of the cumulants, we deduce that $\mathrm{Cum}[x_{i_1}(t),\ldots,x_{i_q}(t),x_{i_{q+1}}(t)^*,\ldots,x_{i_{2q}}(t)^*]=\mathrm{Cum}[x_{i_1}(t),\ldots,x_{i_l}(t),x_{i_{q+1}}(t)^*,\ldots,x_{i_{2q-l}}(t)^*,x_{i_{2q-l+1}}(t)^*,\ldots,x_{i_{2q}}(t)^*,x_{i_{l+1}}(t),\ldots,x_{i_q}(t)]$. Assuming that the latter quantity is the element $[I_l,J_l]$ of the $\boldsymbol{C}_{2q,x}$ matrix, thus noted $\boldsymbol{C}_{2q,x}(l)$, it is straightforward to show, using Equation (8.2) and the cumulant properties of Section 8.2.2, that the $(N^q\times N^q)\boldsymbol{C}_{2q,x}(l)$ matrix can be written as

$$\boldsymbol{C}_{2q,x}(l)=\sum_{g=1}^{G}\boldsymbol{C}_{2q,xg}(l)+\eta_2\boldsymbol{V}(l)\delta(q-1) \tag{8.8}$$

In Equation (8.8), η_2 is the mean power of the noise per sensor, $\boldsymbol{V}(l)$ is the $(N\times N)$ spatial coherence matrix of the noise for the arrangement indexed by l, such that $\mathrm{Tr}[\boldsymbol{V}(l)]=N$; Tr[.] denotes Trace; $\delta(.)$ is the Kronecker symbol; and the $(N^q\times N^q)$ matrix $\boldsymbol{C}_{2q,xg}(l)$ corresponds to the $2q$th-order circular cumulants of $\boldsymbol{x}_g(t)$ for the arrangement indexed by l, which can be written as

$$\boldsymbol{C}_{2q,x_g}(l)=\left[\boldsymbol{A}_g{}^{\otimes l}\otimes\boldsymbol{A}_g{}^{*\otimes(q-l)}\right]\boldsymbol{C}_{2q,m_g}(l)\left[\boldsymbol{A}_g{}^{\otimes l}\otimes\boldsymbol{A}_g{}^{*\otimes(q-l)}\right]^{\dagger} \tag{8.9}$$

In Equation (8.9) $\boldsymbol{C}_{2q,m_g}(l)$ is the $(P_g{}^q\times P_g{}^q)$ matrix of the $2q$th-order circular cumulants of $\boldsymbol{m}_g(t)$ for the arrangement indexed by l; † corresponds to the conjugate transposition; $\otimes$ is the Kronecker product; and $\boldsymbol{A}_g{}^{\otimes l}$ is the $(N^l\times P_g{}^l)$ matrix defined by $\boldsymbol{A}_g{}^{\otimes l}\triangleq\boldsymbol{A}_g\otimes\boldsymbol{A}_g\otimes\ldots\otimes\boldsymbol{A}_g$, with a number of Kronecker products equal to $l-1$. Note that the independence of matrix $\boldsymbol{C}_{2q,x_g}(l)$ with respect to frequency in the reception band requires an HO NB assumption for the sources (i.e., sources with a bandwidth that becomes smaller as q increases; Chevalier et al., 2005). In other words, for a given reception bandwidth and a given array aperture, there exists an integer q_0 such that model Equation (8.9) is valid for $q\leq q_0$ and not valid for $q>q_0$. For HF or GSM links, for example, the HO NB assumption is generally verified up to $q_0=8$ or 10 (i.e., up to a statistical order $m=2q$, which is equal to 16 or 20; Chevalier et al., 2005).

In particular, for P statistically independent sources ($G=P$ and $P_g=1$, $1\le g\le G$), Equation (8.8) takes the form

$$C_{2q,x}(l)=\sum_{i=1}^{P} c_{2q,m_i}\boldsymbol{a}_{q,l}(\boldsymbol{\theta}_i)\boldsymbol{a}_{q,l}(\boldsymbol{\theta}_i)^{\dagger}+\eta_2 V(l)\delta(q-1) \qquad (8.10)$$

where $c_{2q,m_i}\triangleq \text{Cum}[m_{i_1}(t),\ldots,m_{i_q}(t),m_{i_{q+1}}(t)^*,\ldots,m_{i_{2q}}(t)^*]$, with $i_j=i(1\le j\le 2q)$, is the $2q$th-order circular auto-cumulant of $m_i(t)$; and where $\boldsymbol{a}_{q,l}(\boldsymbol{\theta})$ is the $2q$th-order extended steering vector for the DOA $\boldsymbol{\theta}$ and the arrangement l, defined by

$$\boldsymbol{a}_{q,l}(\boldsymbol{\theta})\triangleq \boldsymbol{a}(\boldsymbol{\theta})^{\otimes l}\otimes \boldsymbol{a}(\boldsymbol{\theta})^{*\otimes(q-l)} \qquad (8.11)$$

Equation (8.10) shows that in the absence of noise ($\eta_2=0$) and in the presence of P statistically independent non-Gaussian sources ($c_{2q,m_i}\ne 0, 1\le i\le P$), the rank of $C_{2q,x}(l)$ is equal to the number of sources P as long as the vectors $\boldsymbol{a}_{q,l}(\boldsymbol{\theta}_i)(1\le i\le P)$ remain linearly independent. However, in the presence of $2q$th-order correlated sources, $P_g>1$ for at least one group g and the rank of the associated matrix $C_{2q,x_g}(l)$ is strictly greater than 2. This shows that, contrary to SO source correlation matrices, the rank of $C_{2q,x}(l)$ for $q>1$ is, in the absence of noise, dependent on the correlation of the sources. To illustrate the algebraic structure of $C_{2q,x}(l)$ in the presence of $2q$th-order correlated sources, let us consider the cases $q=1$ and $q=2$.

For $q=1$ and $l=1$, the $(N\times N)$ $C_{2q,x}(l)$ matrix corresponds to the well-known data covariance matrix (since the observations are zero mean) defined by

$$\boldsymbol{R}_x\triangleq C_{2,x}(1)=\text{E}\{\boldsymbol{x}(t)\boldsymbol{x}(t)^{\dagger}\}=\sum_{g=1}^{G}\boldsymbol{A}_g C_{2,m_g}(1)\boldsymbol{A}_g{}^{\dagger}+\eta_2 V(1)$$

$$=\sum_{g=1}^{G}\sum_{i,j=1}^{Pg} C_{2,m_g}(1)[i,j]\,\boldsymbol{a}_{gi}\,\boldsymbol{a}_{gj}{}^{\dagger}+\eta_2 V(1)\triangleq \boldsymbol{R}_s+\eta_2 V(1) \qquad (8.12)$$

where $C_{2,m_g}(1)[i,j]=\text{Cum}[m_i(t),m_j(t)^*]$ is the coefficient $[i,j]$ of the matrix $C_{2,m_g}(1)$; $\boldsymbol{a}_{gi}$ is the steering vector of source i of group g; and $\boldsymbol{R}_s$ is the correlation matrix of the mixed sources only. It is obvious that in the absence of noise, the rank of $\boldsymbol{R}_x$ is equal to P, whatever the correlation of the sources, provided that the sources are not coherent (i.e., not fully correlated) and the vectors $\boldsymbol{a}_{gi}(1\le i\le P)$ are linearly independent.

For $q=2$ and $l=1$, the $(N^2\times N^2)$ $C_{2q,x}(l)$ matrix corresponds to the classical expression of the data quadricovariance matrix, used in most papers dealing with

FO DOA estimation problems, defined by

$$\boldsymbol{Q}_x \triangleq \boldsymbol{C}_{4,x}(1) = \sum_{g=1}^{G} \left[\boldsymbol{A}_g \otimes \boldsymbol{A}_g{}^*\right] \boldsymbol{C}_{4,m_g}(1) \left[\boldsymbol{A}_g \otimes \boldsymbol{A}_g{}^*\right]^\dagger$$
$$= \sum_{g=1}^{G} \sum_{i,j,k,l=1}^{Pg} C_{4,m_g}(1)[i,j,k,l] \left[\boldsymbol{a}_{gi} \otimes \boldsymbol{a}_{gj}{}^*\right] \left[\boldsymbol{a}_{gk} \otimes \boldsymbol{a}_{gl}{}^*\right]^\dagger \quad (8.13)$$

where $C_{4,m_g}(1)[i,j,k,l] = \mathrm{Cum}[m_i(t), m_j(t)^*, m_k(t)^*, m_l(t)]$ is the coefficient $[i,j,k,l]$ of the matrix $\boldsymbol{C}_{4,m_g}(1)$. It is obvious that the rank of $\boldsymbol{Q}_x$ becomes greater than P in the presence of fourth-order correlated sources and for vectors $\left[\boldsymbol{a}_{gi} \otimes \boldsymbol{a}_{gj}{}^*\right] (1 \le i,j \le P_g)$, which are linearly independent.

For $q=2$ and $l=2$, the $(N^2 \times N^2)$ $\boldsymbol{C}_{2q,x}(l)$ matrix corresponds to an alternative expression of the data quadricovariance matrix (not used often) defined by

$$\overline{\boldsymbol{Q}}_x \triangleq \boldsymbol{C}_{4,x}(2) = \sum_{g=1}^{G} \left[\boldsymbol{A}_g \otimes \boldsymbol{A}_g\right] \boldsymbol{C}_{4,m_g}(2) \left[\boldsymbol{A}_g \otimes \boldsymbol{A}_g\right]^\dagger$$
$$= \sum_{g=1}^{G} \sum_{i,j,k,l=1}^{Pg} C_{4,m_g}(2)[i,j,k,l] \left[\boldsymbol{a}_{gi} \otimes \boldsymbol{a}_{gj}\right] \left[\boldsymbol{a}_{gk} \otimes \boldsymbol{a}_{gl}\right]^\dagger \quad (8.14)$$

where $C_{4,m_g}(2)[i,j,k,l] = \mathrm{Cum}[m_i(t), m_j(t), m_k(t)^*, m_l(t)^*]$ is the coefficient $[i,j,k,l]$ of the matrix $\boldsymbol{C}_{4,m_g}(2)$. Again, it is obvious that the rank of $\overline{\boldsymbol{Q}}_x$ becomes greater than P in the presence of fourth-order correlated sources and for vectors $\left[\boldsymbol{a}_{gi} \otimes \boldsymbol{a}_{gj}\right] (1 \le i,j \le P_g)$, which are linearly independent.

Statistics Estimation

In situations of practical interest, the $2q$th-order statistics of the data, $\mathrm{Cum}[x_{i_1}(t), \ldots, x_{i_q}(t), x_{i_{q+1}}(t)^*, \ldots, x_{i_{2q}}(t)^*]$, are not known a priori and have to be estimated from L samples of data, $\boldsymbol{x}(k) \triangleq \boldsymbol{x}(kT_e)$, $1 \le k \le L$, where T_e is the sample period.

For zero-mean stationary observations, using the ergodicity property, an empirical estimator of $\mathrm{Cum}[x_{i_1}(t), \ldots, x_{i_q}(t), x_{i_{q+1}}(t)^*, \ldots, x_{i_{2q}}(t)^*]$, asymptotically unbiased and consistent, may be built from the Leonov-Shiryaev formula (8.4) by replacing in the latter all the moments by their empirical estimate. More precisely, an empirical estimate of Equation (8.4) with $x_{i_j}(t)^{\varepsilon j}$ instead of x_{i_j} is obtained by replacing in Equation (8.4) all the moments $\mathrm{E}\{x_{i_1}(t)^{\varepsilon 1} x_{i_2}(t)^{\varepsilon 2} \ldots x_{i_p}(t)^{\varepsilon p}\} (1 \le p \le n)$ by their empirical estimate, given by

$$\hat{\mathrm{E}}\left\{x_{i_1}(t)^{\varepsilon 1} x_{i_2}(t)^{\varepsilon 2} \ldots x_{i_p}(t)^{\varepsilon p}\right\}(L) \triangleq \frac{1}{L} \sum_{k=1}^{L} x_{i_1}(k)^{\varepsilon 1} x_{i_2}(k)^{\varepsilon 2} \ldots x_{i_p}(k)^{\varepsilon p} \quad (8.15)$$

Explicit expressions of Equation (8.4) for $n = 2q$ with $1 \le q \le 3$ are given in Chevalier et al. (2005).

However, in a radio communications context, most of the sources are no longer stationary but become cyclo-stationary (digital modulations). For zero-mean cyclo-stationary observations, the statistical matrix defined by Equation (8.8) becomes time dependent, noted $\boldsymbol{C}_{2q,x}(l)(t)$, and the theory developed in this chapter can be extended without any difficulties by considering that $\boldsymbol{C}_{2q,x}(l)$ is, in this case, the temporal mean, $< \boldsymbol{C}_{2q,x}(l)(t) >$, over an infinite interval duration, of the instantaneous statistics, $\boldsymbol{C}_{2q,x}(l)(t)$. In these conditions, using a cyclo-ergodicity property, the matrix $\boldsymbol{C}_{2q,x}(l)$ has to be estimated from the sampled data by a nonempirical estimator such as that presented in Ferreol and Chevalier (2000) for $q = 2$, the convergence of which is shown in Dandawate and Giannakis (1995). Note finally that this extension can also be applied to nonzero-mean cyclostationary sources, such as some nonlinearly digitally modulated sources (Ferreol et al., 2004b), provided that a nonempirical statistics estimator, such as that presented in Ferreol et al. (2004b) for $q = 1$ and in Ferreol et al. (2002) for $q = 2$, is used.

8.3 THE 2*q*-MUSIC METHOD

The 2q-MUSIC method is described in this section. To do so, we first introduce some hypotheses in Section 8.3.1 and analyze the algebraic structure of $\boldsymbol{C}_{2q,x}(l)$ matrix in Section 8.3.2. The 2q-MUSIC method is then presented in Section 8.3.3. Finally, Section 8.3.4 is devoted to the possible implementations of the 2q-MUSIC method.

8.3.1 Hypotheses

To develop the 2q-MUSIC algorithm for the arrangement l (Chevalier et al., 2006), we need some hypotheses that correspond to the following:

H1. $P_g < N, 1 \le g \le G$.
H2. Matrix $\boldsymbol{A}_g{}^{\otimes l} \otimes \boldsymbol{A}_g{}^{*\otimes(q-l)}$ has full-rank $P_g{}^q$, $1 \le g \le G$.
H3. $P(G,q) \triangleq \sum_{g=1}^{G} P_g{}^q < N^q$.
H4. Matrix $\overline{\boldsymbol{A}}_{q,l} \triangleq \left[\boldsymbol{A}_1{}^{\otimes l} \otimes \boldsymbol{A}_1{}^{*\otimes(q-l)}, \ldots, \boldsymbol{A}_G{}^{\otimes l} \otimes \boldsymbol{A}_G{}^{*\otimes(q-l)}\right]$ has full-rank $P(G,q)$.

For example, for $(q,l) = (2,1)$, hypothesis H2 reduces to $\boldsymbol{A}_g \otimes \boldsymbol{A}_g{}^*$ and has full-rank $P_g{}^2$. In particular, for sources that are all statistically dependent ($G = 1$), $P(G,q) = P^q$, matrix $\boldsymbol{A}_1$ reduces to $\boldsymbol{A}$ and hypotheses H1 through H4 reduce to

H1′. $P < N$.
H2′. Matrix $\boldsymbol{A}^{\otimes l} \otimes \boldsymbol{A}^{*\otimes(q-l)}$ has full-rank P^q.

However, for statistically independent sources ($G=P$), $P(G,q)=P$, matrix $\boldsymbol{A}_g$ reduces to the vector $\boldsymbol{a}(\boldsymbol{\theta}_g)$ and hypotheses H1 through H4 reduce to

H1''. $P<N^q$.
H2''. matrix $\overline{\boldsymbol{A}}_{q,l} \triangleq \left[\boldsymbol{a}_{q,l}(\boldsymbol{\theta}_1),\ldots,\boldsymbol{a}_{q,l}(\boldsymbol{\theta}_P)\right]$ has full-rank P.

8.3.2 Properties of $C_{2q,x}(l)$

Although components of $\boldsymbol{m}_g(t)$ are statistically dependent, the $(P_g{}^q \times P_g{}^q)$ matrix $\boldsymbol{C}_{2q,m_g}(l)$, which contains the $2q$th-order cumulants of $\boldsymbol{m}_g(t)$ for the arrangement indexed by l, may not be full rank for some couples (q,l). Indeed, assuming for example that $(q,l)=(2,2)$, it is easy to verify that the maximal rank of $\boldsymbol{C}_{4,m_g}(2)$ is 3 (not 4) for $P_g=2$ and 6 (not 9) for $P_g=3$. In this context, noting $r_{2q,m_g}(l)$ as the rank of $\boldsymbol{C}_{2q,m_g}(l)$ ($r_{2q,m_g}(l)\le P_g{}^q$), we deduce from H1 and H2 that matrix $\boldsymbol{C}_{2q,xg}(l)$ for $q>1$ also has rank $r_{2q,m_g}(l)$. Thus, using H4 and for $q>1$, matrix $\boldsymbol{C}_{2q,x}(l)$ has a rank $r_{2q,x}(l)$ equal to

$$r_{2q,x}(l)=\sum_{g=1}^{G} r_{2q,m_g}(l) \tag{8.16}$$

and such that $r_{2q,x}(l)<N^q$ from H3. As matrix $\boldsymbol{C}_{2q,x}(l)$ is Hermitian, we deduce that $\boldsymbol{C}_{2q,x}(l)$ has $r_{2q,x}(l)$ nonzero real-valued eigenvalues and $N^q-r_{2q,x}(l)$ zero eigenvalues for $q>1$.

8.3.3 Construction of 2*q*-MUSIC

To build a MUSIC-like method from the matrix $\boldsymbol{C}_{2q,x}(l)$, for $q>1$, we first compute the eigendecomposition of the latter, given by

$$\boldsymbol{C}_{2q,x}(l)=\boldsymbol{U}_{2q,s}(l)\boldsymbol{\Lambda}_{2q,s}(l)\boldsymbol{U}_{2q,s}(l)^\dagger+\boldsymbol{U}_{2q,n}(l)\boldsymbol{\Lambda}_{2q,n}(l)\boldsymbol{U}_{2q,n}(l)^\dagger \tag{8.17}$$

where $\boldsymbol{\Lambda}_{2q,s}(l)$ is the $(r_{2q,x}(l)\times r_{2q,x}(l))$ diagonal matrix of the nonzero eigenvalues of $\boldsymbol{C}_{2q,x}(l)$; $\boldsymbol{U}_{2q,s}(l)$ is the $(N^q\times r_{2q,x}(l))$ unitary matrix of the associated eigenvectors; $\boldsymbol{\Lambda}_{2q,n}(l)$ is the $((N^q-r_{2q,x}(l))\times(N^q-r_{2q,x}(l)))$ diagonal matrix of the zero eigenvalues of $\boldsymbol{C}_{2q,x}(l)$; and $\boldsymbol{U}_{2q,n}(l)$ is the $(N^q\times(N^q-r_{2q,x}(l)))$ unitary matrix of the associated eigenvectors. As $\boldsymbol{C}_{2q,x}(l)$ is Hermitian, all columns of $\boldsymbol{U}_{2q,s}(l)$ are orthogonal to all columns of $\boldsymbol{U}_{2q,n}(l)$. Moreover, $\text{Span}\{\boldsymbol{U}_{2q,s}(l)\}=\text{Span}\{\overline{\boldsymbol{A}}_{q,l}\}$ when matrices $\boldsymbol{C}_{2q,m_g}(l)$, $1\le g\le G$ are full rank; $\text{Span}\{\boldsymbol{U}_{2q,s}(l)\}\subset\text{Span}\{\overline{\boldsymbol{A}}_{q,l}\}$, otherwise. Denoting by $\boldsymbol{\theta}_{ig}$ the DOA parameters of the ith source in the gth group, it can be easily verified that, in all cases, the vector $\boldsymbol{a}_{q,l}(\boldsymbol{\theta}_{ig})$ always belongs to $\text{Span}\{\boldsymbol{U}_{2q,s}(l)\}$, called the $2q$th-order signal subspace for the arrangement l.

Consequently, all vectors $\{\boldsymbol{a}_{q,l}(\boldsymbol{\theta}_{ig}), 1\le i\le P_g, 1\le g\le G\}$ are orthogonal to the columns of $\boldsymbol{U}_{2q,n}(l)$ and are solutions to

$$\boldsymbol{a}_{q,l}(\boldsymbol{\theta})^\dagger\boldsymbol{\Pi}_{2q,n}(l)\boldsymbol{a}_{q,l}(\boldsymbol{\theta})=0 \tag{8.18}$$

where $\boldsymbol{\Pi}_{2q,n}(l)\triangleq\boldsymbol{U}_{2q,n}(l)\boldsymbol{U}_{2q,n}(l)^\dagger$.

Equation (8.18) corresponds to the heart of the 2q-MUSIC algorithm for the arrangement l (Chevalier et al., 2006), which corresponds to the well-known 2-MUSIC pseudo-spectrum for $(q, l) = (1, 1)$. Because, for an array with identical sensors in the absence of mutual coupling between the sensors, the complex gain of the sensors, $f(\boldsymbol{\theta}, \boldsymbol{\beta})$, has been integrated in the complex envelope of the associated source, we deduce from Equations (8.3) and (8.11) that $\boldsymbol{a}_{q,l}(\boldsymbol{\theta})^{\dagger}\boldsymbol{a}_{q,l}(\boldsymbol{\theta})$ does not depend on $\boldsymbol{\theta}$. However, in the presence of mutual coupling between sensors, a situation for which the array manifold $\boldsymbol{a}(\boldsymbol{\theta})$ has to be estimated by array calibration, $\boldsymbol{a}_{q,l}(\theta)^{\dagger}\boldsymbol{a}_{q,l}(\boldsymbol{\theta})$ may be a function of $\boldsymbol{\theta}$. In this case, in the absence of sources (i.e., when $\boldsymbol{\Pi}_{2q,n}(l) = \mathbf{I}_{N^q}$), to ensure a constant value, independent of parameters $\boldsymbol{\theta}$, of the left side of Equation (8.18), it is necessary to normalize the latter by the quantity $\boldsymbol{a}_{q,l}(\boldsymbol{\theta})^{\dagger}\boldsymbol{a}_{q,l}(\boldsymbol{\theta})$. Moreover, in practical situations, matrix $\boldsymbol{U}_{2q,s}(l)$ has to be estimated from the observations, and the DOAs of the sources may be found by searching for the minima of the left side of Equation (8.18).

The steps of the 2q-MUSIC algorithm for the arrangement l are summarized as follows:

Step 1. Estimation, $\hat{\boldsymbol{C}}_{2q,x}(l)$, of the matrix $\boldsymbol{C}_{2q,x}(l)$ from L snapshots $\mathbf{x}(k)$, $1 \le k \le L$, using a suitable estimator of the 2qth-order cumulants of observations.

Step 2. Eigendecomposition of the matrix $\hat{\boldsymbol{C}}_{2q,x}(l)$ and extraction of an estimate, $\hat{\boldsymbol{U}}_{2q,n}(l)$, of the $\boldsymbol{U}_{2q,n}(l)$ matrix. This step may involve rank determination in cases where the number of sources and/or their mutual statistical dependence are not known a priori.

Step 3. Computation of the estimated pseudo-spectrum

$$\hat{P}_{2q-Music(l)}(\boldsymbol{\theta}) \triangleq \frac{\mathbf{a}_{q,l}(\boldsymbol{\theta})^{\dagger}\hat{\boldsymbol{\Pi}}_{2q,n}(l)\mathbf{a}_{q,l}(\boldsymbol{\theta})}{\mathbf{a}_{q,l}(\boldsymbol{\theta})^{\dagger}\mathbf{a}_{q,l}(\boldsymbol{\theta})} \tag{8.19}$$

over a suitably chosen grid, and search for the local minima (including interpolation at each local minimum), where $\hat{\boldsymbol{\Pi}}_{2q,n}(l) \triangleq \hat{\boldsymbol{U}}_{2q,n}(l)\hat{\boldsymbol{U}}_{2q,n}(l)^{\dagger}$.

Sometimes the number of sources P is known, such that $P < N$, but their statistical dependence is not. In this case, $r_{2q,x}(l) \le P^q$, and a conservative approach is to use only the $(N^q - P^q)$ eigenvectors associated with the smallest eigenvalues to build $\hat{\boldsymbol{U}}_{2q,n}(l)$, which implicitly assumes the statistical dependence of all sources. Note finally that similar to 2-MUSIC, 2q-MUSIC cannot handle perfectly coherent sources. Indeed, in such a case the associated vectors $\boldsymbol{a}_{q,l}(\boldsymbol{\theta}_{ig})$ no longer belong to Span$\{\boldsymbol{U}_{2q,s}(l)\}$ and the corresponding sources become indistinguishable to the algorithm.

8.3.4 Implementations

We note by $\hat{P}$ the estimate of the number of sources P obtained from step 2 of the 2q-MUSIC algorithm. In the standard version of 2q-MUSIC, the $\hat{P}$ local minima of Equation (8.19) are estimated in parallel. However, this parallel search gives rise to interaction between the sources when the spatial correlation coefficient between at least two of the sources is high, which occurs in particular when the sources are close to each other or when $\hat{P}$ approaches the number of sensors N. This interaction between sources generally decreases DOA estimation performance, especially the performance of the weakest sources. To decrease interaction between the sources in the pseudo-spectrum (8.19), and then to increase the quality of their DOA estimation, one may replace the parallel search of local minima in Equation (8.19) with a sequential search.

A sequential search consists of finding the DOA number p after having removed in the pseudo-spectrum (8.19), by some deflation scheme, the contribution of DOA numbers 1 to $p-1$ already estimated. Such sequential schemes were proposed for implementation of 2-MUSIC in Mosher and Leahy (1999), Oh and Un (1993), and Stoica et al. (1995), and have given rise to S-MUSIC (Sequential MUSIC; Oh and Un, 1993, IES-MUSIC (Improved Sequential MUSIC; Stoica et al., 1995, and RAP-MUSIC (Recursively Applied and Projected MUSIC); Mosher and Leahy, 1999, respectively. More recently, sequential versions of the $2q$-MUSIC algorithm were proposed in Albera et al. (2008) and Birot et al. (2007), giving rise to the $2q$-RAP-MUSIC and $2q$-D-MUSIC ($2q$ Deflated MUSIC) algorithms. We limit the following discussion to $2q$-D-MUSIC.

Let us assume, for $p \geq 2$, that $p-1$ DOAs, $\boldsymbol{\theta}_i$ for $1 \leq i \leq p-1$, have already been found, and let us denote by $\tilde{\boldsymbol{A}}_{q,l}(\boldsymbol{\theta}_{1:p-1})$ the matrix of the associated vectors $\boldsymbol{a}_{q,l}(\boldsymbol{\theta}_i)$, defined by $\tilde{\boldsymbol{A}}_{q,l}(\boldsymbol{\theta}_{1:p-1}) \triangleq [\boldsymbol{a}_{q,l}(\boldsymbol{\theta}_1), \ldots, \boldsymbol{a}_{q,l}(\boldsymbol{\theta}_{p-1})]$. We then define, for $1 \leq p \leq \hat{P}$, the orthogonal projector, $\boldsymbol{P}_{q,l}(\boldsymbol{\theta}_{1:p-1})^{\perp}$, on the space that is orthogonal to Span$\{\tilde{\boldsymbol{A}}_{q,l}(\boldsymbol{\theta}_{1:p-1})\}$ by

$$\boldsymbol{P}_{q,l}(\boldsymbol{\theta}_{1:p-1})^{\perp} \triangleq \mathbf{I}_{N^q} - \tilde{\boldsymbol{A}}_{q,l}(\boldsymbol{\theta}_{1:p-1}) \left[\tilde{\boldsymbol{A}}_{q,l}(\boldsymbol{\theta}_{1:p-1})^{\dagger} \tilde{\boldsymbol{A}}_{q,l}(\boldsymbol{\theta}_{1:p-1})\right]^{-1} \tilde{\boldsymbol{A}}_{q,l}(\boldsymbol{\theta}_{1:p-1})^{\dagger} \tag{8.20}$$

For $p=1$, we assume that $\boldsymbol{P}_{q,l}(\boldsymbol{\theta}_{1:0})^{\perp} \triangleq \mathbf{I}_{N^q}$. For $p \geq 1$, it is straightforward to verify that the projected vectors $\boldsymbol{P}_{q,l}(\boldsymbol{\theta}_{1:p-1})^{\perp} \boldsymbol{a}_{q,l}(\boldsymbol{\theta}_i)$ contained in the signal subspace of the matrix $\boldsymbol{P}_{q,l}(\boldsymbol{\theta}_{1:p-1})^{\perp}\ \boldsymbol{C}_{2q,x}(l) \boldsymbol{P}_{q,l}(\boldsymbol{\theta}_{1:p-1})^{\perp}$ correspond to the projected vectors associated with the $\hat{P}-p+1$ DOAs that have yet to be found, while the others have been removed by the projection operation. As a consequence, the implementation of the $2q$-MUSIC algorithm from the matrix $\boldsymbol{P}_{q,l}(\boldsymbol{\theta}_{1:p-1})^{\perp} \boldsymbol{C}_{2q,x}(l) \boldsymbol{P}_{q,l}(\boldsymbol{\theta}_{1:p-1})^{\perp}$ and the $2q$th-order projected array manifold $\boldsymbol{P}_{q,l}(\boldsymbol{\theta}_{1:p-1})^{\perp} \boldsymbol{a}_{q,l}(\boldsymbol{\theta})$ allows us to find the DOA number p with a limited interaction with sources 1 to $p-1$.

The steps of the $2q$-D-MUSIC algorithm for the estimation of $\boldsymbol{\theta}_p (1 \leq p \leq \hat{P})$ and for arrangement l are summarized here from the estimation, $\hat{\boldsymbol{\theta}}_i$, of the DOAs $\boldsymbol{\theta}_i (1 \leq i \leq p-1)$:

Step 1. Estimation, $\hat{\boldsymbol{C}}_{2q,x}(l)$, of the matrix $\boldsymbol{C}_{2q,x}(l)$ from L snapshots, $\boldsymbol{x}(k)$, $1 \leq k \leq L$, using a suitable estimator of the $2q$th-order cumulants of observations.

Step 2. Initialization of parameters: $p=1$ and $\boldsymbol{P}_{q,l}(\hat{\boldsymbol{\theta}}_{1:0})^{\perp} = \mathbf{I}_{N^q}$.

Step 3. Eigendecomposition of the matrix $\boldsymbol{P}_{q,l}(\hat{\boldsymbol{\theta}}_{1:p-1})^{\perp}\hat{\boldsymbol{C}}_{2q,x}(l)\boldsymbol{P}_{q,l}(\hat{\boldsymbol{\theta}}_{1:p-1})^{\perp}$, and extraction of an estimate, $\hat{\boldsymbol{U}}_{2q,n,p-1}(l)$, of the unitary matrix of the noise subspace eigenvectors. This step may involve rank determination in cases where the number of sources and/or their mutual statistical dependence are not known a priori.

Step 4. Computation of the estimated pseudo-spectrum

$$\hat{P}_{2q-D-Music,p(l)}(\boldsymbol{\theta}) \triangleq \frac{\boldsymbol{a}_{q,l}(\boldsymbol{\theta})^{\dagger}\boldsymbol{P}_{q,l}(\hat{\boldsymbol{\theta}}_{1:p-1})^{\perp}\hat{\boldsymbol{\Pi}}_{2q,n,p-1}(l)\boldsymbol{P}_{q,l}(\hat{\boldsymbol{\theta}}_{1:p-1})^{\perp}\boldsymbol{a}_{q,l}(\boldsymbol{\theta})}{\boldsymbol{a}_{q,l}(\boldsymbol{\theta})^{\dagger}\boldsymbol{P}_{q,l}(\hat{\boldsymbol{\theta}}_{1:p-1})^{\perp}\boldsymbol{a}_{q,l}(\boldsymbol{\theta})} \tag{8.21}$$

over a suitably chosen grid, and search for the absolute minimum (including interpolation around this minimum), where $\hat{\boldsymbol{\Pi}}_{2q,n,p-1}(l) \triangleq \hat{\boldsymbol{U}}_{2q,n,p-1}(l)\hat{\boldsymbol{U}}_{2q,n,p-1}(l)^{\dagger}$. The position of this minimum corresponds to $\hat{\boldsymbol{\theta}}_p$.

Step 5. If $p < \hat{P}$, incrementation of $p : p = p+1$.

Step 6. Computation of $\tilde{\boldsymbol{A}}_{q,l}(\hat{\boldsymbol{\theta}}_{1:p-1}) \triangleq [\boldsymbol{a}_{q,l}(\hat{\boldsymbol{\theta}}_1), \ldots, \boldsymbol{a}_{q,l}(\hat{\boldsymbol{\theta}}_{p-1})]$ and the orthogonal projector, $\boldsymbol{P}_{q,l}(\hat{\boldsymbol{\theta}}_{1:p-1})^{\perp}$, defined by Equation (8.20) with $\hat{\boldsymbol{\theta}}_{1:p-1}$ instead of $\boldsymbol{\theta}_{1:p-1}$. To reduce the computational cost of $\boldsymbol{P}_{q,l}(\hat{\boldsymbol{\theta}}_{1:p-1})^{\perp}$, a recursive computation of the latter as a function of $\boldsymbol{P}_{q,l}(\hat{\boldsymbol{\theta}}_{1:p-2})^{\perp}$ is described in Birot et al. (2007) and may be preferred.

Step 7. Reiteration from step 3.

Let us note that in practical situations, the $p-1$ source DOA estimates $\hat{\boldsymbol{\theta}}_i$, $1 \leq i \leq p-1$, have errors and that a contribution of these sources remains present after the orthogonal projection (8.20). However, despite these errors, simulations show that when resolution is required to separate sources, 2q-D-MUSIC still performs better than 2q-MUSIC. To our knowledge, no analytical proof of this result is currently available.

8.4 2*q*-MUSIC IDENTIFIABILITY

Because the maximal number of sources that can be processed by the 2q-MUSIC algorithm is obtained when all sources are statistically independent, we limit the identifiability analysis (i.e., the analysis of the maximal number of sources that can be processed by the 2q-MUSIC algorithm) to the case of statistically independent sources. In such a situation, hypotheses H1 through H4 reduce to H1$''$ and H2$''$, respectively. For this reason, we first describe, in Section 8.4.1, the HO virtual array concept (Chevalier et al., 2005), valid for statistically independent sources, which generalizes to an arbitrary even order the FO virtual array concept initially introduced in Chevalier and Ferreol (1999) and Dogan and Mendel (1995a). Some properties of HO VA, presented in Section 8.4.2, then allow us to show geometrically the increasing resolution of 2q-MUSIC method as q increases. Finally, the identifiability issue of 2q-MUSIC method is described in Section 8.4.3.

8.4.1 Higher-Order Virtual Array

Assuming no noise, statistically independent non-Gaussian sources, and no modeling errors, particularly no mutual coupling between sensors, the matrices $\boldsymbol{C}_{2q,x}(l)$ and $\boldsymbol{R}_x = \boldsymbol{C}_{2,x}(1)$, defined by Equation (8.10), have the same algebraic structure, where the auto-cumulant $c_{2q,mi}$ and the vector $\boldsymbol{a}_{q,l}(\boldsymbol{\theta}_i)$ have for $\boldsymbol{C}_{2q,x}(l)$, the place occupied by the power $c_{2,mi}$ and the steering vector $\boldsymbol{a}(\boldsymbol{\theta}_i)$, respectively, for R_x. Thus, for the $2q$th-order array-processing methods exploiting Equation (8.10), the $(N^q \times 1)$ vector $\boldsymbol{a}_{q,l}(\boldsymbol{\theta}_i)$ can be considered as the *virtual steering vector* of the source i for the *true array* of N identical sensors with coordinates (x_n, y_n, z_n), $1 \le n \le N$. The N^q components of the vector $\boldsymbol{a}_{q,l}(\boldsymbol{\theta}_i)$ correspond to the quantities

$$a_{k_1}(\boldsymbol{\theta}_i)a_{k_2}(\boldsymbol{\theta}_i)\ldots a_{k_l}(\boldsymbol{\theta}_i)a_{k_{l+1}}(\boldsymbol{\theta}_i)^* a_{k_{l+2}}(\boldsymbol{\theta}_i)^* \ldots a_{k_q}(\boldsymbol{\theta}_i)^* (1 \le k_j \le N, 1 \le j \le q)$$

where $a_{k_j}(\boldsymbol{\theta}_i)$ is the component k_j of vector $\boldsymbol{a}(\boldsymbol{\theta}_i)$. Numbering, in a natural way, the N^q values of the q-uplet $(k_1, k_2, \ldots, k_l, k_{l+1}, \ldots, k_q)$ by associating to the latter the integer K defined by

$$K \triangleq \sum_{j=1}^{q} N^{q-j}(k_j - 1) + 1 \tag{8.22}$$

and using Equation (8.3) in the components of $\boldsymbol{a}_{q,l}(\boldsymbol{\theta}_i)$, we find that the Kth component of the vector $\boldsymbol{a}_{q,l}(\boldsymbol{\theta}_i)$, noted $a_{q,l}(\boldsymbol{\theta}_i)_K$, takes the form

$$\begin{aligned} a_{q,l}(\boldsymbol{\theta}_i)_K = \exp\Bigg\{ j2\pi \Bigg[& \left(\sum_{j=1}^{l} x_{k_j} - \sum_{u=1}^{q-1} x_{k_{l+u}} \right) \cos(\theta_i)\cos(\varphi_i) \\ & + \left(\sum_{j=1}^{l} y_{k_j} - \sum_{u=1}^{q-1} y_{k_{l+u}} \right) \sin(\theta_i)\cos(\varphi_i) \\ & + \left(\sum_{j=1}^{l} z_{k_j} - \sum_{u=1}^{q-1} z_{k_{l+u}} \right) \sin(\varphi_i) \Bigg] \Big/ \lambda \Bigg\} \end{aligned} \tag{8.23}$$

Comparing Equation (8.23) to $a_K(\boldsymbol{\theta}_i)$ defined by Equation (8.3) with $n = K$ and $f_n(\boldsymbol{\theta}_i, \boldsymbol{\beta}_i) = 1$, we deduce that the vector $\boldsymbol{a}_{q,l}(\boldsymbol{\theta}_i)$ can also be considered as the *true steering vector* of source i for the VA of N^q identical virtual sensors (VSs), with coordinates $\left(x^l_{k_1,k_2\ldots k_q}, y^l_{k_1,k_2\ldots k_q}, z^l_{k_1,k_2\ldots k_q}\right)(1 \le k_j \le N, 1 \le j \le q)$ given by

$$\begin{aligned} & \left(x^l_{k_1,k_2\ldots k_q}, y^l_{k_1,k_2\ldots k_q}, z^l_{k_1,k_2\ldots k_q}\right) \\ & = \left(\sum_{j=1}^{l} x_{kj} - \sum_{u=1}^{q-l} x_{k_{l+u}}, \sum_{j=1}^{l} y_{k_j} - \sum_{u=1}^{q-l} y_{k_{l+u}}, \sum_{j=1}^{l} z_{k_j} - \sum_{u=1}^{q-l} z_{k_{l+u}} \right) \end{aligned} \tag{8.24}$$

Equation (8.25) introduces, in a very simple way and for arrays with identical sensors, the VA concept for the $2q$th-order DOA estimation problem for the arrangement $\boldsymbol{C}_{2q,x}(l)$.

Thus, we can consider that the $2q$th-order DOA estimation problem of P statistically independent NB non-Gaussian sources from a given array of N identical sensors with coordinates (x_n, y_n, z_n), $1 \le n \le N$, is, for the arrangement $\boldsymbol{C}_{2q,x}(l)$, similar to an SO DOA estimation problem for which these P statistically independent NB sources impinge, with the virtual powers c_{2q,m_i} $(1 \le i \le P)$, on a VA of N^q identical VSs having the coordinates $\left(x^l_{k_1 k_2 \ldots k_q}, y^l_{k_1 k_2 \ldots k_q}, z^l_{k_1 k_2 \ldots k_q}\right)$, $1 \le k_j \le N$ for $1 \le j \le q$, defined by Equation (8.24). Thus, HO array processing may be used to replace sensors and hardware and so decrease the overall cost of a given system.

8.4.2 Properties of Higher-Order Virtual Array

We naturally deduce from the previous theory that, for given values of (q, l), the 3-dB beamwidth of the associated $2q$th-order VA controls the resolution power of $2q$th-order DOA estimation methods for the arrangement l and for a finite observation duration. For further insight into this resolution power, let us compute the spatial correlation coefficient, $\alpha_{\theta,\theta_p}(q,l)$ $(0 \le |\alpha_{\theta,\theta_p}(q,l)| \le 1)$, of DOAs $\boldsymbol{\theta}$ and $\boldsymbol{\theta}_p$ for the HO VA associated with parameters (q, l), defined by

$$\alpha_{\theta,\theta_p}(q,l) \triangleq \frac{\boldsymbol{a}_{q,l}(\boldsymbol{\theta})^\dagger \boldsymbol{a}_{q,l}(\boldsymbol{\theta}_p)}{\left[\boldsymbol{a}_{q,l}(\boldsymbol{\theta})^\dagger \boldsymbol{a}_{q,l}(\boldsymbol{\theta})\right]^{1/2} \left[\boldsymbol{a}_{q,l}(\boldsymbol{\theta}_p)^\dagger \boldsymbol{a}_{q,l}(\boldsymbol{\theta}_p)\right]^{1/2}} \tag{8.25}$$

It is straightforward to verify (Chevalier et al., 2005) that the modulus of $\alpha_{\theta,\theta_p}(q,l)$ can be written as

$$|\alpha_{\theta,\theta_p}(q,l)| = |\alpha_{\theta,\theta_p}(1,1)|^q \tag{8.26}$$

where $\alpha_{\theta,\theta_p}(1,1)$ is the spatial correlation coefficient of DOAs $\boldsymbol{\theta}$ and $\boldsymbol{\theta}_p$ for the true array of N sensors, defined by

$$\alpha_{\theta,\theta_p}(1,1) \triangleq \frac{\boldsymbol{a}(\boldsymbol{\theta})^\dagger \boldsymbol{a}(\boldsymbol{\theta}_p)}{\left[\boldsymbol{a}(\boldsymbol{\theta})^\dagger \boldsymbol{a}(\boldsymbol{\theta})\right]^{1/2} \left[\boldsymbol{a}(\boldsymbol{\theta}_p)^\dagger \boldsymbol{a}(\boldsymbol{\theta}_p)\right]^{1/2}} \tag{8.27}$$

From Equation (8.26), we deduce that $|\alpha_{\theta,\theta_p}(q,l)|$ is independent of l and is a decreasing function of q, showing off the increasing resolution of HO VA as q increases. In particular, it was shown in Chevalier et al. (2005) that the 3-dB beamwidth of the $2q$th-order VA, $\theta^{2q}_{3\mathrm{dB}}$, is equal to $\theta^{2q}_{3\mathrm{dB}} = 0.84\,\theta_{3\mathrm{dB}}, 0.76\,\theta_{3\mathrm{dB}}$, and $0.71\ \theta_{3\mathrm{dB}}$ for $q = 2, 3$, and 4, respectively, where $\theta_{3\mathrm{dB}}$ corresponds to the 3-dB beamwidth of the true array.

Another interesting result deduced from Equation (8.26) is that rank-1 ambiguities (or grating lobes; Compton, 1988) of the $2q$th-order VA coincide with those of the true array whatever the values of q and l, since the directions $\boldsymbol{\theta} \neq \boldsymbol{\theta}_p$ giving rise to $|\alpha_{\theta,\theta_p}(q,l)| = 1$ are exactly the ones that give rise to $|\alpha_{\theta,\theta_p}(1,1)| = 1$.

A consequence of this result is that a necessary and sufficient condition to obtain the $2q$th-order VA without any rank-1 ambiguities is that the true array of N sensors have no rank-1 ambiguities.

8.4.3 Identifiability of the $2q$-MUSIC Method

Following the developments of Section 8.3, we deduce that the $2q$-MUSIC algorithm for the arrangement indexed by l is able to estimate the DOAs of P noncoherent sources from an array of N identical sensors, provided that hypotheses H1 through H4 are verified and that the DOAs of the sources are the only solutions to Equation (8.18).

Analysis

As for statistically independent sources, hypotheses H1 through H4 reduce to H1$''$ and H2$''$, the identifiability analysis of the $2q$-MUSIC method mainly reduces to analysis of the conditions under which H2$''$ is verified. Thus, because some of the N^q VSs of the associated $2q$th-order VA may coincide, we note N_{2q}^l ($N_{2q}^l \leq N^q$) as the number of different VSs of the VA associated with the parameters (q, l). This number, N_{2q}^l, is directly related to the geometry of the true array of N sensors as shown by Equation (8.24). In these conditions, $N^q - N_{2q}^l$ components of all vectors $\boldsymbol{a}_{q,l}(\boldsymbol{\theta})$ are redundant, providing no information. Therefore, $N^q - N_{2q}^l$ rows of the $\bar{\boldsymbol{A}}_{q,l}$ matrix provide no information and are linear combinations of the other rows, which means that the rank of $\bar{\boldsymbol{A}}_{q,l}$ cannot be greater than N_{2q}^l. We then deduce that the $\bar{\boldsymbol{A}}_{q,l}$ matrix may have a rank equal to P only if $P \leq N_{2q}^l$.

Conversely, for a $2q$th-order VA without ambiguities up to order N_{2q}^l, P sources coming from P different directions generate an $\bar{\boldsymbol{A}}_{q,l}$ matrix with a full-rank P as long as $P \leq N_{2q}^l$. Thus, the maximal number of statistically independent sources able to generate a matrix $\bar{\boldsymbol{A}}_{q,l}$ with rank P is N_{2q}^l. However, when $P = N_{2q}^l$, an arbitrary vector $\boldsymbol{a}_{q,l}(\boldsymbol{\theta})$ associated with an arbitrary DOA $\boldsymbol{\theta}$ is necessarily a linear combination of the source steering vectors $\boldsymbol{a}_{q,l}(\boldsymbol{\theta}_i), 1 \leq i \leq N_{2q}^l$, since matrix $\bar{\boldsymbol{A}}_{q,l}$ cannot have a rank greater than N_{2q}^l. All DOAs $\boldsymbol{\theta}$ are thus solutions to Equation (8.18), which does not allow the sources' DOA estimation. A necessary condition for the DOAs of the sources to be the only solutions to Equation (8.18), then, is that $P < N_{2q}^l$ and this condition becomes sufficient for HO VA with no ambiguities. From the previous results we deduce that the $2q$-MUSIC algorithm for the arrangement l is able to process up to $N_{2q}^l - 1$ sources.

This result means that the processing capacity of $2q$-MUSIC depends on the geometry of the true array of N sensors and on the parameters q and l, which show in particular the existence of an optimal arrangement of the $2q$th-order data statistics for a given value of q. It is shown in Chevalier et al. (2005) that for a

TABLE 8.1 $N_{max}[2q, l]$ **as a Function of** N **for Several Values of** q **and** l **and for Arrays with Identical Sensors**

$m = 2q$	l	$N_{max}[2q, l]$
4	2	$N(N+1)/2$
$(q=2)$	1	$N^2 - N + 1$
6	3	$N!/[6(N-3)!] + N(N-1) + N$
$(q=3)$	2	$N!/[2(N-3)!] + N(N-1) + N$
8	4	$N!/[24(N-4)!] + N!/[2(N-3)!] + 1.5N(N-1) + N$
$(q=4)$	3	$N!/[6(N-4)!] + N!/(N-3)! + 1.5N(N-1) + N$
	2	$N!/[4(N-4)!] + N!/(N-3)! + 2N(N-1) + 1$

Source: From Chevalier et al., 2005, © IEEE, with permission.

given value of q, the optimal arrangement is associated with the integer l, which minimizes the quantity $|2l - q|$, finally generating steering vectors $\boldsymbol{a}_{q,l}(\boldsymbol{\theta}_p)$ for which the number of conjugate vectors is the least different from the number of nonconjugate vectors. In particular, for $q = 2$ this corresponds to $l = 1$ (i.e., to steering vectors of the form $[\boldsymbol{a}(\boldsymbol{\theta}_i) \otimes \boldsymbol{a}(\boldsymbol{\theta}_i)^*]$).

Finally, analysis of the symmetries of the HO VA shows (Chevalier et al., 2005) that, for given values of N, q, and l, N_{2q}^l is necessarily upper-bounded by a quantity, noted $N_{max}[2q, l]$, such that $N_{max}[2q, l] < N^q$. Table 8.1 illustrates, for arrays with identical sensors, the expression of $N_{max}[2q, l]$ as a function of N for $2 \leq q \leq 4$ and several values of l. In fact, this upper bound corresponds to N_{2q}^l in most cases of array geometries with no particular symmetries (e.g., a uniform circular array with a number of sensors corresponding to a prime number; Chevalier et al., 2005), but it cannot be reached by N_{2q}^l for arrays with particular symmetries (e.g., a uniform linear array (ULA), with an HO VA that is analyzed in the next section).

Example of the Uniform Linear Array

For a ULA, it is always possible to choose a coordinate system in which the sensor n has the coordinates $(x_n = nd, 0, 0)$, $1 \leq n \leq N$, where d is the inter-element spacing. As a consequence, the VSs of the HO VA for parameters (q, l) are, from Equation (8.24), at coordinates

$$\left(x_{k_1 k_2 \ldots k_q}^l, y_{k_1 k_2 \ldots k_q}^l, z_{k_1 k_2 \ldots k_q}^l\right) = \left(\left(\sum_{j=1}^{l} k_j - \sum_{u=1}^{q-l} k_{l+u}\right) d, 0, 0\right) \tag{8.28}$$

for $1 \le k_j \le N$ and $1 \le j \le q$. This shows that the associated HO VA is also a ULA with the same inter-element spacing, whatever the arrangement l. Moreover, for given values of q, l, and N, the minimum and maximum values of $x^l_{k_1 k_2 \ldots k_q}$, noted $x^l_{q,\min}$ and $x^l_{q,\max}$, respectively, are given by

$$x^l_{q,\min} = [l - (q-l)N]d = [l(1+N) - qN]d \tag{8.29}$$

$$x^l_{q,\max} = [lN - (q-l)]d = [l(1+N) - q]d \tag{8.30}$$

and the number of different VSs, N^l_{2q}, of the associated VA, easily deduced from Equations (8.29) and (8.30), is given by

$$N^l_{2q} = \left(x^l_{q,\max} - x^l_{q,\min}\right)/d + 1 = qN - (q-1) = q(N-1) + 1 \tag{8.31}$$

This is independent of l, and means that, for given values of q and N, the number of VSs is independent of the chosen arrangement l.

Put another way, in terms of processing power, for a given value of q and because of the symmetries of the array, all arrangements $C_{2q,x}(l)$ are equivalent for a ULA. The $2q$-MUSIC algorithm is thus able to process up to $N^l_{2q} - 1 = q(N-1)$ statistically independent non-Gaussian sources from a ULA of $N > 1$ sensors, which is an increasing function of N and q and is strictly lower than $N_{\max}[2q, l] - 1$ as soon as $q > 1$ and $N > 2$.

To illustrate the previous results, Figure 8.2 shows the HO VA, with nine VSs, of a ULA of five sensors for which $d = \lambda/2$, together with the order of

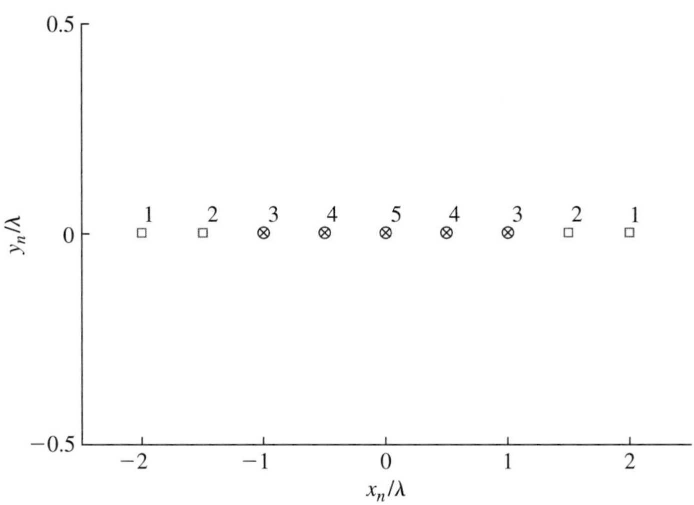

FIGURE 8.2 4th-order VA of a ULA of five sensors with the order of multiplicities of the VSs (*Source:* From Chevalier et al., 2005, © IEEE, with permission.)

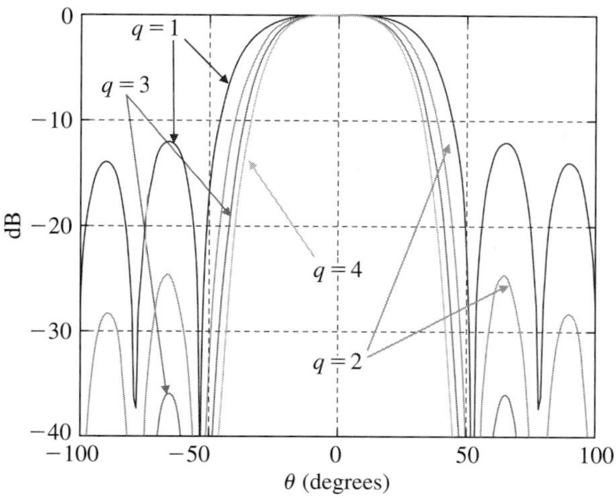

FIGURE 8.3 VA pattern for $q=1,2,3,4$, ULA with five sensors, $d=\lambda/2$; pointing direction: 0° (*Source:* From Chevalier et al., 2005, © IEEE, with permission.)

multiplicity of the VSs (i.e., the number of VSs having the same coordinates), with the x- and y-axes normalized by the wavelength λ.

To complete these results and to illustrate Equation (8.26) in relation to the increasing resolution of HO VA as q increases, Figure 8.3 shows the array pattern (the normalized inner product of two associated steering vectors) of HO VA associated with a ULA of five sensors equi-spaced half a wavelength apart for $q = 1, 2, 3$, and 4, and for a pointing direction equal to 0°. Note the decreasing 3-dB beamwidth and sidelobe level of the array pattern as q increases in proportions given by Equation (8.26).

Example of the Uniform Circular Array

For a UCA of N sensors, it is always possible to choose a coordinate system in which the sensor n has the coordinates $(R\cos\phi_n, R\sin\phi_n, 0) 1 \leq n \leq N$, where R is the radius of the array and where $\phi_n \triangleq (n-1)2\pi/N$. For $q>2$, the analytical computation of the VA (introduced in Chevalier et al., 2005) is a tedious task. It can be verified that, for given values of q and l, the number of different VSs, N_{2q}^{l}, of the associated VA corresponds to the upper bound, $N_{\max}[2q, l]$, when N is a prime number. In this case, the 2q-MUSIC method is able to process up to $N_{2q}^{l} - 1 = N_{\max}[2q, l] - 1$ statistically independent non-Gaussian sources from a UCA of N sensors, where some values of $N_{\max}[2q, l]$ are presented in Table 8.1. Otherwise, N_{2q}^{l} remains smaller than $N_{\max}[2q, l]$.

To illustrate the previous result, Figures 8.4 and 8.5 show the VA of a UCA of five sensors for which $R = 0.8\lambda$, together with the order of multiplicity of the VSs, for $(q, l) = (2, 2)$ and $(q, l) = (2, 1)$, respectively. Note the greater value of N_{2q}^{l}, equal to 21, for the optimal arrangement $l = 1$, whereas $N_{2q}^{l} = 15$ for $l = 2$.

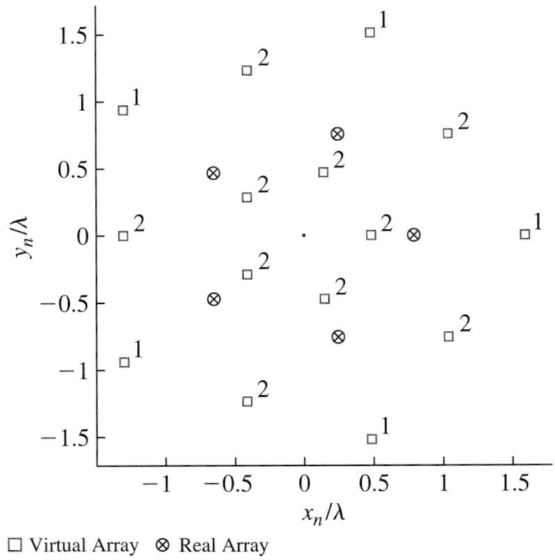

FIGURE 8.4 4th-order VA of a UCA of five sensors with the order of multiplicities of the VSs for $(q, l) = (2, 2)$. (*Source:* From Chevalier et al., 2005, © IEEE, with permission.)

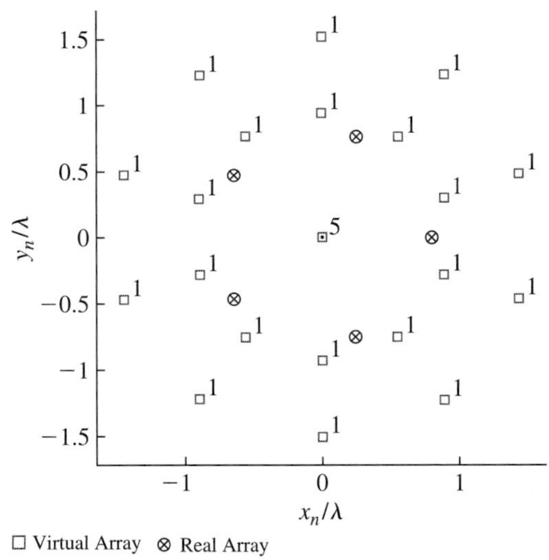

FIGURE 8.5 4th-order VA of a UCA of five sensors with the order of multiplicities of the VSs for $(q, l) = (2, 1)$. (*Source:* From Chevalier et al., 2005, © IEEE, with permission.)

To complete these results and to illustrate Equation (8.26) for a UCA, Figure 8.6 shows the array pattern of HO VA associated with a UCA of five sensors such that $R = 0.8\,\lambda$, for $q = 1, 2, 3$, and 4, and for a pointing direction equal to 0°. Note the decreasing 3-dB beamwidth and sidelobe level of the array pattern as q increases in proportions given by Equation (8.26).

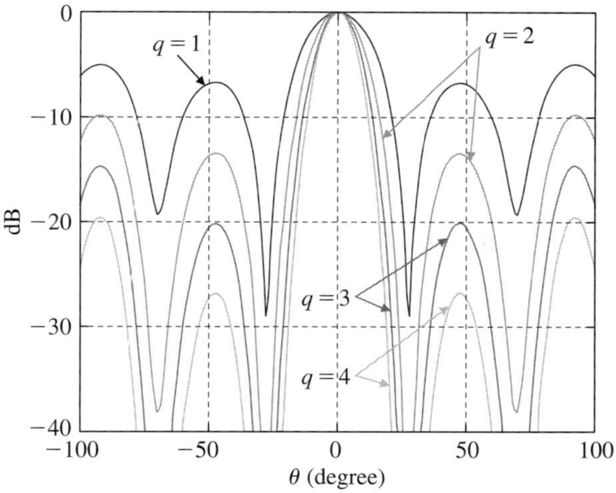

FIGURE 8.6 VA pattern for $q = 1, 2, 3, 4$, UCA with five sensors, $R = 0.8\,\lambda$; pointing direction: 0°.

8.5 2q-MUSIC PERFORMANCE

This section describes the performance of the 2q-MUSIC method in the presence of modeling errors. To this aim, in Section 8.5.1 we present the main source of performance degradation in operational context—that is, the finite-sample effect and modeling errors. In practical situations, modeling errors are generally the dominant source of performance degradation. For this reason, an observation model including the modeling errors is introduced in Section 8.5.2. Performance of 2-MUSIC and 2q-MUSIC for $q > 1$ in the presence of modeling errors are then presented in Sections 8.5.3 and 8.5.4, respectively. Finally, results of previous sections are illustrated in Section 8.5.5.

8.5.1 Finite-Sample Effect and Modeling Errors

In operational contexts, for given choices of sensor array and algorithm, the performance of the latter is mainly controlled by both finite-sample effect, due to the finite observation duration, and modeling errors such as array calibration errors or phase and amplitude residual mismatches between reception chains.

The sensitivity of several FO DOA estimation methods with respect to finite-sample effects was studied analytically in Cardoso and Moulines (1995), Fan and Younan (1995), and Yuen and Friedlander (1996, 1998). Cardoso and Moulines (1995) mainly considered FO-contracted MUSIC methods, whereas Fan and Younan (1995) dealt with the harmonic retrieval problem for $N = 1$. Moreover, Yuen and Friedlander (1996, 1998) analyzed the sensitivity of both 4-ESPRIT and FO virtual ESPRIT for far-field and near-field sources, respectively. Most of these works (Cardoso and Moulines, 1995, in particular) show the higher sensitivity to finite-sample effects of FO versus SO methods, especially for weak sources. Nevertheless, to our knowledge, the sensitivity of $2q$-MUSIC to finite-sample effect has not yet been studied analytically for $q > 1$, which prevents us from describing such results in the next sections.

In contrast, the sensitivity of HO DOA estimation methods with respect to modeling errors, omnipresent in all operational contexts, was analyzed in Liu and Mendel (1999b) for the FO virtual ESPRIT algorithm and in Chevalier et al. (2006) for the $2q$-MUSIC algorithm. We present here some modified results of the analysis described in Chevalier et al. (2006), showing in particular the influence of parameters q and l on robustness with respect to modeling errors of $2q$-MUSIC.

8.5.2 Model and Problem Formulation

In the presence of modeling errors, the observation vector defined by Equation (8.2) becomes

$$\boldsymbol{x}(t) \approx \sum_{i=1}^{P} m_i(t)\tilde{\boldsymbol{a}}(\boldsymbol{\theta}_i) + \boldsymbol{v}(t) \triangleq \tilde{\boldsymbol{A}}\boldsymbol{m}(t) + \boldsymbol{v}(t) \tag{8.32}$$

where $\tilde{\boldsymbol{A}}$ is the $(N \times P)$ matrix of the corrupted source steering vectors $\tilde{\boldsymbol{a}}(\boldsymbol{\theta}_i)$, $(1 \leq i \leq P)$, such that

$$\tilde{\boldsymbol{a}}(\boldsymbol{\theta}_i) = \boldsymbol{a}(\boldsymbol{\theta}_i) + \boldsymbol{e}(\boldsymbol{\theta}_i) \tag{8.33}$$

where $\boldsymbol{e}(\boldsymbol{\theta}_i)$ is the modeling error vector of the source i, whereas vector $\boldsymbol{a}(\boldsymbol{\theta}_i)$ is such that $\boldsymbol{a}(\boldsymbol{\theta}_i)^\dagger \boldsymbol{a}(\boldsymbol{\theta}_i) = N$.

From Equation (8.33) we deduce that

$$\tilde{\boldsymbol{A}} = \boldsymbol{A} + \boldsymbol{E} \triangleq \tilde{\boldsymbol{A}}(\boldsymbol{E}) \tag{8.34}$$

where $\boldsymbol{E}$ is the $(N \times P)$ matrix of the error vectors $\boldsymbol{e}(\boldsymbol{\theta}_i)$, $(1 \leq i \leq P)$. Note that the model (8.33) is well suited for any distortion (mutual coupling between sensors, mismatches between reception chains, sensor position errors) (Ferreol et al., 2006). To simplify the notations, we note in the following that $\boldsymbol{a}_i \triangleq \boldsymbol{a}(\boldsymbol{\theta}_i), \tilde{\boldsymbol{a}}_i \triangleq \tilde{\boldsymbol{a}}(\boldsymbol{\theta}_i)$, and $\boldsymbol{e}_i \triangleq \boldsymbol{e}(\boldsymbol{\theta}_i)$. We also assume that a very large number of sampled observation vectors, $\boldsymbol{x}(kT_e)$, have been collected, resulting in a perfect measurement of the matrix $\boldsymbol{C}_{2q,x}(l)$ given by Equation (8.8) but where $\boldsymbol{C}_{2q,xg}(l)$ is now given by

$$\boldsymbol{C}_{2q,x_g}(l) = \left[\tilde{\boldsymbol{A}}_g^{\otimes l} \otimes \tilde{\boldsymbol{A}}_g^{*\otimes(q-l)}\right] \boldsymbol{C}_{2q,m_g}(l) \left[\tilde{\boldsymbol{A}}_g^{\otimes l} \otimes \tilde{\boldsymbol{A}}_g^{*\otimes(q-l)}\right]^\dagger \tag{8.35}$$

where $\tilde{\boldsymbol{A}}_g$ is the $(N \times P_g)$ submatrix of $\tilde{\boldsymbol{A}}$ corresponding to the gth group of sources.

Assuming that the number of sources P and their correlation properties are known, the problem considered in this section is to find the P DOA $\hat{\boldsymbol{\theta}}_i (1 \leq i \leq P)$, minimizing the left side of Equation (8.18), or the following criterion:

$$P_{2q,l}(\boldsymbol{\theta}, E) \triangleq \boldsymbol{a}_{q,l}(\boldsymbol{\theta})^\dagger \boldsymbol{\Pi}_{2q,n}(l)(\boldsymbol{E}) \boldsymbol{a}_{q,l}(\boldsymbol{\theta}) \tag{8.36}$$

In Equation (8.36), $\boldsymbol{\Pi}_{2q,n}(l)(\boldsymbol{E}) \triangleq \boldsymbol{U}_{2q,n}(l)(\boldsymbol{E})\boldsymbol{U}_{2q,n}(l)(\boldsymbol{E})^{\dagger}$, where $\boldsymbol{U}_{2q,n}(l)(\boldsymbol{E})$ is the $(N^q \times (N^q - r_{2q,x}(l)))$ unitary matrix of the eigenvectors of $\boldsymbol{C}_{2q,x}(l)$ associated with the $(N^q - r_{2q,x}(l))$ zero eigenvalues of $\boldsymbol{C}_{2q,x}(l)$. In the absence of modeling errors ($\boldsymbol{E}=0$), the P minima of Equation (8.36) are P zeros corresponding to the DOA of the sources, $\boldsymbol{\theta}_i(1 \leq i \leq P)$. However, in the presence of modeling errors ($\boldsymbol{E} \neq 0$), the criterion Equation (8.36) is no longer zero for the DOA of the sources but presents P local minima for directions $\hat{\boldsymbol{\theta}}_i(1 \leq i \leq P)$, different from $\boldsymbol{\theta}_i(1 \leq i \leq P)$. The variable $\Delta\boldsymbol{\theta}_i \triangleq \hat{\boldsymbol{\theta}}_i - \boldsymbol{\theta}_i$ defines the estimation error in the DOA of source i. To simplify the mathematical developments, we limit our analysis to a mono-dimensional DOA estimation problem for which $\theta_i, \hat{\theta}_i$ and $\Delta\theta_i$, $(1 \leq i \leq P)$ are scalar quantities.

Using a first-order Taylor expansion of the first partial derivative, $\dot{P}_{2q,l}(\theta, \boldsymbol{E})$, of $P_{2q,l}(\theta, \boldsymbol{E})$ with respect to θ around $\theta = \theta_i$, and exploiting the fact that $\dot{P}_{2q,l}(\hat{\theta}_i, \boldsymbol{E}) = 0$, we obtain an approximated expression of $\Delta\theta_i$ given by

$$\Delta\theta_i \approx -\frac{\dot{P}_{2q,l}(\theta_i, \boldsymbol{E})}{\ddot{P}_{2q,l}(\theta_i, \boldsymbol{E})} \tag{8.37}$$

where $\ddot{P}_{2q,l}(\theta, \boldsymbol{E})$ corresponds to the second derivative of $P_{2q,l}(\theta, \boldsymbol{E})$ with respect to θ at θ. Using Equation (8.36), $\dot{P}_{2q,l}(\theta_i, \boldsymbol{E})$ and $\ddot{P}_{2q,l}(\theta_i, \boldsymbol{E})$ can be written as

$$\dot{P}_{2q,l}(\theta_i, \boldsymbol{E}) = 2\mathrm{Re}[\dot{\boldsymbol{a}}_{q,l}(\theta_i)^{\dagger}\boldsymbol{\Pi}_{2q,n}(l)(\boldsymbol{E})\boldsymbol{a}_{q,l}(\theta_i)] \tag{8.38}$$

$$\ddot{P}_{2q,l}(\theta_i, \boldsymbol{E}) = 2\mathrm{Re}[\ddot{\boldsymbol{a}}_{q,l}(\theta_i)^{\dagger}\boldsymbol{\Pi}_{2q,n}(l)(\boldsymbol{E})\boldsymbol{a}_{q,l}(\theta_i)] + 2\dot{\boldsymbol{a}}_{q,l}(\theta_i)^{\dagger}\boldsymbol{\Pi}_{2q,n}(l)(\boldsymbol{E})\dot{\boldsymbol{a}}_{q,l}(\theta_i) \tag{8.39}$$

where Re[.] indicates the real part and $\dot{\boldsymbol{a}}_{q,l}(\theta_i)$ and $\ddot{\boldsymbol{a}}_{q,l}(\theta_i)$ indicate the first and second derivative of $\boldsymbol{a}_{q,l}(\theta)$ with respect to θ at $\theta = \theta_i$, respectively. Considering that $\boldsymbol{E}$ is a random matrix, the quantities $\Delta\theta_i(1 \leq i \leq P)$ become random variables.

The purpose of this section is to compute the root mean-square error (RMSE) of $\hat{\theta}_i$, defined by $\mathrm{RMSE}_i \triangleq (\mathrm{E}[\Delta\theta_i{}^2])^{1/2}$, as a function of q, l, and the statistics of $\boldsymbol{E}$.

8.5.3 Solution for $q=1$

The performance of the 2-MUSIC method is computed in this section through a first- and a second-order expansion of $\Pi_{2,n}(1)(E) \triangleq \Pi_{2,n}(E)$, respectively.

First-Order Expansion of $\Pi_{2,n}(E)$

The computation of the RMSE of the previous DOA estimates, $\hat{\theta}_i(1 \leq i \leq P)$, were considered in Friedlander (1990), Li and Vaccaro (1992), and Swindlehurst and Kailath (1992) for $q=1$, that is, for the 2-MUSIC algorithm and from an expansion of $\boldsymbol{\Pi}_{2,n}(\boldsymbol{E})$ at the first order in $\boldsymbol{E}$ around $\boldsymbol{E}=0$. More precisely, under these assumptions it was shown in Friedlander (1990) that $\Delta\theta_i$

and $\mathrm{E}[\Delta\theta_i^2]$ are given by

$$\Delta\theta_i \approx -\frac{\mathrm{Re}\left[\boldsymbol{e}_i^{\dagger}\boldsymbol{b}_i\right]}{\dot{\boldsymbol{a}}_i^{\dagger}\boldsymbol{b}_i} \tag{8.40}$$

and

$$\mathrm{E}\left[\Delta\theta_i^2\right] \approx \frac{1}{2}\frac{\boldsymbol{b}_i^{\dagger}\boldsymbol{R}_{ei}\boldsymbol{b}_i + \mathrm{Re}\left[\boldsymbol{b}_i^{\mathrm{T}}\boldsymbol{C}_{ei}^{*}\boldsymbol{b}_i\right]}{\left(\dot{\boldsymbol{a}}_i^{\dagger}\boldsymbol{b}_i\right)^2} \tag{8.41}$$

where $\dot{\boldsymbol{a}}_i \triangleq \dot{\boldsymbol{a}}(\theta_i)$; $\boldsymbol{b}_i \triangleq \boldsymbol{\Pi}_{2,n}(\boldsymbol{E}=\boldsymbol{0})\dot{\boldsymbol{a}}_i$; $\boldsymbol{R}_{ei} \triangleq \mathrm{E}\left[\boldsymbol{e}_i\boldsymbol{e}_i^{\dagger}\right]$ and $\boldsymbol{C}_{ei} \triangleq \mathrm{E}\left[\boldsymbol{e}_i\boldsymbol{e}_i^{\mathrm{T}}\right]$.

Second-Order Expansion of $\Pi_{2,n}(E)$

The first-order expansion of $\boldsymbol{\Pi}_{2,n}(\boldsymbol{E})$ around $\boldsymbol{E}=0$ generates results in agreement with the simulations as long as the sources are not in the resolution limit. However, for sources in the resolution limit, results generated by the first-order expansion do not agree with the simulations, and an SO expansion of $\boldsymbol{\Pi}_{2,n}(\boldsymbol{E})$ around $\boldsymbol{E}=0$ has to be considered. This was done recently in Ferreol et al. (2006) where the associated results were in complete agreement with simulations, even for high spatial correlation between sources. More precisely, using an SO expansion of $\boldsymbol{\Pi}_{2,n}(\boldsymbol{E})$ around $\boldsymbol{E}=0$, it was shown in Ferreol et al. (2006) that $\Delta\theta_i$ and $\boldsymbol{E}[\Delta\theta_i^2]$ are given by

$$\Delta\theta_i \approx -\frac{\boldsymbol{\varepsilon}^{\dagger}\dot{\boldsymbol{Q}}_i\boldsymbol{\varepsilon}}{\boldsymbol{\varepsilon}^{\dagger}\ddot{\boldsymbol{Q}}_i\boldsymbol{\varepsilon}} \tag{8.42}$$

$$\mathrm{E}\left[\Delta\theta_i^2\right] \approx \frac{\mathrm{Tr}\left[\dot{\boldsymbol{Q}}_i^{\otimes 2}\boldsymbol{R}_{\varepsilon}^{(4)}\right]}{\mathrm{Tr}\left[\ddot{\boldsymbol{Q}}_i^{\otimes 2}\boldsymbol{R}_{\varepsilon}^{(4)}\right]} \tag{8.43}$$

where Tr[.] denotes trace; $\boldsymbol{\varepsilon}$ is the $((2NP+1)\times 1)$ vector defined by $\boldsymbol{\varepsilon} \triangleq [1, \mathrm{vec}(\boldsymbol{E})^{\mathrm{T}}, \mathrm{vec}(\boldsymbol{E})^{\dagger}]^{\mathrm{T}}$; $\mathrm{vec}(\boldsymbol{E}) \triangleq [\boldsymbol{e}_1^{\mathrm{T}}, \ldots, \boldsymbol{e}_P^{\mathrm{T}}]^{\mathrm{T}}$; $\boldsymbol{R}_{\varepsilon}^{(4)}$ is the $((2NP+1)^2 \times (2NP+1)^2)$ matrix defined by $\boldsymbol{R}_{\varepsilon}^{(4)} \triangleq \mathrm{E}[\boldsymbol{\varepsilon}^{\otimes 2}\boldsymbol{\varepsilon}^{\otimes 2\dagger}]$; and $\dot{\boldsymbol{Q}}_i$ and $\ddot{\boldsymbol{Q}}_i$ are the $((2NP+1)\times(2NP+1))$ matrices defined by

$$\dot{\boldsymbol{Q}}_i \triangleq \boldsymbol{Q}(\boldsymbol{a}_i, \dot{\boldsymbol{a}}_i) + \boldsymbol{Q}(\dot{\boldsymbol{a}}_i, \boldsymbol{a}_i) \tag{8.44}$$

$$\ddot{\boldsymbol{Q}}_i \triangleq \boldsymbol{Q}(\boldsymbol{a}_i, \ddot{\boldsymbol{a}}_i) + \boldsymbol{Q}(\ddot{\boldsymbol{a}}_i, \boldsymbol{a}_i) + 2\boldsymbol{Q}(\dot{\boldsymbol{a}}_i, \dot{\boldsymbol{a}}_i) \tag{8.45}$$

where $\ddot{\boldsymbol{a}}_i \triangleq \ddot{\boldsymbol{a}}(\theta_i)$. In Equations (8.44) and (8.45), for the $(N\times 1)$ vectors $\boldsymbol{u}$ and $\boldsymbol{v}$, $\boldsymbol{Q}(\boldsymbol{u},\boldsymbol{v})$ is the $((2NP+1)\times(2NP+1))$ matrix defined by

$$\boldsymbol{Q}(\boldsymbol{u},\boldsymbol{v}) \triangleq \begin{pmatrix} \alpha & -\boldsymbol{q}_{12}^{\dagger} & \boldsymbol{0}^{\mathrm{T}} \\ -\boldsymbol{q}_{21} & \boldsymbol{Q}_{22} & \boldsymbol{Q}_{23} \\ \boldsymbol{0} & \boldsymbol{Q}_{32} & \boldsymbol{Q}_{33} \end{pmatrix} \tag{8.46}$$

where $\alpha \triangleq \boldsymbol{v}^{\dagger}\boldsymbol{\Pi}_{2,n}(\boldsymbol{E}=0)\boldsymbol{u}, \boldsymbol{q}_{12}$ and $\boldsymbol{q}_{21}$ are the $(NP\times 1)$ vectors defined by $\boldsymbol{q}_{12} \triangleq \boldsymbol{\phi}(\boldsymbol{u},\boldsymbol{v})$ and $\boldsymbol{q}_{21} \triangleq \boldsymbol{\phi}(\boldsymbol{v},\boldsymbol{u})$, respectively, with $\boldsymbol{\phi}(\boldsymbol{u},\boldsymbol{v}) \triangleq [(\boldsymbol{A}^{\dagger}\boldsymbol{A})^{-1}\boldsymbol{A}^{\dagger}\boldsymbol{u}]^{*} \otimes$

$[\mathbf{\Pi}_{2,n}(\boldsymbol{E}=0)\boldsymbol{v}]$; $\mathbf{0}$ is the $(NP \times 1)$ zero vector; and $\boldsymbol{Q}_{22}$, $\boldsymbol{Q}_{23}$, $\boldsymbol{Q}_{32}$, and $\boldsymbol{Q}_{33}$ are $(NP \times NP)$ matrices defined by

$$\boldsymbol{Q}_{22} \triangleq \boldsymbol{\psi}[(\boldsymbol{A}^{\dagger}\boldsymbol{A})^{-1}\boldsymbol{A}^{\dagger}, (\boldsymbol{A}^{\dagger}\boldsymbol{A})^{-1}\boldsymbol{A}^{\dagger}, \mathbf{\Pi}_{2,n}(\boldsymbol{E}=0)] \tag{8.47}$$

$$\boldsymbol{Q}_{23} \triangleq \boldsymbol{\psi}[(\boldsymbol{A}^{\dagger}\boldsymbol{A})^{-1}\boldsymbol{A}^{\dagger}, \mathbf{\Pi}_{2,n}(\boldsymbol{E}=0), \boldsymbol{A}(\boldsymbol{A}^{\dagger}\boldsymbol{A})^{-1}]\mathbf{\Pi}_p \tag{8.48}$$

$$\boldsymbol{Q}_{32} \triangleq \mathbf{\Pi}_p{}^{\dagger}\, \boldsymbol{\psi}[\mathbf{\Pi}_{2,n}(\boldsymbol{E}=0), (\boldsymbol{A}^{\dagger}\boldsymbol{A})^{-1}\boldsymbol{A}^{\dagger}, (\boldsymbol{A}^{\dagger}\boldsymbol{A})^{-1}\boldsymbol{A}^{\dagger}] \tag{8.49}$$

$$\boldsymbol{Q}_{33} \triangleq -\mathbf{\Pi}_p{}^{\dagger}\, \boldsymbol{\psi}[\mathbf{\Pi}_{2,n}(\boldsymbol{E}=0), \mathbf{\Pi}_{2,n}(\boldsymbol{E}=0), (\boldsymbol{A}^{\dagger}\boldsymbol{A})^{-1}]\mathbf{\Pi}_p \tag{8.50}$$

respectively, where $\boldsymbol{\psi}[\mathbf{X}, \mathbf{Y}, \mathbf{Z}] \triangleq [(\mathbf{X}\boldsymbol{v})^{*}(\mathbf{Y}\boldsymbol{u})^{\mathrm{T}}] \otimes \mathbf{Z}$ and $\mathbf{\Pi}_p$ is the $(NP \times NP)$ permutation matrix such that $\mathrm{vec}(\boldsymbol{E}^{\mathrm{T}}) \triangleq \mathbf{\Pi}_p \mathrm{vec}(\boldsymbol{E})$.

8.5.4 Solution for $q > 1$

The RMSEs of the source DOA estimates were computed in the previous section from Equations (8.37) to (8.39), with $q = l = 1$ and either a first-order or an SO Taylor expansion of $\mathbf{\Pi}_{2,n}(\boldsymbol{E}) \triangleq \mathbf{I}_N - \tilde{\boldsymbol{A}}(\boldsymbol{E})(\tilde{\boldsymbol{A}}(\boldsymbol{E})^{\dagger}\tilde{\boldsymbol{A}}(\boldsymbol{E}))^{-1}\tilde{\boldsymbol{A}}(\boldsymbol{E})^{\dagger}$ around $\boldsymbol{E} = 0$, where $\tilde{\boldsymbol{A}}(\boldsymbol{E}) \triangleq [\tilde{\boldsymbol{a}}_1, \ldots, \tilde{\boldsymbol{a}}_p] \triangleq \boldsymbol{A} + \boldsymbol{E}$. Assuming statistically independent sources and considering arbitrary values of q and l, the projector on the noise subspace, $\mathbf{\Pi}_{2q,n}(l)(\boldsymbol{E})$, takes the form

$$\mathbf{\Pi}_{2q,n}(l)(\boldsymbol{E}) \triangleq \mathbf{I}_{N^q} - \tilde{\bar{\boldsymbol{A}}}_{q,l}(\boldsymbol{E})(\tilde{\bar{\boldsymbol{A}}}_{q,l}(\boldsymbol{E})^{\dagger}\tilde{\bar{\boldsymbol{A}}}_{q,l}(\boldsymbol{E}))^{-1}\tilde{\bar{\boldsymbol{A}}}_{q,l}(\boldsymbol{E})^{\dagger}$$

where $\tilde{\bar{\boldsymbol{A}}}_{q,l}(\boldsymbol{E}) \triangleq [\tilde{\boldsymbol{a}}_{q,l}(\theta_1), \ldots, \tilde{\boldsymbol{a}}_{q,l}(\theta_P)]$ and $\tilde{\boldsymbol{a}}_{q,l}(\theta_i) \triangleq \tilde{\boldsymbol{a}}_i{}^{\otimes l} \otimes \tilde{\boldsymbol{a}}_i{}^{*\otimes(q-l)}$.

The use of an SO Taylor expansion of $\mathbf{\Pi}_{2q,n}(l)(\boldsymbol{E})$ around $\boldsymbol{E} = 0$, for the computation of the RMSE of the sources, requires an SO Taylor expansion of $\tilde{\bar{\boldsymbol{A}}}_{q,l}(\boldsymbol{E})$ around $\boldsymbol{E} = 0$, which dramatically complicates the computations and prevents us from exploiting the results of Section 8.5.3. For this reason, we limit the Taylor expansion of $\tilde{\bar{\boldsymbol{A}}}_{q,l}(\boldsymbol{E})$ around $\boldsymbol{E} = 0$ to the first order, which finally also imposes the limitation to the first order of the Taylor expansion of $\mathbf{\Pi}_{2q,n}(l)(\boldsymbol{E})$ around $\boldsymbol{E} = 0$. In Chevalier et al. (2006), an SO Taylor expansion of $\Pi_{2q,n}(l)(\boldsymbol{E})$ around $\boldsymbol{E} = 0$ was used but only from a first-order expansion of $\tilde{\bar{\boldsymbol{A}}}_{q,l}(\boldsymbol{E})$ around $\boldsymbol{E} = 0$, which is not completely rigorous but still gives good results in agreement with the simulations.

Considering the first-order expansion of $\tilde{\bar{\boldsymbol{A}}}_{q,l}(\boldsymbol{E})$ around $\boldsymbol{E} = 0$, we obtain

$$\tilde{\bar{\boldsymbol{A}}}_{q,l}(\boldsymbol{E}) \approx \bar{\boldsymbol{A}}_{q,l} + \boldsymbol{E}_{q,l} \triangleq \tilde{\bar{\boldsymbol{A}}}_{q,l}(\boldsymbol{E}_{q,l}) \tag{8.51}$$

where $\bar{\boldsymbol{A}}_{q,l}$ is defined by H2″ and where it is easy to show that $\boldsymbol{E}_{q,l}$ is the $(N^q \times P)$ matrix defined by $\boldsymbol{E}_{q,l} \triangleq [\boldsymbol{e}_{q,l}(\boldsymbol{\theta}_1), \ldots, \boldsymbol{e}_{q,l}(\theta_P)]$, where $\boldsymbol{e}_{q,l}(\theta_i)$ is defined by

$$\begin{aligned} \boldsymbol{e}_{q,l}(\theta_i) = & \sum_{u=0}^{l=1} \boldsymbol{a}_i{}^{\otimes u} \otimes \boldsymbol{e}_i \otimes \boldsymbol{a}_i{}^{\otimes(l-u-1)} \otimes \boldsymbol{a}_i{}^{*\otimes(q-l)} \\ & + \sum_{u=0}^{q-l-1} \boldsymbol{a}_i{}^{\otimes l} \otimes \boldsymbol{a}_i{}^{*\otimes u} \otimes \boldsymbol{e}_i{}^{*} \otimes \boldsymbol{a}_i{}^{*\otimes(q-l-u-1)} \end{aligned} \tag{8.52}$$

with the convention $\boldsymbol{a}_i^{\otimes 0} \triangleq 1$. In these conditions, the RMS error of the source DOA estimates can be computed for arbitrary values of q and l from Equations (8.37) to (8.39) and a first-order Taylor expansion of $\boldsymbol{\Pi}_{2q,n}(l)\,(\boldsymbol{E}) \approx \boldsymbol{\Pi}_{2q,n}(l)\,(\boldsymbol{E}_{q,l}) \triangleq \mathbf{I}_{N^q} - \widetilde{\overline{\boldsymbol{A}}}_{q,l}(\boldsymbol{E}_{q,l})\,(\widetilde{\overline{\boldsymbol{A}}}_{q,l}(\boldsymbol{E}_{q,l})^{\dagger}\widetilde{\overline{\boldsymbol{A}}}_{q,l}(\boldsymbol{E}_{q,l}))^{-1}\widetilde{\overline{\boldsymbol{A}}}_{q,l}(\boldsymbol{E}_{q,l})^{\dagger}$ around $\boldsymbol{E}_{q,l} = 0$. This can be done by replacing N with $N^q, \boldsymbol{E}$ with $\boldsymbol{E}_{q,l}, \boldsymbol{A}$ with $\overline{\boldsymbol{A}}_{q,l}$, and $\boldsymbol{a}(\theta_i)$ with $\boldsymbol{a}_{q,l}(\theta_i)$ in (8.41).

8.5.5 Illustrations

After the description of some hypotheses, the previous results are illustrated in this section for both the mono- and bisource cases.

Hypotheses

To illustrate the previous results, we assume that P statistically independent non-Gaussian sources are received by a UCA of $N = 5$ omnidirectional sensors with a radius equal to half the wavelength. The DOA of source 1 is $\theta_1 = 100°$. The modeling error vectors $\boldsymbol{e}_i (1 \leq i \leq P)$ are assumed to be zero-mean statistically independent circular Gaussian vectors such that $\boldsymbol{E}[\boldsymbol{e}_i \boldsymbol{e}_j{}^{\dagger}] = \sigma^2 \delta_{ij} \mathbf{I}_N$.

Performance for $P = 1$

We first consider the mono-source case ($P = 1$). Under the previous assumptions, Figure 8.7 shows, for $q = 1, 2, 3$ and for $l = 1$ (optimal arrangement), the variations of the RMS error of the source as a function of σ, the RMS value of the modeling errors. We note similar performance of 2q-MUSIC methods as q increases, despite the increasing aperture of the HO VA. This can be explained by the fact that the HO virtual modeling errors—that is, the equivalent modeling errors associated with the extended steering vector, $\boldsymbol{a}_{q,l}(\theta_1)$—of the source increase jointly with the HO VA aperture. We note finally a decreasing precision of 2q-MUSIC methods as σ increases.

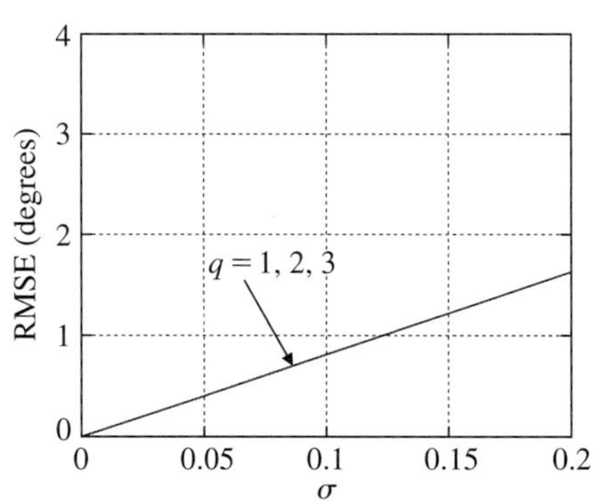

FIGURE 8.7 RMSE of the source as a function of $\sigma, q = 1, 2, 3; N = 5$; UCA; $r/\lambda = 0.5; P = 1$; i.i.d. Gaussian and circular errors. Theoretical results.

Performance for $P=2$

We now consider the case of $P=2$ sources, and we assume that $\sigma=0.12$ (strong errors), which corresponds, for example, to phase and amplitude errors with a standard deviation of 4.6° and 0.7 dB, respectively. Under these assumptions, Figure 8.8 shows, for $q=1,2,3$ and for $l=1$ (optimal arrangement), the variations of the RMS error of source 1 (it is similar for source 2) as a function of the modulus of the spatial correlation coefficient between the two sources, α_{12}, defined by Equation (8.27) with $\theta=\theta_1=100°$ and $\theta_p=\theta_2$. For $q=1$, the RMSEs obtained from both a first-order (solid line) and an SO (dotted line) Taylor expansion of $\mathbf{\Pi}_{2,n}(\boldsymbol{E})$ around $\boldsymbol{E}=0$ have been computed.

Figure 8.8 shows, for $q=1$, similar results obtained from both the first-order and SO Taylor expansion of $\mathbf{\Pi}_{2,n}(\boldsymbol{E})$ around $\boldsymbol{E}=0$ as long as the spatial correlation of the sources is not too high. For high values of the spatial correlation, results obtained from the first-order expansion of $\mathbf{\Pi}_{2,n}(\mathrm{E})$ no longer agree with the simulations, whereas those obtained from the SO expansion do. The latter show an increasing RMSE with $|\alpha_{12}|$ as long as the two sources are resolved by the 2-MUSIC algorithm—in other words, as long as $|\alpha_{12}|\leq 0.9$, and there is a decreasing RMSE with $|\alpha_{12}|$—as soon as the two sources become unresolvable by the algorithm ($|\alpha_{12}|>0.9$).

In all cases, the RMSE obtained from the first-order expansion of $\mathbf{\Pi}_{2,n}(\boldsymbol{E})$ is greater than or approximately equal to that obtained from an SO expansion of $\mathbf{\Pi}_{2,n}(\boldsymbol{E})$. Moreover, for a given value of $|\alpha_{12}|$ such that $|\alpha_{12}|\leq 0.95$, Figure 8.8 shows a decreasing value of the RMSE, obtained from a first-order expansion of $\mathbf{\Pi}_{2q,n}(l)(\boldsymbol{E})$, as q increases. This illustrates the greater robustness of the 2q-MUSIC algorithm to modeling errors as q increases, especially for a small angular separation between the sources. For $|\alpha_{12}|>0.95$, results obtained for $q=2,3$ from the first-order expansion of $\mathbf{\Pi}_{2q,n}(l)(\boldsymbol{E})$ are no longer in agreement

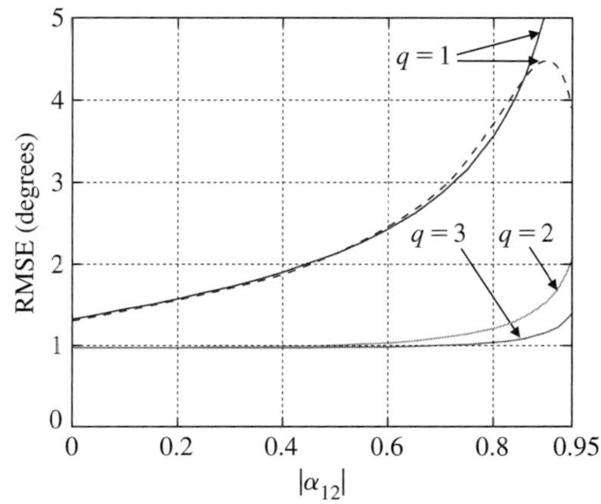

FIGURE 8.8 RMSE of the source 1 as a function of $|\alpha_{12}|, q=1,2,3$; $N=5$; UCA; $r/\lambda=0.5$; P = 2; i.i.d. Gaussian and circular errors, $\boldsymbol{\sigma}=0.12$. Theoretical results.

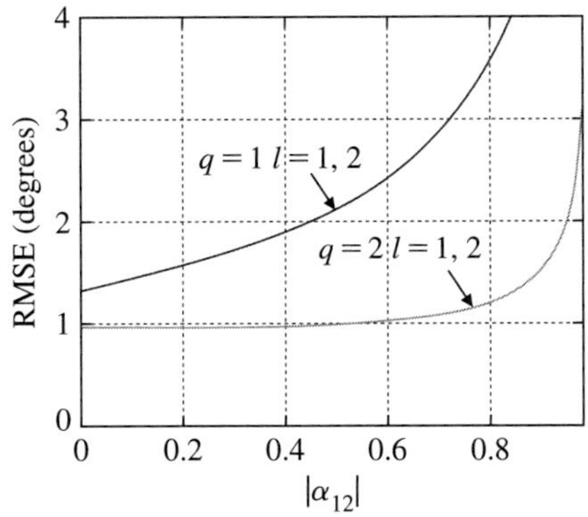

FIGURE 8.9 RMSE of the source 1 as a function of $|\alpha_{12}|, q = 1, 2; l = 1, 2; N = 5$; UCA; $r/\lambda = 0.5$; $P = 2$; i.i.d. Gaussian and circular errors, $\sigma = 0.12$. Theoretical results.

with the simulations and an SO expansion of $\mathbf{\Pi}_{2q,n}(l)(\boldsymbol{E})$ has to be taken into account. For a given value of $q > 1$, such an expansion would indicate an increasing RMSE with $|\alpha_{12}|$ as long as the two sources are resolved by the 2q-MUSIC algorithm—that is, as long as $|\alpha_{12}| \leq |\alpha_{12}|_q$, where $|\alpha_{12}|_q$ is the 2q-MUSIC's resolution threshold and there is a decreasing RMSE with $|\alpha_{12}|$ as soon as the two sources become unresolvable by the algorithm ($|\alpha_{12}| > |\alpha_{12}|_q$).

For high angular separation between sources ($|\alpha_{12}| = 0$), the presence of modeling errors brakes the orthogonality of the sources, which then interact. This explains why, in this case, 4-MUSIC and 6-MUSIC perform better than 2-MUSIC. All of our previous conclusions would have been similar from an SO expansion of $\mathbf{\Pi}_{2q,n}(l)(\boldsymbol{E})$, as was verified by simulations and as illustrated approximately in Chevalier et al. (2006). Such illustrations would have shown that the maximal value of $|\alpha_{12}|$ ensuring resolvability of the two sources increases with q, indicating the increasing resolution of 2q-MUSIC as q increases. These results may be physically interpreted through the HO VA concept introduced in Chevalier et al. (2005) and the results of Section 8.4.2. They definitely show the great interest of 2q-MUSIC methods for $q \geq 2$, even for overdetermined ($P < N$) mixtures of sources, despite the increasing variance and complexity with q.

The influence of the arrangement of the statistics on the robustness to modeling errors of 2q-MUSIC is illustrated in Figure 8.9, which shows the same variations as in Figure 8.8 for the same scenario but for $q = 1, 2$ and for $l = 1, 2$. All the results have been obtained from a first-order expansion of $\mathbf{\Pi}_{2q,n}(l)(\boldsymbol{E})$. We note that performance of 4-MUSIC for $l = 1$ and $l = 2$ is similar since the resolution of the associated VAs is independent of l (indeed $|\alpha_{12}(q, l)|$ is independent of l).

8.6 COMPUTER SIMULATIONS

The performance of 2q-MUSIC and 2q-D-MUSIC methods implemented from a finite number of snapshots is illustrated in this section through computer simulations. We first introduce performance criteria in Section 8.6.1 and then describe the simulations in Sections 8.6.2 and 8.6.3 for both overdetermined ($P < N$) and underdetermined ($P \geq N$) mixtures of sources, respectively. The sources are assumed to have a zero elevation angle φ and to be zero-mean stationary, corresponding to QPSK sources sampled at the symbol rate.

8.6.1 Performance Criteria

For each of the P considered sources and for a given direction-finding method, two criteria are used in the following to quantify the performance of the associated DOA estimation. For a given source, the first criterion is a probability of aberrant results generated by a given method for this source; the second is an averaged RMSE computed from the nonaberrant results generated by a given method for this source.

More precisely, for given values of q and l, a given number of snapshots, L, and a particular realization of the L observation vectors $\boldsymbol{x}(k)(1 \leq k \leq L)$, the estimation, $\hat{\theta}_p$, of the DOA of source $p(1 \leq p \leq P)$ from $2q$-MUSIC or $2q$-D-MUSIC is defined by

$$\hat{\theta}_p \triangleq \underset{\zeta_i}{\mathrm{Arg}}(\underset{i}{\mathrm{Min}}\,|\zeta_i - \theta_p|) \tag{8.53}$$

in which the quantities $\zeta_i(1 \leq i \leq \hat{P})$ correspond to the $\hat{P}$ minima of the pseudo-spectrum $\hat{P}_{2q\text{-}Music(l)}(\theta)$ defined by Equation (8.19) for $\varphi = 0$, or to the $\hat{P}$ absolute minima of the $\hat{P}$ pseudo-spectrum $\hat{P}_{2q\text{-}D\text{-}Music,i(l)}(\theta)$, $1 \leq i \leq \hat{P}$ defined by Equation (8.21) for $p = i$ and $\varphi = 0$. With each estimate $\hat{\theta}_p(1 \leq p \leq P)$, we associate the corresponding value of the pseudo-spectrum, defined by $\eta_p = \hat{P}_{2q\text{-}Music(l)}(\hat{\theta}_p)$ or $\eta_p = \hat{P}_{2q\text{-}D\text{-}Music,p\text{-}1(l)}(\hat{\theta}_p)$. In this context, the estimate $\hat{\theta}_p$ is considered to be aberrant, because of an outlier for example, if $\eta_p > \eta$, where η is a threshold to be defined.

The computation of the optimal threshold requires a deep theoretical analysis beyond the scope of this chapter. This analysis would show that this threshold is a function of several parameters such as the number of sensors, the number of sources, the power of the modeling errors, or the sensor array. In the following we arbitrarily choose $\eta = 0.1$.

We consider M realizations of the L observation vectors $\boldsymbol{x}(k)(1 \leq k \leq L)$. For a given method, the probability of aberrant results for a given source p, $p(\eta_p > \eta)$, is defined by the ratio of the number of realizations for which $\hat{\theta}_p$ is aberrant and the number of realizations M. From the nonaberrant realizations for the source p, we then define the averaged RMSE for source p, RMSE_p, by the quantity

$$\mathrm{RMSE}_p \triangleq \sqrt{\frac{1}{M_p}\sum_{m=1}^{M_p}\left|\hat{\theta}_{pm} - \theta_p\right|^2} \tag{8.54}$$

where M_p is the number of nonaberrant realizations for source p and $\hat{\theta}_{pm}$ is the estimate of θ_p for the nonaberrant realization m.

8.6.2 Overdetermined Mixtures of Sources

To quantify the influence of both the number of independent snapshots L and the parameter q on the performance of $2q$-MUSIC and $2q$-D-MUSIC, we assume that two statistically independent QPSK sources with a raised cosine pulse-shaped filter are received by a ULA of $N=3$ omnidirectional sensors spaced a half-wavelength apart. The two QPSK sources have the same symbol duration, the same roll-off, $\mu=0.3$, and the same input SNR equal to 5 dB. Note that the normalized auto-cumulant of the QPSK symbols is equal to -1 at the FO and $+4$ at the sixth order.

Under these assumptions, Figures 8.10 and 8.11 show the variations, as a function of the number of snapshots, L, of the RMSE for source 1, RMSE_1, and the associated probability of nonaberrant results, $p(\eta_1 \leq \eta)$ (we obtain similar results for source 2), at the output of the 2-MUSIC, 4-MUSIC, and 6-MUSIC

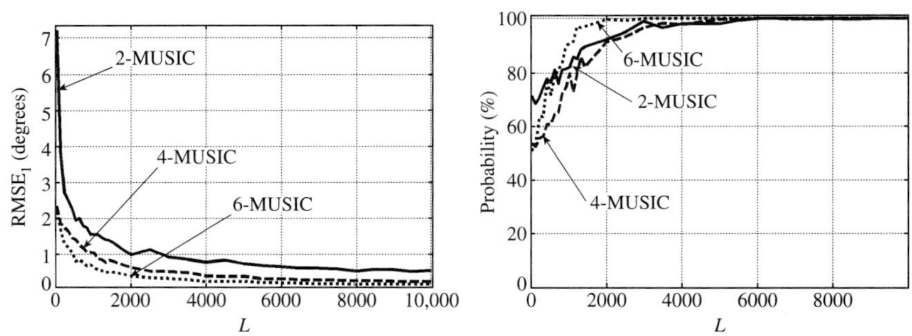

(a) (b)

FIGURE 8.10 (a) RMSE of the source 1 and (b) $p(\eta_1 \leq \eta)$ as a function of L. $l=1, P=2, N=3$, ULA, SNR = 5 dB, $\theta_1=90°$, $\theta_2=97.5°$, no modeling errors.

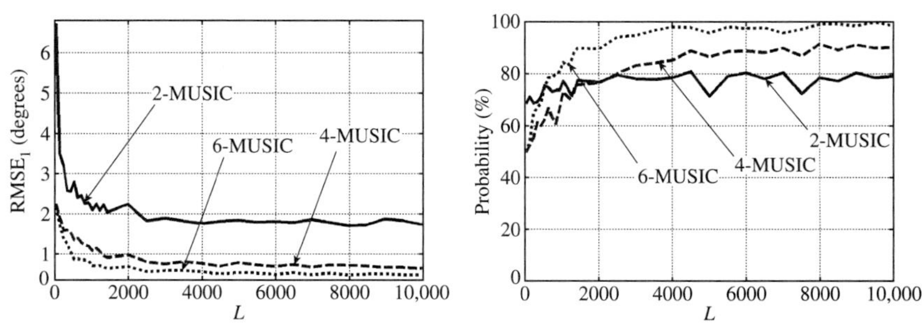

(a) (b)

FIGURE 8.11 (a) RMSE of the source 1 and (b) $p(\eta_1 \leq \eta)$ as a function of L. $l=1, P=2, N=3$, ULA, SNR = 5 dB, $\theta_1=90°$, $\theta_2=97.5°$, with modeling errors, $\sigma=0.03$.

methods for optimal arrangements of the considered statistics, without and with modeling errors, respectively. These performances are estimated from $M = 300$ realizations and the minimum value of L for these figures is $L = 100$. For these figures, the sources, the DOAs of which are equal to $\theta_1 = 90°$ and $\theta_2 = 97.5°$, respectively, are poorly angularly separated for the considered array of sensors, with a 3-dB beamwidth that is around 40°. In the presence of modeling errors, vectors $\boldsymbol{e}_i (1 \leq i \leq 2)$ are assumed to be zero-mean statistically independent circular Gaussian, such that $\mathrm{E}[\boldsymbol{e}_i \boldsymbol{e}_j{}^{\dagger}] = \sigma^2 \delta_{ij} \mathbf{I}_N$, where $\sigma = 0.03$, which corresponds to phase and amplitude errors with standard deviations equal to 1.15° and 0.18 dB, respectively.

In terms of both probability of nonaberrant results and estimation precision, Figures 8.10 and 8.11 show, for poorly angularly separated sources, the best behavior of the 6-MUSIC method with respect to 2-MUSIC and 4-MUSIC as soon as L becomes greater than 660 snapshots without modeling errors and 500 snapshots with them. For such values of L, the resolution gain and the better robustness to modeling errors obtained with 6-MUSIC versus 2-MUSIC and 4-MUSIC, respectively, due to the narrower 3-dB beamwidth of the associated sixth-order VA, are higher than the loss due to a greater variance in the statistics estimates. A similar analysis can be carried out for 4-MUSIC versus 2-MUSIC as soon as L becomes greater than 3500 snapshots without modeling errors and 2000 snapshots with them.

These results confirm that, despite their greater variance and contrary to some generally accepted ideas, HO MUSIC methods may offer better performance than 2-MUSIC or 4-MUSIC methods when some resolution is required—that is, in the presence of several poorly angularly separated sources with or without modeling errors, inherent in operational contexts.

To complete these results, Figures 8.12 and 8.13 show the same variations as in Figure 8.10 and Figure 8.11, respectively, but for sources that are well angularly separated and the DOAs of which are equal to $\theta_1 = 90°$ and $\theta_2 = 130°$, respectively. Note in this case the best behavior of SO versus HO methods, due to a higher variance in the HO statistics estimates, when no particular resolution is required (i.e., in the absence of modeling errors—see Figure 8.12). This result would also be valid in the presence of very weak modeling errors; however, we note a better performance of HO methods as soon as the sources interact with each other, meaning that some resolution is required. This is the case in the presence of higher modeling errors (Figure 8.13), provided the variance in the HO statistics estimate becomes small ($L > 1600$).

To illustrate the significance of the sequential algorithms, we consider again the scenarios of Figure 8.10, but now $\theta_1 = 90°$ and $\theta_2 = 95°$. Under these assumptions, Figure 8.14 shows the same variations as in Figure 8.10 but at the output of 2-MUSIC, 2-D-MUSIC, 4-MUSIC, 4-D-MUSIC, 6-MUSIC, and 6-D-MUSIC. Note, for a given value of q, the best behavior of the sequential algorithms ($2q$-D-MUSIC) with respect to the parallel ($2q$-MUSIC) methods when a strong resolution between sources is required.

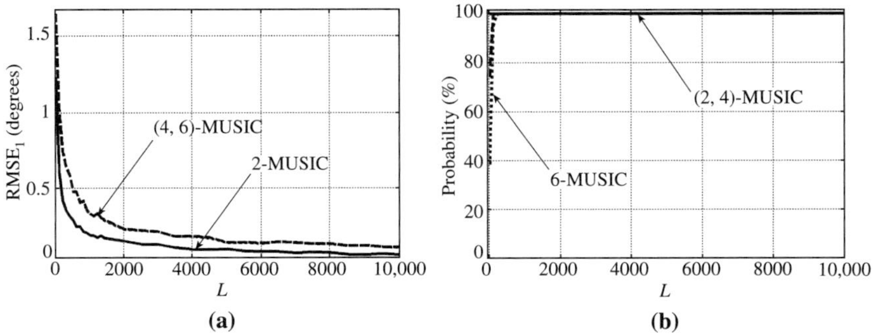

FIGURE 8.12 (a) RMSE of the source 1 and (b) $p(\eta_1 \le \eta)$ as a function of L. $l=1, P=2, N=3$, ULA, SNR $=5$ dB, $\theta_1=90°$, $\theta_2=130°$, no modeling errors.

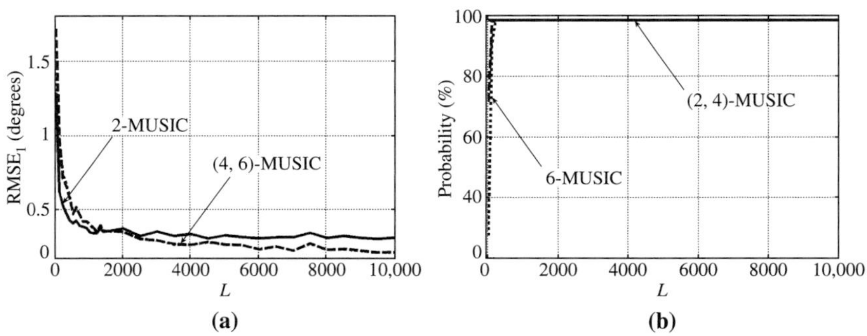

FIGURE 8.13 (a) RMSE of the source 1 and (b) $p(\eta_1 \le \eta)$ as a function of L. $l=1, P=2, N=3$, ULA, SNR $=5$ dB, $\theta_1=90°$, $\theta_2=130°$, with modeling errors, $\sigma=0.03$.

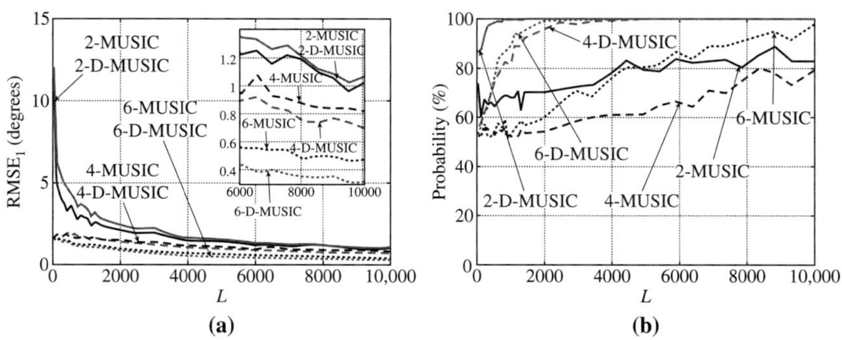

FIGURE 8.14 (a) RMSE of the source 1 and (b) $p(\eta_1 \le \eta)$ as a function of L. $l=1, P=2, N=3$, ULA, SNR $=5$ dB, $\theta_1=90°$, $\theta_2=95°$, no modeling errors.

The influence of the HO statistics arrangement on the performance of $2q$-MUSIC methods was illustrated in Chevalier et al. (2006), which still showed increasing performance with q, whatever the choice of l, when some resolution is required.

8.6.3 Underdetermined Mixtures of Sources

To illustrate the influence of the arrangement of the HO statistics on the number of sources that can be processed by the $2q$-MUSIC algorithm, we assume that six statistically independent QPSK sources with a raised cosine pulse-shaped filter are received by a UCA of $N=3$ omnidirectional sensors with a radius r such that $r=0.3\,\lambda$. The six QPSK sources have the same symbol duration, the same roll-off, $\mu=0.3$, the same input SNR equal to 15 dB, and a DOA equal to $\theta_1=35°$, $\theta_2=80°, \theta_3=110°, \theta_4=140°, \theta_5=230°$, and $\theta_6=304°$, respectively. Under these assumptions, Figure 8.15 shows the variations in the pseudo-spectrum of 4-MUSIC, for $l=1$ and 2, as a function of θ for $L=10,000$ without modeling errors; the good behavior of 4-MUSIC for $l=1$, which succeeds in estimating the DOA of the six sources; and the poor behavior of 4-MUSIC for $l=2$, which fails, since the number of VSs of the associated VA is equal to $N_4^1=7$ for $l=1$ and equal to $N_4^2=6$ for $l=2$.

To complete these results, Figure 8.16 shows the variations, as a function of the number of snapshots L, of the highest RMSE and the lowest probability of nonaberrant results among all sources, estimated from $M=300$ realizations, at the output of the 4-MUSIC algorithm for $l=1$ and without modeling errors. Note the increasing minimum probability of nonaberrant results and the decreasing

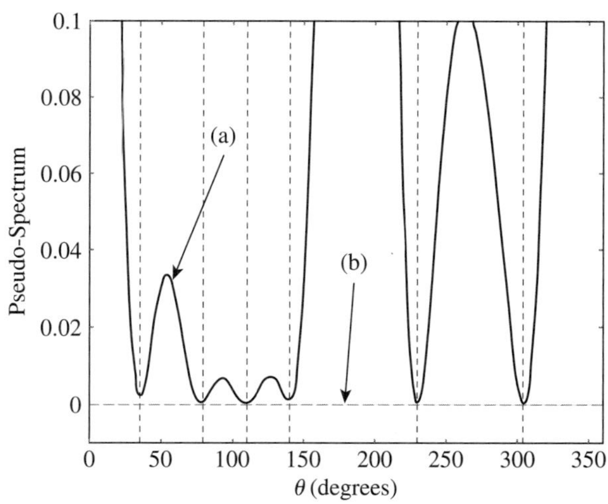

FIGURE 8.15 Pseudo-spectrum of 4-MUSIC as a function of θ: (a) $l=1$; (b) $l=2$, $P=6$, $N=3$, UCA, SNR = 15 dB, $L=10{,}000$, $\theta_1=35°, \theta_2=80°, \theta_3=110°, \theta_4=140°, \theta_5=230°$, $\theta_6=304°$, no modeling errors (*Source:* From Chevalier et al., 2006, © IEEE, with permission.)

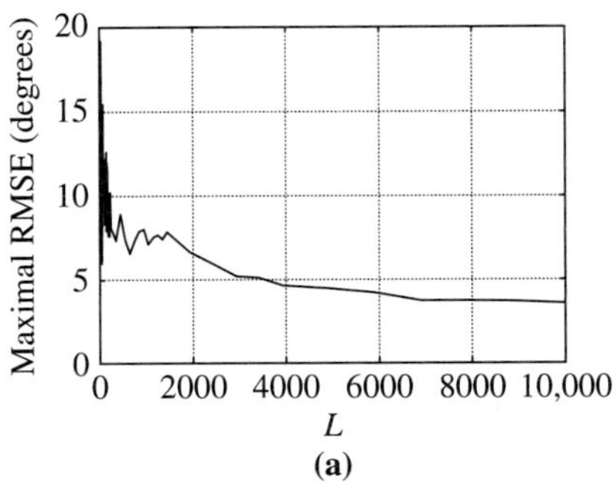

FIGURE 8.16 (a) Maximal RMSE and (b) minimum probability of nonaberrant results as a function of L: $l=1$, $P=6, N=3$, UCA, SNR = 15 dB, $\theta_1=35°, \theta_2=80°, \theta_3=110°, \theta_4=140°, \theta_5=230°, \theta_6=304°$, no modeling errors. (*Source:* From Chevalier et al., 2006, © IEEE, with permission.)

maximal RMSE as L increases, showing the capability of 4-MUSIC to efficiently estimate the DOA of all sources for $l=1$.

Other illustrations of 2q-MUSIC for underdetermined mixtures of sources may be found in Chevalier et al. (2006), particularly for $q=3$.

8.7 EXTENSION TO ARRAYS WITH DIVERSELY POLARIZED ANTENNAS: THE PD-2q-MUSIC METHODS

This section presents extensions of the 2q-MUSIC algorithm to arrays with diversely polarized antennas. These extensions, introduced in Chevalier et al. (2007), give rise to new algorithms called PD-2q-MUSIC. Some hypotheses are described in Section 8.7.1 and PD-2q-MUSIC algorithms are presented in Section 8.7.2. The issue of these algorithms identifiability is described in Section 8.7.3. Finally, performance of PD-2q-MUSIC algorithms are illustrated in Section 8.7.4.

8.7.1 Hypotheses and Notations

The observation model is still assumed to be given by Equation (8.2) where, for an array with diversely polarized sensors and in the absence of coupling between them, the complex response, $f_n(\boldsymbol{\theta}_i, \boldsymbol{\beta}_i)$, of sensor n for DOA $\boldsymbol{\theta}_i$ and polarization $\boldsymbol{\beta}_i$, introduced in Equation (8.3), is potentially a function of n. In these conditions, the steering vector $\boldsymbol{a}(\boldsymbol{\theta}_i, \boldsymbol{\beta}_i)$ actually depends on the parameters $\boldsymbol{\theta}_i$ and $\boldsymbol{\beta}_i$.

Let $\boldsymbol{\beta}_{i1}$ and $\boldsymbol{\beta}_{i2}$ be two distinct polarizations for the ith source (for example, vertical and horizontal) and $\boldsymbol{a}_1(\boldsymbol{\theta}_i) \triangleq \boldsymbol{a}(\boldsymbol{\theta}_i, \boldsymbol{\beta}_{i1})$ and $\boldsymbol{a}_2(\boldsymbol{\theta}_i) \triangleq \boldsymbol{a}(\boldsymbol{\theta}_i, \boldsymbol{\beta}_{i2})$ be the corresponding steering vectors for DOA $\boldsymbol{\theta}_i$. We assume that the vectors $\boldsymbol{a}_1(\boldsymbol{\theta})$ and $\boldsymbol{a}_2(\boldsymbol{\theta})$ can be calculated analytically or measured by calibration whatever the value of $\boldsymbol{\theta}$. Considering an arbitrary polarization $\boldsymbol{\beta}_i$ for the ith source, the complex electric field of the latter can be broken down into the sum of two complex fields, each arriving from the same direction and having the polarizations $\boldsymbol{\beta}_{i1}$ and $\boldsymbol{\beta}_{i2}$ (Compton, 1988). The steering vector $\boldsymbol{a}(\boldsymbol{\theta}_i, \boldsymbol{\beta}_i)$ of source i is then the weighted sum of the steering vectors $\boldsymbol{a}_1(\boldsymbol{\theta}_i)$ and $\boldsymbol{a}_2(\boldsymbol{\theta}_i)$ given by

$$\boldsymbol{a}(\boldsymbol{\theta}_i, \boldsymbol{\beta}_i) = \beta_{i1}\boldsymbol{a}_1(\boldsymbol{\theta}_i) + \beta_{i2}\boldsymbol{a}_2(\boldsymbol{\theta}_i) \triangleq \boldsymbol{A}_{12}(\boldsymbol{\theta}_i)\boldsymbol{\beta}_i \tag{8.55}$$

In Equation (8.55), $\boldsymbol{A}_{12}(\boldsymbol{\theta}_i)$ is the $(N \times 2)$ matrix of vectors $\boldsymbol{a}_1(\boldsymbol{\theta}_i)$ and $\boldsymbol{a}_2(\boldsymbol{\theta}_i)$, whereas β_{i1} and β_{i2} are complex numbers such that $|\beta_{i1}|^2 + |\beta_{i2}|^2 = 1$. Vector $\boldsymbol{\beta}_i$ is the unit norm (2×1) vector with components β_{i1} and β_{i2}. It can be written, to within a phase term, as $\boldsymbol{\beta}_i = [\cos\gamma_i, \mathrm{e}^{\mathrm{j}\phi_i}\sin\gamma_i]^{\mathrm{T}}$, where γ_i and ϕ_i are two angles characterizing the polarization of source i and such that $(0 \le \gamma_i \le \pi/2, -\pi \le \phi_i < \pi)$. Note that for an array with identical sensors, $\boldsymbol{a}_1(\boldsymbol{\theta}_i), \boldsymbol{a}_2(\boldsymbol{\theta}_i)$ and $\boldsymbol{a}(\boldsymbol{\theta}_i, \boldsymbol{\beta}_i)$ are collinear, which means that, to within a constant, $\boldsymbol{a}(\boldsymbol{\theta}_i, \boldsymbol{\beta}_i)$ does not depend on the polarization of source i.

8.7.2 The PD-2q-MUSIC Algorithms

Two kinds of algorithms are defined in the following subsections, depending on whether the polarization of the sources is a priori known or unknown.

Sources with Known Polarization: KP-PD-2q-MUSIC

Assuming that the hypotheses of Section 8.3.1 are verified, and following the same reasoning as in Section 8.3, we deduce that the DOAs of the sources are solutions to Equation (8.18), which can be written, for an array with diversely polarized sensors, as

$$\boldsymbol{a}_{q,l}(\boldsymbol{\theta}, \boldsymbol{\beta})^{\dagger}\boldsymbol{\Pi}_{2q,n}(l)\boldsymbol{a}_{q,l}(\boldsymbol{\theta}, \boldsymbol{\beta}) = 0 \tag{8.56}$$

where $\boldsymbol{a}_{q,l}(\boldsymbol{\theta}, \boldsymbol{\beta})$ is defined by Equation (8.11) with $\boldsymbol{a}(\boldsymbol{\theta}, \boldsymbol{\beta})$ instead of $\boldsymbol{a}(\boldsymbol{\theta})$. Equation (8.56) corresponds to the heart of the PD-2q-MUSIC algorithms for the arrangement l and can also be written, using Equations (8.55) and (8.11), as

$$\boldsymbol{\beta}_{q,l}{}^{\dagger}\boldsymbol{A}_{12,q,l}(\boldsymbol{\theta})^{\dagger}\boldsymbol{\Pi}_{2q,n}(l)\boldsymbol{A}_{12,q,l}(\boldsymbol{\theta})\boldsymbol{\beta}_{q,l} = 0 \tag{8.57}$$

where $\boldsymbol{\beta}_{q,l}$ and $\boldsymbol{A}_{12,q,l}(\boldsymbol{\theta})$ are, respectively, the $(2^q \times 1)$ vector and the $(N^q \times 2^q)$ matrix defined by

$$\boldsymbol{\beta}_{q,l} \triangleq \left[\boldsymbol{\beta}^{\otimes l} \otimes \boldsymbol{\beta}^{*\otimes(q-l)}\right] \tag{8.58}$$

$$\boldsymbol{A}_{12,q,l}(\boldsymbol{\theta}) \triangleq \left[\boldsymbol{A}_{12}(\boldsymbol{\theta})^{\otimes l} \otimes \boldsymbol{A}_{12}(\boldsymbol{\theta})^{*\otimes(q-l)}\right] \tag{8.59}$$

For some values of (q, l), some components of $\boldsymbol{\beta}_{q,l}$ are equal. It is then useful not only to at least reduce the complexity of the computation of the left side of Equation (8.57) but also, for improving the performance of the algorithms presented in the next section, to remove the redundant components of $\boldsymbol{\beta}_{q,l}$.

This can be done by removing the redundant components of both $\boldsymbol{\beta}^{\otimes l}$ and $\boldsymbol{\beta}^{*\otimes(q-l)}$. It is straightforward to show that $\boldsymbol{\beta}^{\otimes l}$ can be written as

$$\boldsymbol{\beta}^{\otimes l} = \boldsymbol{B}_l \tilde{\boldsymbol{\beta}}_l \tag{8.60}$$

where $\boldsymbol{B}_l$ is the $(2^l \times (l+1))$ real matrix such that

$$\boldsymbol{B}_1 = \mathbf{I}_2 \tag{8.61a}$$

$$\begin{aligned}\boldsymbol{B}_{l+1} = {} & [\boldsymbol{B}_l \otimes \mathbf{I}_2]\{[\mathbf{I}_{l+1} \otimes \boldsymbol{c}_1][\mathbf{I}_{l+1}\ \mathbf{0}_{l+1}] \\ & + [\mathbf{I}_{l+1} \otimes \boldsymbol{c}_2][\mathbf{0}_{l+1} \mathbf{I}_{l+1}]\}, \quad (l > 1)\end{aligned} \tag{8.61b}$$

In both parts of Equation (8.61), $\mathbf{I}_r$ is the $(r \times r)$ identity matrix; $\boldsymbol{c}_1$ and $\boldsymbol{c}_2$ are the (2×1) vectors defined by $\boldsymbol{c}_1 \triangleq [1\ \ 0]^{\mathrm{T}}$ and $\boldsymbol{c}_2 \triangleq [0\ \ 1]^{\mathrm{T}}$; $\mathbf{0}_{l+1}$ is the $((l+1) \times 1)$ zero vector; and $\tilde{\boldsymbol{\beta}}_l$ is the $((l+1) \times 1))$ vector with components $\tilde{\boldsymbol{\beta}}_l[j]$ defined by

$$\tilde{\beta}_l[j] = {\beta_1}^{l-j+1}\, {\beta_2}^{j-1}; \quad 1 \leq j \leq l+1 \tag{8.62}$$

where β_1 and β_2 are the components of the polarization vector $\boldsymbol{\beta}$. From Equations (8.58) and (8.60), we deduce that

$$\begin{aligned}\boldsymbol{\beta}_{q,l} &= (\boldsymbol{B}_l \tilde{\boldsymbol{\beta}}_l) \otimes (\boldsymbol{B}_{q-l} \tilde{\boldsymbol{\beta}}_{q-l})^* \\ &= [\boldsymbol{B}_l \otimes \boldsymbol{B}_{q-l}][\tilde{\boldsymbol{\beta}}_l \otimes \tilde{\boldsymbol{\beta}}_{q-l}{}^*] \triangleq [\boldsymbol{B}_l \otimes \boldsymbol{B}_{q-l}]\tilde{\boldsymbol{\beta}}_{q,l}\end{aligned} \tag{8.63}$$

where $\tilde{\boldsymbol{\beta}}_{q,l} \triangleq [\tilde{\boldsymbol{\beta}}_l \otimes \tilde{\boldsymbol{\beta}}_{q-l}{}^*]$ is an $((l+1)(q-l+1) \times 1)$ vector.

Note a dimension reduction of $\tilde{\boldsymbol{\beta}}_{q,l}$ with respect to $\boldsymbol{\beta}_{q,l}$ for most values of (q, l). To ensure, in the absence of sources (i.e., when $\boldsymbol{\Pi}_{2q,n}(l) = \mathbf{I}_{N^q}$), a constant value independent of parameters $\boldsymbol{\theta}$ and $\boldsymbol{\beta}$ of the left side of Equation (8.57), it is necessary to normalize the latter by the quantity $\boldsymbol{\beta}_{q,l}{}^{\dagger} \boldsymbol{A}_{12,q,l}(\boldsymbol{\theta})^{\dagger} \boldsymbol{A}_{12,q,l}(\boldsymbol{\theta}) \boldsymbol{\beta}_{q,l}$.

Using Equation (8.63) in Equation (8.57), the problem of source DOA estimation by the PD-2q-MUSIC algorithm for the arrangement l then consists of finding the P sets of parameters $(\hat{\boldsymbol{\theta}}_i, \hat{\boldsymbol{\beta}}_i) = (\hat{\theta}_i, \hat{\varphi}_i, \hat{\gamma}_i, \hat{\phi}_i)$, $(1 \leq i \leq P)$, which either are solutions to Equation (8.64) or minimize its left side, where Equation (8.64) is defined by

$$\frac{\tilde{\boldsymbol{\beta}}_{q,l}{}^{\dagger} \boldsymbol{Q}_{q,l,1}(\boldsymbol{\theta}) \tilde{\boldsymbol{\beta}}_{q,l}}{\tilde{\boldsymbol{\beta}}_{q,l}{}^{\dagger} \boldsymbol{Q}_{q,l,2}(\boldsymbol{\theta}) \tilde{\boldsymbol{\beta}}_{q,l}} = 0 \tag{8.64}$$

where the $((l+1)(q-l+1) \times (l+1)(q-l+1))$ matrices $\boldsymbol{Q}_{q,l,1}(\boldsymbol{\theta})$ and $\boldsymbol{Q}_{q,l,2}(\boldsymbol{\theta})$ are defined by

$$\boldsymbol{Q}_{q,l,1}(\boldsymbol{\theta}) \triangleq [\boldsymbol{B}_l \otimes \boldsymbol{B}_{q-l}]^{\dagger} \boldsymbol{A}_{12,q,l}(\boldsymbol{\theta})^{\dagger} \boldsymbol{\Pi}_{2q,n}(l) \boldsymbol{A}_{12,q,l}(\boldsymbol{\theta}) [\boldsymbol{B}_l \otimes \boldsymbol{B}_{q-l}] \tag{8.65}$$

$$\boldsymbol{Q}_{q,l,2}(\boldsymbol{\theta}) \triangleq [\boldsymbol{B}_l \otimes \boldsymbol{B}_{q-l}]^{\dagger} \boldsymbol{A}_{12,q,l}(\boldsymbol{\theta})^{\dagger} \boldsymbol{A}_{12,q,l}(\boldsymbol{\theta}) [\boldsymbol{B}_l \otimes \boldsymbol{B}_{q-l}] \tag{8.66}$$

For sources with known polarization, the set of parameters for a given source reduces to the set of its DOA parameters, and the complexity of PD-2q-MUSIC, called in this case KP-PD-2q-MUSIC (Known Polarization PD-2q-MUSIC), corresponds to that of $2q$-MUSIC. However, for sources with unknown polarization, the set of parameters for a given source has to take into account polarization parameters in addition to DOA parameters, causing the complexity of the searching procedure of PD-$2q$-MUSIC to dramatically increase beyond what is generally and practically reasonable. For this reason, Chevalier et al. (2007) proposed using the KP-PD-$2q$-MUSIC algorithm only when the polarization of sources is known.

In practical situations, matrices $\boldsymbol{\Pi}_{2q,n}(l)$ and $\boldsymbol{U}_{2q,n}(l)$ must be estimated from the observations; assuming sources with known polarization, the sources' DOA may be found by searching for the minima of the estimated left side of Equation (8.64). The different steps of the KP-PD-$2q$-MUSIC algorithm for the arrangement l are summarized as follows:

Step 1. Estimation, $\hat{\boldsymbol{C}}_{2q,x}(l)$, of the matrix $\boldsymbol{C}_{2q,x}(l)$ from L snapshots $\boldsymbol{x}(k)$, $1 \leq k \leq L$, using a suitable estimator of the $2q$th-order cumulants of observations.

Step 2. Eigendecomposition of the matrix, $\hat{\boldsymbol{C}}_{2q,x}(l)$, and extraction of an estimate, $\hat{\boldsymbol{U}}_{2q,n}(l)$, of the $\boldsymbol{U}_{2q,n}(l)$ matrix. This step may involve rank determination in cases where the number of sources and/or their mutual statistical dependence are not known a priori.

Step 3. Computation, for each known vector $\boldsymbol{\beta} = \boldsymbol{\beta}_i (1 \leq i \leq P)$, of the estimated pseudo-spectrum,

$$\hat{P}_{KP\text{-}PD\text{-}2q\text{-}Music(l)}(\boldsymbol{\theta}, \boldsymbol{\beta}) \triangleq \frac{\tilde{\boldsymbol{\beta}}_{q,l}^{\dagger} \hat{\boldsymbol{Q}}_{q,l,1}(\boldsymbol{\theta}) \tilde{\boldsymbol{\beta}}_{q,l}}{\tilde{\boldsymbol{\beta}}_{q,l}^{\dagger} \boldsymbol{Q}_{q,l,2}(\boldsymbol{\theta}) \tilde{\boldsymbol{\beta}}_{q,l}} \tag{8.67}$$

over a suitably chosen grid. Then search for the local minima (including interpolation at each local minimum), where the $((l+1)(q-l+1) \times (l+1)(q-l+1))$ matrix $\hat{\boldsymbol{Q}}_{q,l,1}(\boldsymbol{\theta})$ is defined by

$$\hat{\boldsymbol{Q}}_{q,l,1}(\boldsymbol{\theta}) \triangleq [\boldsymbol{B}_l \otimes \boldsymbol{B}_{q-l}]^{\dagger} \boldsymbol{A}_{12,q,l}(\boldsymbol{\theta})^{\dagger} \hat{\boldsymbol{\Pi}}_{2q,n}(l) \boldsymbol{A}_{12,q,l}(\boldsymbol{\theta}) [\boldsymbol{B}_l \otimes \boldsymbol{B}_{q-l}] \tag{8.68}$$

In Equation (8.68), $\hat{\boldsymbol{\Pi}}_{2q,n}(l) \triangleq \hat{\boldsymbol{U}}_{2q,n}(l) \hat{\boldsymbol{U}}_{2q,n}(l)^{\dagger}$. Note that, similar to $2q$-MUSIC, PD-$2q$-MUSIC cannot handle perfectly coherent sources.

Sources with Unknown Polarization: UP-PD-2q-MUSIC

For sources with unknown polarization, the complexity of the PD-$2q$-MUSIC searching procedure, described in the previous section, with respect to DOA and polarization parameters is dramatically high. A simple way to remove the searching procedure for the polarization parameter consists, for any fixed DOA, of minimizing the left side of Equation (8.64) with respect to this parameter, as was proposed in Ferrara and Parks (1983) for $q = 1$. This gives rise to UP-PD-$2q$-MUSIC (Unknown Polarization PD-$2q$-MUSIC), of which the pseudo-spectrum, for the arrangement indexed by l, is given by

$$P_{UP\text{-}PD\text{-}2q-Music(l)}(\boldsymbol{\theta}) \triangleq \underset{\tilde{\boldsymbol{\beta}}_{q,l}}{\text{Min}} \left\{ \frac{\tilde{\boldsymbol{\beta}}_{q,l}^{\dagger} \boldsymbol{Q}_{q,l,1}(\boldsymbol{\theta}) \tilde{\boldsymbol{\beta}}_{q,l}}{\tilde{\boldsymbol{\beta}}_{q,l}^{\dagger} \boldsymbol{Q}_{q,l,2}(\boldsymbol{\theta}) \tilde{\boldsymbol{\beta}}_{q,l}} \right\} \tag{8.69}$$

It is well known (Ferrara and Parks, 1983) that the right side of Equation (8.69) corresponds to the minimum eigenvalue, $\lambda_{q,l,\min}(\boldsymbol{\theta})$, of the $((l+1)(q-l+1) \times (l+1)(q-l+1))$ matrix $\boldsymbol{Q}_{q,l,1}(\boldsymbol{\theta})$ in the metric $\boldsymbol{Q}_{q,l,2}(\boldsymbol{\theta})$, and that the minimizing vector $\tilde{\boldsymbol{\beta}}_{q,l}$, denoted $\tilde{\boldsymbol{\beta}}_{q,l,\min}(\boldsymbol{\theta})$, corresponds to the associated eigenvector. In other words, $\lambda_{q,l,\min}(\boldsymbol{\theta})$ and $\tilde{\boldsymbol{\beta}}_{q,l,\min}(\boldsymbol{\theta})$ satisfy

$$\boldsymbol{Q}_{q,l,1}(\boldsymbol{\theta})\tilde{\boldsymbol{\beta}}_{q,l,\min}(\boldsymbol{\theta}) = \lambda_{q,l,\min}(\boldsymbol{\theta})\boldsymbol{Q}_{q,l,2}(\boldsymbol{\theta})\tilde{\boldsymbol{\beta}}_{q,l,\min}(\boldsymbol{\theta}) \tag{8.70}$$

Thus, a first version of UP-PD-2q-MUSIC for the arrangement indexed by l, called UP-PD-2q-MUSIC(l)-1, consists of finding the P sets of parameters $\hat{\boldsymbol{\theta}}_i = (\hat{\theta}_i, \hat{\varphi}_i)$, $(1 \le i \le P)$, for which the pseudo-spectrum

$$P_{UP\text{-}PD\text{-}2q\text{-}Music(l)\text{-}1}(\boldsymbol{\theta}) \triangleq \lambda_{q,l,\min}(\boldsymbol{\theta}) \tag{8.71}$$

is zero.

This algorithm corresponds to a 2qth-order extension, for the arrangement indexed by l, of the algorithm proposed in Ferrara and Parks (1983) for $q=1$. It was shown in Chevalier et al. (2007) that it thus becomes possible to estimate the polarization of each source i from the associated eigenvector $\tilde{\boldsymbol{\beta}}_{q,l,\min}(\hat{\boldsymbol{\theta}}_i)$, which is the solution to Equation (8.70) for $\boldsymbol{\theta} = \hat{\boldsymbol{\theta}}_i$. Note that one way in which the eigenvalue $\lambda_{q,l,\min}(\boldsymbol{\theta})$ can be computed is by determining the minimum root of

$$\det[\boldsymbol{Q}_{q,l,1}(\boldsymbol{\theta}) - \lambda \boldsymbol{Q}_{q,l,2}(\boldsymbol{\theta})] = 0 \tag{8.72}$$

where det[$\boldsymbol{X}$] denotes the determinant of $\boldsymbol{X}$. Thus, for each value of $\boldsymbol{\theta}$, searching in polarization space has been avoided by finding the roots of an equation of order $(l+1)(q-l+1)$, which corresponds to a substantial reduction in computation, at least for small values of q. We deduce from Equation (8.72) and from Chevalier et al. (2007) and Ferreol et al. (2004a) that for invertible matrix $\boldsymbol{Q}_{q,l,2}(\boldsymbol{\theta})$, finding $\boldsymbol{\theta}$ such that $\lambda = \lambda_{q,l,\min}(\boldsymbol{\theta})$ is zero is equivalent to finding $\boldsymbol{\theta}$ such that $\det[\boldsymbol{Q}_{q,l,2}(\boldsymbol{\theta})^{-1}\boldsymbol{Q}_{q,l,1}(\boldsymbol{\theta})] = \det[\boldsymbol{Q}_{q,l,1}(\boldsymbol{\theta})]/\det[\boldsymbol{Q}_{q,l,2}(\boldsymbol{\theta})] = 0$.

A second version of the UP-PD-2q-MUSIC algorithm for the arrangement indexed by l, called UP-PD-2q-MUSIC(l)-2, consists of finding the P sets of parameters $\hat{\boldsymbol{\theta}}_i = (\hat{\theta}_i, \hat{\varphi} i)$, $(1 \le i \le P)$, for which the pseudo-spectrum

$$P_{UP\text{-}PD\text{-}2q\text{-}Music(l)\text{-}2}(\boldsymbol{\theta}) \triangleq \frac{\det[\boldsymbol{Q}_{q,l,1}(\boldsymbol{\theta})]}{\det[\boldsymbol{Q}_{q,l,2}(\boldsymbol{\theta})]} \tag{8.73}$$

is zero, which allows a decrease in complexity with respect to the computation of Equation (8.71).

In practical situations, matrix $\boldsymbol{U}_{2q,n}(l)$ has to be estimated from observations. Assuming sources with unknown polarization, their DOA may be found by searching for the minima of the right side of Equation (8.71) or (8.73). The steps of the two versions of the UP-PD-2q-MUSIC algorithm for the arrangement l are summarized as follows:

Step 1. Estimation, $\hat{C}_{2q,x}(l)$, of the matrix $C_{2q,x}(l)$ from L snapshots $x(k)$, $1 \le k \le L$, using a suitable estimator of the 2qth-order cumulants of observations.

Step 2. Eigendecomposition of the matrix, $\hat{C}_{2q,x}(l)$, and extraction of an estimate, $\hat{U}_{2q,n}(l)$, of the $U_{2q,n}(l)$ matrix. This step may involve rank determination in cases where the number of sources and/or their mutual statistical dependence are not known a priori.

Step 3. Computation of matrices $Q_{q,l,2}(\theta), \hat{Q}_{q,l,1}(\theta)$ and one of the two estimated pseudo-spectrums,

$$\hat{P}_{UP\text{-}PD\text{-}2q\text{-}Music(l)\text{-}1}(\theta) \triangleq \hat{\lambda}_{q,l,\min}(\theta)$$

$$\hat{P}_{UP\text{-}PD\text{-}2q\text{-}Music(l)\text{-}2}(\theta) \triangleq \frac{\det[\hat{Q}_{q,l,1}(\theta)]}{\det[Q_{q,l,2}(\theta)]}$$

over a suitably chosen grid. Then a search for the local minima of $\hat{P}_{UP\text{-}PD\text{-}2q\text{-}Music(l)\text{-}1}(\theta)$ or $\hat{P}_{UP\text{-}PD\text{-}2q\text{-}Music(l)\text{-}2}(\theta)$ (including interpolation at each local minimum), where $\hat{\lambda}_{q,l,\min}(\theta)$ is the minimum eigenvalue of $\hat{Q}_{q,l,1}(\theta)$ in the metric $Q_{q,l,2}(\theta)$.

Step 4. If needed, computation of both the associated estimated vectors $\hat{\hat{\beta}}_{q,l,\min}(\hat{\theta}_i)$ and the polarization vector of the sources.

8.7.3 Identifiability

As for arrays with identical sensors, we assume an absence of coupling between the sensors and we limit the identifiability analysis of PD-2q-MUSIC algorithms to the case of statistically independent sources.

Higher-Order Virtual Array

As in Section 8.4.1, we deduce that, for the 2qth-order array-processing methods exploiting Equation (8.10) with vectors $a_{q,l}(\theta_i, \beta_i)$ instead of $a_{q,l}(\theta_i)$, the $(N^q \times 1)$ vector $a_{q,l}(\theta_i, \beta_i)$ can be considered as the *true steering vector* of source i for the *VA* of N^q VSs, with coordinates, $(x^l_{k_1k_2,\dots,k_q}, y^l_{k_1k_2,\dots,k_q}, z^l_{k_1k_2,\dots,k_q})$, defined by Equation (8.24) and complex amplitude patterns, $f^l_{k_1k_2,\dots,k_q}(\theta,\beta)$, $1 \le k_j \le N$ for $1 \le j \le q$, given by

$$f^l_{k_1k_2,\dots,k_q}(\theta,\beta) = \prod_{j=1}^{l}\prod_{u=1}^{q-l} f_{k_j}(\theta,\beta) f_{k_{l+u}}(\theta,\beta)^* \tag{8.74}$$

This introduces in a very simple way, and for arrays with diversely polarized sensors, the VA concept for the 2qth-order DOA estimation problem for the arrangement indexed by l.

KP-PD-2q-MUSIC

The KP-PD-2q-MUSIC algorithm for the arrangement indexed by l can estimate the DOA of P statistically independent sources from an array of N sensors provided that hypotheses H1$''$ and H2$''$ are verified, with $\boldsymbol{a}_{q,l}(\boldsymbol{\theta}_i, \boldsymbol{\beta}_i)$ instead of $\boldsymbol{a}_{q,l}(\boldsymbol{\theta}_i)$, and that the DOA and the polarization of the sources are the only solutions to Equation (8.56). Noting N^l_{2q} $(N^l_{2q} \leq N^q)$ as the number of different VSs of the associated VA, and following a reasoning similar to that for identical sensors, we deduce that the KP-PD-2q-MUSIC algorithm for the arrangement indexed by l is able to process up to

$$P_{\max} \triangleq P^l_{2q,K} = N^l_{2q} - 1 \tag{8.75}$$

sources, provided the associated 2qth-order VA has no ambiguities up to order $N^l_{2q} - 1$. The rth-order ambiguities of HO VA are an important problem that was partially discussed in Chevalier et al. (2007), where it was found, in particular, that, for a VA without colocalized sensors, $P_{\max}$ is effectively given by Equation (8.75), whereas for a VA with colocalized sensors, $P_{\max}$ is generally lower than Equation (8.75).

For given values of N, q, and l and for general arrays of N sensors with diversely polarized sensors, it was shown in Chevalier et al. (2005) that N^l_{2q} is necessarily upper-bounded by a quantity, noted $N_{\max}[2q, l]$, such that $N_{\max}[2q, l] \leq N^q$. Table 8.2 shows, for a general array with diversely polarized sensors, the expression of $N_{\max}[2q, l]$ as a function of N for $2 \leq q \leq 4$ and several values of l. This upper bound corresponds to N^l_{2q} in most cases of sensor responses and array geometry.

TABLE 8.2 $N_{\max}[2q, l]$ **as a Function of** N **for Several Values of** q **and** l **and for Arrays with Diversely Polarized Sensors**

$m=2q$	l	$N_{\max}[2q, l]$
4	2	$N(N+1)/2$
$(q=2)$	1	N^2
6	3	$N!/[6(N-3)!]+N(N-1)+N$
$(q=3)$	2	$N!/[2(N-3)!]+2N(N-1)+N$
8	4	$N!/[24(N-4)!]+N!/[2(N-3)!]+1.5N(N-1)+N$
$(q=4)$	3	$N!/[6(N-4)!]+1.5N!/(N-3)!+3N(N-1)+N$
	2	$N!/[4(N-4)!]+2N!/(N-3)!+3.5N(N-1)+N$

Source: From Chevalier et al., 2005, © IEEE, with permission.

UP-PD-2q-MUSIC

The developments of the previous section are still valid for UP-PD-2q-MUSIC algorithms. In particular, the maximal number of statistically independent sources that may be processed by these algorithms for the arrangement indexed by l cannot exceed $N_{2q}^{l}-1$ if the associated VA has no ambiguities up to order $N_{2q}^{l}-1$. Moreover, we deduce from the HO VA theory (Chevalier et al., 2005) that $(N^q - N_{2q}^{l})$ components of $\boldsymbol{a}_{q,l}(\boldsymbol{\theta}, \boldsymbol{\beta})$ produce no information. This means that $\boldsymbol{a}_{q,l}(\boldsymbol{\theta}, \boldsymbol{\beta})$ can be written as

$$\boldsymbol{a}_{q,l}(\boldsymbol{\theta}, \boldsymbol{\beta}) \triangleq \boldsymbol{G}\boldsymbol{a}_{q,l,nr}(\boldsymbol{\theta}, \boldsymbol{\beta}) \tag{8.76}$$

$\boldsymbol{G}$ is a full-rank and constant $(N^q \times N_{2q}^{l})$ matrix, and $\boldsymbol{a}_{q,l,nr}(\boldsymbol{\theta}, \boldsymbol{\beta})$ is the nonredundant $(N_{2q}^{l} \times 1)$ steering vector of a source coming from DOA $\boldsymbol{\theta}$ with polarization $\boldsymbol{\beta}$ for the VA associated with parameters (q, l). Equation (8.76) shows that an arbitrary steering vector $\boldsymbol{a}_{q,l}(\boldsymbol{\theta}, \boldsymbol{\beta})$ necessarily belongs to the space spanned by the N_{2q}^{l} columns of $\boldsymbol{G}$, noted Span$\{\boldsymbol{G}\}$. Regarding the $\boldsymbol{G}^{\perp}$ a $(N^q \times (N^q - N_{2q}^{l}))$ full-rank matrix with columns that span the space orthogonal to Span$\{\boldsymbol{G}\}$, we deduce that all the vectors, $\boldsymbol{u} = \boldsymbol{G}^{\perp}\boldsymbol{v}$, of Im$\{\boldsymbol{G}^{\perp}\}$, where $\boldsymbol{v}$ is an arbitrary $\left((N^q - N_{2q}^{l}) \times 1\right)$ vector, are orthogonal to $\boldsymbol{a}_{q,l}(\boldsymbol{\theta}, \boldsymbol{\beta})$ for arbitrary values of $(\boldsymbol{\theta}, \boldsymbol{\beta})$. A direct consequence of this result is that whatever the number, P, of statistically independent sources such that $P \leq N_{2q}^{l}-1$, and whatever their DOA and polarization, Im$\{\boldsymbol{G}^{\perp}\} \subset$ Span$\{\boldsymbol{U}_{2q,n}(l)\}$, which means that $\left(N^q - N_{2q}^{l}\right)$ columns of $\boldsymbol{U}_{2q,n}(l)$ are not discriminant. In other words, we deduce from Equation (8.76) and the previous results that only $\left(N_{2q}^{l} - P\right)$ columns of $\boldsymbol{U}_{2q,n}(l)$ are discriminant, while Equation (8.56) takes the form

$$\boldsymbol{a}_{q,l,nr}(\boldsymbol{\theta}, \boldsymbol{\beta})^{\dagger}\boldsymbol{G}^{\dagger}\boldsymbol{\Pi}_{2q,n}(l)\boldsymbol{G}\boldsymbol{a}_{q,l,nr}(\boldsymbol{\theta}, \boldsymbol{\beta}) = \boldsymbol{a}_{q,l}(\boldsymbol{\theta}, \boldsymbol{\beta})^{\dagger}\boldsymbol{\Pi}_{2q,n,G}(l)\boldsymbol{a}_{q,l}(\boldsymbol{\theta}, \boldsymbol{\beta}) = 0 \tag{8.77}$$

where $\boldsymbol{\Pi}_{2q,n,G}(l)$ is the $(N^q \times (N_{2q}^{l} - P))$ orthogonal projector on span$\{\boldsymbol{U}_{2q,n}(l)\}$ $\cap$Span$\{\boldsymbol{G}\}$.

Replacing Equation (8.56) with Equation (8.77) in the developments of Section 8.7.2, we deduce that, for given values of (q, l), $\lambda_{q,l,\min}(\boldsymbol{\theta})$ and $\tilde{\boldsymbol{\beta}}_{q,l,\min}(\boldsymbol{\theta})$ are also solutions to Equation (8.70) where $\boldsymbol{Q}_{q,l,1}(\boldsymbol{\theta})$ has been replaced with $\boldsymbol{Q}_{q,l,1,G}(\boldsymbol{\theta})$, defined by

$$\boldsymbol{Q}_{q,l,1,G}(\boldsymbol{\theta}) \triangleq [\boldsymbol{B}_l \otimes \boldsymbol{B}_{q-l}]^{\dagger}\boldsymbol{A}_{12,q,l}(\boldsymbol{\theta})^{\dagger}\boldsymbol{\Pi}_{2q,n,G}(l)\boldsymbol{A}_{12,q,l}(\boldsymbol{\theta})[\boldsymbol{B}_l \otimes \boldsymbol{B}_{q-l}] \tag{8.78}$$

Because the quantity $\lambda_{q,l,\min}(\boldsymbol{\theta})$, defined by Equation (8.70) with $\boldsymbol{Q}_{q,l,1,G}(\theta)$ instead of $\boldsymbol{Q}_{q,l,1}(\boldsymbol{\theta})$, has to be nulled only for the DOA of the sources and not for other DOAs, the $((l+1)(q-l+1) \times (l+1)(q-l+1))$ matrix $\boldsymbol{Q}_{q,l,1,G}(\boldsymbol{\theta})$ has to be full rank when $\boldsymbol{\theta}$ does not correspond to a source's DOA. Using Equation (8.78), this means that the rank of $\boldsymbol{\Pi}_{2q,n,G}(l)$ cannot be lower than $(l+1)$ $(q-l+1)$. Thus, the number of columns of $\boldsymbol{\Pi}_{2q,n,G}(l)$ must be greater than or equal to $(l+1)(q-l+1)$. Moreover, in the presence of P statistically independent sources such that $P \leq N_{2q}^{l}-1$, the number of columns of $\mathbf{\Pi}_{2q,n,G}(l)$ is equal to $N_{2q}^{l}-P$ for the associated VA with no ambiguities up to order $N_{2q}^{l}-1$. As a consequence, the maximal number of sources, $P_{\max}$, that may be processed by UP-PD-$2q$-MUSIC algorithms for the arrangement indexed by l must, for such VAs, verify $P_{\max} \leq N_{2q}^{l}-(l+1)(q-l+1)$.

Conversely, for a $2q$th-order VA with no ambiguities up to order $N_{2q}^{l}-1$, the P sources coming from P different directions with different polarizations and such that $P \leq N_{2q}^{l}-(l+1)(q-l+1)$, are such that their DOAs are the only solutions to $\lambda_{q,l,\min}(\boldsymbol{\theta})=0$. From the previous results, assuming a $2q$th-order VA for the arrangement indexed by l with N_{2q}^{l} different VSs and with no ambiguities up to order $N_{2q}^{l}-1$, we deduce that UP-PD-$2q$-MUSIC algorithms for the arrangement l are able to process up to

$$P_{\max} \triangleq P_{2q,U}^{l} = N_{2q}^{l}-(l+1)(q-l+1) \tag{8.79}$$

sources. This is strictly lower than Equation (8.75), and gives $P_{\max}=N-2$ for $q=1$ and arrays with scalar sensors—a result obtained in Ferrara and Parks (1983). Note that for VA with HO ambiguities, $P_{\max}$ is lower than Equation (8.79) (Chevalier et al., 2007).

8.7.4 Computer Simulations

Performance of PD-$2q$-MUSIC algorithms are illustrated in this section through computer simulations for both overdetermined and underdetermined mixtures of sources.

Overdetermined Mixtures of Sources

To show the interest of taking into account both the polarization of the sources and the HO statistics for direction finding, we consider a UCA of $N=6$ crossed dipoles with a radius r such that $r=0.3\,\lambda$. One dipole is parallel to the x-axis; the other is parallel to the z-axis. Three of these crossed dipoles are combined to generate, in the y-axis, a right-sense circular polarization while the other three are combined to generate, in the y-axis, a left-sense circular polarization. The array is thus composed of two orthogonally polarized overlapped (noncolocated) circular subarrays of $M=3$ sensors so that adjacent sensors always have different polarizations, as depicted in Figure 8.17(a).

Under these assumptions, the sensors of the first and second subarrays have a complex response to a unit electric field coming from DOA $\boldsymbol{\theta}$ with polarizations $\boldsymbol{\beta}$ equal to $f(\boldsymbol{\theta},\boldsymbol{\beta})=g_1(\boldsymbol{\theta},\boldsymbol{\beta})+\mathrm{j}g_2(\boldsymbol{\theta},\boldsymbol{\beta})$ and $f'(\boldsymbol{\theta},\boldsymbol{\beta})=g_1(\boldsymbol{\theta},\boldsymbol{\beta})-\mathrm{j}g_2(\boldsymbol{\theta},\boldsymbol{\beta})$, respectively. In these expressions, $g_1(\boldsymbol{\theta},\boldsymbol{\beta})$ and $g_2(\boldsymbol{\theta},\boldsymbol{\beta})$, which correspond to the complex responses of the two dipoles, are given by

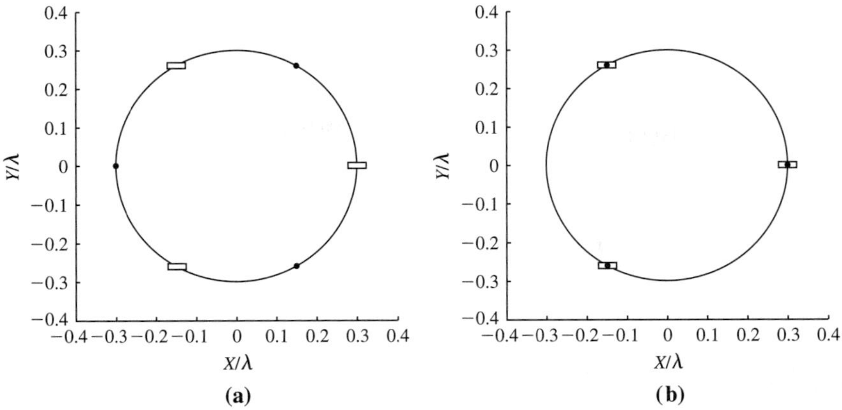

FIGURE 8.17 Circular array of six equi-spaced sensors composed of two overlapped orthogonally polarized subarrays of three sensors: (a) non-collocated subarrays; (b) collocated subarrays. (*Source:* From Chevalier et al., 2007, © IEEE, with permission.)

Chevalier and Ferreol (1999) and Compton (1988):

$$g_1(\boldsymbol{\theta}, \boldsymbol{\beta}) = g_1(\theta, \varphi, \gamma, \phi) = \sin\gamma \sin\varphi \cos\theta \mathrm{e}^{\mathrm{j}\phi} - \cos\gamma \sin\theta \tag{8.80}$$

$$g_2(\boldsymbol{\theta}, \boldsymbol{\beta}) = g_2(\theta, \varphi, \gamma, \phi) = -\sin\gamma \cos\varphi \mathrm{e}^{\mathrm{j}\phi} \tag{8.81}$$

In other words, the complex response, $f_n(\boldsymbol{\theta}, \boldsymbol{\beta})$, of sensor $n(1 \le n \le N)$ to a unit electric field coming from DOA $\boldsymbol{\theta}$ with polarization $\boldsymbol{\beta}$ is given by

$$f_n(\boldsymbol{\theta}, \boldsymbol{\beta}) = (\sin\varphi \cos\theta + \mathrm{j}(-1)^n \cos\varphi) \sin\gamma \mathrm{e}^{\mathrm{j}\phi} - \cos\gamma \sin\theta \tag{8.82}$$

We then deduce from Equations (8.3), (8.55), and (8.82) that, in this case, the 2N coefficients of matrix $\boldsymbol{A}_{12}(\boldsymbol{\theta})$ are defined by

$$A_{12}(\boldsymbol{\theta})[n, 1] = -\sin\theta \mathrm{e}^{\mathrm{j}\zeta_n} \tag{8.83a}$$

$$A_{12}(\boldsymbol{\theta})[n, 2] = (\sin\varphi \cos\theta + \mathrm{j}(-1)^n \cos\varphi) \mathrm{e}^{\mathrm{j}\zeta_n} \tag{8.83b}$$

where $1 \le n \le N$ and where $\zeta_n (1 \le n \le N)$ is defined by

$$\zeta_n \triangleq 2\pi[x_n \cos(\theta)\cos(\varphi) + y_n \sin(\theta)\cos(\varphi) + z_n \sin(\varphi)]/\lambda \tag{8.84}$$

Note that this corresponds to choosing vectors $\boldsymbol{a}_1(\boldsymbol{\theta}) \triangleq \boldsymbol{a}(\boldsymbol{\theta}, \boldsymbol{\beta}_1)$ and $\boldsymbol{a}_2(\boldsymbol{\theta}) \triangleq \boldsymbol{a}(\boldsymbol{\theta}, \boldsymbol{\beta}_2)$ such that $\boldsymbol{\beta}_1 = [1, 0]^{\mathrm{T}}$ and $\boldsymbol{\beta}_2 = [0, 1]^{\mathrm{T}}$. Note also that the chosen array of sensors presents ambiguities for $\boldsymbol{\theta}_0 = (0, k\pi)$, where k is an integer, and thus prevents estimating the DOA of sources coming from $\boldsymbol{\theta}_0$. Indeed, $\boldsymbol{a}_1(\boldsymbol{\theta}_0) = 0$, $\boldsymbol{Q}_{q,l,1}(\boldsymbol{\theta}_0)$, and $\boldsymbol{Q}_{q,l,2}(\boldsymbol{\theta}_0)$ are not full rank and $\boldsymbol{\theta}_0$ is always a solution to Equations (8.57) and (8.70) and $\lambda_{q,l,\min}(\boldsymbol{\theta}_0) = 0$.

In this context, two QPSK sources with the same symbol duration, the same raised cosine pulse-shaped filter with a roll-off of 0.3, and the same input SNR (which would be received by an omnidirectional sensor) equal to 5 dB

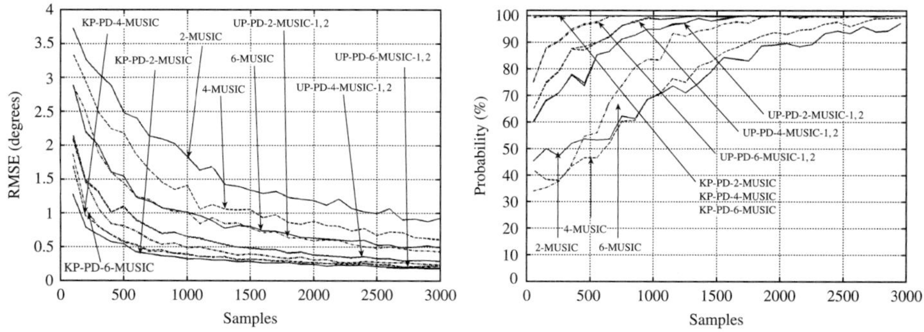

(a) (b)

FIGURE 8.18 (a) RMSE and (b) probability of nonaberrant results of source 2 as a function of L. $P=2, N=6$, UCA, SNR $=5$ dB, $(\theta_1, \gamma_1, \phi_1)=(50°, 45°, 0°)$, $(\theta_2, \gamma_2, \phi_2)=(60°, 45°, 10°)$, no modeling errors. (*Source:* From Chevalier et al., 2007, © IEEE, with permission.)

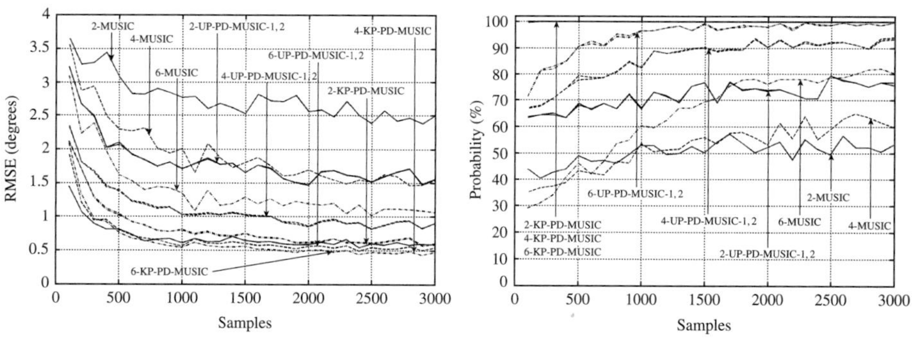

(a) (b)

FIGURE 8.19 (a) RMSE and (b) probability of nonaberrant results of source 2 as a function of L. $P=2, N=6$, UCA, SNR $=5$ dB, $(\theta_1, \gamma_1, \phi_1)=(50°, 45°, 0°)$, $(\theta_2, \gamma_2, \phi_2)=(60°, 45°, 10°)$, with modeling errors $\sigma=0.017$. (*Source:* From Chevalier et al., 2007, © IEEE, with permission.)

are assumed to be received by the array. They are first assumed to be weakly separated in both space and polarization and such that $(\theta_1, \gamma_1, \phi_1)=(50°, 45°, 0°)$ and $(\theta_2, \gamma_2, \phi_2)=(60°, 45°, 10°)$, respectively. Under these assumptions, Figures 8.18 and 8.19 show, as a function of the number of snapshots L, the variations of the RMSE for source 2, RMSE_2, and the associated probability of nonaberrant results, for several methods with and without modeling errors, respectively. For arrangement of the considered statistics indexed by $l=1$, the analyzed methods correspond to 2-MUSIC, KP-PD-2-MUSIC, UP-PD-2-MUSIC-1, UP-PD-2-MUSIC-2, 4-MUSIC, KP-PD-4-MUSIC, UP-PD-4-MUSIC-1, UP-PD-4-MUSIC-2, 6-MUSIC, KP-PD-6-MUSIC, UP-PD-6-MUSIC-1, and UP-PD-6-MUSIC-2. Performance is computed from 300 realizations, and similar results are obtained for source 1.

In the presence of modeling errors, the steering vector of source i at the output of the sensors becomes

$$\tilde{\boldsymbol{a}}(\boldsymbol{\theta}_i, \boldsymbol{\beta}_i) = \mu(\boldsymbol{\theta}_i, \boldsymbol{\beta}_i)[\overline{\boldsymbol{a}}(\boldsymbol{\theta}_i, \boldsymbol{\beta}_i) + \boldsymbol{e}_i]$$

where $\overline{\boldsymbol{a}}(\boldsymbol{\theta}_i, \boldsymbol{\beta}_i)$ is the normalized steering vector $\boldsymbol{a}(\boldsymbol{\theta}_i, \boldsymbol{\beta}_i)$ such that $\boldsymbol{a}(\boldsymbol{\theta}_i, \boldsymbol{\beta}_i) = \mu(\boldsymbol{\theta}_i, \boldsymbol{\beta}_i)\overline{\boldsymbol{a}}(\boldsymbol{\theta}_i, \boldsymbol{\beta}_i)$, with $\overline{\boldsymbol{a}}(\boldsymbol{\theta}_i, \boldsymbol{\beta}_i)^\dagger \overline{\boldsymbol{a}}(\boldsymbol{\theta}_i, \boldsymbol{\beta}_i) = N$, and the vectors $\boldsymbol{e}_i (1 \leq i \leq 2)$ are assumed to be zero-mean statistically independent circular Gaussian vectors such that $\mathrm{E}[\boldsymbol{e}_i \boldsymbol{e}_j^\dagger] = \sigma^2 \delta_{ij} \mathbf{I}_N$. For the simulations, $\sigma^2 = 0.0003$ (weak modeling errors), which corresponds, for example, to a phase error with a standard deviation of 0.6° jointly having an amplitude error with a standard deviation of 0.1 dB. For 2-MUSIC, 4-MUSIC, and 6-MUSIC, the six sensors of the UCA are assumed to be identical with complex responses $f_n(\boldsymbol{\theta}, \boldsymbol{\beta}) = f(\boldsymbol{\theta}, \boldsymbol{\beta}) = g_1(\boldsymbol{\theta}, \boldsymbol{\beta}) + \mathrm{j}g_2(\boldsymbol{\theta}, \boldsymbol{\beta})$ $(1 \leq n \leq N)$.

Figures 8.18 and 8.19 show, for sources that are weakly separated both in DOA and polarization, with and without modeling errors, and for a given value of $q(=1,2,3)$, the best behavior of DP-2q-MUSIC versus 2q-MUSIC as soon as the polarization of sources is different. This shows the better resolution and greater robustness to modeling errors, whatever the value of q, of methods that discriminate in both space and polarization over methods that discriminate the sources only in space. Moreover, we note, for a given value of q, the similar performance of UP-PD-2q-MUSIC-1 and UP-PD-2q-MUSIC-2, which seem to differ only in complexity.

We note as well increasing performances with q of UP-PD-2q-MUSIC for situations where resolution is required. This is due to an increasing resolution in both DOA and polarization of the associated 2qth-order VA and shows the interest of exploiting both polarization diversity and HO statistics for direction finding with poorly separated sources.

For a given value of q, the better performance of KP-PD-2q-MUSIC over UP-PD-2q-MUSIC, due to the exploitation of true a priori knowledge of the source polarization, is clear. Also clear, for two sources with known polarizations, is the improving performance of KP-PD-2q-MUSIC as q increases in the presence of modeling errors as soon as the number of snapshots grows beyond 1300. This result seems to be directly related to the degree of coupling of the two estimated pseudo-spectra (one for each polarization), computed by a given KP-PD-2q-MUSIC method, which increases with modeling errors and when the polarization separation of the sources decreases.

More precisely, when this coupling is high (weak polarization separation with modeling errors), the two sources interact in each of the two computed pseudo-spectra and resolution is required to separate them—thus, the improved performance with q of KP-PD-2q-MUSIC. However, when this coupling is weak (strong polarization separation or absence of modeling errors), sources no longer interact in a given pseudo-spectrum. Then only one source has to be found for a given pseudo-spectrum and no resolution is required, leading

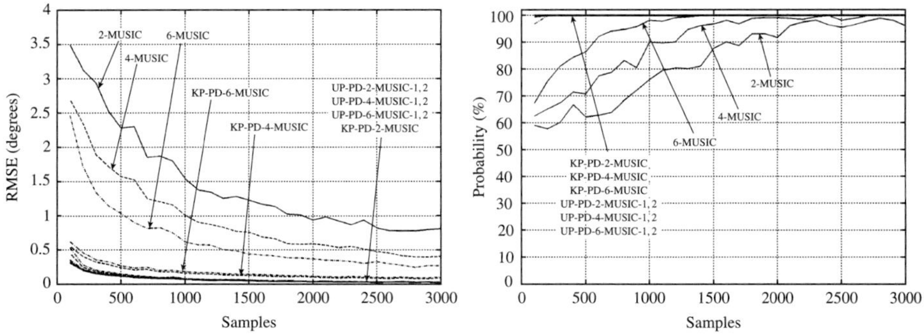

(a) (b)

FIGURE 8.20 (a) RMSE and (b) probability of nonaberrant results of source 2 as a function of L. $P=2, N=6$, UCA, $SNR=5$ dB, $(\theta_1, \gamma_1, \phi_1)=(50°, 45°, 0°)$, $(\theta_2, \gamma_2, \phi_2)=(60°, 45°, 180°)$, no modeling errors. (*Source:* From Chevalier et al., 2007, © IEEE, with permission.)

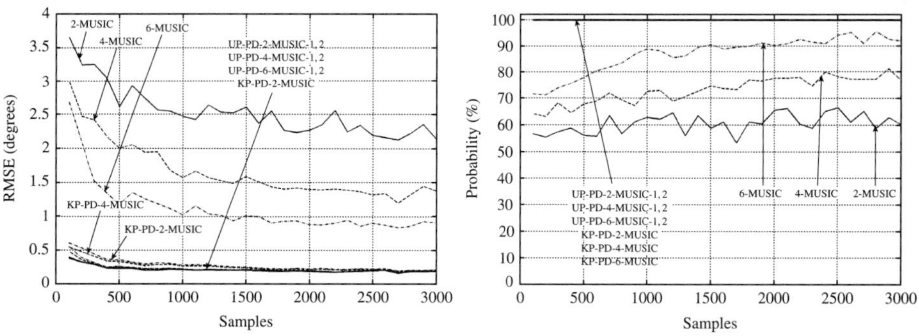

(a) (b)

FIGURE 8.21 (a) RMSE and (b) probability of nonaberrant results of source 2 as a function of L. $P=2$, $N=6$, UCA, $SNR=5$ dB, $(\theta_1, \gamma_1, \phi_1)=(50°, 45°, 0°)$, $(\theta_2, \gamma_2, \phi_2)=(60°, 45°, 180°)$, with modeling errors, $\sigma=0.017$. (*Source:* From Chevalier et al., 2007, © IEEE, with permission.)

to lower performance due to a higher variance in the statistics estimation as q increases.

We now consider the scenario of Figures 8.18 and 8.19, but we assume that the two sources are still poorly angularly separated but well separated in polarization, such that $(\theta_1, \gamma_1, \phi_1)=(50°, 45°, 0°)$ and $(\theta_2, \gamma_2, \phi_2)=(60°, 45°, 180°)$, respectively. Under these assumptions, Figures 8.20 and 8.21 show variations similar to those in Figures 8.18 and 8.19. Whatever the value of q, we still see better performances with PD-2q-MUSIC methods over 2q-MUSIC methods because of polarization discrimination in addition to space discrimination. And we remark the very close performances of UP-PD-2q-MUSIC-1 and UP-PD-2q-MUSIC-2, as well as the decreasing performance of UP-PD-2q-MUSIC-1,2 and KP-PD-2q-MUSIC, as q increases. This is due to a higher variance in the statistics estimates given that no resolution is required for direction finding because of a high separation of sources in polarization. Finally, we note that, for $q>2$, KP-PD-2q-MUSIC may perform surprisingly less well than UP-PD-2q-MUSIC.

Underdetermined Mixtures of Sources

To illustrate the capability of PD-4-MUSIC and PD-6-MUSIC to process underdetermined mixtures of sources, we limit the number of sensors in the previous circular array to $N=3$. These sensors are such that $f_1(\boldsymbol{\theta},\boldsymbol{\beta})=f_3(\boldsymbol{\theta},\boldsymbol{\beta})=g_1(\boldsymbol{\theta},\boldsymbol{\beta})+\mathrm{j}g_2(\boldsymbol{\theta},\boldsymbol{\beta})$ and $f_2(\boldsymbol{\theta},\boldsymbol{\beta})=g_1(\boldsymbol{\theta},\boldsymbol{\beta})-\mathrm{j}g_2(\boldsymbol{\theta},\boldsymbol{\beta})$ for PD-2q-MUSIC and $f_1(\boldsymbol{\theta},\boldsymbol{\beta})=f_2(\boldsymbol{\theta},\boldsymbol{\beta})=f_3(\boldsymbol{\theta},\boldsymbol{\beta})=g_1(\boldsymbol{\theta},\boldsymbol{\beta})+\mathrm{j}g_2(\boldsymbol{\theta},\boldsymbol{\beta})$ for 2q-MUSIC. We also assume that $l=1$. Under these assumptions, we deduce from Tables 1 and 2 in Chevalier et al. (2005), and Equations (8.75) and (8.79) that $\left(N_4^1, P_{4,K}^1, P_{4,U}^1\right)=(8,7,4)$ and $\left(N_6^1, P_{6,K}^1, P_{6,U}^1\right)=(15,14,9)$ for PD-2q-MUSIC methods; we deduce from Tables 6 and 7 in Chevalier et al. (2005) that $\left(N_4^1, P_{\max}\right)=(7,6)$ and $\left(N_6^1, P_{\max}\right)=(12,11)$ for 2q-MUSIC.

We then assume that four statistically independent QPSK sources with a raised cosine pulse-shaped filter are received by the array. These sources have the same symbol duration, the same roll-off $\mu=0.3$, the same input SNR equal to 15 dB and DOA, and polarization parameters equal to $(\theta_1,\gamma_1,\phi_1)=(15°,45°,-75°)$, $(\theta_2,\gamma_2,\phi_2)=(45°,45°,0°)$, $(\theta_3,\gamma_3,\phi_3)=(95°,22.5°,75°)$, and $(\theta_4,\gamma_4,\phi_4)=(122.5°,45°,150°)$, respectively.

Under these assumptions, Figure 8.22 shows the variations, as a function of the number of snapshots L, of the highest RMSE and the lowest probability of nonaberrant results, among all the sources, at the output of several methods without modeling errors. These methods correspond to 4-MUSIC, KP-PD-4-MUSIC, UP-PD-4-MUSIC-1, UP-PD-4-MUSIC-2, 6-MUSIC, KP-

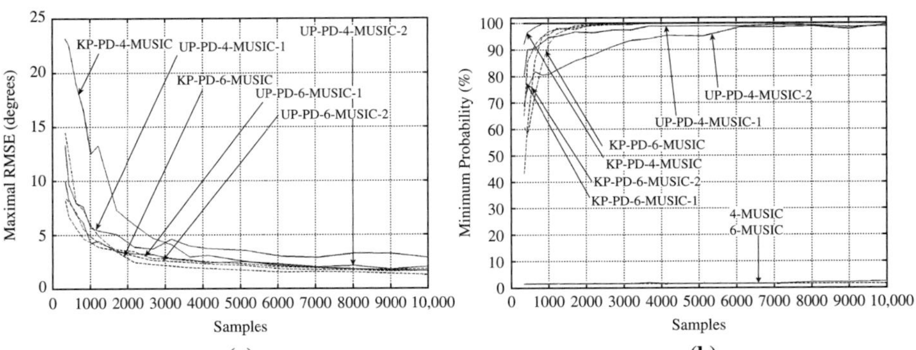

FIGURE 8.22 (a) Maximal RMSE and (b) minimal probability of nonaberrant results of sources as a function of L. $P=4$, $N=3$, UCA, $SNR=15$ dB, $(\theta_1,\gamma_1,\phi_1)=(15°,45°,-75°)$, $(\theta_2,\gamma_2,\phi_2)=(45°,45°,0°)$, $(\theta_3,\gamma_3,\phi_3)=(95°,22.5°,75°)$, $(\theta_4,\gamma_4,\phi_4)=(122.5°,45°,150°)$, 150°), no modeling errors. (*Source:* From Chevalier et al., 2007, © IEEE, with permission.)

PD-6-MUSIC, UP-PD-6-MUSIC-1, and UP-PD-6-MUSIC-2, respectively, with performances computed from 300 realizations. Note the capability of PD-2q-MUSIC to process underdetermined mixtures of sources provided that $P \leq P_{\max}$ given by Equation (8.75) or (8.79). Note also the poor performance of 2q-MUSIC for the considered scenario due to the low input power of the weakest source at the sensor outputs. Better performance would be obtained for higher values of L.

8.8 CONCLUSION

In this chapter, the philosophy, properties, implementations, both parallel and sequential, and performance of HO HR DOA estimation methods were presented through description of the 2qMUSIC and PD-2q-MUSIC algorithms families ($q \geq 1$). These are well suited for any kind of array with identical and diversely polarized sensors, respectively, for several arrangements of the 2qth-order data statistics.

It was shown that HO cumulant-based HR DOA estimation methods have many advantages over SO methods. They are asymptotically robust with respect to a Gaussian background noise with spatial coherence that is unknown, and, because of a virtual increase in both the effective aperture and the number of array sensors, they allow for increasing resolution, increasing robustness to modeling errors in multiple-source contexts, and a greater number of sources to be processed as q increases.

For a given value of q, the maximal number of sources that may be processed by an HO cumulant-based method is directly related to both the considered array of sensors and the way the 2qth-order data statistics are arranged in the exploited 2qth-order statistical matrix, illustrating the existence of an optimal arrangement. In particular, the results may allow minimization of the number of sensors and reception chains for a given number of sources to be processed and may allow relaxation of some constraints on antenna calibration or receiver chain equalization.

The main drawbacks of HO cumulant-based HR DOA versus SO methods are their numerical complexity and their higher sensitivity to finite-sample effect, especially for weak sources and when no resolution is required to resolve the sources. Numerical complexity may be decreased by implementing contracted methods, which exploit only a part of the HO information and are intermediate between SO and HO methods. However, this comes at the expense of the number of sources to be processed. Sensitivity to finite-sample effect may be decreased by exploiting both SO and HO data statistics through implementation of a direction-finding process from a blind estimation of source steering vectors, but this lessens both robustness with respect to an unknown colored Gaussian noise and the number of sources to be processed.

Acknowledgement This work was carried out in part within the framework of the Decotes research project (ANR06-BLAN-0074).

REFERENCES

Albera, L., Ferreol, A., Cosandier-Rimele, D., Merlet, I., Wendling, F., 2008. Brain source localization using a fourth-order deflation scheme. IEEE Trans. Biomed. Eng. 55 (2), 490–501.

Amblard, P.O., Gaeta, M., Lacoume, J.L., 1996. Statistics for complex variables and signs—parts I and II. Signal Process. 53 (1), 1–25.

Bienvenu, G., Kopp, L., 1983. Optimality of high resolution array processing using the eigensystem approach. IEEE Trans. Acoust. Speech Signal Process. 3 (5), 1235–1247.

Birot, G., Albera, L., Ferreol, A., Chevalier, P., 2007. DOA estimation based on an even order deflation scheme. In: Proc. DSP'07.

Cardoso, J.F., 1990. Localisation et Identification par la quadricovariance. Trait. Sign. 7 (5), 397–406.

Cardoso, J.F., 1994. How much more DOA information in higher order statistics. In: Proc. 7th Workshop on Statistical Signal and Array Processing, pp. 199–202.

Cardoso, J.F., Moulines, E., 1995. Asymptotic performance analysis of direction finding algorithms based on fourth-order cumulants. IEEE Trans. Signal Process. 43 (1), 214–224.

Cardoso, J.F., Souloumiac, A., 1993. Blind beamforming for non-Gaussian signals. IEE Proc.-F 140 (6), 362–370.

Challa, R.N., Shamsunder, S., 1998. Passive near-field localization of multiple non-Gaussian sources in 3D using cumulants. Signal Process. 65, 39–53.

Chen, Y.H., Lin, Y.S., 1994a. Cumulant-based method for bearing estimation in the presence of non-Gaussian noise. IEEE Trans. Antennas Propag. 42 (4), 548–552.

Chen, Y.H., Lin, Y.S., 1994b. A modified cumulant matrix for DOA estimation. IEEE Trans. Signal Process. 42 (11), 3287–3291.

Chevalier, P., Ferreol, A., 1999. On the virtual array concept for the fourth-order direction finding problem. IEEE Trans. Signal Process. 47 (9), 2592–2595.

Chevalier, P., Albera, L., Ferreol, A., Comon, P., 2005. On the virtual array concept for higher order array processing. IEEE Trans. Signal Process. 53 (4), 1254–1271.

Chevalier, P., Benoit, G., Ferreol, A., 1996. Direction finding after blind identification of sources steering vectors: the blind-MAXCOr and blind-MUSIC methods. In: Proc. EUSIPCO, pp. 2097–2100.

Chevalier, P., Ferreol, A., Albera, L., 2006. High resolution direction finding from higher order statistics: the $2q$-MUSIC algorithm. IEEE Trans. Signal Process. 54 (8), 2986–2997.

Chevalier, P., Ferreol, A., Albera, L., Birot, G., 2007. Higher order direction finding from arrays with diversely polarized antennas: the PD-2q-MUSIC algorithms. IEEE Trans. Signal Process. 55 (11), 5337–5350.

Chiang, H.H., Nikias, C.L., 1989. The ESPRIT algorithm with high order statistics. In: Proc. Workshop on Higher Order Statistics, pp. 163–168.

Comon, P., 1994. Independent component analysis—a new concept? Signal Process. 36 (3), 287–314 (special issue on higher-order statistics).

Compton Jr., R.T., 1988. Adaptive Antenna—Concepts and Performance. Prentice-Hall.

Dandawate, A.V., Giannakis, G.B., 1995. Asymptotic theory of mixed time averages and Kth-order cyclic-moment and cumulant statistics. IEEE Trans. Inf. Theory 41(1), 216–232.

Demeure, C., Chevalier, P., 1998. The smart antennas at Thomson-CSF communications: concepts, implementations, performances, applications. Ann. Telecommun. 53 (11–12), 466–482.

Dogan, M.C., Mendel, J.M., 1995a. Applications of cumulants to array processing—part I: aperture extension and array calibration. IEEE Trans. Signal Process. 43 (5), 1200–1216.

Dogan, M.C., Mendel, J.M., 1995b. Applications of cumulants to array processing—part II: non-Gaussian noise suppression. IEEE Trans. Signal Process. 43 (7), 1663–1676.

Fan, X., Younan, N.H., 1995. Asymptotic analysis of the cumulant-based MUSIC method in the presence of sample cumulant errors. IEEE Trans. Signal Process. 43 (3), 799–802.

Ferrara Jr., E.R., Parks, T.M., 1983. Direction finding with an array of antennas having diverse polarizations. IEEE Trans. Antennas Propag. 31 (2), 231–236.

Ferreol, A., Chevalier, P., 2000. On the behavior of current second and higher order blind source separation methods for cyclostationary sources. IEEE Trans. Signal Process. 48 (6), 1712–1725.

Ferreol, A., Chevalier, P., Albera, L., 2002. Higher order blind separation of non zero-mean cyclostationary sources. In: Proc. EUSIPCO 02, pp. 103–106.

Ferreol, A., Boyer, E., Larzabal, P., 2004a. Low-cost algorithm for some bearing estimation methods in presence of separable nuisance parameters. Electron. Lett. 4015, 966–967.

Ferreol, A., Chevalier, P., Albera, L., 2004b. Second order blind separation of first and second order cyclostationary sources—application to AM, FSK, CPFSK and deterministic sources. IEEE Trans. Signal Process. 52 (4), 845–861.

Ferreol, A., Larzabal, P., Viberg, M., 2006. On the asymptotic performance analysis of subspace DOA estimation in the presence of modeling errors: case of MUSIC. IEEE Trans. Signal Process. 54 (3), 907–920.

Forster, P., Nikias, C., 1991. Bearing estimation in the bispectrum domain. IEEE Trans. Signal Process. 39 (9), 1994–2006.

Friedlander, B., 1990. A sensitivity analysis of the MUSIC algorithm. IEEE Trans. Acoust. Speech Signal Process. 38 (10), 1740–1751.

Germain, P., Maguer, A., Kopp, L., 1989. Comparison of resolving power of array processing methods by using an analytical criterion. In: Proc. ICASSP.

Gönen, E., Mendel, J.M., 1999. Applications of cumulants to array processing—part VI: polarization and direction of arrival estimation with minimally constrained arrays. IEEE Trans. Signal Process. 47 (9), 2589–2592.

Gönen, E., Mendel, J.M., Dogan, M.C., 1997. Applications of cumulants to array processing—part IV: direction finding in coherent signals case. IEEE Trans. Signal Process. 45 (9), 2265–2276.

Jacovitti, G., Scarano, G., 1994. Hybrid nonlinear moments in array processing and spectrum analysis. IEEE Trans. Signal Process. 42 (7), 1708–1718.

Kaveh, M., Barabell, A.J., 1986. The statistical performance of the MUSIC and the minimum norm algorithms in resolving plane waves in noise. IEEE Trans. Acoust. Speech Signal Process. 34 (4), 331–341.

Li, F., Vaccaro, R.J., 1992. Sensitivity analysis of DOA estimation algorithms to sensor errors. IEEE Trans. Aerosp. Electron. Syst. 28 (3), 708–717.

Liu, T.H., Mendel, J.M., 1999a. Cumulant-based subspace tracking. Signal Process. 76 (3), 237–252.

Liu, T.H., Mendel, J.M., 1999b. Applications of cumulants to array processing—part V: sensitivity issues. IEEE Trans. Signal Process. 47 (3), 746–759.

Liu, J., Huang, Z., Zhou, Y., 2008. Extended 2q-MUSIC algorithm for noncircular signals. Signal Process. 88, 1327–1339.

McCullagh, P., 1987. Tensor Methods in Statistics. Monographs on Statistics and Applied Probability. Chapman and Hall.

Mendel, J.M., 1991. Tutorial on higher-order statistics (spectra) in signal processing and system theory: theoretical results and some applications. Proc. IEEE 79 (3), 278–305.

Mosher, J.C., Leahy, R.M., 1999. Source localization using recursively applied and projected (RAP) MUSIC. IEEE Trans. Signal Process. 47 (2), 332–340.

Oh, S.K., Un, C.K., 1993. A sequential estimation approach for performance improvement of eigenstructure-based methods in array processing. IEEE Trans. Signal Process. 41, 457–463.

Paulraj, A., Kailath, T., 1986. Eigenstructure methods for direction of arrival estimation in the presence of unknown noise field. IEEE Trans. Acoust. Speech Signal. Process. 34 (1), 13–20.

Paulraj, A., Roy, R., Kailath, T., 1986. A subspace rotation approach to signal parameter estimation. Proc. IEEE 74 (7), 1044–1045.

Picinbono, B., 1994. On circularity. IEEE Trans. Signal Process. 42 (12), 3473–3482.

Porat, B., Friedlander, B., 1988. Analysis of the asymptotic relative efficiency of the MUSIC algorithm. IEEE Trans. Acoust. Speech Signal Process. 36 (4), 532–544.

Porat, B., Friedlander, B., 1991. Direction finding algorithms based on higher order statistics. IEEE Trans. Signal Process. 39 (9), 2016–2024.

Priestley, M.B., 1981. Spectral Analysis and Time. Academic Press.

Proakis, J.G., 1995. Digital Communications, Third ed. McGraw-Hill.

Roy, R., Kailath, T., 1989. ESPRIT–Estimation of signal parameters via rotational invariance techniques. IEEE Trans. Acoust. Speech Signal Process. 37 (7), 984–995.

Scarano, G., Jacovitti, G., 1996. Applications of generalized cumulants to array processing. Signal Process. 53, 179–193.

Schmidt, R.O., 1986. Multiple emitter location and signal parameter estimation. IEEE Trans. Antennas Propag. 34 (3), 276–280.

Shamsunder, S., Giannakis, G.B., 1993. Modeling of non-Gaussian array data using cumulants: DOA estimation of more sources with less sensors. Signal Process. 30 (3), 279–297.

Shamsunder, S., Giannakis, G.B., 1994. Signal selective localization of non-Gaussian cyclostationary sources. IEEE Trans. Signal Process. 42 (10), 2860–2864.

Stoica, P., Nehorai, A., 1989. MUSIC, maximum likelihood and Cramer Rao bound. IEEE Trans. Acoust. Speech Signal Process. 37 (5), 720–741.

Stoica, P., Nehorai, A., 1990. MUSIC, maximum likelihood and Cramer Rao bound: further results and comparisons. IEEE Trans. Acoust. Speech Signal Process. 38 (12), 2140–2150.

Stoica, P., Handel, P., Nehorai, A., 1995. Improved sequential MUSIC. IEEE Trans. Aerosp. Electron. Syst. 31 (4), 1230–1239.

Swami, A., Mendel, J., 1991. Cumulant-based approach to the harmonic retrieval and related problems. IEEE Trans. Signal Process. 39 (5), 1099–1109.

Swindlehurst, A.L., Kailath, T., 1992. A performance analysis of subspaced-based methods in the presence of model errors, part I: the MUSIC algorithm. IEEE Trans. Signal Process. 40 (3), 1758–1773.

Yuen, N., Friedlander, B., 1996. Asymptotic performance analysis of ESPRIT, higher order ESPRIT and virtual ESPRIT algorithms. IEEE Trans. Signal Process. 44 (10), 2537–2550.

Yuen, N., Friedlander, B., 1997. DOA estimation in multipath: an approach using fourth-order cumulants. IEEE Trans. Signal Process. 45 (5), 1253–1263.

Yuen, N., Friedlander, B., 1998. Performance analysis of higher order ESPRIT for localization of near-field sources. IEEE Trans. Signal Process. 46 (3), 709–719.

Chapter 9

Source and Node Localization in Sensor Networks

Chiao-En Chen, Kung Yao

9.1 INTRODUCTION

The technologies of information processing in the last 50 years that made the Internet and the Web possible, including modern microelectronics, have resulted in low-cost personal computers (PCs), servers, worldwide telecommunication, and computer-networking infrastructure. In the last 10 years, sensor networking has combined the technology of modern microelectronic sensors, embedded computational processing systems, and modern computer and wireless networking methodologies. It is believed that sensor networking in the 21st century will be equally significant by providing measurement of spatial–temporal physical phenomena, leading to a better understanding and utilization of this information in a wide range of applications. Sensor networking will be able to bring a finer-grained and fuller measurement (using acoustic, seismic, magnetic, infrared (IR), imaging, and video data) to characterize the world to be processed and communicated, so that decision makers can utilize the information to take actions in near real time.

Localization is the determination of the location of an object, whether that of one or more source(s) emitting some energy of interest (e.g., radio frequency (RF), acoustic, ultrasonic, optical, IR, seismic, thermal) or a sensor node in the sensor network (SN). It is a major issue of interest for many applications. Often, the localization of a source is of primary interest, while the localization of the sensor node is needed to support other functions, including one or more source(s). For example, the Federal Communications Commission (FCC) has mandated the E911 system (Europe has the E112 system), requiring cellular telephone providers to determine the location of a cell phone user (in an emergency mode) to tens of meters. In various military scenarios, the localization of a vehicle; in automated manufacturing, the location of a robotic device; and in bio-complexity studies, the locations of birds and animals, are of great interest.

Copyright © 2009 by Academic Press, Inc. All rights of reproduction in any form reserved.

Before discussing source and node localizations in an SN, it is important to review some history. The SN as a concept and in realization appeared only around 2000 as the result of the accumulation of enabling technologies in the last 50 years. The concept of a programmable digital computer was originated in the 1940s. In the 1950s, mainframe electronic digital computers were built—they were expensive and were only available in a few educational, governmental, and commercial research organizations. At this time, basic concepts of digital communication also became known. In the 1960s, mini-computers became popular and digital computations were made available to more users. In that period, satellite and terrestrial microwave communication made the transmission of large amounts of digital data possible. The concept of data transmission over a network of many nodes distributed over large areas was pioneered by researchers of the Arpanet. In the 1970s, microprocessors significantly reduced the cost of digital computations, and the availability of low-cost DSP chips made digital processing possible for many applications. Commercial and military communication and computer networks spread around the world.

In the 1980s, PCs appeared, and the beginning of the Internet allowed researchers at a few research and large commercial organizations to easily communicate. In the 1990s, optical communication networks and the availability of the Web browser allowed the explosive growth of worldwide communications among individuals through the Internet. In this period, advances in embedded processors and wireless communication technology led to the creation of ad hoc networks and explosive worldwide usage of cellular telephony. Around 2000, with all of these available technologies, sensor networking was made possible. In recent years, the interest in SN at the research and development level has resulted in many papers (Gharavi and Kumar, 2003; Iyengar and Kumar, 2002; Kumar et al., 2002; Lu et al., 2006; Pottie et al., 2005; Yao et al., 2003, 2005), books (Bulusu and Jha, 2005; Estrin et al., 2001; Ilyas and Mahgoub, 2005; Iyengar and Brooks, 2004; Karl and Willig, 2005; Krishnamachari, 2006; Lesser, 2003; Pottie and Kaiser, 2005; Raghavendra et al., 2004; Sohraby et al., 2007; Stojmenovic, 2005; Zhao and Guilas, 2004), conferences/workshop proceedings (ACM International Workshop on Wireless Sensor Networks; IEEE and ACM International Symposium on Information Processing in Sensor Networks (IPSN), International Conference on Embedded Networked Sensor Systems (SenSys)), and new SN journals (*ACM Transactions on Sensor Networks*; *International Journal of Ad Hoc and Sensor Wireless Networks*; *International Journal of Distributed Sensor Networks*; *International Journal of Sensor Networks*).

A sensor network consists of dozens/hundreds/thousands of nodes (possibly randomly distributed), each with a sensor (acoustic, seismic, magnetic, chemical, image, video, temperature, etc.), a low-power embedded processor (of varying processing capability), a radio (e.g., a low-power transceiver of varying capability and range), a battery, often of limited energy and size, and a program controlling one or more nodes and possibly some parts of the network to perform some given task. Constraints from all factors (distribution of the sensors in unfortunate locations, limited processing capability, bad propagation conditions, nonuniform energy use among the nodes resulting in the early death of some, nonrobust networking protocols and network management software, etc.) can limit the proper operation of an SN and its ability to perform source and node localization.

Another important factor in source localization is the physical nature of the source. In radar and wireless communications, the information signal is modulated on some high RF frequency, f_0, for efficient transmission. In general, the bandwidth of the signal over $[0, f_s]$ is much less than the RF frequency. Thus, the ratio of the highest to lowest transmitted frequency, $(f_0 + f_s)/(f_0 - f_s)$, is typically near unity. For the 802.11b ISM wireless local area network (LAN) system, for example, the ratio is 2.4835 GHz/2 GHz $= 1.03$. These waveforms are denoted as narrowband. Narrowband waveforms have a well-defined nominal wavelength, and time delays can be compensated by simple phase shifts. The conventional narrowband beamformer operating on these waveforms is merely a spatial extension of the matched filter. In classical time-domain filtering, the time-domain signal is linearly combined with filtering weight to achieve desired high/low/bandpass filtering. The narrowband beamformer also combines the spatially distributed collected sensor array data linearly with the beamforming weight to achieve spatial filtering. Beamforming enhances the signal from the desired spatial direction and reduces the signal(s) from other direction(s) in addition to possible time–frequency filtering.

The movement of persons, cars, trucks, wheeled/tracked vehicles, and vibrating machinery can all generate acoustic or seismic waveforms. The processing of seismic/vibrational sensor data is similar to that of acoustic sensors except for the propagation's medium and unknown speed. For acoustic/seismic waveforms, the ratio of the highest to lowest frequencies can be several octaves. For audio waveforms (i.e., 30 Hz–15 KHz), the ratio is about 500, and these waveforms are denoted as wideband. Dominant acoustical waveforms generated from wheeled/tracked vehicles may range from 20 Hz to 2 KHz, resulting in a ratio of about 100. Similarly, dominant seismic waveforms generated from wheeled vehicles may range from 5 Hz to 500 Hz, also resulting in a ratio of about 100. Thus, the acoustic and seismic signals of interest are generally wideband. However, even for certain RF applications, the ratio of the highest to lowest frequencies can also be considerably greater than unity. For wideband waveforms there is no characteristic wavelength and time delays must be obtained by interpolation.

When an acoustic or seismic source is located close to the sensors, the wavefront of the received signal is curved and the curvature depends on distance; the source is thus said to be in the near field. As the distances become large, the wavefront is planar and parallel; the source is then said to be in the far field. For a far-field source, only the direction-of-arrival (DOA) angle in the coordinate system of the sensors is observable to characterize the source. Consider an ideal (noise-free) linear beamforming array with uniform inter-sensor spacing. For a far-field source, any adjacent sensors receive identical waveforms with only an equal time delay, which can estimate the DOA of the source. For a near-field source, the collection of all relative time delays and the propagation speed of the source can be used to determine the source location.

In general, wideband beamforming is considerably more complex than narrowband beamforming. Thus, acoustic source localization and beamforming problem is challenging because of its wideband nature, near- and far-field geometry (relatively near/far distance of the source from the sensor array), and arbitrary array shape.

The presentation in this chapter is as follows. The source localization problem is treated in Section 9.2, in which localization using trilateration and multilateration is first introduced in Section 9.2.1. The time-difference-of-arrival (TDOA) source localization method using the maximum-power blind beamforming and least squares is presented subsequently in Section 9.2.2. In Section 9.2.3, we briefly introduce the maximum likelihood formulation of the source localization problem and its theoretical performance bound. In Section 9.2.4, we introduce the energy-based source localization methodology.

In Section 9.3, we introduce a few approaches that address the node localization problem. Noise-free node localization using SDP relaxation is first introduced in Section 9.3.1. Sections 9.3.2 and 9.3.3 present two formulations for cases where the distance measurement is contaminated by noise. Classical multidimensional scaling is introduced in Section 9.3.4. In Section 9.3.5, we describe a distributed algorithm based on the nonlinear Gauss-Seidel algorithm and present a few numerical simulation results.

In Section 9.4, we demonstrate a few experimental results of source localization and compare the performance of several algorithms using the implemented acoustic testbed.

9.2 SOURCE LOCALIZATION METHODS APPLIED TO SENSOR NETWORKS

9.2.1 Trilateration/Multilateration

Trilateration is conceptually the simplest method for performing source localization of a single source based on the range estimates from several sensors. Here we assume each sensor node can estimate the range from the source to itself. There are various ways to make this range estimate in practical scenarios. To find the location of the source, we need to reference it with respect to some coordinate system. Furthermore, we must assume that the locations of the sensor nodes are known with respect to the same coordinate system. The problem of determining the locations of the sensor nodes is referred to as the node localization problem and will be discussed in Section 9.3.

In trilateration, we assume all range-sensing nodes and the source are situated on the x-y plane: range-sensing nodes A, B, and C are located at (x_A, y_A), (x_B, y_B), and (x_C, y_C), respectively, as shown in Figure 9.1. If node A estimates the source to be at range d_A and node B estimates the source at range d_B, then the intersection

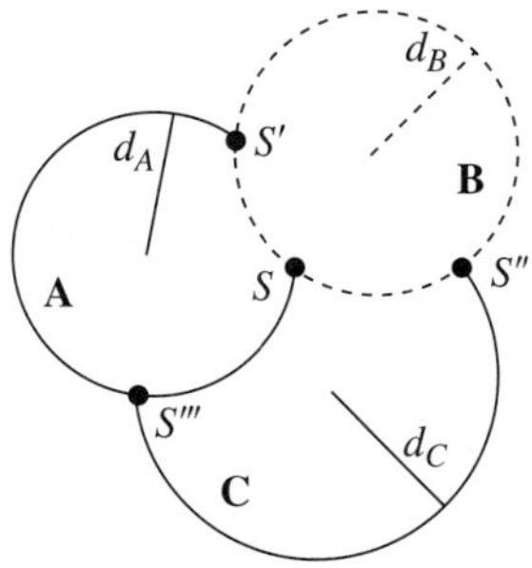

FIGURE 9.1 Trilateration based on three circles.

of these two circles yields two possible ambiguous source locations marked by S and S'. Similarly, if node C estimates the source at range d_C, then the intersection between node B and node C yields two possible ambiguous source locations S and S'', while the intersection of node C and node A yields two possible ambiguous source locations S and S'''. Clearly, the true source location (x, y) is at the intersection of all three circles at S, which satisfies the following equations:

$$(x-x_A)^2+(y-y_A)^2=d_A^2 \tag{9.1}$$

$$(x-x_B)^2+(y-y_B)^2=d_B^2 \tag{9.2}$$

$$(x-x_C)^2+(y-y_C)^2=d_C^2 \tag{9.3}$$

Of course, in the presence of noisy estimations of d_A, d_B, d_C, the three circles will not intersect at a single point to yield a unique location S. As a result, the source location estimate is obtained through a least-squares (LS) fit given by

$$\begin{bmatrix} x \\ y \end{bmatrix} = \begin{bmatrix} 2(x_A-x_C) & 2(y_A-y_C) \\ 2(x_B-x_C) & 2(y_B-y_C) \end{bmatrix}^{-1} \begin{bmatrix} x_A^2-x_C^2+y_A^2-y_C^2+d_C^2-d_A^2 \\ x_B^2-x_C^2+y_B^2-y_C^2+d_C^2-d_B^2 \end{bmatrix} \tag{9.4}$$

In multilateration, N range-sensing nodes with known locations (x_1, y_1), $(x_2, y_2), \ldots, (x_N, y_N)$ are used. Let the distance measurement obtained at each node be $d_1, d_2, \ldots, d_N$. Then the least-squares solution can be obtained similarly by

$$\mathbf{r}_s = \left(\mathbf{A}^T\mathbf{A}\right)^{-1}\mathbf{A}^T\mathbf{b}, \tag{9.5}$$

where

$$\mathbf{A} = \begin{bmatrix} 2(x_1-x_N) & 2(y_1-y_N) \\ \vdots & \vdots \\ 2(x_{N-1}-x_N) & 2(y_{N-1}-y_N) \end{bmatrix}$$

$$\mathbf{b} = \begin{bmatrix} x_1^2-x_N^2+y_1^2-y_N^2+d_N^2-d_1^2 \\ \vdots \\ x_{N-1}^2-x_N^2+y_{N-1}^2-y_N^2+d_N^2-d_{N-1}^2 \end{bmatrix} \tag{9.6}$$

9.2.2 TDOA Source Localization Methods

The problem of determining the location of a source given a set of differential time delays between sensors has been studied for many years. Localization techniques include a two-stage procedure: TDOA estimation and TDOA-based source localization. Closed-form solutions can be derived for the second stage using spherical interpolation (SI) (Schau and Robinson, 1987; Smith and Abel, 1987), hyperbolic intersection (Chan and Ho, 1994), or linear intersection (Brandstein et al., 1997). However, these techniques all require knowledge of the speed of propagation. Chen et al. (2002c) and Yao et al. (1998) derived both the closed-form LS and constrained least-squares (CLS) solutions for this. It can be shown that when the speed of propagation is known, the LS method of Chen et al. (2002c) and Yao et al. (1998) coincides with the method independently derived later by Huang et al. (2000). Their method is mathematically equivalent to the SI technique with less computational complexity. In theory, these two-stage methods can be used for multiple-source localization as long as the relative time delays can be estimated from the data. However, this remains challenging in practice. As a result, we will only consider single-source TDOA and source localization schemes in this subsection.

In the following paragraphs, we present the maximum-power beamforming method (Yao et al., 1998) for TDOA estimation and the LS and CLS methods for source localization.

Maximum-Power Blind Beamforming

Let there be M sources impinging on an array composed of P sensor elements. Denote the wavefront generated by the Mth source by $s_m(t)$, $m=1,\ldots,M$; then the waveform received by the pth sensor can be expressed as

$$x_p(t)=\sum_{m=1}^{M} s_m\left(t-t_p^{(m)}\right)+\eta_p(t), \quad p=1,\ldots,P \tag{9.7}$$

Here $t_p^{(m)}$ denotes the propagation time from the mth source to the pth sensor, and $\eta_p(t)$ is additive noise with zero mean and variance σ^2. Figure 9.2 shows a simple example where $M=2$ and $P=3$. It is assumed that each sensor has a memory of L taps.

We first consider the impact where only the wavefront from the first source (solid circles) is present. The impact of the second wavefront (dashed circles) is considered later. Let the received waveform be sampled at the rate of $1/T$ samples per second. Without loss of generality, we can let $T=1$. We also assume sensor 1 as our reference point and the furthest away, followed by sensor 2, with sensor 3 closest to the first source in Figure 9.2. It follows that the sampled data vector at the three sensors can be denoted as

$$\mathbf{x}^{(1)}=[x_1(n-n_1),x_1(n-n_1-1),\ldots,x_1(n-n_1-L+1)]^T \tag{9.8}$$

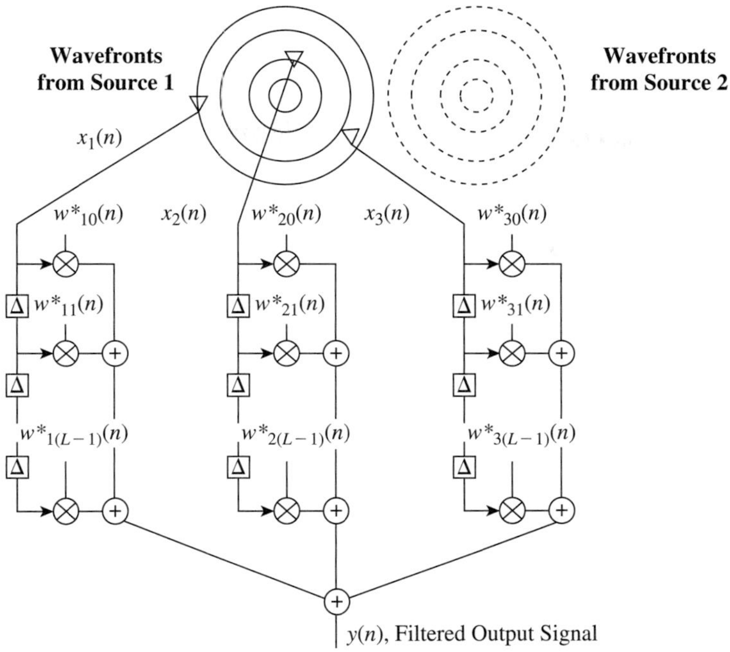

FIGURE 9.2 Beamforming using three randomly distributed sensors with L taps. (*Source:* From Yao et al., 1998, © 1998 IEEE, with permission.)

$$\mathbf{x}^{(2)} = [x_2(n-n_2), x_2(n-n_2-1), \ldots, x_2(n-n_2-L+1)]^T \tag{9.9}$$

$$\mathbf{x}^{(3)} = [x_3(n-n_3), x_3(n-n_3-1), \ldots, x_3(n-n_3-L+1)]^T \tag{9.10}$$

where $n_1 = 0$ and $0 < n_3 < n_2 < L-1$ from our earlier assumptions. Define

$$\mathbf{x} = \left[\mathbf{x}^{(1)T}, \mathbf{x}^{(2)T}, \mathbf{x}^{(3)T}\right]^T \tag{9.11}$$

Then the space–time correlation matrix can be expressed as

$$\mathbf{R}_{3L} = \mathrm{E}\{\mathbf{x}\mathbf{x}^H\} = \begin{bmatrix} \mathbf{R}_L^{11} & \mathbf{R}_L^{12} & \mathbf{R}_L^{13} \\ \mathbf{R}_L^{12H} & \mathbf{R}_L^{11} & \mathbf{R}_L^{23} \\ \mathbf{R}_L^{13H} & \mathbf{R}_L^{23H} & \mathbf{R}_L^{11} \end{bmatrix} \tag{9.12}$$

In general, we want to find an algorithm that generates the beamformer output

$$y(n) = \sum_{p=1}^{P} \sum_{l=0}^{L-1} w_{pl}^{*} x_p\left(n - n_p - l\right) \tag{9.13}$$

which satisfies some desired criterion. The maximum-power beamformer (Yao et al., 1998) chooses the weights such that the power of its output is maximized under the constraint $\|\mathbf{w}_{3L}\|^2 = 1$, where

$$\mathbf{w}_{3L} = \left[w_{10}, w_{11}, \ldots, w_{1(L-1)}, w_{20}, \ldots, w_{2(L-1)}, \ldots, w_{30}, \ldots, w_{3(L-1)}\right]^T \quad (9.14)$$

It is reasonable to expect that the combination corresponding to the largest output power is the one that sums coherently the strongest of the signals, to the disadvantage of the weaker ones. Putting the optimality criterion into matrix form, we have the following maximization problem:

$$\text{maximize } \mathbf{w}_{3L}^H \mathbf{R}_{3L} \mathbf{w}_{3L}, \text{ subject to } \|\mathbf{w}_{3L}\| = 1 \quad (9.15)$$

The unity constraint on the norm of the weights $\mathbf{w}_{3L}$ ensures that the array output noise power is the same as the input noise power, and therefore the maximization of Equation (9.15) is equivalent to the maximization of the signal-to-noise ratio (SNR) at the output of the array. The solution to Equation (9.15) is then given by the eigenvector $\boldsymbol{\phi}_{3L}^{(3L)}$ corresponding to the largest eigenvalue of $\mathbf{R}_{3L}$ in the following eigenvalue problem:

$$\mathbf{R}_{3L} \boldsymbol{\phi}_k^{(3L)} = \lambda_k^{(3L)} \boldsymbol{\phi}_k^{(3L)} \quad (9.16)$$

where $0 \leq \lambda_1^{(3L)} \ldots \leq \lambda_k^{(3L)} \ldots \leq \lambda_{3L}^{(3L)}$. In practice, $\mathbf{R}_{3L}$ is implemented by the time average sample covariance matrix of $\mathbf{x}$. The maximum-power criterion discussed here can be justified from the point of view of Szegö's asymptotic eigenvalue distribution theory (Grenander and Szegö, 1958). A corresponding singular value decomposition (SVD) formulation of this problem has been used but is omitted here. An efficient QLP approximation of the SVD, proposed by Stewart (1999), can also be used to find the weights of the maximum-power (MP) beamformer.

The MP blind beamformer can be used to estimate the relative time delays from the dominant source with and without interference with the sensors. Consider a simulation using a field-measured track vehicle acoustic source with a spectral peak near 100 Hz plus an interferer modeled as a second-order autoregressive (AR) source of coefficients $a_1 = -2 * 0.989 \cos(2\pi * 0.12)$ and $a_2 = 0.989^2$, resulting in a spectral peak near 120 Hz. The x-y coordinates of the three sensors, the tracked vehicle, and the interferer are set to $\{(12, 0), (0, 12), (-9, 0)\}$, $(7, -12)$, and $(6.08, -8.438)$, respectively. The true time delays of the vehicle are at 12, 7, and 5 samples, while those of the interferer are at 11, 7, and 4 samples, among all the sensors. Figure 9.3(a) shows the time delay estimates based on the dominant eigenvector method using the MP blind beamformer; Figure 9.3(b) shows these estimates based on the classical correlation method (Carter, 1987) operating directly on the sensor data. For a signal-to-interference ratio (SIR) higher than approximately 4 dB, the eigenvector method finds the delays of the strongest source (the vehicle) with essentially no error. For an SIR of less than approximately 3 dB, the strongest source is the interferer and its delays are also found with essentially no errors. The estimated delay uncertainty region is only

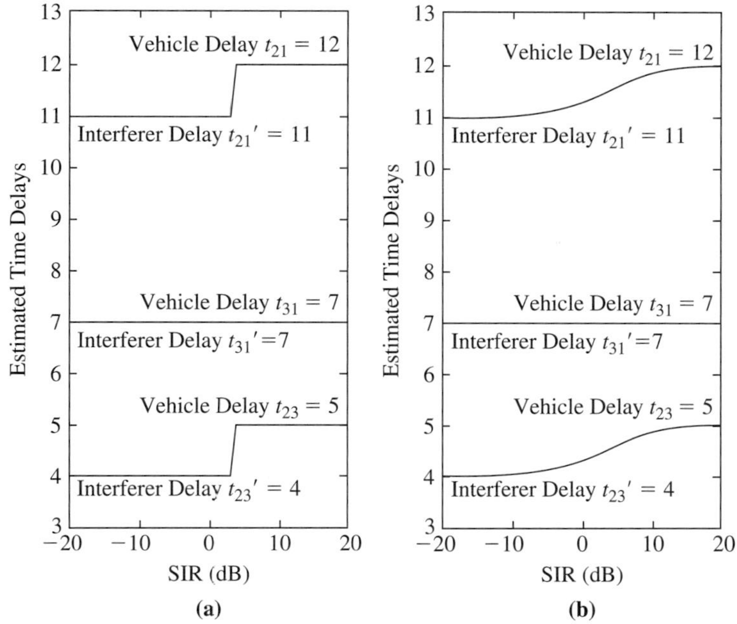

FIGURE 9.3 Comparison of eigenvector (a) and classical (b) correlation methods on time-delay estimation. (*Source:* From Yao et al., 1998, © 1998 IEEE, with permission.)

about 1 dB. Figure 9.3(b) shows that the classical correlation method yields much less accurate estimation results and that the estimated delay uncertainty region is about 20 dB.

Least-Squares and Constrained Least-Squares Source Localization

The problem of determining the location of a source based on a set of TDOA measurements is equivalent to the problem of determining the position of a receiver given equally biased range measurements to a set of satellites (Chafee and Abel, 1994). There are several ways to solve the problem (Bancroft, 1985), but no closed-form solution appears to be available when the propagation speed is unknown. This section introduces an algebraic solution to the problem of estimating both propagation speed and position.

Denote the source location in the Cartesian coordinate of $\mathbf{r}_s = [x_s, y_s, z_s]^T$ and the pth sensor location at $\mathbf{r}_p = [x_p, y_p, z_p]^T$. Without loss of generality, we choose the first sensor as the reference sensor for differential time delays and let the reference sensor be the origin of the coordinate system for simplicity. The speed of propagation v in this formulation can also be estimated from the data. In some problems, v may be considered partially known (e.g., acoustic applications), while in others it may be considered to be unknown (e.g., seismic applications).

Let the relative time delay between the pth sensor and the first sensor be t_{p1}; then

$$t_{p1} = t_p - t_1 = \frac{\|\mathbf{r}_s - \mathbf{r}_p\| - \|\mathbf{r}_s - \mathbf{r}_1\|}{v} \tag{9.17}$$

where $p = 2, \ldots, P$. This is a set of $(P-1)$ nonlinear equations that makes finding the solution for $\mathbf{r}_s$ nontrivial.

To simplify the estimation to a linear problem, we make use of the following relationship:

$$\|\mathbf{r}_s - \mathbf{r}_p\|^2 - \|\mathbf{r}_s\|^2 = \|\mathbf{r}_p\|^2 - 2\,(x_s x_r + y_s y_r + z_s z_r) \tag{9.18}$$

The left side of Equation (9.18) is equivalent to

$$\left(\|\mathbf{r}_s - \mathbf{r}_p\| - \|\mathbf{r}_s\|\right)\left(\|\mathbf{r}_s - \mathbf{r}_p\| + \|\mathbf{r}_s\|\right) = v t_{p1}\left(2\|\mathbf{r}_s\| + v t_{p1}\right) \tag{9.19}$$

in the case of $\mathbf{r}_1 = \mathbf{0}$. By combining both expressions, we have the following linear relation:

$$\mathbf{r}_p^T \mathbf{r}_s + v t_{p1}\|\mathbf{r}_s\| + v^2 t_{p1}^2/2 = \|\mathbf{r}_p\|^2/2, \quad p = 2, \ldots, P \tag{9.20}$$

With P sensors, we formulate the least-squares solution by putting the $(P-1)$ linear equations of Equation (9.20) into the matrix form of

$$\mathbf{A}_1 \mathbf{y}_1 = \mathbf{b} \tag{9.21}$$

where

$$\mathbf{A}_1 = \begin{bmatrix} \mathbf{r}_2^T & t_{21} & t_{21}^2/2 \\ \mathbf{r}_3^T & t_{31} & t_{31}^2/2 \\ \vdots & \vdots & \vdots \\ \mathbf{r}_P^T & t_{P1} & t_{P1}^2/2 \end{bmatrix} \tag{9.22}$$

and

$$\mathbf{y}_1 = \begin{bmatrix} \mathbf{r}_s \\ v\|\mathbf{r}_s\| \\ v^2 \end{bmatrix}, \; \mathbf{b} = \frac{1}{2}\begin{bmatrix} \|\mathbf{r}_2\|^2 \\ \|\mathbf{r}_3\|^2 \\ \vdots \\ \|\mathbf{r}_P\|^2 \end{bmatrix} \tag{9.23}$$

The LS solution for the unknown vector $\mathbf{y}_1$ is simply $\hat{\mathbf{y}}_1 = \mathbf{A}_1^{\dagger}\mathbf{b}$, where $\mathbf{A}_1^{\dagger} = (\mathbf{A}_1^T\mathbf{A}_1)^{-1}\mathbf{A}_1^T$ is the pseudo-inverse of $\mathbf{A}_1$. The source location estimate is then given by the first three elements of $\hat{\mathbf{y}}_1$, and the speed-of-propagation estimate is given by the square root of the last element of $\hat{\mathbf{y}}_1$.

In the three-dimensional localization problem, there are five unknowns in $\mathbf{y}_1$ and therefore at least six sensors are required to obtain an overdetermined solution. However, placing sensors randomly does not provide much assurance against ill-conditioned solutions. The preferred approach is to use seven or more sensors, yielding six or more relative delays, and to perform a least-squares fitting to the data. For a two-dimensional localization problem, the minimum number of sensors can be reduced by one. If the propagation speed v is known, then the minimum number of sensors can be further reduced by one.

Note that the speed of propagation can also be estimated by

$$\hat{v} = \frac{\widehat{v\|\mathbf{r}_s\|}}{\|\hat{\mathbf{r}}_s\|} \tag{9.24}$$

using the fourth and first three elements of $\hat{\mathbf{y}}_1$. To exploit this relationship, we can add another nonlinear constraint to ensure the equivalence between the speed of propagation estimates from the fourth and fifth elements of $\hat{\mathbf{y}}_1$. We first rewrite Equation (9.21) as

$$\mathbf{A}_2\mathbf{y}_2 = \mathbf{b} + v^2\mathbf{d} \tag{9.25}$$

where

$$\mathbf{A}_2 = \begin{bmatrix} \mathbf{r}_2^T & t_{21} \\ \mathbf{r}_3^T & t_{31} \\ \vdots & \vdots \\ \mathbf{r}_P^T & t_{P1} \end{bmatrix}, \quad \mathbf{y}_2 = \begin{bmatrix} \mathbf{r}_s \\ v\|\mathbf{r}_s\| \end{bmatrix} \tag{9.26}$$

and

$$\mathbf{b} = \frac{1}{2}\begin{bmatrix} \|\mathbf{r}_2\|^2 \\ \|\mathbf{r}_3\|^2 \\ \vdots \\ \|\mathbf{r}_P\|^2 \end{bmatrix}, \quad \mathbf{d} = -\frac{1}{2}\begin{bmatrix} t_{21}^2 \\ t_{31}^2 \\ \vdots \\ t_{P1}^2 \end{bmatrix} \tag{9.27}$$

The CLS solution for the unknown vector is then given by $\hat{\mathbf{y}}_2 = \mathbf{A}_2^\dagger\mathbf{b} + v^2\mathbf{A}_2^\dagger\mathbf{d}$. Define $\mathbf{p} = \mathbf{A}_2^\dagger\mathbf{b}$ and $\mathbf{q} = \mathbf{A}_2^\dagger\mathbf{d}$; then the estimates for the source location and propagation speed can be expressed as

$$x_s = p_1 + v^2 q_1 \tag{9.28}$$

$$y_s = p_2 + v^2 q_2 \tag{9.29}$$

$$z_s = p_3 + v^2 q_3 \tag{9.30}$$

$$v\|\mathbf{r}_s\| = p_4 + v^2 q_4 \tag{9.31}$$

where p_i and q_i are the ith entries of $\mathbf{p}$ and $\mathbf{q}$, respectively. Using the relation $\|\mathbf{r}_s\|^2 = x_s^2 + y_s^2 + z_s^2$ and Equation (9.31), we obtain the following third-order constraint equation:

$$\alpha(v^2)^3 + \beta(v^2)^2 + \gamma(v^2) + \delta = 0 \tag{9.32}$$

where

$$\alpha = q_1^2 + q_2^2 + q_3^2 \tag{9.33}$$

$$\beta = 2(p_1 q_1 + p_2 q_2 + p_3 q_3) - q_4^2 \tag{9.34}$$

$$\gamma = p_1^2 + p_2^2 + p_3^2 - 2p_4 q_4 \tag{9.35}$$

$$\delta = -p_4^2 \tag{9.36}$$

There exist at most three solutions to the third-order equation. The estimate of propagation speed is then given by the positive square root of the real positive solution of Equation (9.32). If there are more than one positive and real roots, only one of the estimates in the range of feasible propagation speed consistent with the known propagation medium is used. From real-life data, we have not encountered multiple estimates within the range of feasible propagation speed.

Once the speed of propagation is estimated, the source location can be obtained by setting

$$\hat{x}_s = p_1 + \hat{v}^2 q_1 \tag{9.37}$$

$$\hat{y}_s = p_2 + \hat{v}^2 q_2 \tag{9.38}$$

$$\hat{z}_s = p_3 + \hat{v}^2 q_3 \tag{9.39}$$

Compared to the LS method, the minimum required number of sensors in the CLS method is reduced by one.

Cramer-Rao Bound Analysis

The Cramer-Rao bound (CRB) is one of the most well-known theoretical lower bounds on the variances among all unbiased estimators. Here we derive the CRB for the estimation of source location and speed of propagation with respect to the relative time-delay estimation error.

We first denote the unknown parameter vector by $\Psi = [\mathbf{r}_s^T, v]^T$, and model the time-delay measurement as

$$\tilde{\mathbf{t}} = \mathbf{t}(\Psi) + \boldsymbol{\xi} \tag{9.40}$$

where $\mathbf{t} = [t_{21}, \ldots, t_{R1}]^T$ is the true relative time-delay vector and $\boldsymbol{\xi}$ is the estimation error vector, which is assumed to be zero-mean white Gaussian with covariance $\mathbf{R}_{\xi} = \sigma^2 \mathbf{I}$. Define the gradient matrix $\mathbf{H} = \frac{\partial \mathbf{t}}{\partial \Psi^T} = [\frac{\partial \mathbf{t}}{\partial \mathbf{r}_s^T}, \frac{\partial \mathbf{t}}{\partial v}]$; then $\mathbf{H}$ can be expressed as

$$\mathbf{H} = \frac{1}{v}[\mathbf{B}, -\mathbf{t}] \tag{9.41}$$

where $\mathbf{B} = [\mathbf{u}_2 - \mathbf{u}_1, \ldots, \mathbf{u}_P - \mathbf{u}_1]^T$, and $\mathbf{u}_p = (\mathbf{r}_s - \mathbf{r}_p)/\|\mathbf{r}_s - \mathbf{r}_p\|$ denotes the direction of the source from the pth sensor. The Fisher information matrix (Kay, 1993) can then be derived as

$$\mathbf{F}_\Psi = \mathbf{H}^T \mathbf{R}_\xi^{-1} \mathbf{H} \tag{9.42}$$

$$= \frac{1}{\sigma^2 v^2} \begin{bmatrix} \mathbf{B}^T\mathbf{B} & -\mathbf{B}^T\mathbf{t} \\ -\mathbf{t}^T\mathbf{B} & \|\mathbf{t}\|^2 \end{bmatrix} \tag{9.43}$$

The theoretical lower bound of the variances in estimating $\mathbf{r}_s$ is given by the diagonal elements of the leading 3×3 submatrix of the inverse Fisher information matrix

$$\left[\mathbf{F}_\Psi^{-1}\right]_{11:33} = \sigma^2 v^2 \left[\mathbf{B}^T \mathbf{P}_\mathbf{t}^\perp \mathbf{B}\right]^{-1} \tag{9.44}$$

where $\mathbf{P}_\mathbf{t}^\perp = \mathbf{I} - \mathbf{t}\mathbf{t}^T / \|\mathbf{t}\|^2$. The variance bound of the distance d between the estimated source location and the actual location is given by $\sigma_d^2 \geq \text{trace}[\mathbf{F}_\Psi^{-1}]_{11:33}$. For the estimation of propagation speed, the variance bound is given by

$$\sigma_v^2 \geq \left[\mathbf{F}_\Psi^{-1}\right]_{44} = \sigma^2 v^2 \|\mathbf{P}_\mathbf{B}^\perp \mathbf{t}\|^2 \tag{9.45}$$

where $\mathbf{P}_\mathbf{B}^\perp = \mathbf{I} - \mathbf{B}(\mathbf{B}^T\mathbf{B})^{-1}\mathbf{B}^T$.

Analysis shows that the CRBs of both the source location and the propagation speed estimates grow linearly with the variance of the location estimate and v^2. When the speed of propagation is known, the CRB for the source location estimate becomes $\sigma^2 v^2 \text{trace}[(\mathbf{B}^T\mathbf{B})^{-1}]$, which is always smaller than that of the unknown speed of propagation case. We can define $\mathbf{A}_{\mathbf{r}_s} = \mathbf{B}^T\mathbf{B}$ as the *array matrix*, which provides a measure of geometric relations between the source and the sensor array. Poor array geometry may lead to a degeneration in the rank of $\mathbf{A}_{\mathbf{r}_s}$ and thus result in a large CRB.

By transforming to the polar coordinate system in the $2-D$ case, the CRB for the source range and the DOA estimate can also be obtained. Denote $r_s = \sqrt{x_s^2 + y_s^2}$ as the distance between the source and the reference position such as the array centroid; then the DOA is given by $\phi_s = \tan^{-1}(x_s/y_s)$ with respect to the y-axis. The time delay from the source to the pth sensor can be expressed as $t_p = \sqrt{r_s^2 + r_p^2 - 2r_s r_p \cos(\phi_s - \phi_p)}/v$, where (r_p, ϕ_p) is the location of the pth sensor in the polar coordinate system. The gradient matrix $\mathbf{H}$ can be computed as

$$\mathbf{H} = \left[\frac{\partial \mathbf{t}}{\partial r_s}, \frac{\partial \mathbf{t}}{\partial \phi_s}, \frac{\partial \mathbf{t}}{\partial v}\right] = \frac{1}{v}\left[\mathbf{B}_{pol}, -\mathbf{t}\right] \tag{9.46}$$

where

$$\mathbf{B}_{pol} = v\left[\frac{\partial \mathbf{t}}{\partial r_s}, \frac{\partial \mathbf{t}}{\partial \phi_s}\right] \tag{9.47}$$

$$\frac{\partial t_{p1}}{\partial r_s} = \frac{1}{v}\left[\frac{r_s - r_p\cos(\phi_s - \phi_p)}{\|\mathbf{r}_s - \mathbf{r}_p\|} - \frac{r_s - r_1\cos(\phi_s - \phi_1)}{\|\mathbf{r}_s - \mathbf{r}_1\|}\right] \tag{9.48}$$

$$\frac{\partial t_{p1}}{\partial \phi_s} = \frac{1}{v}\left[\frac{r_s r_p \sin(\phi_s - \phi_p)}{\|\mathbf{r}_s - \mathbf{r}_p\|} - \frac{r_s r_1 \sin(\phi_s - \phi_1)}{\|\mathbf{r}_s - \mathbf{r}_1\|}\right] \tag{9.49}$$

The leading 2×2 submatrix of the inverse polar Fisher information matrix is then given by

$$\left[\mathbf{F}_{\Psi_{pol}}^{-1}\right]_{11:22} = \sigma^2 v^2 \left[\mathbf{B}_{pol}^T \mathbf{P}_{\mathbf{t}}^{\perp} \mathbf{B}_{pol}\right]^{-1} \tag{9.50}$$

and the CRBs of the range and DOA estimates can then be obtained as $[\mathbf{F}_{\Psi_{pol}}^{-1}]_{11}$ and $[\mathbf{F}_{\Psi_{pol}}^{-1}]_{22}$, respectively.

To evaluate the performance of the LS and CLS source localization algorithms, a simulated scenario, depicted in Figure 9.4(a), is used. A randomly distributed array of seven sensors collects the signal generated by a source moving on a straight line. By perturbing the actual time delays by a white Gaussian noise with zero mean and standard deviation $\sigma_{td} = 10\,\mu$sec, the LS and CLS algorithms are applied to estimate the source location at each time frame. As depicted in Figure 9.4(b), the CLS yields much better range and DOA estimates than the LS on an average of 10, 000 random realizations. The CLS estimates are also very close to the corresponding CRB.

The speed of propagation estimation results for $\sigma_{td} = 10\,\mu$ and $50\,\mu$sec are compared in Figure 9.5(a). The CLS is again dramatically better than the LS because of the additional nonlinear constraint equation. At a fixed source location (10, 10), the range and DOA estimation errors are plotted for different levels of relative time-delay measurement error. We consider the same array of seven randomly distributed sensors, and then take away one sensor and use only the remaining six for comparison. The range and DOA estimation errors are linearly proportional to the relative time-delay error, as depicted in Figure 9.5(b). The CLS is again shown to attain the CRB and to be much better than the LS. Performance also improves when an additional sensor is added to the overdetermined solution.

9.2.3 Maximum Likelihood Source Localization

Maximum Likelihood Formulation

In this subsection, we present the maximum likelihood solution (Bohme, 1986; Chen et al., 2002a; and Ziskind and Wax, 1988) for both near-field and far-field source localization. Consider the problem of localizing M wideband sources using a P-element arbitrarily distributed array ($M < P$). The sampled waveform received at the pth sensor can be described by the following time-domain model:

$$x_p(n) = \sum_{m=1}^{M} a_p^{(m)} s_0^{(m)}\left(n - n_p^{(m)}\right) + \eta_p(n), \quad n = 1, \ldots, L,\ p = 1, \ldots, P \tag{9.51}$$

Here $a_p^{(m)}$ denotes the signal level of the mth source at the pth sensor, $s_0^{(m)}(n)$ is the signal of the mth source at the reference point, and $n_p^{(m)}$ is the relative time delay (in samples) of the mth source at the pth sensor with respect to the reference point.

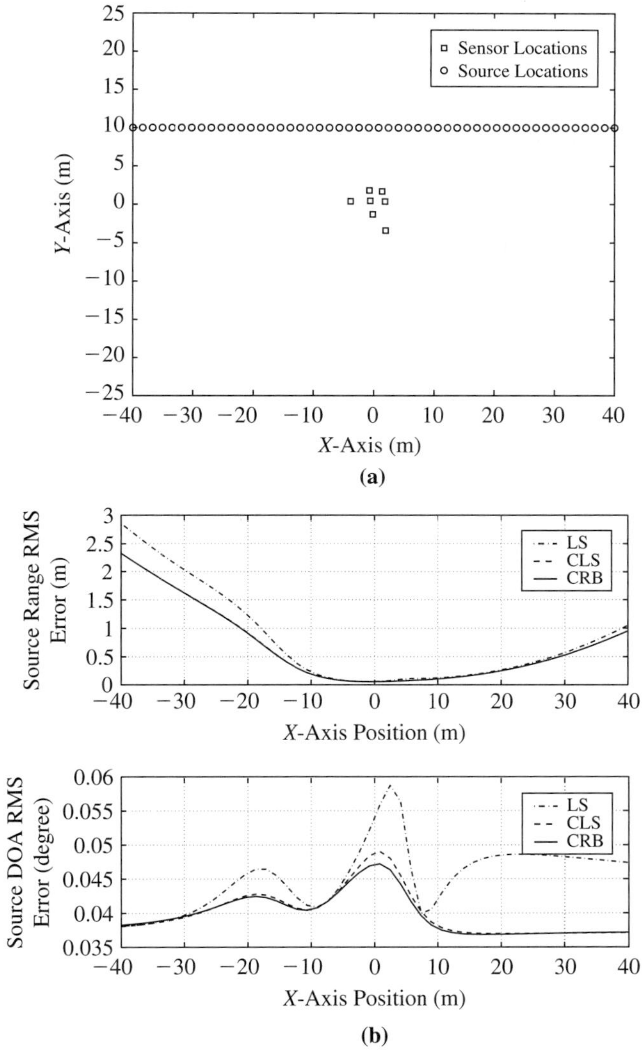

FIGURE 9.4 (a) Traveling source scenario and sensor scenario. (b) Range estimate (*top*) and DOA estimate (*bottom*) versus source locations ($\sigma_{td} = 10\,\mu$sec). (*Source:* From Chen et al., 2002c, © 2002 Sage, with permission.)

The sensor noise $\eta_p(n)$ is modeled as a zero-mean, i.i.d. Gaussian process with variance σ^2. A block of L samples in each sensor's data can be transformed to the frequency domain using the discrete-time Fourier transform, which results in the following array signal model:

$$\mathbf{X}(\omega) = \mathbf{D}(\omega)\mathbf{S}_0(\omega) + \boldsymbol{\eta}(\omega), \quad -\pi < \omega < \pi \tag{9.52}$$

Here $\mathbf{D}(\omega)$ is the steering matrix at frequency ω, and $\boldsymbol{\eta}(\omega)$ is the noise spectrum vector. Under certain assumptions (Chen et al., 2002a) and using the discrete Fourier transform (DFT)/fast Fourier transform (FFT) implementation,

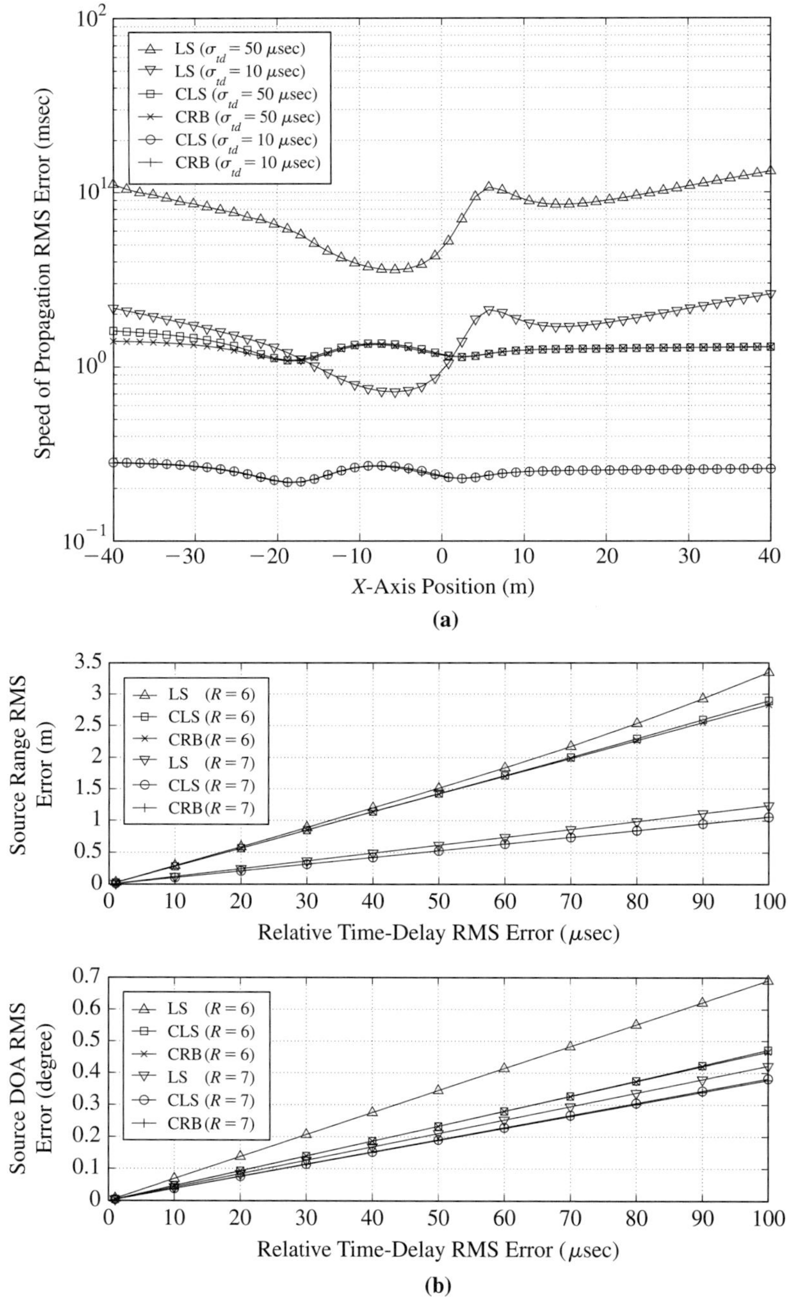

FIGURE 9.5 (a) RMSE speed of propagation versus source locations. (b) Range estimate (*top*) and DOA estimate (*bottom*) versus source locations ($\mathbf{r}_S = [10, 10]^T$). (*Source:* From Chen et al., 2002c, © 2002 Sage, with permission.)

the maximum likelihood estimate for the considered source localization problem can be obtained by solving

$$\hat{\mathbf{r}}_s = \arg\max_{\mathbf{r}_s} J(\mathbf{r}_s) = \arg\max_{\mathbf{r}_s} \sum_k \|\mathbf{P}_{\mathbf{D}(\omega_k)}\mathbf{X}(\omega_k)\|^2 \tag{9.53}$$

where $\mathbf{r}_s$ is the source location vector, and $\mathbf{P}_{\mathbf{D}(\omega_k)} = \mathbf{D}(\omega_k)[\mathbf{D}^H(\omega_k)\mathbf{D}(\omega_k)]^{-1}$ $\mathbf{D}^H(\omega_k)$.

Figure 9.6(a) shows the simulation result of the near-field normalized $J(\mathbf{r}_s)$ metric of a vehicle source having high values about its location inside the convex hull of an array of five sensors. Thus, for near-field scenarios this metric is capable of estimating the source location. Figure 9.6(b) shows the simulation result of the far-field normalized $J(\mathbf{r}_s)$ metric of a vehicle source having high values in

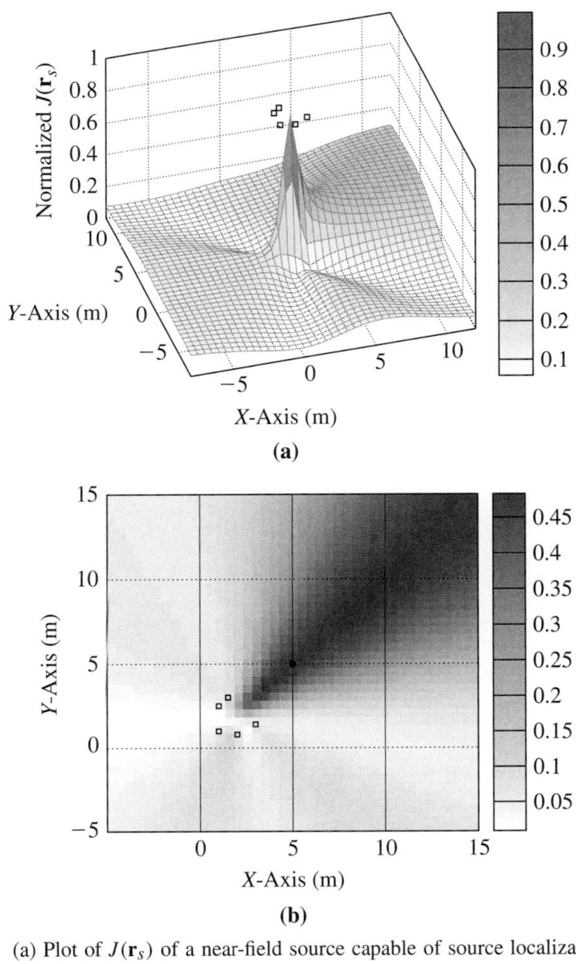

FIGURE 9.6 (a) Plot of $J(\mathbf{r}_s)$ of a near-field source capable of source localization. (b) Plot of $J(\mathbf{r}_s)$ of a far-field source capable of DOA source estimation.

the angular sector in the direction of the source. Thus, for far-field scenarios, this metric is capable of estimating the source DOA. The details of this algorithm as well as other simulation and field-measured estimated results can be found in Chen et al. (2002b). Note that the set of parameters to estimate can be augmented with the propagation velocity and noise variance, if these are unknown.

This algorithm is fairly computationally complex, mostly because of the maximization of a function, $J(\mathbf{r}_s)$, that is usually nonconvex and displays a large number of local maxima. Moreover, all the sensors that participate in the estimate need to be synchronized and fully collaborative. The algorithm is centralized—that is, the relevant information has to be sent to a central node provided with sufficient computational power. These are issues in networks of low-power sensors, but they may not be a problem in environment-monitoring networks where nodes are more complex and may have higher available energy supplies.

Cramer-Rao Bound Analysis

In this subsection, we derive the Cramer-Rao bound for the DOA estimation of wideband sources in the far-field scenario. The Cramer-Rao bound for source location estimation can be derived under the same concept, but the analytical expression is more complicated and therefore will not be presented here.

Consider the following far-field array signal model

$$\mathbf{X}(\omega_k) = \mathbf{D}(\omega_k)\mathbf{S}_0(\omega_k) + \boldsymbol{\eta}(\omega_k), \quad k = 1, \ldots, L \tag{9.54}$$

where $\mathbf{D}(\omega_k) = [\mathbf{d}^{(1)}(\omega_k), \ldots, \mathbf{d}^{(M)}(\omega_k)]$, and $\boldsymbol{\eta}(\omega_k)$ is modeled as i.i.d. zero-mean Gaussian with unknown covariance $\mathbf{R}_\eta = \sigma^2\mathbf{I}$. Define $\boldsymbol{\psi} = [\boldsymbol{\theta}^T, \mathbf{S}_0^T, \sigma^2]^T$ as the vector of unknown parameters, where $\boldsymbol{\theta} = [\theta_1, \ldots, \theta_M]^T$ denotes the DOAs of the M sources, and $\mathbf{S}_0 = [\mathbf{S}_0^T(\omega_1), \ldots, \mathbf{S}_0^T(\omega_L)]^T$. Using Equation (9.54), the (i,j) element of the Fisher information matrix for the estimation of $\boldsymbol{\psi}$ can be computed as

$$[\mathbf{F}_{\boldsymbol{\psi}}]_{i,j} = \frac{N}{2} tr\left\{\mathbf{R}_\eta^{-1}\frac{\partial \mathbf{R}_\eta}{\partial \psi_i}\mathbf{R}_\eta^{-1}\frac{\partial \mathbf{R}_\eta}{\partial \psi_j}\right\} + 2\Re\left\{\frac{\partial \mathbf{b}^H}{\partial \psi_i}\left(\mathbf{I}_{N/2} \otimes \mathbf{R}_\eta^{-1}\right)\frac{\partial \mathbf{b}}{\partial \psi_j}\right\} \tag{9.55}$$

where

$$\mathbf{b} = \left[\mathbf{S}_0(\omega_1)^T\mathbf{D}(\omega_1)^T, \ldots, \mathbf{S}_0(\omega_L)^T\mathbf{D}(\omega_L)^T\right]^T \tag{9.56}$$

After some mathematical manipulations, the Cramer-Rao bound expression (Chen et al., 2007)

$$\mathbf{CRB}_{\boldsymbol{\theta}\boldsymbol{\theta}} = \frac{\sigma^2}{2}\Re\left[\sum_k \left[\mathbf{E}(\omega_k)^H \mathbf{P}^{\perp}_{\mathbf{D}(\omega_k)}\mathbf{E}(\omega_k)\right] \odot \mathbf{R}_{\mathbf{S}}(\omega_k)^T\right]^{-1} \tag{9.57}$$

can be obtained, where $\mathbf{E}(\omega_k) = [\frac{d}{d\theta_1}\mathbf{d}^{(1)}(\omega_k), \ldots, \frac{d}{d\theta_M}\mathbf{d}^{(M)}(\omega_k)]$, and $\mathbf{R}_{\mathbf{S}}(\omega_k) = \mathbf{S}_0(\omega_k)\mathbf{S}_0(\omega_k)^H$.

The CRB expression in Equation (9.57) contains contributions from all frequency bins through a direct summation. The contribution from each frequency bin is an element-wise matrix product of the geometry factor, $\mathbf{E}(\omega_k)^H \mathbf{P}^{\perp}_{\mathbf{D}(\omega_k)} \mathbf{E}(\omega_k)$, and the spectral factor, $\mathbf{R_S}(\omega_k)$. The geometry factor provides a measure of geometric relations between the sources and the sensor array, while the spectral factor provides a measure of correlation among M sources. For frequency bins where no signals are present, the spectral factors are just zero matrices and thus do not contribute to $\mathbf{CRB}_{\theta\theta}$, as is intuitively expected. We note that the CRB in Friedlander and Weiss (1993) is a stochastic CRB, in which the narrowband signal at each frequency bin is characterized by its covariance function; in the CRB of Chen et al. (2007), as shown in Equation (9.57), we treated the narrowband signal at each frequency bin as deterministic and unknown. A general CRB method for source localization based on the one-step approach (jointly process all the sensor data) and the two-step approach (measure signal parameters, then infer source location) were proposed (Moore et al., 2008).

9.2.4 Energy-Based Source Localization

One of the key requirements of TDOA-based and many other source localization algorithms is synchronization among sensor nodes. However, accurate synchronization requires extra energy consumption and communication bandwidth, which can be costly for wireless sensor networks. Recently, energy-based approaches were proposed for acoustic source localization to relax such constraints (Li et al., 2002). This is possible because the acoustic power emitted by targets such as vehicles usually varies slowly with time and therefore a coarser sampling time can be used to relieve the burden of accurate synchronization. In this subsection, we summarize the main results of Sheng and Hu (2005). For more details, refer to Li et al. (2002), Li and Hu (2003), and Sheng and Hu (2005).

Let P be the number of the sensors and M the number of acoustic sources in the sensor field. Then the received signal measured on the pth sensor over the time interval n can be expressed as

$$x_p(n) = s_p(n) + \nu_p(n) \tag{9.58}$$

where

$$s_p(n) = \gamma_p \sum_{m=1}^{M} \frac{a^{(m)}\left(n - t_p^{(m)}\right)}{\|\boldsymbol{\rho}^{(m)}\left(n - t_p^{(m)}\right) - \mathbf{r}_p\|} \tag{9.59}$$

is the signal intensity; $\nu_p(n)$ is the background noise modeled as a zero-mean additive white Gaussian noise with variance ς_p^2; $t_p^{(m)}$ is the propagation delay from the mth source to the pth sensor; and $a^{(m)}(n - t_p^{(m)})$ represents the intensity of the mth acoustic source measured 1 m away from that source and modeled as a random variable uncorrelated with other the intensities of sources. The notations $\boldsymbol{\rho}^{(m)}$, $\mathbf{r}_p$, and γ_p denote the position vector of the mth source, the position vector of the pth sensor, and the sensor gain factor of the pth sensor, respectively.

$s_p(n)$ and $v_p(n)$ are assumed to be uncorrelated, and as a result the acoustic energy received at the pth sensor can be represented as

$$\mathrm{E}\left[s_p^2(n)\right] = \gamma_p^2 \sum_{m=1}^{M} \frac{\mathrm{E}\left[a^{(m)2}\left(n - t_p^{(m)}\right)\right]}{\|\boldsymbol{\rho}^{(m)}\left(n - t_p^{(m)}\right) - \mathbf{r}_p\|^2} \tag{9.60}$$

$$= g_p \sum_{m=1}^{M} \frac{\mathcal{E}_s^{(m)}\left(n - t_p^{(m)}\right)}{\|\boldsymbol{\rho}^{(m)}\left(n - t_p^{(m)}\right) - \mathbf{r}_p\|^2} \tag{9.61}$$

where $g_p = \gamma_p^2$, and $\mathcal{E}_s^{(m)}(n - t_p^{(m)}) = \mathrm{E}[a^{(m)2}(n - t_p^{(m)})]$. In practice, the ensemble average is implemented using time average over a time window $T = N/F_s$, where N is the number of samples and F_s is the sampling frequency. Under some practical assumptions (Sheng and Hu, 2005), the average energy measurement over the time window $[t - \frac{N}{2F_s}, t + \frac{N}{2F_s}]$ can be modeled as

$$y_p(t) = g_p \sum_{m=1}^{M} \frac{\mathcal{E}_s^{(m)}(t)}{\left[d_p^{(m)}(t)\right]^2} + \epsilon_p(t) \tag{9.62}$$

where $d_p^{(m)}(t) = \|\boldsymbol{\rho}^{(m)}(t) - \mathbf{r}_p\|$. Note that v_p^2 has an χ^2 distribution with mean ς_p^2 and variance $2\varsigma_p^4$. When N is sufficiently large, ($N \gg 30$), $\epsilon_p(t)$ can be approximated by a normal distribution, $\mathcal{N}(\mu_p, \sigma_p^2)$, where $\mu_p = \varsigma_p^2$ and $\sigma_p^2 = 2\varsigma_p^4/N$. Using Equation (9.62) and omitting the time index t for brevity, the energy-based source localization problem can be formulated using the maximum likelihood criterion

$$\hat{\boldsymbol{\psi}} = \arg\min_{\boldsymbol{\psi}} \ell(\boldsymbol{\psi}) \tag{9.63}$$

where

$$\boldsymbol{\psi} = \left[\boldsymbol{\rho}^T, \boldsymbol{\mathcal{E}}_s{}^T\right]^T \tag{9.64}$$

$$\ell(\boldsymbol{\psi}) = \|\mathbf{Z} - \mathbf{G}\mathbf{D}\boldsymbol{\mathcal{E}}_s\|^2 \tag{9.65}$$

and

$$\mathbf{Z} = \left[\frac{(\mathcal{E}_1 - \mu_1)}{\sigma_1}, \ldots, \frac{(\mathcal{E}_P - \mu_P)}{\sigma_P}\right]^T \tag{9.66}$$

$$\mathbf{G} = \mathrm{diag}\left[\frac{g_1}{\sigma_1}, \ldots, \frac{g_P}{\sigma_P}\right] \tag{9.67}$$

$$\mathbf{D} = \begin{bmatrix} \frac{1}{|d_1^{(1)}|^2} & \frac{1}{|d_1^{(2)}|^2} & \cdots & \frac{1}{|d_1^{(M)}|^2} \\ \frac{1}{|d_2^{(1)}|^2} & \frac{1}{|d_2^{(2)}|^2} & \cdots & \frac{1}{|d_2^{(M)}|^2} \\ \vdots & \vdots & \ddots & \vdots \\ \frac{1}{|d_P^{(1)}|^2} & \frac{1}{|d_P^{(2)}|^2} & \cdots & \frac{1}{|d_P^{(M)}|^2} \end{bmatrix} \tag{9.68}$$

$$\mathcal{E}_s = \left[\mathcal{E}_s^{(1)}, \ldots, \mathcal{E}_s^{(M)}\right]^T \tag{9.69}$$

$$\rho = \left[\rho^{(1)T}, \ldots, \rho^{(M)T}\right]^T \tag{9.70}$$

Besides the maximum likelihood estimator (9.63), many other energy-based source localization methods have been presented. The closest point approach (CPA), generally used for single-source localization and probably the simplest method, assigns the source location as the location of the sensor with the largest energy reading. It works reasonably well only when a sufficient number of sensors are deployed. Other energy-based source localization methods such as energy-ratio nonlinear least-squares (ER-NLS) and energy-ratio least-squares (ER-LS) have also been proposed (Li and Hu, 2003). The basic idea of ER-NLS and ER-LS is to determine target locations by intersecting the target location hyperspheres of every pair of sensors. Both can be considered as approximations of the ML solution (9.63) in a single-source scenario (Sheng and Hu, 2005).

9.3 NODE LOCALIZATION METHODS APPLIED TO SENSOR NETWORKS

A typical sensor network consists of a large number of small, inexpensive, collaborative, and relatively autonomous sensor nodes. Theses sensors are deployed to collect local information and respond to changes in the environment by communicating with neighboring nodes or the central station, making the sensor network a huge distributed computation machine. Applications featuring sensor network capabilities include earthquake prediction, agricultural monitoring, battlefield surveillance, and traffic control, among others.

In many situations, it is useful for sensors to be aware of their relative or even absolute positions in the network. Since the number of sensors is usually large and the nodes are often randomly distributed, exactly locating all sensor positions is a nontrivial task. One possible solution is to equip each sensor node with a global positioning systems (GPS) device. However, this is not practical for the following reasons (Savvides et al., 2001):

- GPS cannot work indoors or in the presence of dense vegetation, foliage, or other obstacles that block the line-of-sight signals from the GPS satellite.
- GPS devices introduce extra power consumption that shortens the lifetime of both the sensor nodes and the network as a whole.
- The production of GPS devices is expensive, especially when large numbers of nodes are to be produced.
- The size of GPS devices can be an issue when the nodes are required to be miniaturized.

As a result, alternative solutions for node localization are required.

Techniques for locating sensor nodes in a sensor network generally fall into two categories: localization with anchor nodes (or beacons) and anchor-free localization (Sun et al., 2005). Anchor nodes are equipped with special positioning devices (e.g., GPS) that are aware of their locations. For anchor-based localization techniques, the sensor nodes with unknown locations try to determine their positions relative to the anchor nodes. Once their locations are estimated, they can be treated as anchor nodes and used to locate other nodes. Clearly, anchor-based localization requires another positioning system to initialize the estimation process. For anchor-free localization techniques, all nodes collaborate with each other (within the communication range) and then construct a relative map. To convert the relative map into absolute position information, postprocessing is required.

Another classification of node localization can be made between centralized and distributed algorithms (Patwari et al., 2005). In centralized algorithms, all the information collected in the sensor network is transferred to a central node where the information is processed and the location of every node is estimated. These algorithms are often associated with some constrained optimization problems, and many of them are nonconvex (Moses et al., 2003). Finding the global optimum without being trapped in local optimum solutions is in general challenging, especially when the size of the sensor network increases. Convex formulations using myriad relaxation techniques (Biswas et al., 2006; Biswas and Ye, 2004; Doherty et al., 2001; Tseng, 2007) have been developed and approximate solutions are sought. In favorable conditions, these approximate solutions perform reasonably well or even solve the original problem (Biswas and Ye, 2004). They can also be used as initializations for other iterative localization algorithms to further improve localization accuracy.

Distributed algorithms are motivated by the following observations. First, there might not be a central node or the central node may not have the capability or computational power to handle a huge number of calculations. Second, forwarding all the information in the network to the central node can result in a communication bottleneck at the central processor. In distributed algorithms, each node runs a subset of the operations to obtain localization estimates and then passes them to neighboring nodes. The neighboring nodes then recalculate the localization estimates and pass the updates to their neighbors until the algorithm converges. To make an algorithm distributed, the convergence issue needs to be addressed to ensure that the global optimum is obtained. Exchanging information locally alleviates the communication bottleneck at the central node, but may increase total energy consumption compared to what is required to forward the information to the central node. It is conceivable that a trade-off can be achieved.

Many localization algorithms rely on distance measurements between the sensor nodes (Gustafsson and Gunnarsson, 2005). These can be estimated based on signal characteristics such as received signal strength (RSS), time of arrival (TOA), or TDOA. Network connectivity information can also be used to estimate distance. For example, the distance between nodes can be estimated by the number of hops required for the packets to travel from the source to the destination nodes (Nicolescu and Nath, 2001). For the ease of later discussions, it will be assumed that the distance measurement between a pair of nodes lying within the communication range can be obtained using any of these approaches.

In the following sections, we first review formulations of the node localization problem and then introduce some of the important techniques such as multidimensional scaling (MDS) and various types of constraint relaxations that have been successfully applied. In the last part of this section, we present a new distributed algorithm based on the nonlinear Gauss-Seidel algorithm (Cheng, 2006; Cheng et al., 2009) and a few simulation results.

Without loss of generality, let us consider a two-dimensional(2D) node localization problem where all sources and sensors are located on a single plane. Let there be m anchor nodes with known locations $\mathbf{a}_k \in \mathbb{R}^2$, $k=1,\dots,m$, and n nodes with unknown locations $\mathbf{x}_i \in \mathbb{R}^2$, $i=1,\dots,n$. The true Euclidean distances between $\mathbf{x}_i$ and $\mathbf{x}_j$ and $\mathbf{x}_i$ and $\mathbf{a}_k$ are denoted as d_{ij} and d_{ik}, respectively. It is assumed that measurements of the distance to all neighboring nodes are available only if these nodes are within radio range, R. Denote w_{ij} and w_{ik} as the measurement errors for d_{ij} and d_{ik}, respectively; then the distance measurements can be expressed as

$$\tilde{d}_{ij} = d_{ij} + w_{ij} \tag{9.71}$$

$$\tilde{d}_{ik} = d_{ik} + w_{ik} \tag{9.72}$$

9.3.1 Noise-Free Node Localization

In the noise-free case, $\tilde{d}_{ij} = d_{ij}$ and $\tilde{d}_{ik} = d_{ik}$. Thus, the node localization problem can be formulated as

$$\begin{aligned} \text{Find} \quad & \mathbf{x}_1, \dots, \mathbf{x}_n, \\ \text{s.t.} \quad & \|\mathbf{x}_i - \mathbf{x}_j\| = d_{ij}, \quad \forall (i,j) \in N_x \\ & \|\mathbf{x}_i - \mathbf{a}_k\| = d_{ik}, \quad \forall (i,k) \in N_a \end{aligned} \tag{9.73}$$

where $N_x = \{(i,j) | i<j; \|\mathbf{x}_i^{true} - \mathbf{x}_j^{true}\| < R\}$; $N_a = \{(i,k) | \|\mathbf{x}_i^{true} - \mathbf{a}_k\| < R\}$; and $\mathbf{x}_i^{true}$ is the true location of the ith node. Define $\mathbf{X} = [\mathbf{x}_1, \dots, \mathbf{x}_n] \in \mathbb{R}^{2\times n}$. Then Equation (9.73) can be rewritten (Biswas et al., 2006) as

$$\begin{aligned} \text{Find} \quad & \mathbf{X} \in \mathbb{R}^{2\times n}, \mathbf{Y} \in \mathbb{R}^{n\times n} \\ \text{s.t.} \quad & (\mathbf{e}_i - \mathbf{e}_j)^T \mathbf{Y} (\mathbf{e}_i - \mathbf{e}_j) = d_{ij}^2, \quad \forall (i,j) \in N_x \\ & \begin{bmatrix} \mathbf{a}_k \\ -\mathbf{e}_i \end{bmatrix}^T \begin{bmatrix} \mathbf{I} & \mathbf{X} \\ \mathbf{X}^T & \mathbf{Y} \end{bmatrix} \begin{bmatrix} \mathbf{a}_k \\ -\mathbf{e}_i \end{bmatrix} = d_{ik}^2, \quad \forall (i,k) \in N_a \\ & \mathbf{Y} = \mathbf{X}^T \mathbf{X} \end{aligned} \tag{9.74}$$

where $\mathbf{e}_i$ is the vector of all zeros except the ith element, which is 1. The semidefinite programming (SDP) formulation of Equation (9.74) can be obtained by first changing the constraint $\mathbf{Y} = \mathbf{X}^T\mathbf{X}$ to $\mathbf{Y} \succeq \mathbf{X}^T\mathbf{X}$, which is equivalent (Boyd et al., 1994) to

$$\mathbf{Z} := \begin{bmatrix} \mathbf{I} & \mathbf{X} \\ \mathbf{X}^T & \mathbf{Y} \end{bmatrix} \succeq \mathbf{0} \tag{9.75}$$

The notation $\mathbf{A} \succeq \mathbf{B}$ means that $\mathbf{A} - \mathbf{B}$ is a positive semi-definite matrix. Using the new variable $\mathbf{Z}$, the original problem is converted into a standard SDP feasibility problem (Biswas et al., 2006):

$$\begin{aligned} \text{Find} \quad & \mathbf{Z} \in \mathbb{R}^{(n+2)\times(n+2)} \\ \text{s.t.} \quad & [1; 0; \mathbf{0}]^T \mathbf{Z} [1; 0; \mathbf{0}] = 1 \\ & [0; 1; \mathbf{0}]^T \mathbf{Z} [0; 1; \mathbf{0}] = 1 \\ & [1; 1; \mathbf{0}]^T \mathbf{Z} [1; 1; \mathbf{0}] = 2 \\ & \left[\mathbf{0}; \mathbf{e}_i - \mathbf{e}_j\right]^T \mathbf{Z} \left[\mathbf{0}; \mathbf{e}_i - \mathbf{e}_j\right] = d_{ij}^2, \quad \forall (i,j) \in N_x \\ & [\mathbf{a}_k; -\mathbf{e}_i]^T \mathbf{Z} [\mathbf{a}_k; -\mathbf{e}_i] = d_{ik}^2, \quad \forall (i,k) \in N_a \\ & \mathbf{Z} \succeq \mathbf{0} \end{aligned} \tag{9.76}$$

The notation $[\mathbf{x}; \mathbf{y}]$ is defined as the column vector constructed by stacking $\mathbf{x}$ on top of $\mathbf{y}$. Biswas et al. (2006) further showed that, if there are $2n + n(n+1)/2$ distance pairs, each of which has an accurate distance measure, then Equation (9.76) solves the original problem exactly.

When the distance measurement is contaminated by noise, Equation (9.73) is not applicable since no feasible solution can be found in general that satisfies all constraints. One way of solving this problem is to assign certain distribution functions to the measurement errors w_{ij} and w_{ik} and reformulate the problem using the maximum likelihood criterion (Biswas et al., 2006; Liang et al., 2004).

9.3.2 Maximum Likelihood Formulation

Suppose the measurement errors follow the following distributions: $w_{ij} \sim \mathcal{N}(0, \sigma_{ij}^2)$ and $w_{ik} \sim \mathcal{N}(0, \sigma_{ik}^2)$, where $\mathcal{N}(0, \sigma^2)$ denotes the Gaussian p.d.f. with zero mean and variance σ^2. Assuming all the measurement errors are independent, the maximum likelihood estimate for $\mathbf{X}$ can be obtained by solving

$$\begin{aligned} \text{Minimize} \quad & \sum_{(i,k)\in N_a} \frac{1}{\sigma_{ik}^2} \epsilon_{ik} + \sum_{(i,j)\in N_x} \frac{1}{\sigma_{ij}^2} \epsilon_{ij} \\ \text{s.t.} \quad & \left(\|\mathbf{x}_i - \mathbf{x}_j\| - \tilde{d}_{ij}\right)^2 = \epsilon_{ij}^2, \quad \forall (i,j) \in N_x \\ & \left(\|\mathbf{x}_i - \mathbf{a}_k\| - \tilde{d}_{ik}\right)^2 = \epsilon_{ik}^2, \quad \forall (i,k) \in N_a \end{aligned} \tag{9.77}$$

Equation (9.77) is a nonconvex optimization problem, but again admits an SDP relaxation problem (Biswas et al., 2006) of

$$
\begin{aligned}
\text{Minimize} \quad & \sum_{(i,k)\in N_a} \frac{1}{\sigma_{ik}^2}\epsilon_{ik} + \sum_{(i,j)\in N_x} \frac{1}{\sigma_{ij}^2}\epsilon_{ij} \\
\text{s.t.} \quad & \left[-\tilde{d}_{ij};1\right]^T \mathbf{D}_{ij}\left[-\tilde{d}_{ij};1\right] = \epsilon_{ij}, \quad \forall (i,j)\in N_x \\
& \left[-\tilde{d}_{ik};1\right]^T \mathbf{D}_{ik}\left[-\tilde{d}_{ik};1\right] = \epsilon_{ik}, \quad \forall (i,k)\in N_a \\
& \left[\mathbf{0};\mathbf{e}_i-\mathbf{e}_j\right]^T \mathbf{Z}\left[\mathbf{0};\mathbf{e}_i-\mathbf{e}_j\right] = v_{ij}, \quad \forall (i,j)\in N_x \\
& [\mathbf{a}_k;-\mathbf{e}_i]^T \mathbf{Z}[\mathbf{a}_k;-\mathbf{e}_i] = v_{ik}, \quad \forall (i,j)\in N_a \\
& \mathbf{D}_{ij} \succcurlyeq \mathbf{0}, \quad \forall (i,j)\in N_x \\
& \mathbf{D}_{ik} \succcurlyeq \mathbf{0}, \quad \forall (i,k)\in N_a \\
& \mathbf{Z} \succcurlyeq \mathbf{0}
\end{aligned} \tag{9.78}
$$

where

$$
\mathbf{D}_{ij} = \begin{bmatrix} 1 & u_{ij} \\ u_{ij} & v_{ij} \end{bmatrix}, \quad \forall (i,j)\in N_x, \quad \mathbf{D}_{ik} = \begin{bmatrix} 1 & u_{ik} \\ u_{ik} & v_{ik} \end{bmatrix}, \quad \forall (i,k)\in N_a
$$

The original problem (9.77) can then be solved by solving and rounding its SDP relaxation (9.78) (Biswas et al., 2006; Liang et al., 2004).

9.3.3 Minimization on the Sum of Absolute Errors

Other than the maximum likelihood formulation, the node localization problem can also be solved by minimizing the sum of the absolute errors (Biswas and Ye, 2004):

$$
\sum_{(i,k)\in N_a} \left| \|\mathbf{x}_i - \mathbf{x}_j\|^2 - \tilde{d}_{ij}^2 \right| + \sum_{(i,j)\in N_x} \left| \|\mathbf{x}_i - \mathbf{a}_k\|^2 - \tilde{d}_{ik}^2 \right| \tag{9.79}
$$

Introducing the slack variable α's, Equation (9.79) can then be put into the following form:

$$
\begin{aligned}
\text{Minimize} \quad & \sum_{(i,j)\in N_x} \left(\alpha_{ij}^+ + \alpha_{ij}^-\right) + \sum_{(i,k)\in N_a} \left(\alpha_{ik}^+ + \alpha_{ik}^-\right) \\
\text{s.t.} \quad & \left(\mathbf{e}_i - \mathbf{e}_j\right)^T \mathbf{Y}\left(\mathbf{e}_i - \mathbf{e}_j\right) - \tilde{d}_{ij}^2 = \alpha_{ij}^+ - \alpha_{ij}^-, \quad \forall\,(i,j)\in N_x \\
& [\mathbf{a}_k;-\mathbf{e}_i]^T \begin{bmatrix} \mathbf{I} & \mathbf{X} \\ \mathbf{X}^T & \mathbf{Y} \end{bmatrix} [\mathbf{a}_k;-\mathbf{e}_i] - \tilde{d}_{ik}^2 = \alpha_{ik}^+ - \alpha_{ik}^-, \quad \forall\,(i,k)\in N_a \\
& \mathbf{Y} = \mathbf{X}^T\mathbf{X}
\end{aligned} \tag{9.80}
$$

If we change the equality constraint $\mathbf{Y}=\mathbf{X}^T\mathbf{X}$ of Equation (9.80) to $\mathbf{Y}\succcurlyeq\mathbf{X}^T\mathbf{X}$ and use the inequality constraint (9.75) as before, the SDP relaxation of Equation

(9.80) is obtained (Biswas and Ye, 2004) as

$$\begin{aligned}
\text{Minimize} \quad & \sum_{(i,j)\in N_x} \left(\alpha_{ij}^{+} + \alpha_{ij}^{-}\right) + \sum_{i,k\in N_a} \left(\alpha_{ik}^{+} + \alpha_{ik}^{-}\right) \\
\text{s.t.} \quad & [1;0;\mathbf{0}]^T \mathbf{Z}[1;0;\mathbf{0}] = 1 \\
& [0;1;\mathbf{0}]^T \mathbf{Z}[0;1;\mathbf{0}] = 1 \\
& [1;1;\mathbf{0}]^T \mathbf{Z}[1;1;\mathbf{0}] = 2 \\
& \left[\mathbf{0};\mathbf{e}_i - \mathbf{e}_j\right]^T \mathbf{Z}\left[\mathbf{0};\mathbf{e}_i - \mathbf{e}_j\right] - \tilde{d}_{ij}^2 = \alpha_{ij}^{+} - \alpha_{ij}^{-}, \quad \forall i,j \in N_x \\
& [\mathbf{a}_k; -\mathbf{e}_i]^T \mathbf{Z}[\mathbf{a}_k; -\mathbf{e}_i] - \tilde{d}_{ik}^2 = \alpha_{ik}^{+} - \alpha_{ik}^{-}, \quad \forall i,k \in N_a \\
& \alpha_{ij}^{+}, \alpha_{ij}^{-}, \alpha_{ik}^{+}, \alpha_{ik}^{-} \geq 0 \\
& \mathbf{Z} \succcurlyeq \mathbf{0}
\end{aligned} \tag{9.81}$$

The challenge in solving SDP relaxation problems lies in the fact that existing SDP solvers can only deal with small- to medium-size localization problems ($n < 100$) (Biswas and Ye, 2006). To overcome such difficulties, a distributed SDP method followed by a gradient search approach has been proposed (Biswas and Ye, 2006). Other relaxation techniques have been considered in the literature, such as second-order-cone programming (SOCP) relaxation (Doherty et al., 2001). The SOCP relaxation of Equation (9.79) can be formulated as

$$\begin{aligned}
\text{Minimize} \quad & \sum_{(i,j)\in N_x} \left(y_{ij} - \tilde{d}_{ij}^2\right) + \sum_{(i,k)\in N_a} \left(y_{ik} - \tilde{d}_{ik}^2\right) \\
\text{s.t.} \quad & y_{ij} \geq \|\mathbf{x}_i - \mathbf{x}_j\|^2, \quad \forall i,j \in N_x \\
& y_{ik} \geq \|\mathbf{x}_i - \mathbf{a}_k\|^2, \quad \forall i,k \in N_a
\end{aligned} \tag{9.82}$$

Theoretical and numerical studies of SOCP relaxation in sensor network localization were investigated in Tseng (2007).

9.3.4 Classical Multidimensional Scaling

Another family of node location algorithms is MDS, which aims to find a low-dimensional representation of a group of objects such that the distances between the objects fit a given set of measured pairwise "dissimilarities" as well as possible. Here we introduce classical multidimensional scaling using Euclidean distance as the proximity measure.

Let $\mathbf{P} = [\mathbf{p}_1, \ldots, \mathbf{p}_{n+m}] = [\mathbf{x}_1, \ldots, \mathbf{x}_n, \mathbf{a}_1, \ldots, \mathbf{a}_m]$ be the coordinate matrix of all sensor nodes and δ_{ij} and $\tilde{\delta}_{ij}$ be the true Euclidean distance and the distance measurement between $\mathbf{p}_i$ and $\mathbf{p}_j$; then the squared distance matrices, $\mathbf{D}$ and $\tilde{\mathbf{D}}$, can be defined as $[\mathbf{D}]_{i,j} = \delta_{ij}^2$ and $[\tilde{\mathbf{D}}]_{i,j} = \tilde{\delta}_{ij}^2$, respectively, $\forall\, 0 < i,j < n+m$. Since $\delta_{ij}^2 = \mathbf{p}_i^T \mathbf{p}_i - 2\mathbf{p}_i^T \mathbf{p}_j + \mathbf{p}_j^T \mathbf{p}_j$, the matrix $\mathbf{D}$ can be expressed as

$$\mathbf{D} = \boldsymbol{\psi}\mathbf{1}^T - 2\mathbf{P}^T\mathbf{P} + \mathbf{1}\boldsymbol{\psi}^T \tag{9.83}$$

where $\psi = [\mathbf{p}_1^T\mathbf{p}_1, \ldots, \mathbf{p}_{n+m}^T\mathbf{p}_{n+m}]^T$, and $\mathbf{1}$ is the vector of all ones. In classical MDS, the matrix $\mathbf{D}$ is then shifted to the center, resulting in a centered squared distance matrix $\mathbf{B}$,

$$\mathbf{B} = -\frac{1}{2}\left(\mathbf{I}-\mathbf{1}\mathbf{1}^T/N\right)\mathbf{D}\left(\mathbf{I}-\mathbf{1}\mathbf{1}^T/N\right) \tag{9.84}$$

$$= \left(\mathbf{I}-\mathbf{1}\mathbf{1}^T/N\right)\mathbf{P}^T\mathbf{P}\left(\mathbf{I}-\mathbf{1}\mathbf{1}^T/N\right) \tag{9.85}$$

The node localization problem is then formulated as

$$\begin{aligned} &\text{minimize } \|\mathbf{B}-\mathbf{Q}^T\mathbf{Q}\|_F \\ &\text{s.t.} \quad \text{Rank}(\mathbf{Q})=2, \quad \forall \mathbf{Q} \in \mathbb{R}^{2\times(n+m)} \end{aligned} \tag{9.86}$$

The solution to Equation (9.86) can be obtained by performing the SVD on $\mathbf{B}$:

$$\mathbf{B} = \mathbf{U}\text{diag}\{\lambda_1, \ldots, \lambda_{n+m}\}\mathbf{U}^T, \quad \lambda_1 \geq \lambda_2 \geq \cdots \geq \lambda_{n+m} \tag{9.87}$$

and then setting $\mathbf{Q} = \text{diag}\{\lambda_1, \lambda_2, 0, \ldots, 0\}\mathbf{U}^T$. $\mathbf{P}$ is then recovered by the obtained solution $\mathbf{Q}$ up to a translation and orthogonal transformation. Given a sufficient number of anchors nodes (three or more in a 2D problem), this relative map can then be transformed back to the absolute map based on the absolute positions of the anchor nodes. For real applications, $\mathbf{D}$ of Equation (9.84) is replaced by the measured squared distance matrix $\tilde{\mathbf{D}}$.

One of the major drawbacks of classical MDS is the required pairwise distance measurements, which is not realistic for most power/bandwidth-constrained sensor network applications. Also, classical MDS minimizes the squared distances between δ_{ij}^2 and $\tilde{\delta}_{ij}^2$ instead of the distances themselves. Because the distance measurement error increases proportionally with distance, minimizing the squared distance can amplify measurement errors and may result in poor performance (Costa et al., 2006).

To overcome these difficulties, many improved MDS localization schemes have been proposed. MDS-MAP(P) (Shang and Ruml, 2004) decentralizes classical MDS by using patches of local maps that can be built up in a distributed fashion and then merged to form a global map. The distributed weighted-MDS (dwMDS) algorithm minimizes a local cost function and is able to adaptively emphasize the most accurate distance measurements (Costa et al., 2006). Other variants are abundant in the literature.

9.3.5 Distributed Node Localization Based on the Nonlinear Gauss-Seidel Algorithm

In this section, we present a new distributed algorithm (Cheng, 2006; Cheng et al., 2005) based on the nonlinear Gauss-Seidel algorithm (Bertsekas and Tsitsiklis, 1989), in which each node is responsible for computing and updating its own location estimate based on the estimated locations of neighboring nodes and the locations of anchor nodes (when they are within radio range). Once the location estimate is updated, each node broadcasts it to all neighboring nodes. It stops updating when the magnitude of the updating step is less than a certain threshold. Additionally, a simple scheduling algorithm updates the node localization estimates in a consistent order without the use of a global coordinator or a predefined routing loop.

In this scheme node localization is first formulated as the following nonlinear least-squares problem:

$$\begin{aligned} &\text{Minimize} \quad F(\mathbf{x}_1,\ldots,\mathbf{x}_n) \\ &\text{where} \quad F(\mathbf{x}_1,\ldots,\mathbf{x}_n) = \sum_{(i,k)\in N_a} |s_{ik}(\mathbf{x}_i)|^2 + \sum_{(i,j)\in N_x} |r_{ij}(\mathbf{x}_i,\mathbf{x}_j)|^2 \\ &r_{ij}(\mathbf{x}_i,\mathbf{x}_j) = \|\mathbf{x}_i-\mathbf{x}_j\|^2 - \tilde{d}_{ij}^2, \quad \forall (i,j)\in N_x \\ &s_{ik}(\mathbf{x}_i) = \|\mathbf{x}_i-\mathbf{a}_k\|^2 - \tilde{d}_{ik}^2, \quad \forall (i,k)\in N_a \end{aligned} \tag{9.88}$$

which is commonly solved using the centralized Gauss-Newton method (Moses et al., 2003). The main advantage of this formulation is that it admits a simple distributed implementation and guaranteed convergence because the Jacobian and Hessian of the cost function F are continuous in the $\mathbb{R}^2$ plane. We also show in the simulation that when the measurement noise is small, the distributed implementation of Equation (9.88) yields better performance in terms of accuracy for a given number of updates.

From Bertsekas and Tsitsiklis (1989), the nonlinear Gauss-Seidel algorithm of the ith node for the problem in Equation (9.88) is described by

$$\mathbf{x}_i^{(t+1)} = \arg\min_{\mathbf{x}_i} F\left(\mathbf{x}_1^{(t+1)},\ldots,\mathbf{x}_{i-1}^{(t+1)},\mathbf{x}_i^{(t)},\mathbf{x}_{i+1}^{(t)},\ldots,\mathbf{x}_n^{(t)}\right), \quad i=1,\ldots,n \tag{9.89}$$

where t is the number of iterations. Note that, if $(i,j)\notin N_x$ or $(j,i)\notin N_x$, the solution to Equation (9.89) will not depend on $\mathbf{x}_j^{(t)}$ or $\mathbf{x}_j^{(t+1)}$, respectively. Therefore, we can simplify the optimization problem in Equation (9.89) as

$$\begin{aligned} &\min_{\mathbf{x}_i} \quad F^{(t)}(\mathbf{x}_i), \\ &\text{where} \quad F^{(t)}(\mathbf{x}_i) = \sum_{(j,i)\in N_x} \left| r_{ji}\left(\mathbf{x}_j^{(t+1)},\mathbf{x}_i\right)\right|^2 + \sum_{(i,j)\in N_x} \left| r_{ij}\left(\mathbf{x}_i,\mathbf{x}_j^{(t)}\right)\right|^2 \\ &\qquad + \sum_{(i,k)\in N_a} |s_{ik}(\mathbf{x}_i)|^2 \end{aligned} \tag{9.90}$$

Given $\mathbf{x}_i^{(t)}$, the estimate of node i's location at iteration t, we can approximate Equation (9.90) as a linear least-squares problem using the Gauss-Newton method:

$$\begin{aligned} \min_{\mathbf{g}_i} \Bigg\{ &\sum_{(j,i)\in N_x} \left| r_{ji} + \left(\nabla_{\mathbf{x}_i} r_{ji}\right)^T \mathbf{g}_i \right|^2 + \sum_{(i,j)\in N_x} \left| r_{ij} + \left(\nabla_{\mathbf{x}_i} r_{ij}\right)^T \mathbf{g}_i \right|^2 \\ &+ \sum_{(i,k)\in N_a} \left| s_{ik} + \left(\nabla_{\mathbf{x}_i} s_{ik}\right)^T \mathbf{g}_i \right|^2 \Bigg\} \end{aligned} \tag{9.91}$$

where $\mathbf{g}_i = \mathbf{x}_i - \mathbf{x}_i^{(t)}$; $r_{ji} = r_{ji}(\mathbf{x}_j^{(t+1)}, \mathbf{x}_i^{(t)})$; $r_{ij} = r_{ij}(\mathbf{x}_i^{(t)}, \mathbf{x}_j^{(t)})$; $s_{ik} = s_{ik}(\mathbf{x}_i^{(t)})$; and $\nabla_{\mathbf{x}_i}$ is the gradient with respect to $\mathbf{x}_i$. Note that Equation (9.91) is now a least-squares problem with only two variables. Let $\mathbf{g}_i^{(t)}$ be the solution to Equation (9.91); the update process estimate of sensor i is given by $\mathbf{x}_i^{(t+1)} = \mathbf{x}_i^{(t)} + \alpha_i \mathbf{g}_i^{(t)}$, where $\alpha_i \mathbf{g}_i^{(t)}$ is the step length.

The value of α_i can be found using a backtracking line search (Boyd and Vandenberghe, 2004). That is, find an α_i such that

$$F_i^{(t)}\left(\mathbf{x}_i^{(t)} + \alpha_i \mathbf{g}_i^{(t)}\right) \le F_i^{(t)}\left(\mathbf{x}_i^{(t)}\right) + c\alpha_i \nabla_{\mathbf{x}_i} F_i^{(t)}\left(\mathbf{x}_i^{(t)}\right)^T \mathbf{g}_i^{(t)} \tag{9.92}$$

where $0 < c < 0.5$ is a predetermined constant that can be interpreted as the fraction of the decrease in $F_i^{(t)}$ predicted by a linear extrapolation that we will accept. This is known as the damped Gauss-Newton method (Dennis and Schnabel, 1996). Note that only local information is required to compute the step $\mathbf{g}_i^{(t)}$ and find the step length α_i. After the location estimate is updated, node i broadcasts it to all neighboring nodes. For each node, the update terminates if $\|\mathbf{g}_i^{(t)}\| < \epsilon$, where ϵ is a predetermined threshold. Once the node decides to terminate the updated process, it sends out a message to its neighboring nodes so that the rest of the nodes can keep updating using the broadcasting node's last location estimate $\mathbf{x}_i^{(final)}$ without waiting for an update. The updating process for each node from iteration t to iteration $t+1$ is summarized in Algorithm 9.1. Since $\nabla_{\mathbf{x}_i} r_{ji} = 2(\mathbf{x}_i^{(t)} - \mathbf{x}_j^{(t+1)})$, $\nabla_{\mathbf{x}_i} r_{ij} = 2(\mathbf{x}_i^{(t)} - \mathbf{x}_j^{(t)})$, $\nabla_{\mathbf{x}_i} s_{ik} = 2(\mathbf{x}_i^{(t)} - \mathbf{a}_k)$, and $\mathbf{A}^T\mathbf{A}$ is a 2×2 matrix, the computation required for each node is minimal and thus the algorithm can be implemented on sensor nodes that have basic processing capabilities.

ALGORITHM 9.1 Gauss-Newton Update Process

Let $\mathbf{A} = [\nabla_{\mathbf{x}_i} r_{ji}, \ldots, \nabla_{\mathbf{x}_i} r_{ij}, \ldots, \nabla_{\mathbf{x}_i} s_{ik}, \ldots]^T$, $\mathbf{b} = -[r_{ji}, \ldots, r_{ij}, \ldots, s_{ik}, \ldots]$.
Compute the step $\mathbf{g}_i^{(t)} = (\mathbf{A}^T\mathbf{A})^{-1}\mathbf{A}^T\mathbf{b}$.
Compute the gradient $\nabla_{\mathbf{x}_i} F_i^{(t)}(\mathbf{x}_i^{(t)}) = \mathbf{A}^T\mathbf{b}$.
If $\|\mathbf{g}_i^{(t)}\| > \epsilon$
    While $F_i^{(t)}(\mathbf{x}_i^{(t)} + \alpha_i \mathbf{g}_i^{(t)}) > F_i^{(t)}(\mathbf{x}_i^{t}) + c\alpha_i \nabla_{\mathbf{x}_i} F_i^{(t)}(\mathbf{x}_i^{(t)})^T \mathbf{g}_i^{(t)}$
        $\alpha_i = \alpha_i / 2$;
    end
    Update the estimate $\mathbf{x}_i^{(t+1)} = \mathbf{x}_i^{(t)} + \alpha_i \mathbf{g}_i^{(t)}$.
    Broadcast the updated estimate $\mathbf{x}_i^{(t+1)}$ to the neighboring nodes.
else
    Stop update.
    Broadcast $\mathbf{x}_i^{(final)}$ to the neighboring nodes.
end

It is possible to further reduce the amount of time for all nodes to update from t to $t+1$ while satisfying the Gauss-Seidel algorithm by exploiting the network structure. Note that in Equation (9.90) only the location estimates of the neighboring nodes $\{\mathbf{x}_j^{(t+1)} | (j,i) \in N_x\}$ and $\{\mathbf{x}_j^{(t)} | (i,j) \in N_x\}$ are needed for updating node i. Therefore, node i can determine locally when to update itself without waiting for other nonneighboring nodes. We propose a simple scheduling algorithm that satisfies the Gauss-Seidel updating rule as follows.

Scheduling Algorithm. Node i starts to update its estimate $\mathbf{x}_i^{(t)}$ if one of the following criteria are met:

- Node i receives $\{\mathbf{x}_j^{(t+1)}$ or $\mathbf{x}_j^{(final)} | j < i,\ (j,i) \in N_x\}$ and $\{\mathbf{x}_j^{(t)}$ or $\mathbf{x}_j^{(final)} | j > i,\ (i,j) \in N_x\}$.
- The timeout threshold is reached.

In the ideal situation where all updated messages are received by neighboring nodes, the first criterion ensures that the nodes update according to the Gauss-Seidel algorithm. However, if some of the updated messages are not received by the intended neighboring nodes, applying only the first criterion will halt the entire updating process. Therefore, the second criterion is used to ensure that the updating process resumes in the event of outage in the wireless transmission. In this case, a timer is used to record the idling time from the last update. Once the idling time reaches the timeout threshold, the node will stop waiting for update messages from its neighboring nodes and start the update using the last received information.

Two properties result from this scheduling algorithm. First, each node waits only for its neighboring nodes, instead of $\{j | j < i\}$ as in the sequential case. Thus, a subset of nodes in the network can update simultaneously if they are not each other's neighbors. Second, it can be seen that two connected nodes cannot be updated at the same time. Thus, when one node is updating and broadcasting, its neighbors are not. This reduces the probability of message collision.

The proposed scheduling algorithm can be demonstrated by the simple example shown in Figure 9.7 (Cheng, 2006). Consider a network of six nodes and seven edges (or connections). Assume that $\{\mathbf{x}_i^{(t)} | (i,j) \in N_x\}$ are available to node i for $i = 1, \ldots, 6$. Denote $T = 1, 2, \ldots$ as the instance of time. The processing time for each node to update from iteration t to iteration $t+1$ is one unit. At $T = 1$, node

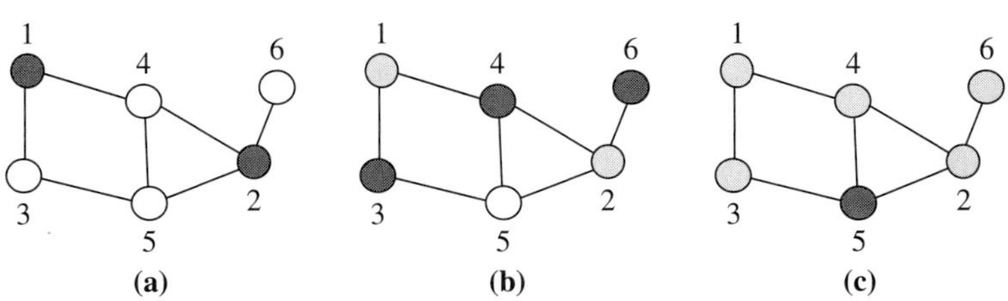

FIGURE 9.7 Proposed scheduling algorithm: (a) $T = 1$, (b) $T = 2$, and (c) $T = 3$.

1 and node 2 will start to update simultaneously since both satisfy the updating rule. This is shown in Figure 9.7(a). After nodes 1 and 2 have updated and broadcast their estimates $\mathbf{x}_1^{(t+1)}$ and $\mathbf{x}_2^{(t+1)}$ to their neighbors, nodes 3, 4, and 6 will start to update simultaneously at $T=2$, as shown in Figure 9.7(b). Finally, at $T=3$, node 5 will start to update after it receives the updated estimates from nodes 3 and 4. Notice that in Figure 9.7(c), node 1 will also start to update again to iteration $t+2$ since it received the updated estimates $\mathbf{x}_3^{(t+1)}$ and $\mathbf{x}_4^{(t+1)}$. In this simple example, it only takes three units of time to update all six nodes from iteration t to iteration $t+1$, while in the sequential case it would be six units.

Figure 9.8 (Cheng, 2006) shows the average total processing time for the proposed scheduling rule to update all nodes in randomly generated networks as the size of the network increases. In this simulation, the node locations are uniformly distributed over an area of $1.6\sqrt{n/100} \times 1.6\sqrt{n/100}$, where n is the network size (total number of nodes) and 100 realizations are used for each network to compute the average processing time. For simplicity and demonstration, we assume that all nodes have the same processing time in this simulation. In practice, the scheduling algorithm does not require such an assumption.

Simulation Results

Consider a network with n and $n/10$ randomly placed sensor nodes and anchor nodes, respectively. If the distance between two nodes or a node and an anchor node is less than radio range R, a noisy measurement of the distance is available and is given by $\tilde{d}_{ij} = d_{ij} + w_{ij}$, where w_{ij} is the noise modeled as zero-mean

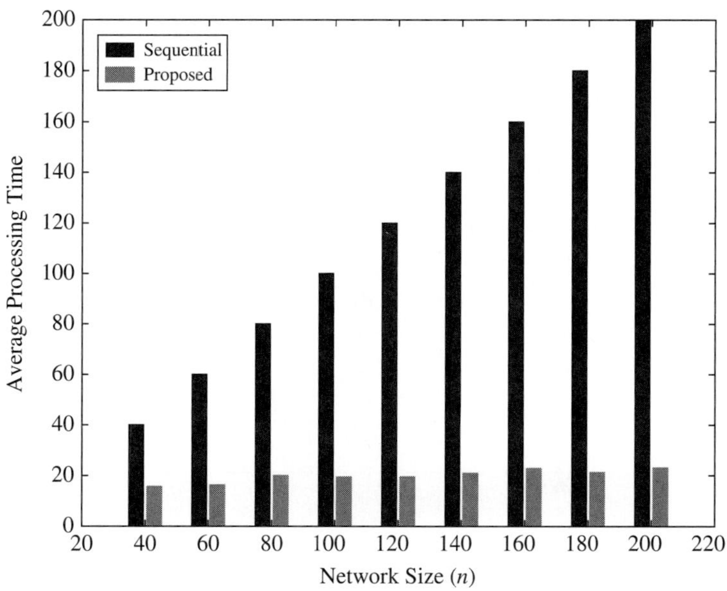

FIGURE 9.8 Average processing time for the proposed scheduling rule.

Gaussian noise with standard deviation σ_d related to the accuracy of the distance measurement. We evaluate the performance of the algorithm by the average normalized error (with respect to radio range) of the estimated location to the true location: $\frac{1}{nR}\sum_{i=1}^{n}\|\mathbf{x}_i^{(t)}-\mathbf{x}_i\|$.

Figure 9.9 shows examples of the algorithm with 60 sensors and 6 anchors placed randomly in a 0.6×0.6 region. The radio range is set to 0.35 and $\sigma_d=0.02$. The starting points are chosen as random perturbations from the true sensor locations, where the perturbation is given by a uniformly distributed random variable in $[-0.2, 0.2]$. The normalized error for the starting point is 0.4628. The x represents the estimate of the true node location represented by the o. Anchor nodes are represented by ◇. The normalized error after 10 iterations is 0.0456.

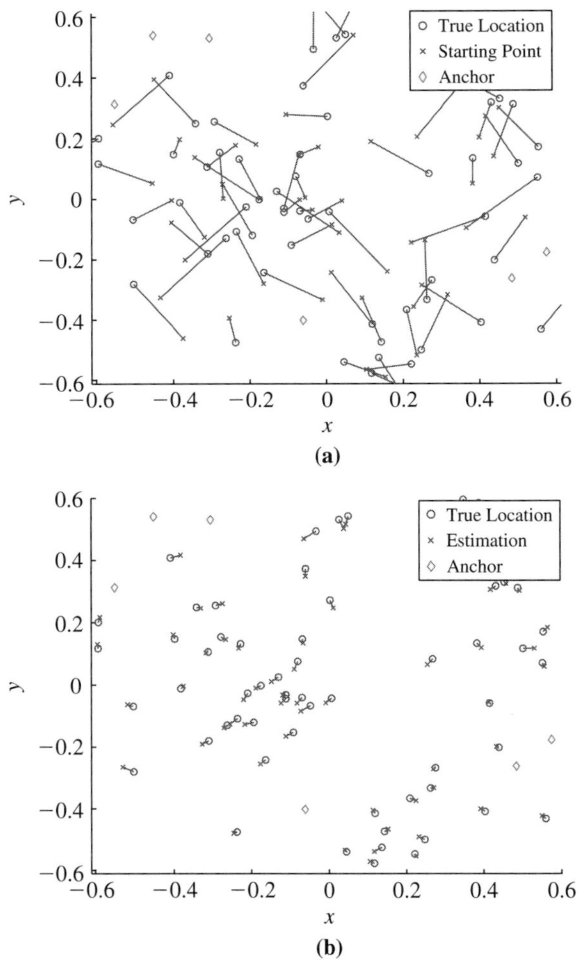

FIGURE 9.9 Realization of node localization for 60 sensor nodes and 6 anchor nodes: (a) initial estimates and (b) estimation results.

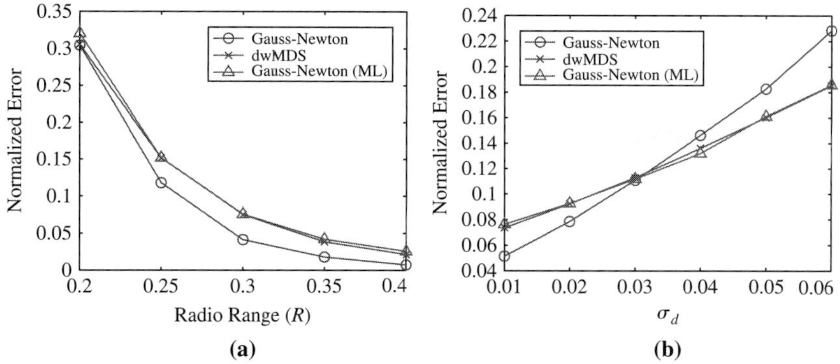

FIGURE 9.10 Comparisons of normalized errors of radio range (a) and noise level (b) for three node localization algorithms.

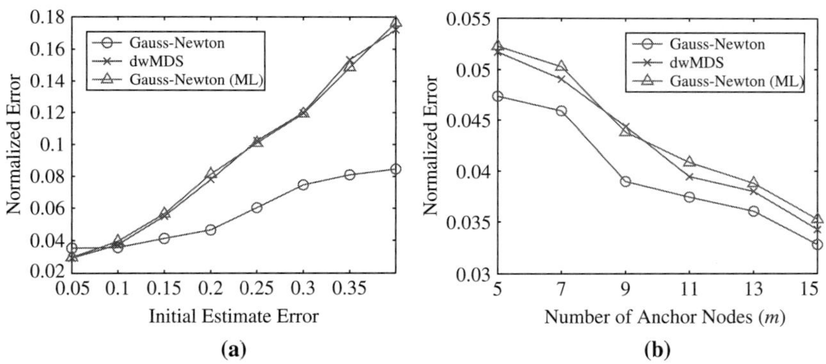

FIGURE 9.11 Comparisons of normalized errors of initial estimate (a) and number of anchor nodes (b) for three node localization algorithms.

Figures 9.10 and 9.11 compare the proposed distributed algorithm with two other localization methods: distributed Gauss-Newton and the maximum likelihood formulation and dwMDS (Costa et al., 2005) methods. In this simulation, 100 nodes and 10 anchors are randomly placed in a 0.6×0.6 region. Each node in the network updates 20 times, $t = 20$, and the normalized errors are averaged over 20 random realizations. Figure 9.10(a) shows the comparisons for different radio ranges (R) in the absence of measurement noise. Figure 9.10(b) shows the comparisons for different measurement noise levels (σ_d) with radio range R set to 0.3. Notice that in situations where the noise levels are small, the proposed algorithm yields better estimation accuracy for a given number of iterations. However, as the noise level increases, the nonlinear least-squares formulation becomes suboptimal, as expected.

Figure 9.11 shows the sensitivity of the localization algorithms to initial estimate error and total number of anchor nodes. In both simulations, the radio range R is set to 0.3 and the noise level σ_d is set to 0.01. It can be seen that the proposed distributed algorithm is less sensitive to the initial estimate error and yields better estimation results for a given number of anchor nodes when compared with the two other methods. An interesting observation from the previous four simulation results is that the distributed algorithm with the maximum likelihood formulation behaves almost identical by to the dwMDS algorithm.

9.4 SOURCE LOCALIZATION APPLICATIONS USING SENSOR NETWORKS

In this section, we describe some experimental results from using the DOAs to perform source localization in the far field. In Chen et al. (2003), we built a wireless acoustical testbed using Compaq iPAQ 3760 Pocket PCs as the testbed nodes. Each iPAQ had a microphone, an ADC, a codec, and an 802.11b wireless card. The combination of COTS hardware and open-source operating system was used to implement both the TDOA-CLS and AML (Chan and Ho, 1994) DOA estimation algorithms. The choice of these devices, which admittedly may be quite different from the nodes of a sensor network, was motivated by the desire to set up a working testbed that would allow the generation of significant data as well as the development and testing of localization algorithms.

In the first experimental setting, the source (a computer speaker) was placed in the middle of a wirelessly connected square array (each side of length ℓ) of the iPAQs. In this near-field case (relative to the array), the four nodes acted as one array with inter-sensor spacing ℓ of 20 ft (6.1 m). The sound of a moving light-wheeled vehicle was played through the speaker and collected by the microphone array embedded in the iPAQs. Figure 9.12(a) shows the direct localization results of the speaker using the AML method. A similar experiment was conducted using the same configuration except that ℓ was 40 ft (12.2 m). This time the loudspeaker played prerecorded organ music, the spectrum of which had a 2-kHz bandwidth with a central frequency at 1.75 kHz. The AML source localization result is shown in Figure 9.12(b). From the RMS errors of the previous two experiments, both algorithms show comparably promising results.

The second experimental setting is depicted in Figure 9.13(a), where three triangular subarrays, each with three iPAQs, formed the sensor network. In this far-field case (relative to each subarray), the DOA of the source was independently estimated in each subarray using a far-field version of the ML estimator described in Section 9.2.3. The location estimate was subsequently determined by crossing these bearing estimates and performing a least-squares fit. In this experiment, the speaker was placed at four distinct source locations $S1, \ldots, S4$, simulating the movement of the source, while the same vehicle sound was played each time. Figure 9.13(b) depicts one snapshot (for clear illustration) of the AML and TDOA-CLS results at the four locations. We note that better results were clearly obtained when the source was inside the convex hull of the

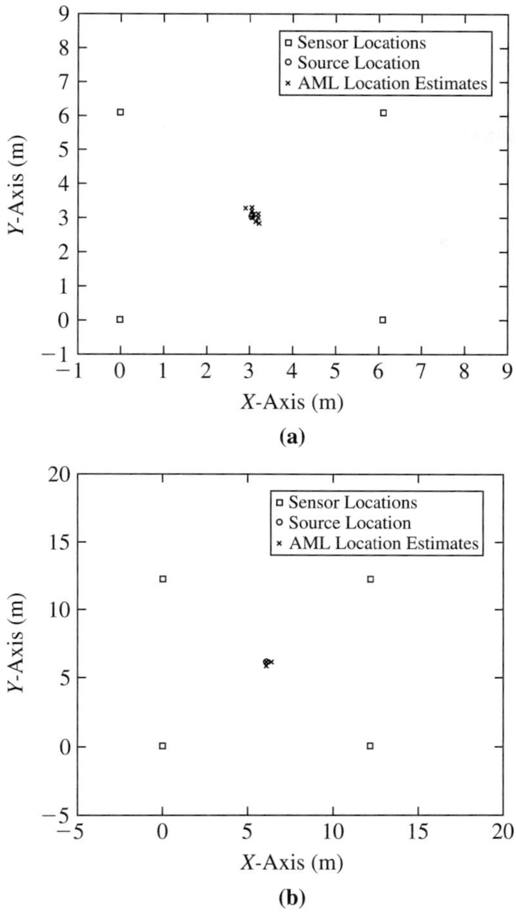

FIGURE 9.12 AML source localization: (a) vehicle source and (b) music source. (*Source:* From Chen et al., 2003, © 2003 IEEE, with permission.)

overall array. Moreover, the RMS errors of Figure 9.12 are much less than the RMS errors here: The RMS errors of direct source localization are much less because of the favorable geometry, the shorter ranges, and the fully coherent process in contrast to coherent DOA estimation and the noncoherent cross-bearing process.

A few more far-field cases are considered. In the third experimental setting, depicted in Figure 9.14(a), three linear subarrays, each with three iPAQs, formed one sensor network. The speaker, this time playing the organ music, was placed at six distinct locations. Figure 9.14(b) shows the one-snapshot results of the two algorithms at the six locations. The RMS error calculation shows performance similar to the second experiment, which demonstrates that both AML and TDOA-CLS can locate different sources.

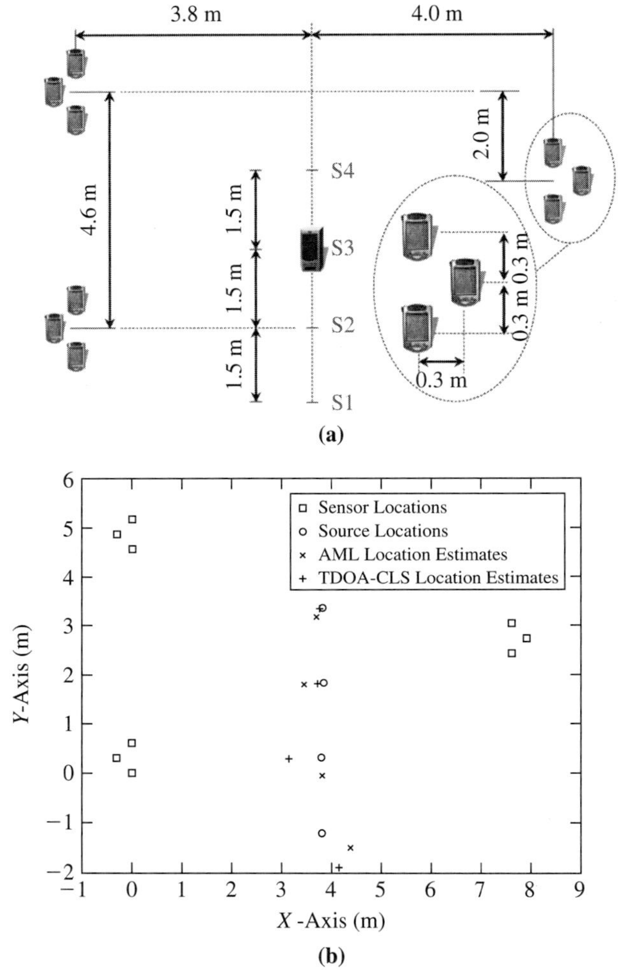

FIGURE 9.13 Source localization under the second experimental setting. (a) Experimental setting using triangular subarrays; (b) localization results of a vehicle source. (*Source:* From Chen et al., 2003, © 2003 IEEE, with permission.)

In the fourth experimental setting, depicted in Figure 9.15, four square subarrays, each with four iPAQs, formed a single network. Two speakers, one playing the vehicle sound and the other one playing the organ music simultaneously, were placed inside the convex hull of the sensors. Figure 9.16(a) shows the AML localization results when only the vehicle source was active, while Figure 9.16(b) shows the AML localization results when both sources were active. Comparing Figure 9.16(a) and 9.16(b), we found that good multisource localization results can be obtained by AML, but not as good as the performance with only a single source. When the number of subarray elements increases, localization accuracy

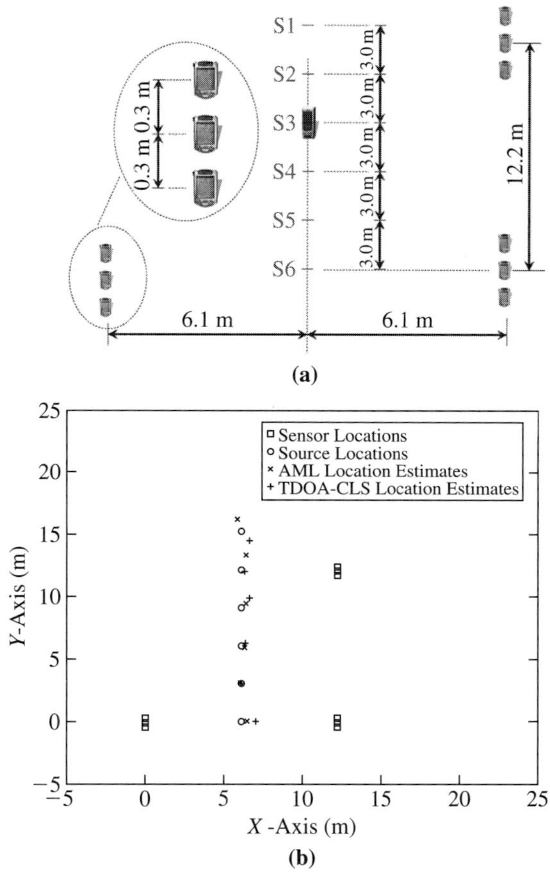

FIGURE 9.14 Source localization under the third experimental setting. (a) Experimental setting using linear subarrays; (b) localization results of a music source. (*Source:* From Chen et al., 2003, © 2003 IEEE, with permission.)

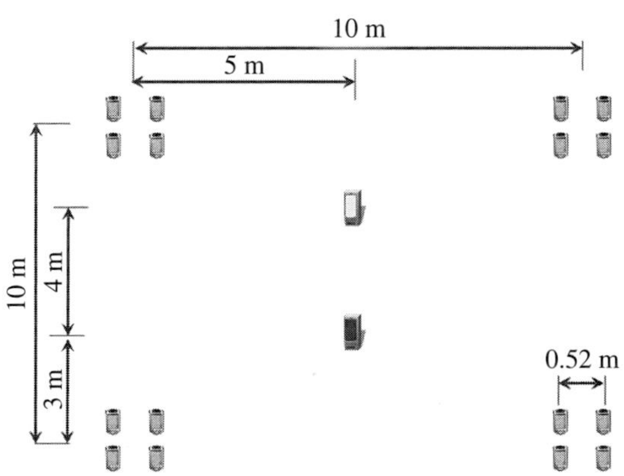

FIGURE 9.15 Experimental setting using square subarrays. (*Source:* From Chen et al., 2003, © 2003 IEEE, with permission.)

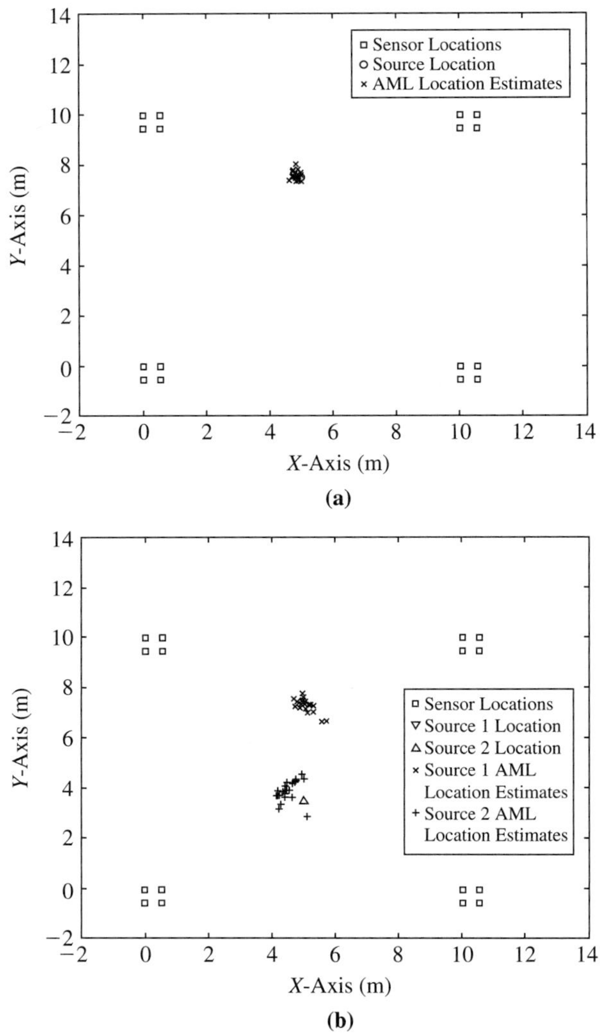

FIGURE 9.16 Source localization under the fourth experimental setting: (a) vehicle source; (b) vehicle and organ music sources. (*Source:* From Chen et al., 2003, © 2003 IEEE, with permission.)

also improves, which agrees with the Cramer-Rao bound analysis reported in Stoica and Nehorai (1989).

Acknowledgement The authors would like to express their appreciations for the technical contributions of many UCLA students and colleagues who were co-authors of papers cited in the references. This work was partially supported by NSC grant NSC-97-2218-E-194-007, an NSF CENS program grant, NSF grant EF-0410438, and NSF grant DBI0754247/F.

REFERENCES

Bancroft, S., 1985. An algebraic solution of the GPS equations. IEEE Trans. Aerosp. Electron. Syst. 21 (1), 56–59.

Bertsekas, D.B., Tsitsiklis, J.N., 1989. Parallel and Distributed Computation. Prentice-Hall.

Biswas, P., Lian, T.-C., Wang, T.-C., Ye, Y., 2006. Semidefinite programming based algorithm for sensor network localization. ACM Trans. Sens. Net. 2 (2), 188–220.

Biswas, P., Ye, Y., 2004. Semidefinite programming for ad hoc wireless sensor network localization. In: Proc. 3rd International Symposium on Information Processing in Sensor Networks, pp. 46–54.

Biswas, P., Ye, Y., 2006. A distributed method for solving semidefinite programs arising from ad hoc wireless sensor network localization. In: Multiscale Optimization Methods and Applications, Nonconvex Optimization and Its Applications, pp. 69–84, Springer-Verlag.

Bohme, J.F., 1986. Separated estimation of wave parameters and spectral parameters by maximum likelihood. In: Proc. IEEE International Conference on Acoustics, Speech, and Signal Processing.

Boyd, S., Ghaoui, L.E., Feron, E., Balakrishnan, V., 1994. Linear Matrix Inequalities in System and Control Theory. SIAM.

Boyd, S., Vandenberghe, L., 2004. Convex Optimization. Cambridge University Press.

Brandstein, M.S., Adcock, J.E., Silverman, H.F., 1997. A closed-form location estimator for use with room environment microphone arrays. IEEE Trans. Speech Audio Process. 5 (1), 45–50.

Bulusu, N., Jha, S. (Eds.), 2005. Wireless Sensor Networks. Artech House.

Carter, G.C., 1987. Coherence and time delay estimation. Proc. IEEE 75, 236–255.

Chafee, J., Abel, J.S., 1994. On the exact solutions of pseudorange equations. IEEE Trans. Aerosp. Electron. Syst. 30 (4), 1021–1030.

Chan, Y.T., Ho, K.C., 1994. A simple and efficient estimator for hyperbolic location. IEEE Trans. Signal Process. 42 (8), 1905–1915.

Chen, C.E., Lorenzelli, F., Hudson, R.E., Yao, K., 2007. Maximum likelihood DOA estimation of multiple wideband sources in the presence of nonuniform sensor noise. EURASIP J. Adv. Signal Process. 2008, Article ID 835079, pp. 1–12.

Chen, J., Hudson, R., Yao, K., 2002a. Maximum-likelihood source localization and unknown sensor location estimation for wideband signals in the near-field. IEEE Trans. Signal Process. 50 (8), 1843–1854.

Chen, J., Yao, K., Hudson, R., 2002b. Source localization and beamforming. IEEE Signal Process. Mag. 19 (2), 30–39.

Chen, J.C., Yao, K., Tung, T.L., Reed, C.W., Chen, D., 2002c. Source localization and tracking of a wideband source using a randomly distributed beamforming sensor array. Int. J. High Perform. Comput. Appli. 16 (3), 259–272.

Chen, J.C., Yip, L., Elson, J., Wang, H., Maniezzo, D., Hudson, R.E., Yao, K., Estrin, D., 2003. Coherent acoustic array processing and localization on wireless sensor networks. In: Proc. IEEE 91, 1154–1185.

Cheng, B.H., Hudson, R.E., Lorenzelli, F., Vandenberghe, L., Yao, K., 2005. Distributed Gauss-Newton method for node loclaization in wireless sensor networks. In: Proc. IEEE 6th Workshop on Signal Processing Advances in Wireless Communications, pp. 915–919.

Cheng, B.W., 2006. Distributed and Cooperative Processing in Wireless Sensor Networks. Ph.D. Thesis, University of California, Los Angeles.

Cheng, B.W., Vandenberghe, L., Yao, K., Distributed algorithm for node localization in wireless Ad-Hoc networks. ACM Trans. Sens. Net. In press, Dec. 2009.

Costa, J., Patwari, N., Hero, A.O., 2005. Achieving high-accuracy distributed localization in sensor networks. In: Proc. IEEE International Conference on Acoustics, Speech, and Signal Processing, pp. 641–644.

Costa, J.A., Patwari, N., Hero III, A.O., 2006. Distributed weighted-multidimensional scaling for node localization in sensor networks. ACM Trans. Sens. Netw. 2 (1), 39–64.

Dennis, Jr., J.E., Schnabel, R.B., 1996. Numerical Methods for Unconstrained Optimization and Nonlinear Equations. Society for Industrial and Applied Mathematics.

Doherty, L., Pister, D.S.J., Ghaoui, L.E., 2001. Convex position estimation in wireless sensor networks. In: Proc. IEEE INFOCOM, pp. 1655–1663.

Estrin, D., et al., 2001. Embedded Everywhere. Committee on Networked Systems of Embedded Computers, National Research Council. National Academy Press.

Friedlander, B., Weiss, A.J., 1993. Direction finding for wideband signals using an interpolated array. IEEE Trans. Signal Process. 41 (4), 1618–1634.

Gharavi, H., Kumar, S.P. (Eds.), 2003. Special issue on sensor networks and applications. Proc. IEEE. 91 (8), 1151–1259.

Grenander, U., Szegö, G. 1958. Toeplitz Forms and Their Applications. University of California Press.

Gustafsson, F., Gunnarsson, F., 2005. Mobile positioning using wireless networks: possibilities and fundamental limitations based on available wireless network measurements. IEEE Signal Process. Mag. 22 (4), 41–53.

Huang, Y., Benesty, J., Elko, G.W., 2000. Passive acoustic source localization for video camera steering. In: Proc. IEEE International Conference on Acoustics, Speech, and Signal Processing, vol. 2, pp. 909–912.

Ilyas, M., Mahgoub, I. (Eds.), 2005. Handbook of Sensor Networks: Compact Wireless and Wired Sensing Systems. CRC Press.

Iyengar, S.S., Brooks, R.R. (Eds.), 2004. Distributed Sensor Networks. CRC Press.

Iyengar, S.S., Kumar, S. (Eds.), 2002. Special issue on advances in information technology for high performance and computationally intensive distributed sensor networks. Int. J. High Perform. Comput. Appl. 16 (3), 203–353.

Karl, H., Willig, A. (Eds.), 2005. Protocols and Architectures for Wireless Sensor Networks. John Wiley & Sons.

Kay, S.M., 1993. Fundamentals of Statistical Signal Processing: Estimation Theory. Prentice-Hall.

Krishnamachari, B. (Ed.), 2006. Networking Wireless Sensors. Cambridge University Press.

Kumar, S., Zhao, F., Shepherd, D. (Eds.), 2002. Special issue on collaborative signal and information processing in microsensor networks. IEEE Signal Process. Mag. 19 (2), 13–94.

Lesser, V., 2003. Distributed Sensor Networks: A Multiagent Perspective. Kluwer.

Li, D., Hu, Y., 2003. Energy-based collaborative source localization using acoustic microphone array. EURASIP J. Appl. Signal Process. 2003 (4), 321–337.

Li, D., Wong, K.D., Hu, Y., Sayeed, A.M., 2002. Detection, classification, and tracking of targets. IEEE Signal Process. Mag. 19 (2), 17–29.

Liang, T.-C., Wang, T.-C., Ye, Y., 2004. A gradient search method to round the semidefinite programming relaxation solution for ad hoc wireless sensor network localization. Technical Report Department of Management Science and Engineering, Stanford University.

Lu, Z., Gastpar, M., Liu, J., Swami, A. (Eds.), 2006. Special issue on distributed signal processing in sensor networks. IEEE Signal Process Mag. 23 (4), 14–118.

Moore, T., Kozick, R.J., Sadler, B.M., 2008. Reverse parametric transformations and the cramer-rao bound. IEEE Signal Process. Lett. 2, 909–912.

Moses, R., Krishnamurthy, D., Patterson, R., 2003. Self-localization for wireless networks. EURASIP J. Appl. Signal Process. 2003 (4), 348–358.

Nicolescu, D., Nath, B., 2001. Ad-hoc positioning system. In: Proc. IEEE GLOBECOM, pp. 2926–2931.

Patwari, N., Ash, J.N., Kyperountas, S., Hero, III, A.O., Moses, R.L., Correal, N.S., 2005. Locating the nodes: cooperative localization in wireless sensor networks. IEEE Signal Process. Mag. 22 (4), 54–69.

Pottie, G.J., Kaiser, W. (Eds.), 2005. Principles of Embedded Networked Systems Design. Cambridge University Press.

Pottie, G.J., Sayeed, A.M., Estrin, D., Ramchandran, K. (Eds.), 2005. Special issue on self-organizing distributed collaborative sensor networks. IEEE J. Sel. Areas Commun. 23 (4), 689–872.

Raghavendra, C.S., Sivalingam, K.M., Znati, T. (Eds.), 2004. Wireless Sensor Networks. Springer.

Savvides, A., Han, C., Srivastava, M.B., 2001. Dynamic fine-grained localization in ad-hoc networks of sensors. In: Proc. 7th Annual International Conference on Mobile Computing and Networking, pp. 166–179.

Schau, H., Robinson, A., 1987. Passive source localization employing intersecting spherical surfaces from time-of-arrival differences. IEEE Trans. Acoust. Speech Signal Process. 35 (8), 1223–1225.

Shang, Y., Ruml, W., 2004. Improved MDS-based localization. In: Proc. IEEE INFOCOM, pp. 2640–2651.

Sheng, X., Hu, Y., 2005. Maximum likelihood multiple-source localization using acoustic energy measurements with wireless sensor networks. IEEE Trans. Signal Process. 53 (1), 44–53.

Smith, J., Abel, J., 1987. Closed-form least-squares source location estimation from range-difference measurements. IEEE Trans. Acoust. Speech Signal Process. 35 (12), 1661–1669.

Sohraby, K., Minoli, D., Znati, T. (Eds.), 2007. Wireless Sensor Networks. John Wiley & Sons.

Stewart, G.W., 1999. The QLP approximation to the singular value decomposition. SIAM J. Sci. Comput. 20 (4), 1336–1348.

Stoica, P., Nehorai, A., 1989. Music, maximum likelihood and Cramer-Rao bound. IEEE Trans. Acoust. Speech Signal Process. 37, 720–741.

Stojmenovic, I. (Ed.), 2005. Handbook of Sensor Networks: Algorithms and Architectures. John Wiley & Sons.

Sun, G., Chen, J., Liu, K.J.R., 2005. Signal processing techniques in network-aided positioning: a survey of state-of-the-art positioning designs. IEEE Signal Process. Mag. 22 (4), 12–23.

Tseng, P., 2007. Second-order cone programming relaxation of sensor network localization. SIAM J. Optim. 18 (1), 156–185.

Yao, K., Estrin, D., Hu, Y.H. (Eds.), 2003. Special issue on sensor networks. EURASIP J. Appl. Signal Process.

Yao, K., Hudson, R.E., Reed, C.W., Chen, D., Lorenzelli, F., 1998. Blind beamforming on a randomly distributed sensor array system. IEEE J. Sel. Areas Commun. 16 (8), 1555–1567.

Yao, K., Kohno, R., Mitra, U., Tong, L., Vanzago, L. (Eds.), 2005. Special section on communication and signal processing in sensor networks. J. Commun. Network. 7 (4), 401–536.

Zhao, F., Guilas, L. (Eds.), 2004. Wireless Sensor Networks: An Information Processing Approach. Elsevier.

Ziskind, I., Wax, M., 1988. Maximum likelihood localization of multiple sources by alternating projection. IEEE Trans. Acoust Speech Signal Process. 36 (10), 1553–1560.

Chapter 10

Direct Position Determination: A Single-Step Emitter Localization Approach

Alon Amar, Anthony J. Weiss

10.1 BACKGROUND

Emitter localization attracts significant interest in the signal processing, radar, sonar, bioengineering, seismology, and astronomy literature. Emitter location techniques are currently used for many purposes, such as emergency cellular phone location, radio spectrum monitoring, medical imaging, law enforcement (and law breaking), fraud detection, and homeland security (Gezici et al., 2005; Gustafsson and Gunnarsson, 2005; Patwari et al., 2005; Sayed et al., 2005; Sun et al., 2005; Zagami et al., 1998).

The localization process is based on the exchange of signals between the emitter and a number of reference stations. There are mainly two types of positioning situations: (1) self-positioning, where emitter position is determined based on the transmitted signals from the stations; and (2) remote positioning, where the system determines emitter position. The main concern in this chapter is remote positioning. However, the methods discussed here can be easily applied to self-positioning.

The traditional methods for determining remote position of a radio frequency (RF) emitter are based on a two-step procedure (see Figure 10.1):

Step 1 Estimation of position-related parameters such as angle of arrival (AOA), time of arrival (TOA), time difference of arrival (TDOA), received signal strength (RSS), and Doppler frequency shift.

Step 2 Determination of position based on the estimates that were obtained in step 1. The unit performing this step is referred to as a location processing unit (LPU).

Classical and Modern Direction-of-Arrival Estimation
Copyright © 2009 by Academic Press, Inc. All rights of reproduction in any form reserved.

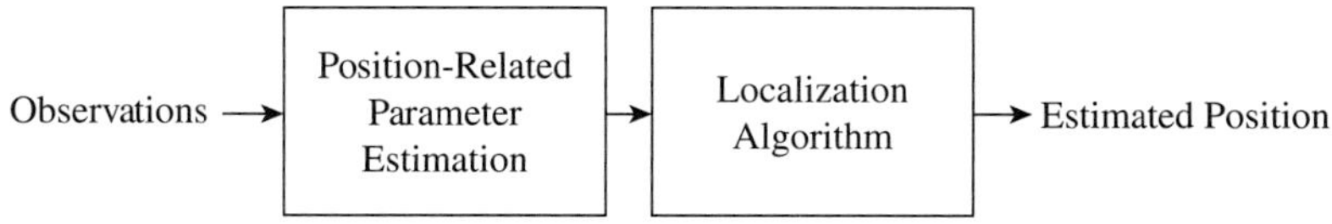

FIGURE 10.1 Two-step position determination scheme.

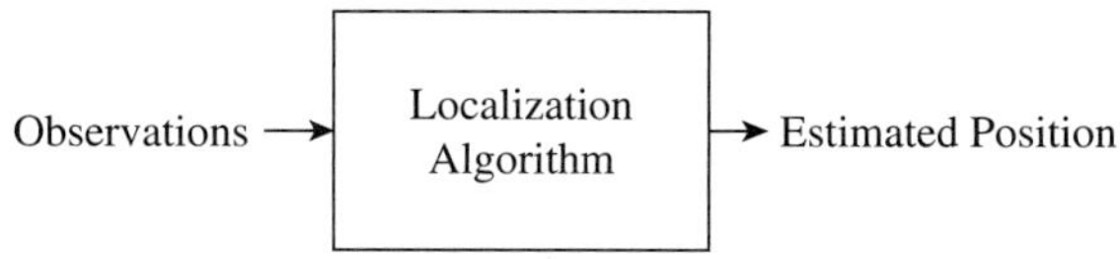

FIGURE 10.2 One-step position determination scheme.

Another approach for remote position determination is based on a one-step procedure where position is estimated directly from observations (see Figure 10.2).

Surprisingly, the two-step approach is common practice in most systems using spatially distributed stations. This trend is a result of a five factors:

- Tradition.
- The transmission of the first step results to the LPU requires a small amount of bandwidth.
- The (unjustified) belief that all information on position is included in the parameters of the first step and therefore nothing better can be done.
- The (unjustified) claim that the maximum likelihood (ML) invariance theorem (Scharf, 1990) supports the two-step approach.
- In many cases of interest the variance of the two-step error approaches the Cramer-Rao lower bound (CRLB) as the data set increases. This leads to the false belief that the two-step approach cannot be outperformed even for short data sets or low signal-to-noise ratio (SNR).

10.2 LOCALIZATION FOR STATIONARY GEOMETRY

Perhaps the first paper on the mathematics of emitter location, using AOA, was Stansfield's (1947). Many other publications followed, including a fine review by Torrieri (1984). Krim and Viberg (1996) and Wax (1996) presented comprehensive reviews on antenna array processing for location by AOA. Recently, Trees (2002) published a book fully devoted to array processing.

Positioning by TOA is used extensively in cellular phone localization (Liberti and Rappaport, 1999), radar systems (Skolnik, 1980), and underwater acoustics (Carter, 1993). Its traditional techniques can be classified as decentralized processing (Kozick and Sadler, 2004; Stoica et al., 1995; Wax and Kailath, 1985; Weinstein, 1981). Wax and Kailath (1985) discussed eigenstructure algorithms for narrowband signals observed by multiple arrays, assuming perfect spatial coherence across each array but no coherence between arrays. Stoica et al.

(1995) proposed variants of the direction estimation algorithm for decentralized processing. Weinstein (1981) discussed pairwise processing as an alternative to centralized processing of a wideband single array. His results indicated that pairwise processing may be used at high SNR without significant loss of performance. Recently, Kozick and Sadler (2004) presented a performance analysis of localization of a single source using multiple distributed arrays. They assumed perfect spatial coherence over each array and frequency-selective coherence between arrays. Their proposed method is based on bearing estimation at each array and delay estimation.

As indicated in Wax and Kailath (1983), it is rather obvious that measuring AOA/TOA at each base station separately and independently is suboptimal, since this approach ignores the constraint that the measurements must correspond to the same source position. Moreover, the base stations are geographically separated, and therefore the desired signal is often weak or absent in some of them. Thus, the system must somehow ensure that all AOA/TOA measurements used to locate a specific source correspond to the same source. In the case of cochannel simultaneous sources, the localization system confronts an association problem of deciding which of the multiple AOA/TOA estimates reported by the base stations correspond to which source.

We now discuss the direct position determination (DPD) approach, which solves the localization problem using the same data collected at all base stations together. The DPD method takes advantage of the rather simple propagation assumptions usually used for RF signals. We assume line-of-sight (LOS) propagation with unknown complex attenuation at each base station.

According to the DPD approach, each base station transfers the intercepted signals to the LPU, where the set of positions that best matches all collected data simultaneously is determined. Although there are many stray parameters, only a two-dimensional (2D) search is required for a 2D geometry. Positioning with DPD is performed in a single step. For illustration, Figure 10.3 shows the cost function of a planar geometry (receivers and emitters are all in a single plane) consisting of four base stations, each equipped with an antenna array, and three emitters. DPD determines the emitter's positions by selecting the peaks of the cost function in a single grid search. Performance comparison of DPD and the AOA-based technique for unknown as well as known waveforms shows that, for high SNR, both methods converge to the CRLB while, for low SNR, DPD achieves better accuracy. The DPD estimator does not use the first step of the conventional methods and therefore does not encounter the requirement of associating measurements to sources.

10.2.1 Problem Formulation

Consider L stations located at geographically separated sites within a plane. Each station is equipped with an antenna array that consists of M sensors. Denote by $\mathbf{p}_q \triangleq [x_q, y_q]^T$, $q = 1, \ldots, Q$ the position of the qth source. The vector of signal

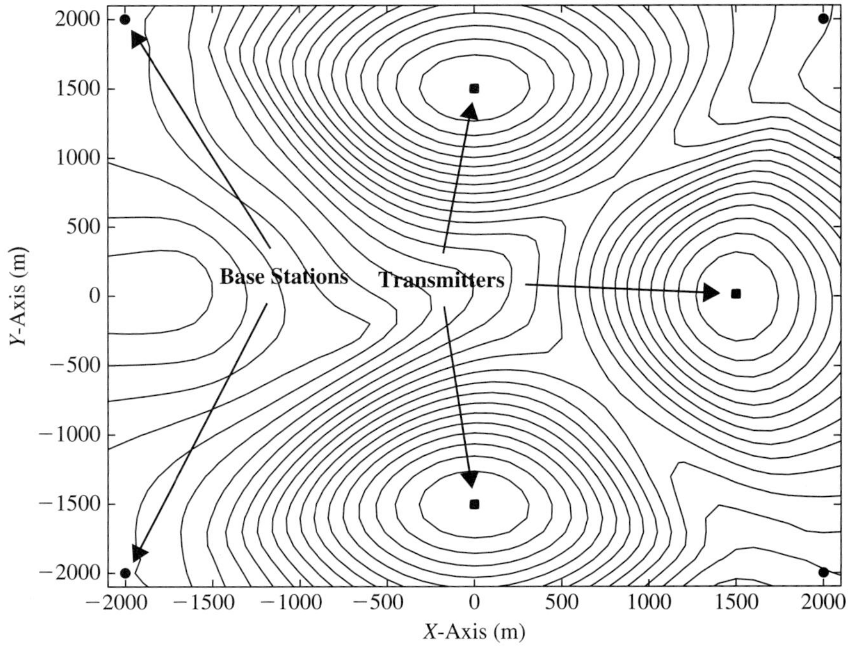

FIGURE 10.3 Contour description of the DPD cost function for four base stations, each with an antenna array, and three emitters.

complex envelopes observed by the elements of the ℓth array is given by

$$\mathbf{r}_\ell(t) = \sum_{q=1}^{Q} b_{q,\ell}\mathbf{a}_{q,\ell}s_q(t-\tau_{q,\ell}) + \mathbf{n}_\ell(t), \quad 0 \le t \le T \tag{10.1}$$

where T is the observation time, $b_{q,\ell}$ is an unknown complex scalar representing the attenuation of the qth signal as observed by the ℓth array, $s_q(t)$ is the qth wideband signal waveform, $\tau_{q,\ell}$ is the propagation delay from the qth emitter to the reference point of the ℓth array, $\mathbf{n}_\ell(t)$ represents the noise and interference observed by the array, and $\mathbf{a}_{q,\ell}$ is the $M \times 1$ column vector that denotes the ℓth array response to the qth source. We assume that the noise at the output of each of the array elements is a wide-sense stationary, zero-mean, complex Gaussian process, uncorrelated with the noise at the other antenna elements and uncorrelated with the signals.

The first stage of the processing consists of partitioning the observation interval into K sections, each of length $T' \triangleq T/K$. In Amar and Weiss (2006, Appendix G), it is shown that this procedure holds if $T' \gg \max\{\Delta t_{\max}, \tau_{\max}\}$, where $\Delta t_{\max}$ is the propagation time between the two most spatially separated stations, and $\tau_{\max}$ is the maximal correlation time of the processes. Denote by $\bar{\mathbf{r}}_\ell(k,u)$, $\bar{s}_q(k,u)$, and $\bar{\mathbf{n}}_\ell(k,u)$ the uth Fourier coefficients of the kth section of

$\mathbf{r}_\ell(t)$, $s_q(t)$, and $\mathbf{n}_\ell(t)$, respectively. Because the observed signal is a stationary process, it can be represented by a set of Fourier coefficients:

$$\bar{\mathbf{r}}_\ell(k,u) \triangleq \frac{1}{\sqrt{T'}} \int_{(k-1)T'}^{kT'} \mathbf{r}_\ell(t) e^{-j\omega_u t} dt \tag{10.2}$$

where $k=1,\ldots,K$, and $\omega_u \triangleq \frac{2\pi}{T}u, u=1,\ldots,J$. The Fourier coefficients of Equation (10.1) satisfy

$$\bar{\mathbf{r}}_\ell(k,u) = \sum_{q=1}^{Q} b_{q,\ell} \bar{\mathbf{a}}_{q,\ell}(u) \bar{s}_q(k,u) + \bar{\mathbf{n}}_\ell(k,u) \tag{10.3}$$

where we define

$$\bar{\mathbf{a}}_{q,\ell}(u) \triangleq \mathbf{a}_{q,\ell} e^{-j\omega_u \tau_{q,\ell}} \tag{10.4}$$

The noise vectors $\bar{\mathbf{n}}_\ell(k,u)$ are independent and identically distributed (i.i.d.), zero-mean complex Gaussian vectors such that $\mathbf{R}_{\bar{\mathbf{n}}}(u) = \eta_u \mathbf{I}$, where η_u is the noise spectral level at frequency ω_u. Define

$$\begin{aligned} \bar{\mathbf{r}}(k,u) &\triangleq \left[\bar{\mathbf{r}}_1^T(k,u),\ldots,\bar{\mathbf{r}}_L^T(k,u)\right]^T \\ \bar{\mathbf{n}}(k,u) &\triangleq \left[\bar{\mathbf{n}}_1^T(k,u),\ldots,\bar{\mathbf{n}}_L^T(k,u)\right]^T \\ \bar{\mathbf{s}}(k,u) &\triangleq \left[\bar{s}_1(k,u),\ldots,\bar{s}_Q(k,u)\right]^T \end{aligned} \tag{10.5}$$

and

$$\begin{aligned} \bar{\mathbf{A}}(u) &\triangleq \left[\bar{\mathbf{A}}_1(u),\ldots,\bar{\mathbf{A}}_Q(u)\right] \\ \bar{\mathbf{A}}_q(u) &\triangleq \text{Diag}\left\{\bar{\mathbf{a}}_{q,1}(u),\ldots,\bar{\mathbf{a}}_{q,L}(u)\right\} \\ \mathbf{B} &\triangleq \text{Diag}\left\{\mathbf{b}_1,\ldots,\mathbf{b}_Q\right\} \\ \mathbf{b}_q &\triangleq [b_{q,1}, b_{q,2},\ldots,b_{q,L}]^T \end{aligned} \tag{10.6}$$

where Diag stands for block diagonal. We can now rewrite Equation (10.3) as

$$\bar{\mathbf{r}}(k,u) = \bar{\mathbf{A}}(u)\mathbf{B}\bar{\mathbf{s}}(k,u) + \bar{\mathbf{n}}(k,u) \tag{10.7}$$

We are mainly interested in estimating the positions, and therefore all the other unknown parameters are nuisances.

The transmitted signals may be considered deterministic or random with known distribution (usually zero-mean complex Gaussian). Obviously, the probability distribution function (p.d.f.) of the observations in Equation (10.7) is affected by the assumption on the signals. However, in both cases the log likelihood function (LLF) of the collection of all observations in Equation (10.7) is

$$\ell(\boldsymbol{\mu}) = \sum_{k=1}^{K} \sum_{u=1}^{J} \ell(\bar{\mathbf{r}}(k,u);\boldsymbol{\mu}) \tag{10.8}$$

where $\boldsymbol{\mu}$ is the parameter vector determined according to the assumption on the signals, and $\ell(\bar{\mathbf{r}}(k,u);\boldsymbol{\mu})$ is the LLF of the observation vector $\bar{\mathbf{r}}(k,u)$ parameterized by $\boldsymbol{\mu}$.

In case of deterministic signals, the LLF of $\bar{\mathbf{r}}(k,u)$ in Equation (10.7) is

$$\ell(\bar{\mathbf{r}}(k,u);\boldsymbol{\mu}) = -\frac{1}{\eta_u}\left\|\bar{\mathbf{r}}(k,u) - \bar{\mathbf{A}}(u)\mathbf{B}\bar{\mathbf{s}}(k,u)\right\|^2 - ML\ln(\pi\eta_u) \tag{10.9}$$

where the parameter vector is defined as

$$\boldsymbol{\mu} \triangleq \left[\mathbf{p}^T, \mathbf{b}^T, \bar{\mathbf{s}}^T, \boldsymbol{\eta}^T\right]^T \tag{10.10}$$

with

$$\begin{aligned}
\mathbf{p} &\triangleq \left[\mathbf{p}_1^T, \ldots, \mathbf{p}_Q^T\right]^T \\
\mathbf{b} &\triangleq \left[\mathbf{b}_1^T, \ldots, \mathbf{b}_Q^T\right]^T \\
\bar{\mathbf{s}} &\triangleq \left[\bar{\mathbf{s}}^T(1,1), \ldots, \bar{\mathbf{s}}^T(K,J)\right]^T \\
\boldsymbol{\eta} &\triangleq [\eta_1, \ldots, \eta_J]^T
\end{aligned} \tag{10.11}$$

If the signal waveforms are known, the vector $\bar{\mathbf{s}}$ is omitted from $\boldsymbol{\mu}$. Note that, if the signals are unknown, the size of $\boldsymbol{\mu}$ increases with the data, which may result in inefficient estimation.

If the signals are assumed to be random zero-mean Gaussian processes, then the LLF of $\bar{\mathbf{r}}(k,u)$ in Equation (10.7) is

$$\ell(\bar{\mathbf{r}}(k,u);\boldsymbol{\mu}) = -\ln|\mathbf{R}_{\bar{\mathbf{r}}}(u)| - \mathrm{tr}\{\hat{\mathbf{R}}_{\bar{\mathbf{r}}}(u)\mathbf{R}_{\bar{\mathbf{r}}}^{-1}(u)\} - ML\ln(\pi) \tag{10.12}$$

where

$$\mathbf{R}_{\bar{\mathbf{r}}}(u) \triangleq E\{\bar{\mathbf{r}}(k,u)\bar{\mathbf{r}}^H(k,u)\} = \bar{\mathbf{A}}(u)\mathbf{B}\mathbf{R}_{\bar{\mathbf{s}}}(u)\mathbf{B}^H\bar{\mathbf{A}}^H(u) + \eta_u\mathbf{I} \tag{10.13}$$

$$\hat{\mathbf{R}}_{\bar{\mathbf{r}}}(u) \triangleq \frac{1}{K}\sum_{k=1}^{K}\bar{\mathbf{r}}(k,u)\bar{\mathbf{r}}^H(k,u) \tag{10.14}$$

$$\mathbf{R}_{\bar{\mathbf{s}}}(u) \triangleq E\{\bar{\mathbf{s}}(k,u)\bar{\mathbf{s}}^H(k,u)\} \tag{10.15}$$

and the parameter vector is defined as

$$\boldsymbol{\mu} \triangleq \left[\mathbf{p}^T, \mathbf{b}^T, \boldsymbol{\theta}^T, \boldsymbol{\eta}^T\right]^T \tag{10.16}$$

where the vector $\boldsymbol{\theta}$ represents the independent parameters of $\mathbf{R}_{\bar{\mathbf{s}}}(u)$ for all values of u.

10.2.2 Uniqueness Conditions

In this section we focus on stochastic signals. We determine conditions that ensure unique estimation of the emitter's parameters. The conditions identify the maximum number of uniquely resolvable emitters, known in the literature as resolution capacity (RC). We consider two cases: no prior knowledge of the correlation between signals, and uncorrelated signals. We derive two types of conditions for uniqueness. The first is a necessary condition based on the requirement that the number of unknown parameters will be equal to or smaller than the number of independent equations. The second guarantees that the observation covariance matrix is unique for almost any parameter vector.

Necessary Condition for Uniqueness

A necessary condition for unique estimation stems from the requirement that the number of unknown real parameters not be larger than the number of independent equations. In mathematical notation the necessary condition is given as

$$\dim(\mathbf{p})+\dim(\mathbf{b})+\dim(\{\mathbf{R}_{\bar{\mathbf{s}}}(u)\})+\dim(\boldsymbol{\eta})\leq\dim(\{\hat{\mathbf{R}}_{\bar{\mathbf{r}}}(u)\}) \tag{10.17}$$

where $\dim(\mathbf{x})$ is the number of real variables needed to describe $\mathbf{x}$.

Consider first the unknown's dimensionality. Assuming planar geometry, it is clear that $\dim(\mathbf{p})=2Q$. Note that only the product of the signal vector and the attenuation coefficient is observed. Thus, a unique estimation of the signal vector and the attenuation coefficient requires some constraints on the attenuation vector. Without loss of generality, we assume that $b_{q,1}=1$, and therefore for Q sources $\dim(\mathbf{b})=2Q(L-1)$. Also, since $\mathbf{R}_{\bar{\mathbf{n}}}(u)=\eta_u\mathbf{I}$, it is clear that $\dim(\boldsymbol{\eta})=J$. It can further be shown that

$$\dim(\{\mathbf{R}_{\bar{\mathbf{s}}}(u)\})=\begin{cases}JQ^2, & \text{General case}\\ JQ, & \text{Uncorrelated signals}\end{cases} \tag{10.18}$$

Consider now the dimensionality of Equation (10.17). Define a baseline vector as the vector from one sensor to another. Define by Y the number of distinct baseline vectors in an array. The minimal number of baseline vectors is associated with a uniform linear array (ULA). A baseline of length 0 is also counted. The maximal number of baseline vectors is associated with a random array. Given that the number of sensors in the array is M, Y is related to the array geometry and is restricted to the range

$$M\leq Y\leq\frac{M(M-1)}{2}+1 \tag{10.19}$$

It is shown in Amar and Weiss (2007) that given $\hat{\mathbf{R}}_{\bar{\mathbf{r}}}(u)$ with a rank $r=\min\{K,LM\}$ for all u, then

$$\dim(\{\hat{\mathbf{R}}_{\bar{\mathbf{r}}}(u)\})=\begin{cases}J(2LMr-r^2), & \text{General case}\\ \min\{J(2LMr-r^2),J\varphi(Y)\}, & \text{Uncorrelated signals}\end{cases} \tag{10.20}$$

where

$$\varphi(Y)\triangleq L(2Y-1+(L-1)M^2) \tag{10.21}$$

Substituting the previous results in Equation (10.17) yields in the general case

$$Q \leq \sqrt{2LMr - r^2 - 1 + \left(\frac{L}{J}\right)^2} - \frac{L}{J} \tag{10.22}$$

The minimal and maximal values of the right side of Equation(10.22) are obtained for $r=1$ and $r=LM$, respectively. Assuming $J>L$,

$$\begin{cases} Q < \sqrt{2LM}, & r=1 \\ Q < LM, & r=LM \end{cases} \tag{10.23}$$

Thus, the maximum number of uniquely resolvable emitters is $LM-1$, where LM is the total number of antenna elements in the system. Further, assuming $r=LM$ and narrowband signals ($J=1$), we get from Equation (10.22) that $Q \leq L(M-1)$—that is, L times the corresponding AOA bound (Wax and Ziskind, 1989). As expected, more wideband signals than narrowband signals can be resolved.

For uncorrelated signals we obtain, by substituting Equation (10.20) in Equation (10.17),

$$Q \leq \frac{J}{J+2L}(\min\{J(2LMr - r^2), J\varphi(Y)\} - 1) \tag{10.24}$$

It is shown in Amar and Weiss (2007) that the lowest upper bound is obtained for a ULA for which $Y=M$, and the highest upper bound is obtained for an arbitrary array for which $Y=\frac{M(M-1)}{2}+1$. That is,

$$\begin{cases} Q \leq \frac{J}{J+2L}((LM)^2 - L(M-1)^2 - 1), & \text{ULA } (Y=M) \\ Q \leq \frac{J}{J+2L}((LM)^2 - L(M-1) - 1), & \text{Random array } \left(Y=\frac{M(M-1)}{2}+1\right) \end{cases} \tag{10.25}$$

Thus, a ULA is not recommended for achieving large RC. Furthermore, for ULA configuration it is shown in Amar and Weiss (2007) that, for narrowband ($J=1$) and wideband ($J \gg 1$) signals, we obtain from Equation (10.24) that the RC is

$$\begin{cases} Q < M + M^2(L-1)/2, & \text{Narrowband signals} \\ Q < L(2M-1) + M^2L(L-1) - 1, & \text{Wideband signals} \end{cases} \tag{10.26}$$

For narrowband signals this result is exactly $M^2(L-1)/2$ above the corresponding AOA result; for wideband signals it should be compared with $2M-1$, the corresponding AOA result. The difference is significant.

Uniqueness with Probability One

We derive expressions for the RC that guarantee a unique solution for every parameter vector, except for a set of parameter vectors with measure zero. This condition ensures that, except for a set of vectors with measure zero, for any parameter vector $\boldsymbol{\mu}$ that is associated with the true covariance matrix, there is no other parameter vector $\boldsymbol{\mu}'$ that corresponds to an identical covariance matrix.

Formally define the "legitimate set" and the "ambiguity set" as all Hermitian matrices satisfying

$$\mathcal{F} \triangleq \{\mathbf{R}(u) : \mathbf{R}(u) = \mathbf{R}_{\tilde{\mathbf{r}}}(u; \boldsymbol{\mu}), u = 1, \ldots, J\} \tag{10.27}$$

$$\mathcal{G} \triangleq \left\{\mathbf{R}(u) : \mathbf{R}(u) = \mathbf{R}_{\tilde{\mathbf{r}}}(u; \boldsymbol{\mu}) = \mathbf{R}_{\tilde{\mathbf{r}}}(u; \boldsymbol{\mu}'); \boldsymbol{\mu} \neq \boldsymbol{\mu}', u = 1, \ldots, J\right\} \tag{10.28}$$

For this condition to hold, the dimensionality of the legitimate set must be strictly smaller than the dimensionality of the ambiguity set. Using the results presented in the previous subsection, it is clear that

$$\dim(\mathcal{F}) = \begin{cases} 2Q + 2Q(L-1) + JQ^2 + J, & \text{General case} \\ 2Q + 2Q(L-1) + JQ + J, & \text{Uncorrelated signals} \end{cases} \tag{10.29}$$

To determine the dimensionality of $\mathcal{G}$, we first define

$$\mathbf{H}(u; \boldsymbol{\mu}, \boldsymbol{\mu}') \triangleq \mathbf{R}_{\tilde{\mathbf{r}}}(u; \boldsymbol{\mu}) - \mathbf{R}_{\tilde{\mathbf{r}}}(u; \boldsymbol{\mu}') \tag{10.30}$$

The set $\mathcal{G}$ can thus be defined by

$$\mathcal{G} \triangleq \left\{\mathbf{H}(u; \boldsymbol{\mu}, \boldsymbol{\mu}') : \mathbf{H}(u; \boldsymbol{\mu}, \boldsymbol{\mu}') = \mathbf{0}; \boldsymbol{\mu} \neq \boldsymbol{\mu}', u = 1, \ldots, J\right\} \tag{10.31}$$

The dimension of set $\mathcal{G}$ is determined by the dimension of $\{\mathbf{H}(u; \boldsymbol{\mu}, \boldsymbol{\mu}')\}$ minus the number of constraints associated with the equation $\{\mathbf{H}(u; \boldsymbol{\mu}, \boldsymbol{\mu}') = \mathbf{0}; \boldsymbol{\mu} \neq \boldsymbol{\mu}'\}$. The dimension of $\{\mathbf{H}(u; \boldsymbol{\mu}, \boldsymbol{\mu}')\}$ is

$$\dim(\{\mathbf{H}(u; \boldsymbol{\mu}, \boldsymbol{\mu}')\}) = \begin{cases} 2(2Q + 2Q(L-1) + JQ^2 + J), & \text{General case} \\ 2(2Q + 2Q(L-1) + JQ + J), & \text{Uncorrelated signals} \end{cases} \tag{10.32}$$

The number of constraints imposed by $\{\mathbf{H}(u; \boldsymbol{\mu}, \boldsymbol{\mu}') = \mathbf{0}\}$ is $J(ML)^2$ in the general case. For uncorrelated signals there is a functional dependency between the entries of $\mathbf{R}_{\tilde{\mathbf{r}}}(u; \mu)$, and in this case the number of constraints is $J\varphi(Y)$. Thus,

$$\dim(\mathcal{G}) = \begin{cases} 2(2Q + 2Q(L-1) + JQ^2 + J) - J(LM)^2, & \text{General case} \\ 2(2Q + 2Q(L-1) + JQ + J) - J\varphi(Y), & \text{Uncorrelated signals} \end{cases} \tag{10.33}$$

The uniqueness condition holds if Equation (10.29) is smaller than Equation (10.33). That is,

$$\begin{cases} Q < \sqrt{(LM)^2 + (L/J)^2 - 1} - \frac{L}{J}, & \text{General case} \\ Q < \frac{J}{J+2L}(\varphi(Y) - 1), & \text{Uncorrelated signals} \end{cases} \tag{10.34}$$

These bounds are similar to the necessary bounds without the equalities.

Numerical Examples

We check the upper bounds by examining the singularity of the Fisher information matrix (FIM). Normally, when the FIM is singular it is impossible to get a reasonable estimate. Thus, we expect the FIM to become singular if the number of sources is too large for obtaining reasonable location estimates. We should emphasize here that we know of certain estimation problems in which the FIM becomes singular and it is still possible to get well-behaved estimates.

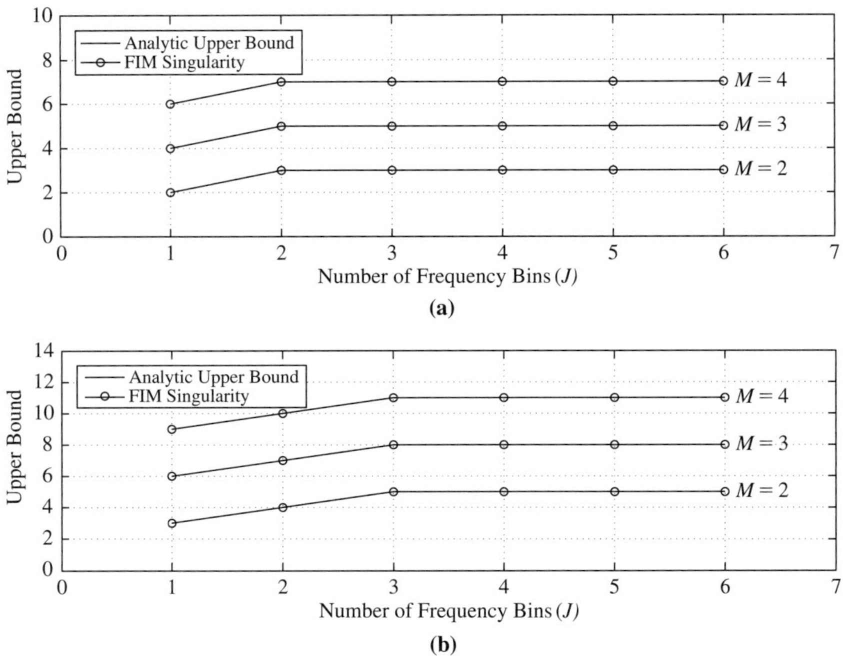

FIGURE 10.4 Largest number of resolvable emitters in the general case: (a) $L=2$ and (b) $L=3$.

In Figure 10.4 we plotted the upper bound on the RC as given in Equation (10.22) for the general case, and for the largest number of emitters for which the FIM (Weiss and Friedlander, 1993) does not become singular, versus the number of frequency bins. We examined the cases of $L=2$ with $M=2,3,4$, and $L=3$ with $M=2,3,4$. In each case the station positions were randomly chosen and the arrays were ULAs. As can be seen, the analytical upper bound on the RC agrees for any given value of M,L,J with the FIM singularity.

We compare the analytic upper bounds for $Y=M$ (corresponding to a ULA) and the upper bound for $Y=M(M-1)/2+1$ (corresponding to an arbitrary array) with the FIM singularity condition (Weiss and Amar, 2005) assuming uncorrelated signals. We use three stations, each equipped with an array of three elements. As can be seen in Figure 10.5, the upper bound related to an arbitrary array is higher than the bound related to a ULA. For $J=1$ (narrowband signals) we see that the upper bound for an arbitrary array is larger than the upper bound for a ULA by a single emitter. However, for $J \gg 1$ (wideband signals) the gap between the two upper bounds for a large number of frequency bins is increased.

10.2.3 DPD for Unknown Transmitted Waveforms

We discuss potential location algorithms for the common case where the signal waveforms are unknown to the receivers. We start by assuming that the

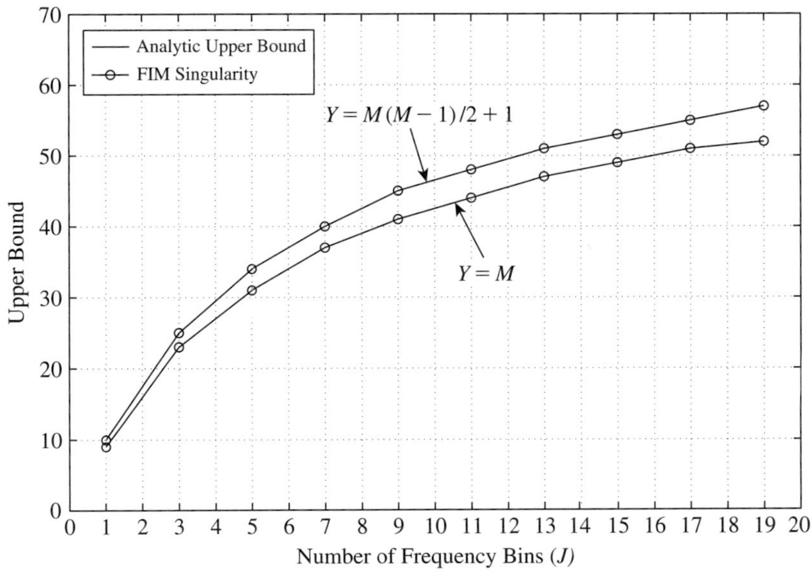

FIGURE 10.5 Comparison of the upper bounds on the number of resolvable emitters for uncorrelated signals with $Y = M$ and $Y = M(M-1)/2+1$.

signals are random, and we apply a Multiple-Signals Classification (MUSIC) approach.

MUSIC Approach

To avoid a multidimensional search for the multisource case, we follow the steps leading to the MUSIC algorithm (Schmidt, 1986). We assume that the noise spectral level is identical at all frequencies (flat). Recall the definition of the covariance matrix of the intercepted signals in Equation (10.13). The qth column vector of $\bar{\mathbf{A}}(u)\mathbf{B}$ is $\bar{\mathbf{A}}_q(u)\mathbf{b}_q$. The column vectors of $\bar{\mathbf{A}}(u)\mathbf{B}$ are orthogonal to the noise subspace and are contained in the signal subspace. Define the matrix $\bar{\mathbf{A}}_{\tilde{\mathbf{p}}}(u)$, which has the same structure as $\bar{\mathbf{A}}_q(u)$ for a source located at some point $\tilde{\mathbf{p}}$. We also use $\tilde{\mathbf{b}}$ as a general vector having the structure of $\mathbf{b}_q$.

Let $\mathbf{U}_s(u)$ be the $ML \times Q$ matrix consisting of the eigenvectors of $\hat{\mathbf{R}}_{\tilde{\mathbf{r}}}(u)$ corresponding to the Q largest eigenvalues. Following the MUSIC algorithm we propose the cost function (Weiss and Amar, 2005)

$$
\begin{aligned}
F(\tilde{\mathbf{p}},\tilde{\mathbf{b}}) &= \sum_{u=1}^{J} (\bar{\mathbf{A}}_{\tilde{\mathbf{p}}}(u)\tilde{\mathbf{b}})^H \mathbf{U}_s(u)\mathbf{U}_s^H(u)\bar{\mathbf{A}}_{\tilde{\mathbf{p}}}(u)\tilde{\mathbf{b}} \\
&= \tilde{\mathbf{b}}^H \left[\sum_{u=1}^{J} \bar{\mathbf{A}}_{\tilde{\mathbf{p}}}^H(u)\mathbf{U}_s(u)\mathbf{U}_s^H(u)\bar{\mathbf{A}}_{\tilde{\mathbf{p}}}(u)\right]\tilde{\mathbf{b}}
\end{aligned}
\tag{10.35}
$$

Note that, to facilitate a unique solution, we assume that the norm of $\tilde{\mathbf{b}}$ is one.

In the sequel we use the Rayleigh-Ritz theorem for maximization of a quadratic form (Rao, 2002), which we display next for easy reference.

Theorem 10.1. Given an $N \times N$ matrix $\mathbf{A}$ and an $N \times 1$ vector $\mathbf{x}$, define the Rayleigh quotient $R(\mathbf{x}) \triangleq \frac{\mathbf{x}^H \mathbf{A}\mathbf{x}}{\mathbf{x}^H \mathbf{x}}$. The vector $\mathbf{x}$ that maximizes $R(\mathbf{x})$ is the eigenvector of $\mathbf{A}$ that corresponds to the largest eigenvalue of $\mathbf{A}$. The maximum of $R(\mathbf{x})$ is then $\lambda_{\max}\{\mathbf{A}\}$.

Using Theorem 10.1 we obtain that, for any assumed position $\tilde{\mathbf{p}}$, the maximum of $F(\tilde{\mathbf{p}}, \tilde{\mathbf{b}})$ corresponds to the maximal eigenvalue of the $L \times L$ matrix $\mathbf{D}(\tilde{\mathbf{p}})$ defined by

$$\mathbf{D}(\tilde{\mathbf{p}}) \triangleq \sum_{u=1}^{J} \bar{\mathbf{A}}_{\tilde{\mathbf{p}}}^H(u)\mathbf{U}_s(u)\mathbf{U}_s^H(u)\bar{\mathbf{A}}_{\tilde{\mathbf{p}}}(u) \tag{10.36}$$

Therefore, Equation (10.35) reduces to

$$F(\tilde{\mathbf{p}}) = \lambda_{\max}\{\mathbf{D}(\tilde{\mathbf{p}})\} \tag{10.37}$$

The matrix $\mathbf{D}(\tilde{\mathbf{p}})$ is a function of the observed data (i.e., $\mathbf{U}_s(u)$) and the array response at each base station to an emitter located at $\tilde{\mathbf{p}}$. It is clear that the maximization of Equation (10.37) requires only a 2D search for emitters confined to a plane or a three-dimensional (3D) search in a general setting. This is a DPD approach. It is interesting to note that the dimensions of the matrix $\mathbf{D}(\tilde{\mathbf{p}})$ are $L \times L$ and are usually rather small.

Numerical Examples

We compared the performance of the DPD method with the traditional AOA approach by Monte Carlo (MC) simulations. The performance evaluation was based on the position root mean-square error (RMSE) defined as

$$\text{RMSE} \triangleq \sqrt{\frac{1}{N_{exp}} \sum_{i=1}^{N_{exp}} \|\hat{\mathbf{p}}_0(i) - \mathbf{p}_0\|^2} \tag{10.38}$$

where N_{exp} is the number of MC trials, and $\hat{\mathbf{p}}_0(i)$ is the estimated emitter position at the ith trial. We performed 100 MC trials in order to obtain the statistical properties of the performance.

Consider three base stations placed at coordinates $(2, -2), (2, 0), (2, 2)$ (in km) and two emitters placed at $(0, +1.5)$ and $(0, -1.5)$ (in km). Each transmitted waveform is a carrier amplitude modulated by a narrowband random Gaussian waveform. The signal is unknown to the receivers. Each base station is equipped with a ULA of only three antenna elements. Each location determination is based on 200 snapshots, each of 4.5 msec at a single frequency ($K = 200$, $J = 1$). The snapshot length ensures that the errors introduced by the finite-length fast Fourier transform (FFT) are 30 dB below the signal level. The SNR (at the base station

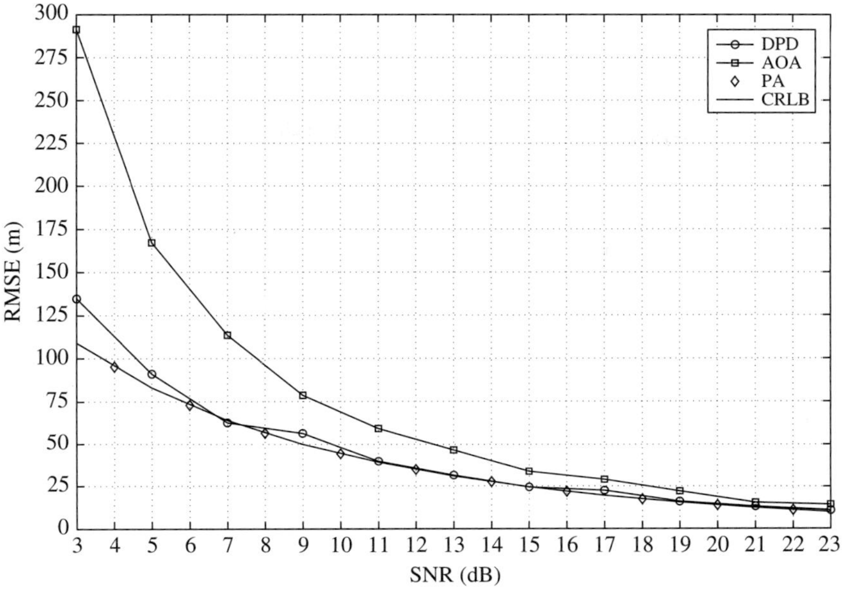

FIGURE 10.6 RMSE of DPD and traditional AOA, CRLB, and PA results for three base stations, two sources, and unknown waveforms.

receiving the strongest signal) is varied between 3 and 23 dB. The attenuation vector is selected as $\mathbf{b} = [1, 0.8, 0.4]^T$.

The results for each of the sources are shown in Figure 10.6. Also plotted is the theoretical performance analysis (PA) obtained by small error analysis of the DPD method. (For details see Weiss and Amar, 2005.) Notice that the theoretical results almost coincide with the CRLB. It is again clear that DPD outperforms AOA at low SNR, whereas the two methods are equivalent at high SNR. Consider now that three transmitters are placed at (0, 1.5), (0, −1.5), (−1, 0) (in km). Each base station collects 1000 snapshots, and attenuation is equal at all base stations. Since each base station is equipped with an array of only three elements, traditional AOA based on MUSIC fails. However, DPD works well, as shown in Figure 10.7.

Iterative Maximum Likelihood Approach

We now propose a solution for deterministic signals. The maximum likelihood estimation (MLE) of the parameter vector given by the observation model in Equation (10.9), assuming the known and flat noise spectral level, is

$$\hat{\boldsymbol{\mu}} = \underset{\boldsymbol{\mu}}{\operatorname{argmin}} \left\{ \sum_{k=1}^{K} \sum_{u=1}^{J} \left\| \bar{\mathbf{r}}(k,u) - \bar{\mathbf{A}}(u)\mathbf{B}\bar{\mathbf{s}}(k,u) \right\|^2 \right\} \tag{10.39}$$

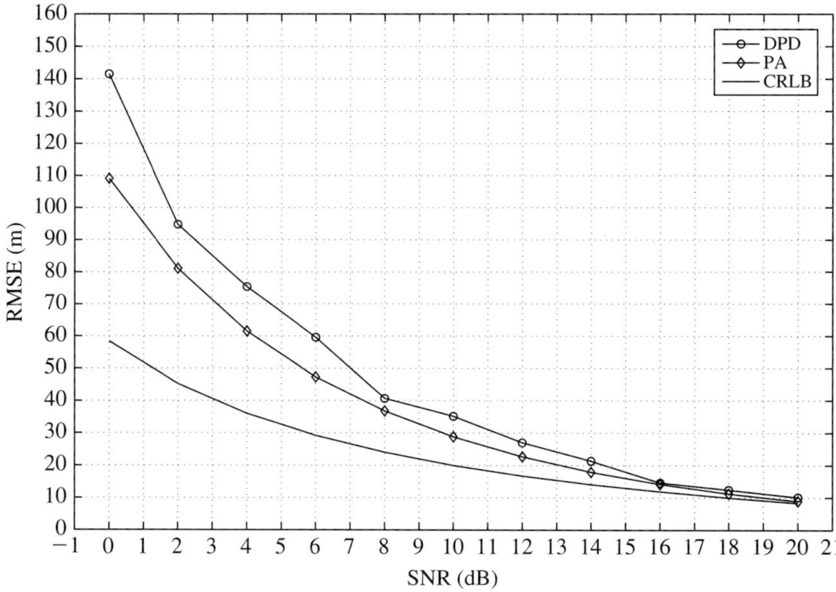

FIGURE 10.7 RMSE of DPD, CRLB, and traditional AOA, CRLB, and PA results for three base stations, three sources, and unknown waveforms.

To reduce the multidimensional search of the MLE, we resort to an iterative least-squares (LS) algorithm by modifying the technique presented in Weiss et al. (1988) to the geolocation model. The proposed Decoupled Geolocation ML (DGML) algorithm estimates the positions of Q emitters in a plane by a 2D search in each iteration instead of by a multidimensional search over all positions.

Assume we have an initial rough estimate of the vectors $\{\mathbf{p}_q, \mathbf{b}_q\}_{q=1}^{Q}$. As in many iterative algorithms, the convergence to the global minimum of the cost function depends on the initial estimate. Such an initial estimate can be provided by the DPD method previously presented. Let $\hat{\bar{\mathbf{A}}}(u)$, $\hat{\mathbf{B}}$ be the matrices $\bar{\mathbf{A}}(u)$, $\mathbf{B}$ computed with the initial estimate of $\{\mathbf{p}_q, \mathbf{b}_q\}_{q=1}^{Q}$. Now we minimize Equation (10.39) w.r.t. $\bar{\mathbf{s}}(k,u)$:

$$\hat{\bar{\mathbf{s}}}(k,u) = [(\hat{\bar{\mathbf{A}}}(u)\hat{\mathbf{B}})^H \hat{\bar{\mathbf{A}}}(u)\hat{\mathbf{B}}]^{-1} (\hat{\bar{\mathbf{A}}}(u)\hat{\mathbf{B}})^H \bar{\mathbf{r}}(k,u) \tag{10.40}$$

Define the $Q \times Q$ diagonal matrix, $\mathbf{E}_q$, where the entries on the main diagonal are all 1 except for the (q,q)-th entry, and let all other entries equal 0. Define the vector,

$$\hat{\bar{\mathbf{s}}}^{(q)}(k,u) \triangleq \mathbf{E}_q \hat{\bar{\mathbf{s}}}(k,u) \tag{10.41}$$

Also define a vector associated with the qth emitter,

$$\mathbf{w}_q(k,u) \triangleq \bar{\mathbf{r}}(k,u) - \hat{\bar{\mathbf{A}}}(u)\hat{\mathbf{B}}\hat{\bar{\mathbf{s}}}^{(q)}(k,u) \tag{10.42}$$

Assume that the positions and attenuations of all emitters, except those of the qth emitter, are equal to their true values. In such a case, using the definitions in Equation (10.6), Equation (10.42) reduces to

$$\mathbf{w}_q(k,u) = \bar{\mathbf{A}}_q(u)\mathbf{b}_q\bar{s}_q(k,u) + \bar{\mathbf{n}}(k,u) \tag{10.43}$$

This is the qth column vector of $\bar{\mathbf{A}}(u)\mathbf{B}$ multiplied by the qth signal $\bar{s}_q(k,u)$, and added with the noise. This motivates a decoupled search. Instead of minimizing Equation (10.39) we minimize

$$F = \sum_{k=1}^{K}\sum_{u=1}^{J}\left\|\mathbf{w}_q(k,u) - \bar{\mathbf{A}}_q(u)\mathbf{b}_q\hat{\bar{s}}_q(k,u)\right\|^2 \tag{10.44}$$

where $\hat{\bar{s}}_q(k,u)$ is the qth element of $\hat{\bar{\mathbf{s}}}(k,u)$. Minimizing Equation (10.44) w.r.t. $\hat{\bar{s}}_q(k,u)$ yields

$$\hat{\bar{s}}_q(k,u) = [(\bar{\mathbf{A}}_q(u)\mathbf{b}_q)^H\bar{\mathbf{A}}_q(u)\mathbf{b}_q]^{-1}(\bar{\mathbf{A}}_q(u)\mathbf{b}_q)^H\mathbf{w}_q(k,u) \tag{10.45}$$

By substituting Equation (10.45) in Equation (10.44) we see that the estimate of $\mathbf{p}_q$ and $\mathbf{b}_q$ is given by

$$(\hat{\mathbf{p}}_q,\hat{\mathbf{b}}_q) = \underset{\mathbf{p}_q,\mathbf{b}_q}{\operatorname{argmax}}\left\{\|\mathbf{b}_q\|^{-2}\sum_{k=1}^{K}\sum_{u=1}^{J}\left|\mathbf{b}_q^H\bar{\mathbf{A}}_q^H(u)\mathbf{w}_q(k,u)\right|^2\right\} \tag{10.46}$$

Recall that we assume that $\|\mathbf{b}_q\|^2 = 1$. Define the sample correlation matrix associated with the qth signal and the uth Fourier coefficient:

$$\hat{\mathbf{W}}_q(u) \triangleq \frac{1}{K}\sum_{k=1}^{K}\mathbf{w}_q(k,u)\mathbf{w}_q^H(k,u) \tag{10.47}$$

and the $L \times L$ Hermitian matrix:

$$\mathbf{D}_q \triangleq \sum_{u=1}^{J}\bar{\mathbf{A}}_q(u)\hat{\mathbf{W}}_q(u)\bar{\mathbf{A}}_q(u) \tag{10.48}$$

Substituting Equation (10.48) in Equation (10.46) yields

$$(\hat{\mathbf{p}}_q,\hat{\mathbf{b}}_q) = \underset{\mathbf{p}_q,\mathbf{b}_q}{\operatorname{argmax}}\{\mathbf{b}_q^H\mathbf{D}_q\mathbf{b}_q\} \tag{10.49}$$

Using Theorem 10.1 we obtain that, for any given emitter position, the vector $\hat{\mathbf{b}}_q$ that maximizes the expression is equal to the eigenvector $\mathbf{u}_{\max}$ corresponding to the largest eigenvalue of $\mathbf{D}_q$—that is, $\hat{\mathbf{b}}_q = \mathbf{u}_{\max}(\mathbf{D}_q)$—and the estimated position is

$$\hat{\mathbf{p}}_q = \underset{\mathbf{p}_q}{\operatorname{argmax}}\ \lambda_{\max}\{\mathbf{D}_q\} \tag{10.50}$$

Therefore, the maximum of the cost function is found by performing a 2D grid search over the area of interest. For each grid point the largest eigenvalue of $\mathbf{D}_q$

is the local value of the cost function. The grid point associated with the largest value of the cost function is the temporary estimate of the qth source position.

Numerical Examples

We demonstrate the performance of the proposed iterative algorithm via MC simulations. In all simulations (unless stated otherwise) we used the following parameters. Four base stations ($L=4$), located at the corners of a square at positions $(X_1, Y_1)=[0,0]$, $(X_2, Y_2)=[1,0]$, $(X_3, Y_3)=[1,1]$, and $(X_4, Y_4)=[0,1]$, with all units in kilometers. The rotation angles of the arrays, measured between the array baseline and the west–east axis, denoted by ϕ_ℓ, are $\phi_1=135°$, $\phi_2=45°$, $\phi_3=135°$, and $\phi_4=45°$. The signals and the noise are random, complex Gaussian vectors with covariance matrices $\sigma_s^2\mathbf{I}$ and $\sigma^2\mathbf{I}$, respectively. We assume the signals are narrowband ($J=1$). We generated 50 independent Fourier coefficients ($K=50$) of the signals and noise processes. Our definition of the SNR is $\text{SNR}[dB] \triangleq 10\log_{10}(\sigma_s^2/\sigma^2)$. The initial emitter positions were set randomly as $\mathbf{p}_q^{(0)}=\mathbf{p}_q+100\times\boldsymbol{\xi}_q$, $q=1,\ldots,Q$, where $\boldsymbol{\xi}_q$ is a 2×1 Gaussian vector with zero mean and covariance equal to the identity matrix. The initial values of the attenuations were chosen randomly. We performed a coarse search with 10-m resolution over the full area of interest and a fine search with 1-m resolution near the peaks found in the course search. The convergence threshold of the algorithm was set to $\varepsilon=1$ m.

In the following examples we consider several values of SNR. For each SNR we performed 50 independent trials. In each the initial positions and initial attenuations were randomly chosen. We then estimated the positions and attenuations of the emitters in each iteration step (the number of iteration steps in each trial was not necessarily equal). We evaluated the performance using the following RMSE for each iteration step:

$$\text{RMSE}_q(i) \triangleq \frac{1}{N}\sum_{n=1}^{N}\sqrt{(\hat{x}_{n,q}(i)-x_q)^2+(\hat{y}_{n,q}(i)-y_q)^2}, \quad i=1,\ldots,I$$

where $(\hat{x}_{n,q}(i), \hat{y}_{n,q}(i))$ is the estimated position of the qth emitter at the nth experiment and at the ith iteration step; I is the maximum number of iterations in all trials; and N is the number of experiments with at least i iterations. Denote by N_{exp} the number of experiments. As indicated previously, $N_{exp}=50$. The value of N/N_{exp} (in percent) is shown in Figures 10.8 and 10.10

Assume now that we have two emitters ($Q=2$) when each of the stations is equipped with a uniform circular array (UCA). Each array consists of five elements—four equally distributed on a circle with radius $\lambda/2$ and the fifth (the reference) located at the circle center, where λ is the wavelength of the transmitted signals. In Figures 10.8 we show the RMSE of the first emitter using circular arrays in all stations. Each array consists of five elements—four equally distributed on a circle with radius $\lambda/2$ and the fifth (the reference) located at

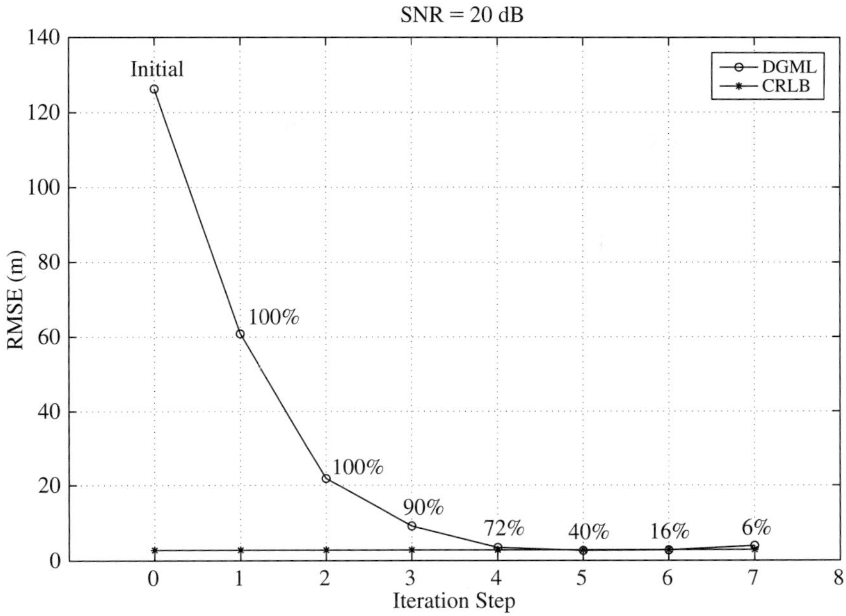

FIGURE 10.8 Algorithm convergence for two emitters using UCA.

the circle center. The SNR is 20 dB. The numbers indicate N/N_{exp} in percent. As can be seen, after only four iterations the CRLB is reached.

Also assume we have two emitters ($Q=2$). A ULA with three sensors ($M=3$) is located at each station. The spacing between adjacent sensors is $\lambda/2$. In Figure 10.9 we show the RMSE of the first emitter versus the SNR. The SNR varies between -4 dB and 4 dB with a 1-dB step. The number of sections is $K=50$. As expected, the RMSE converges to the CRLB as the SNR increases. Also, an SNR of -2 dB is sufficient for approaching the CRLB.

Finally, assume we have three emitters ($Q=3$) and a ULA with three sensors ($M=3$) located at each station. In Figure 10.10 we plot the RMSE of the three emitters for SNR = 20 dB. As can be seen, after five iterations the CRLB is reached in 92% of the trials. Only in 2% of the cases are 10 iterations required for convergence.

10.2.4 DPD for Known Transmitted Waveforms

In certain applications the transmitted waveforms are known to the location system. In cellular systems, for example, synchronization and training sequences are transmitted periodically and are known a priori. Moreover, it is possible to detect the data sequence of a digitally modulated signal and then restore the complex signal envelope based on the known modulation scheme.

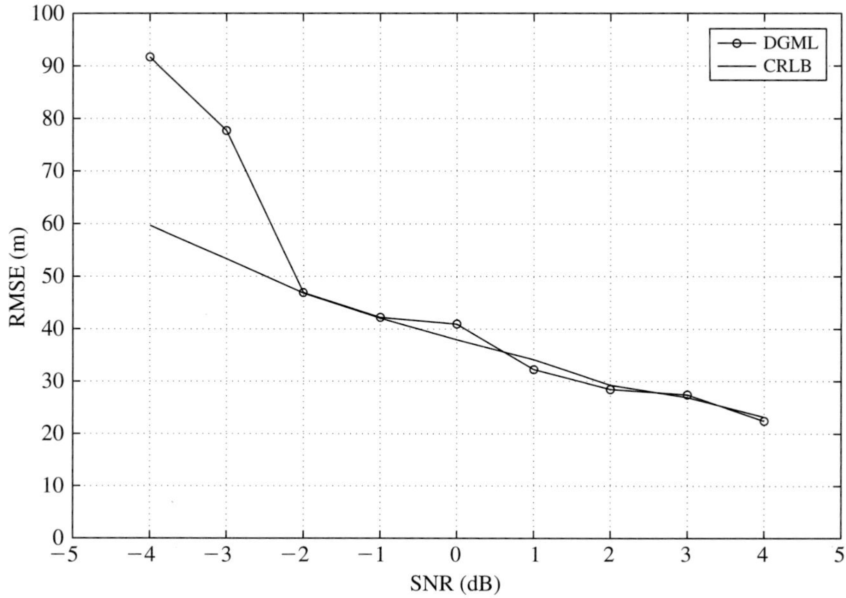

FIGURE 10.9 RMSE versus SNR.

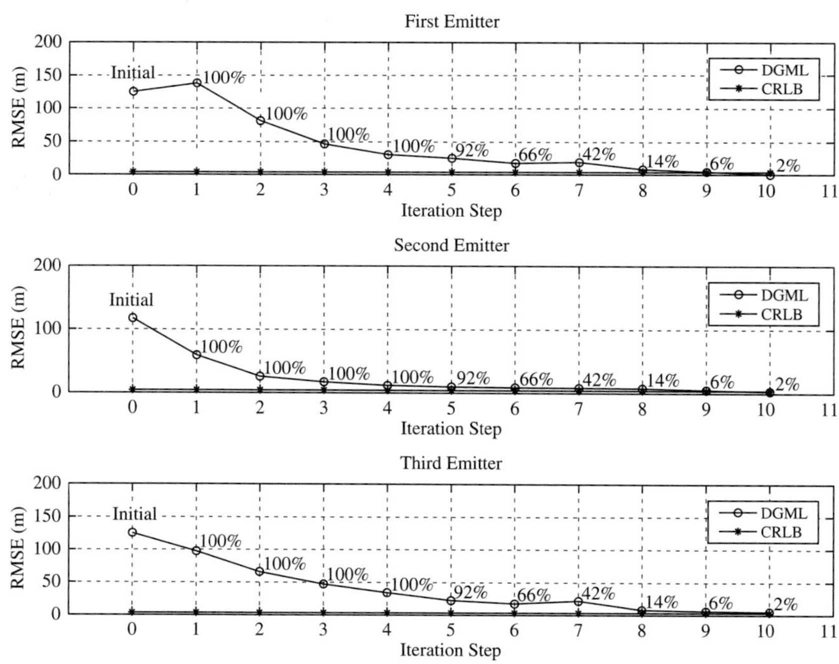

FIGURE 10.10 Convergence of the algorithm for three emitters.

In Amar and Weiss (2006) it is shown that the LLF in Equation (10.9) is given under the assumption of a known flat noise spectral level as

$$F_1 = \mathrm{tr}\left\{\sum_{u=1}^{J}\left[\bar{\mathbf{A}}(u)\mathbf{B} - \hat{\mathbf{C}}(u)\right]^H\left[\bar{\mathbf{A}}(u)\mathbf{B} - \hat{\mathbf{C}}(u)\right]\hat{\mathbf{R}}_{\bar{\mathbf{s}}\bar{\mathbf{s}}}(u)\right\} \tag{10.51}$$

where

$$\hat{\mathbf{C}}(u) \triangleq \hat{\mathbf{R}}^H_{\bar{\mathbf{s}}\bar{\mathbf{r}}}(u)\hat{\mathbf{R}}_{\bar{\mathbf{s}}\bar{\mathbf{s}}}(u) \tag{10.52}$$

and

$$\hat{\mathbf{R}}_{\bar{\mathbf{s}}\bar{\mathbf{s}}}(u) \triangleq \frac{1}{K}\sum_{k=1}^{K}\bar{\mathbf{s}}(k,u)\bar{\mathbf{s}}^H(k,u) \tag{10.53}$$

$$\hat{\mathbf{R}}_{\bar{\mathbf{s}}\bar{\mathbf{r}}}(u) \triangleq \frac{1}{K}\sum_{k=1}^{K}\bar{\mathbf{s}}(k,u)\bar{\mathbf{r}}^H(k,u) \tag{10.54}$$

Assuming that the signals are uncorrelated, $\hat{\mathbf{R}}_{\bar{\mathbf{s}}\bar{\mathbf{s}}}(u)$ is asymptotically diagonal. Therefore, the cost function F_1 can be written as

$$F_1 = \sum_{q=1}^{Q} F_1(q) \tag{10.55}$$

where $F_1(q)$ is the cost function associated with the qth emitter. That is, the estimation of the position and attenuations of each emitter is decoupled from the estimation of the positions and attenuations of the others.

Define the matrix $\bar{\mathbf{A}}_{\tilde{\mathbf{p}}}(u)$, which has the same structure as $\bar{\mathbf{A}}_q(u)$ for a source located at some point $\tilde{\mathbf{p}}$. We also use $\tilde{\mathbf{b}}$ as a general vector having the structure $\mathbf{b}_q$. It can be shown that the cost function $F_1(q)$ is given by

$$F_1(q) = \sum_{u=1}^{J}\|\hat{\mathbf{c}}_q(u) - \bar{\mathbf{A}}_{\tilde{\mathbf{p}}}(u)\tilde{\mathbf{b}}\|^2 \tag{10.56}$$

where $\hat{\mathbf{c}}_q(u)$ represents the qth column of $\hat{\mathbf{C}}(u)$. The vector $\tilde{\mathbf{b}}$ that minimizes the cost function is given by

$$\hat{\tilde{\mathbf{b}}} = \left[\sum_{u=1}^{J}\bar{\mathbf{A}}^H_{\tilde{\mathbf{p}}}(u)\bar{\mathbf{A}}_{\tilde{\mathbf{p}}}(u)\right]^{-1}\sum_{u=1}^{J}\bar{\mathbf{A}}^H_{\tilde{\mathbf{p}}}(u)\hat{\mathbf{c}}_q(u) \tag{10.57}$$

Substituting Equation (10.57) in Equation (10.56), we get a cost function that depends on $\tilde{\mathbf{p}}$ only. It is then easy to verify that minimizing Equation (10.56) is equivalent to maximizing

$$F_1(q) = \left\|\sum_{u=1}^{J}\bar{\mathbf{A}}^H_{\tilde{\mathbf{p}}}(u)\hat{\mathbf{c}}_q(u)\right\|^2 = \sum_{\ell=1}^{L}\left|\sum_{u=1}^{J}e^{j\omega_u\tau_\ell(\tilde{\mathbf{p}})}\mathbf{a}^H_\ell(\tilde{\mathbf{p}})\hat{\mathbf{c}}^{(\ell)}_q(u)\right|^2 \tag{10.58}$$

where $\hat{\mathbf{c}}_q^{(\ell)}(u)$ is the ℓth subvector of $\hat{\mathbf{c}}_q(u)$ associated with the ℓth base station. The estimated position of the qth emitter is given as

$$\hat{\tilde{\mathbf{p}}} = \underset{\tilde{\mathbf{p}}}{\operatorname{argmax}} \, F_1(q) \tag{10.59}$$

The estimation of each Q emitter is performed as described in Equation (10.59) by substituting the associated vector $\hat{\mathbf{c}}_q^{(\ell)}(u)$, $q = 1, \ldots, Q$ in Equation (10.58).

Note that Equation (10.58) indicates that $F_1(q)$ is a sum of L distinct cost functions, each associated with a distinct base station. This is the reason that DPD outperforms methods that maximize the cost function at each base station independently.

An important aspect of an algorithm is its computation load. In Amar and Weiss (2006, Appendix H), it is shown that the ratio between the total number of operations required for AOA localization and for DPD localization is

$$\frac{(\mathrm{AOA})_{op}}{(\mathrm{DPD})_{op}} = \frac{2\sum_{q=1}^{Q} q^L}{MQ} > \frac{2}{M} Q^{L-1} \tag{10.60}$$

As the number of sources increases, this ratio increases exponentially, due to the complexity involved in the "angle source" association required in AOA localization of several sources simultaneously. However, for a single emitter no association is required and then the number of AOA operations is fewer than the number of DPD operations.

Model Errors Analysis

Errors in DPD estimates can arise from no LOS propagation, array calibration errors, sensor-positioning errors, and so forth. For example, in practice the assumption that sensor locations are known precisely does not always hold. Sensors mounted on moving platforms such as cars, ships, and airplanes are prone to location errors. In Amar and Weiss (2006) we discussed imprecise knowledge of sensor locations and scattering environment. Here we consider the former case only.

Assume that L_e out of L arrays suffer from position errors and in each array only M_e of M elements is displaced. Assume that sensor position errors are Gaussian random variables with zero mean and variance ε^2. In Amar and Weiss (2006) it is shown that the effect of sensor position errors is approximately on the same order as the effect of observation noise if

$$\varepsilon^2 \cong \frac{2}{(2\pi/\lambda)^2} \ln\left(\frac{1}{1 - \frac{ML}{2M_e L_e} \gamma^2} \right) \tag{10.61}$$

where

$$\gamma^2 \triangleq \frac{\sigma^2}{K[\hat{\mathbf{R}}_{\bar{\mathbf{s}}\bar{\mathbf{s}}}]_{q,q}} \tag{10.62}$$

which can be interpreted as the SNR.

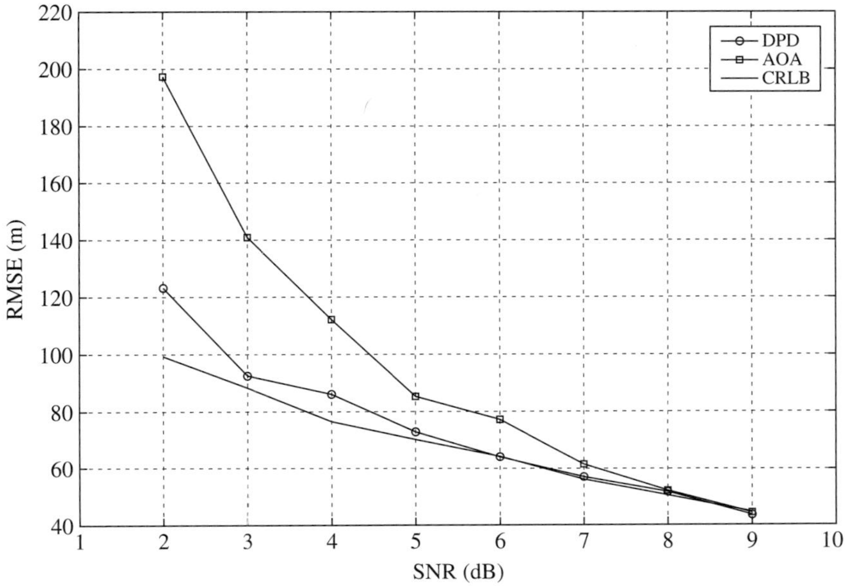

FIGURE 10.11 RMSE of DPD, traditional AOA, and CRLB for known waveforms. Four base stations, each equipped with a 5-element circular array, are used.

Numerical Examples

We compare the performance of the DPD method with that of the traditional AOA approach for known signals. Each location determination is based on a specified number, K, of 4.5-msec snapshots using a single-frequency bin ($J = 1$). The snapshot length ensures that the errors introduced by the finite-length FFT are 30 dB below the signal level. The transmitted waveforms are realizations of a normal Gaussian in random process.

We used four base stations located at $(-2, -2)$, $(-2, +2)$, $(2, -2)$, $(2, +2)$ (in km) and a single source at $(1, 1)$ (in km). Each base station is equipped with a UCA of five elements. The waveforms are known, and the number of snapshots is 1000. The attenuation to two base stations is 0 dB; to the other two, -10 dB. The accuracy results are plotted in Figure 10.11. The plots indicate that DPD is superior to the traditional approach of independent AOA estimates at each base station.

We also consider displaced sensors. Each of the four stations is equipped with a ULA of 11 elements. Only the array elements of the first station are displaced. The standard deviation of the location error is ϵ. We discuss the performance of DPD and AOA as a function of ε/λ and the number of displaced elements, M_e, with that number changed between 1 and 11. The RMSE of both DPD and AOA is shown in Figure 10.12, where we plot DPD and AOA error as a function of the standard deviation for three cases: 1, 6, and 11 displaced elements. Note that for a small number of displaced elements, the difference between DPD and AOA is small. However, as that number increases, so does the difference.

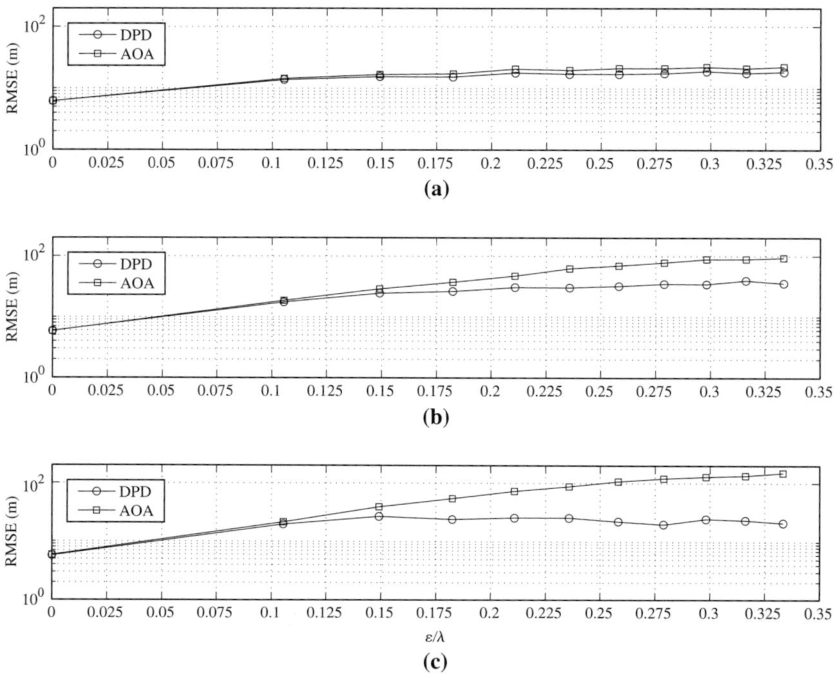

FIGURE 10.12 RMSE of AOA and DPD for (a) 1, (b) 6, and (c) 11 displaced elements.

10.3 LOCALIZATION FOR NONSTATIONARY GEOMETRY

The use of the Doppler effect for localization is seen in radar, sonar, and satellites. Here, focus on locating a stationary radio emitter by moving receivers. The motion induces a frequency shift that is proportional to the signal frequency and to the radial velocity of each receiver toward (or away from) the emitter. It is assumed that receiver location and velocity are known and therefore that emitter location can be estimated.

One of the localization methods discussed in the literature is differential Doppler (DD), also known as frequency difference of arrival, which consists of measuring frequency differences between receivers since it eliminates the need to know the exact transmit frequency. DD can be obtained by estimating the frequency at each receiver and then computing the difference (Chestnut, 1982, Section III) or by directly measuring the frequency difference using signal cross-correlation (Knapp and Carter, 1977; Stein, 1993). The latter approach requires the transfer of raw data to the LPU and therefore is associated with higher transmission rates compared with the former approach. A data compression technique was recently proposed to reduce the transmission load (Fowler, 2000).

Differential Doppler was proposed for locating a moving emitter using stationary receivers in sonar (Schultheiss and Weinstein, 1979; Weinstein, 1982) and artillery (Weinstein and Levanon, 1980) applications. A similar problem of localizing a moving tone source was discussed in Chan and Jardine (1990) and Chan and Towers (1992a,b) without employing DD.

A system for localization of a stationary emitter can be implemented by at least a single platform carrying at least two receivers. Several platforms, each carrying a single receiver, are more common. Becker (1992) investigated the positioning error of a radar transmitting a pure, stable, unknown tone based on multiple measurements of frequency and bearing taken along a single platform trajectory. Becker (1999) extended his previous contribution by considering the drift and frequency hopping of the transmitted signal. Considering measurements collected by a single receiver, Fowler (2001) investigated the accuracy of 3D localization using terrain data. Levanon (1989) discussed a DD location system based on two receivers on a single platform and compared its performance with interferometer measurements.

The Doppler effect has also been used for stationary emitter geolocation by satellites. The SARSAT/COSPAS (Scales and Swanson, 1984) and ARGOS (Bessis, 1981) systems determine the emitter's position using a single satellite receiver. The satellite relays the observed signal to an Earth station where the instance of zero Doppler shift is determined. Zero Doppler shift is associated with the point where the satellite is closest to the emitter, also known as the point of closest approach (PCA). The emitter's position is then determined from the PCA and the known satellite location and velocity. An error analysis of these systems was presented by Levanon and Zaken (1985).

An interesting and related application of localization based on the Doppler effect was demonstrated by the U.S. Navy TRANSIT system (Levanon, 1980). The localization of a stationary emitter with multiple moving platforms was discussed by Haworth et al. (1997), who presented a system for localizing satellite interference sources based on DD and TDOA. Ho and Chan (1997) considered the same system and provided an analytical solution for source location. Later, Pattison and Chou (2000) examined the effect of satellite position and velocity errors on this solution.

All of the just mentioned approaches use two steps for localization. In the first step the Doppler frequency shifts, or their differences, are estimated without the constraint that all estimates must correspond to the same emitter location and the same transmitted frequency. Only in the second step is the location estimated based on the results of the first step. For this reason, these methods are not guaranteed to yield optimal localization results. The objective of this section is to apply the ideas of the DPD approach to this location problem.

10.3.1 Problem Formulation

Consider a stationary narrowband radio emitter and L moving receivers. The receivers are assumed to be synchronized in frequency and time. The emitter's position is denoted by the vector of coordinates $\mathbf{p}_0$. Each receiver intercepts the transmitted signal at K short intervals along its trajectory. Let $\mathbf{p}_{\ell,k}$ and $\mathbf{v}_{\ell,k}$, where $k=1,\ldots,K$ and $\ell=1,\ldots,L$, denote the position and velocity vectors of the ℓth receiver at the kth interception interval, respectively.

The complex signal observed by the ℓth receiver at the kth interception interval at time t is

$$r_{\ell,k}(t)=b_{\ell,k}s_k(t)e^{j2\pi f_{\ell,k}t}+w_{\ell,k}(t), \quad 0\leq t\leq T \tag{10.63}$$

where T is the observation time interval; $b_{\ell,k}$ is an unknown complex scalar representing the path attenuation at the kth interception interval observed by the ℓ-th receiver; $s_k(t)$ is the observed narrowband signal envelope during the kth interception interval, which may be known or unknown depending on the application; $w_{\ell,k}(t)$ is a white zero-mean complex Gaussian noise; and $f_{\ell,k}$ is the frequency observed by the ℓth receiver during the kth interception interval given by

$$f_{\ell,k} \triangleq [f_c+\nu_k][1+\mu_{\ell,k}(\mathbf{p}_0)] \tag{10.64}$$

$$\mu_{\ell,k}(\mathbf{p}_0) \triangleq \frac{1}{c}\frac{\mathbf{v}_{\ell,k}^T[\mathbf{p}_0-\mathbf{p}_{\ell,k}]}{\|\mathbf{p}_0-\mathbf{p}_{\ell,k}\|} \tag{10.65}$$

where f_c is the nominal frequency of the transmitted signal, assumed known; ν_k is the unknown transmitted frequency shift due to source instability during the kth interception interval; and c is the signal's propagation speed. Since $\mu_{\ell,k} \ll 1$ and $\nu_k \ll f_c$, Equation (10.64) can be approximated as

$$f_{\ell,k} \cong \nu_k+f_c[1+\mu_{\ell,k}(\mathbf{p}_0)] \tag{10.66}$$

where the term $\nu_k\mu_{\ell,k}$, which is negligible w.r.t. all other terms, is omitted.

Assume that each receiver performs a down-conversion of the intercepted signal by f_c. Thus, Equation (10.64) is replaced by

$$\bar{f}_{\ell,k}=f_{\ell,k}-f_c \cong \nu_k+f_c\mu_{\ell,k}(\mathbf{p}_0) \tag{10.67}$$

The transmitted frequency is assumed to be constant during the interception interval, T. The down-converted signal is sampled at times $t_n=nT_s$, where $n=0,\ldots,N-1$ and $T_s=T/(N-1)$. Denote the sampled signal at the kth interception interval by $r_{\ell,k}[n]\triangleq r_{\ell,k}(nT_s)$. Then Equation (10.63) can be written in vector form as

$$\mathbf{r}_{\ell,k}=b_{\ell,k}\mathbf{A}_{\ell,k}\mathbf{C}_k\mathbf{s}_k+\mathbf{w}_{\ell,k} \tag{10.68}$$

where

$$\mathbf{r}_{\ell,k} \triangleq \left[r_{\ell,k}[0], \ldots, r_{\ell,k}[N-1]\right]^T \tag{10.69}$$

$$\mathbf{w}_{\ell,k} \triangleq \left[w_{\ell,k}[0], \ldots, w_{\ell,k}[N-1]\right]^T \tag{10.70}$$

$$\mathbf{s}_k \triangleq \left[s_k[0], \ldots, s_k[N-1]\right]^T \tag{10.71}$$

$$\mathbf{A}_{\ell,k} \triangleq \text{diag}\left\{1, e^{j2\pi f_c \mu_{\ell,k} T_s}, \ldots, e^{j2\pi f_c \mu_{\ell,k}(N-1)T_s}\right\} \tag{10.72}$$

$$\mathbf{C}_k \triangleq \text{diag}\left\{1, e^{j2\pi \nu_k T_s}, \ldots, e^{j2\pi \nu_k (N-1)T_s}\right\} \tag{10.73}$$

Note that the preceding equations are accurate only if all receivers are synchronized in frequency and time. Moreover, the signal complex envelope $\mathbf{s}_k$ is the same at all spatially separated receivers provided that the signal bandwidth (rate of signal change) is small compared to the inverse of the propagation time between receivers. This places a restriction on the receiver spatial separation for a given signal bandwidth.

The problem can be briefly stated as follows: Given the observation vectors $\{\mathbf{r}_{\ell,k}\}$ in Equation (10.68), estimate the position of the emitter.

10.3.2 Differential Doppler Localization Method

To simplify our description of the DD positioning approach, we assume two receivers. Denote by $\widehat{\Delta f}_k$ the estimated frequency difference between the first receiver and the second receiver during the kth interception interval. For example, the MLE of the frequency difference f between two sequences $r_{1,k}[n], r_{2,k}[n]$ of length N can be obtained by converting Equation 14 in Stein (1993) into the discrete time domain:

$$\widehat{\Delta f}_k = \underset{f}{\text{argmax}} \left| \sum_{n=0}^{N-1} r_{1,k}[n] r_{2,k}^*[n] e^{j2\pi f n} \right| \tag{10.74}$$

In our case, $r_{1,k}[n]$ represents the sampled output of a receiver at the the kth interception interval, and $r_{2,k}[n]$ may represent either the sampled output of a second receiver or samples of the known waveform.

The frequency difference is used to eliminate the unknown transmitted frequency offset ν_k. Using Equation (10.67) the frequency difference associated with the kth interception interval is given by

$$\widehat{\Delta f}_k = f_c \Delta m_k(\mathbf{p}_0) + \epsilon_k \tag{10.75}$$

where

$$\Delta m_k(\mathbf{p}_0) \triangleq \mu_{1,k}(\mathbf{p}_0) - \mu_{2,k}(\mathbf{p}_0) \tag{10.76}$$

and ϵ_k is an error reflecting the measurement errors and all other model errors. Using the LS principle, the position estimator is given by

$$\hat{\mathbf{p}}_0 = \arg\min_{\mathbf{p}} \sum_{k=1}^{K} |\widehat{\Delta f}_k - f_c \Delta m_k(\mathbf{p})|^2 \tag{10.77}$$

Note that for accurate DD estimates the receivers should be frequency-synchronized with high precision. Otherwise, these estimates will affect the measurements.

10.3.3 DPD Localization Approach

Consider the observation vectors in Equation (10.68). The information on the emitter's position is embedded in each of the matrices $\mathbf{A}_{\ell,k}$. This position is common to all observations at all interception intervals, so we estimate it as the position that best explains all the data together. Because of its excellent asymptotic properties (consistency and efficiency), we focus on the MLE.

The LLF of the observation vectors is given (up to an additive constant) by

$$L_1 = -\frac{1}{\sigma_n^2} \sum_{k=1}^{K} \sum_{\ell=1}^{L} \|\mathbf{r}_{\ell,k} - b_{\ell,k} \mathbf{A}_{\ell,k} \mathbf{C}_k \mathbf{s}_k\|^2 \tag{10.78}$$

The path attenuation scalars that maximize Equation (10.78) are

$$\begin{aligned} \hat{b}_{\ell,k} &= [(\mathbf{A}_{\ell,k}\mathbf{C}_k\mathbf{s}_k)^H \mathbf{A}_{\ell,k}\mathbf{C}_k\mathbf{s}_k]^{-1} (\mathbf{A}_{\ell,k}\mathbf{C}_k\mathbf{s}_k)^H \mathbf{r}_{\ell,k} \\ &= (\mathbf{A}_{\ell,k}\mathbf{C}_k\mathbf{s}_k)^H \mathbf{r}_{\ell,k} \end{aligned} \tag{10.79}$$

where we assume, without loss of generality, that $\|\mathbf{s}_k\|^2 = 1$ and use the special structure of $\mathbf{A}_{\ell,k}$ and $\mathbf{C}_k$. Substitution of Equation (10.79) in Equation (10.78) yields

$$L_1 = -\frac{1}{\sigma_n^2} \left[\sum_{k=1}^{K} \sum_{\ell=1}^{L} \|\mathbf{r}_{\ell,k}\|^2 - |(\mathbf{A}_{\ell,k}\mathbf{C}_k\mathbf{s}_k)^H \mathbf{r}_{\ell,k}|^2 \right] \tag{10.80}$$

Since $\|\mathbf{r}_{\ell,k}\|^2$ is independent of the parameters, instead of maximizing Equation (10.80) we can now maximize the cost function L_2 given by

$$L_2 = \sum_{k=1}^{K} \sum_{\ell=1}^{L} |(\mathbf{A}_{\ell,k}\mathbf{C}_k\mathbf{s}_k)^H \mathbf{r}_{\ell,k}|^2 = \sum_{k=1}^{K} (\mathbf{C}_k\mathbf{s}_k)^H \mathbf{Q}_k \mathbf{C}_k \mathbf{s}_k \tag{10.81}$$

where we define the $N \times N$ Hermitian matrix $\mathbf{Q}_k$ as

$$\mathbf{Q}_k \triangleq \mathbf{V}_k \mathbf{V}_k^H \tag{10.82}$$

and where the $N \times L$ matrix $\mathbf{V}_k$ is given as

$$\mathbf{V}_k \triangleq \left[\mathbf{A}_{1,k}^H \mathbf{r}_{1,k}, \ldots, \mathbf{A}_{L,k}^H \mathbf{r}_{L,k}\right] \tag{10.83}$$

We now consider the two cases of unknown and a priori known transmitted signals.

Unknown Transmitted Signals

Define the unknown $N \times 1$ vector $\mathbf{u}_k \triangleq \mathbf{C}_k \mathbf{s}_k$. The cost function in Equation (10.81) is maximized by maximizing each of the K quadratic forms w.r.t. $\mathbf{u}_k$. Thus, according to Theorem 10.1 the vector $\mathbf{u}_k$ should be selected as the eigenvector corresponding to the largest eigenvalue of the matrix $\mathbf{Q}_k$. The cost function in Equation (10.81) therefore reduces to

$$L_{us} = \sum_{k=1}^{K} \lambda_{\max}\{\mathbf{Q}_k\} \tag{10.84}$$

The dimension of $\mathbf{Q}_k$ is $N \times N$, and thus it increases with the number of data samples. Determining the eigenvalues of $\mathbf{Q}_k$ can in turn result in high computation effort. Instead, define the $L \times L$ matrix

$$\bar{\mathbf{Q}}_k \triangleq \mathbf{V}_k^H \mathbf{V}_k \tag{10.85}$$

According to Rao (2002, pp. 42–43) we see that the nonzero eigenvalues of $\mathbf{Q}_k$ and $\bar{\mathbf{Q}}_k$ are identical. Therefore, recalling the definition in Equation (10.82), $\mathbf{Q}_k$ can be replaced with $\bar{\mathbf{Q}}_k$. This substantially reduces the computation load since $L \ll N$.

The estimated emitter's position $\hat{\mathbf{p}}_0$ is determined by a grid search. For any grid point $\mathbf{p}$ in the position space, evaluate Equation (10.84) with $\mathbf{Q}_k$ replaced by $\bar{\mathbf{Q}}_k$ and obtain

$$\bar{L}_{us}(\mathbf{p}) = \sum_{k=1}^{K} \lambda_{\max}\{\bar{\mathbf{Q}}_k\} \tag{10.86}$$

The estimated emitter's position is then given by

$$\hat{\mathbf{p}}_0 = \underset{\mathbf{p}}{\operatorname{argmax}}\{\bar{L}_{us}(\mathbf{p})\} \tag{10.87}$$

An algorithm that uses Equation (10.87) to estimate the emitter position is usually more precise than a two-step procedure.

Known Transmitted Signals

Define the vector $\mathbf{c}_k \triangleq \operatorname{diag}\{\mathbf{C}_k\}$, the matrix $\mathbf{S}_k \triangleq \operatorname{diag}\{\mathbf{s}_k\}$, and the $N \times N$ matrix:

$$\mathbf{G}_k \triangleq \mathbf{S}_k^H \mathbf{Q}_k \mathbf{S}_k \tag{10.88}$$

Now Equation (10.81) can be simplified:

$$L_2 = \sum_{k=1}^{K} \mathbf{c}_k^H \mathbf{G}_k \mathbf{c}_k \tag{10.89}$$

Equation (10.89) is a polynomial in $z \triangleq e^{j2\pi\nu T_s}$. Define by

$$\alpha_n \triangleq \mathbf{1}^T \text{diag}\{\mathbf{G}_k; n\}, \quad n = 0, \pm 1, \ldots, \pm(N-1) \tag{10.90}$$

where $\text{diag}(\mathbf{G}_k; n)$ is a column vector containing the nth diagonal elements of the matrix $\mathbf{G}_k$, α_0 is the sum of the main diagonal elements, and $\alpha_{N-1}, \alpha_{-(N-1)}$ are the $(1, N)$-th and $(N, 1)$-th elements of $\mathbf{G}_k$, respectively. Since $\mathbf{G}_k$ is Hermitian, $\alpha_n = \alpha^*_{-n}$. Let

$$\beta_m \triangleq \begin{cases} \alpha_0, & m = 0 \\ 2\alpha_m, & m = 1, \ldots, N-1 \end{cases} \tag{10.91}$$

It can be shown that

$$\mathbf{c}_k^H \mathbf{G}_k \mathbf{c}_k = \sum_{i,j=1}^{N-1} \mathbf{G}_k[i,j] z^{i-j} = \sum_{n=-N+1}^{N-1} \alpha_n z^m = \Re\left\{\sum_{m=0}^{N-1} \beta^*_m z^{-m}\right\} \tag{10.92}$$

Thus, in order to find the ν_k that maximizes $\mathbf{c}_k^H \mathbf{G}_k \mathbf{c}_k$, compute the FFT of the sequence $\{\beta^*_m\}_{m=0}^{N-1}$ and take the real part of the result. The FFT length should satisfy $M \geq N$. If the largest FFT coefficient is the m_0th coefficient, then

$$\hat{\nu}_k = \begin{cases} \frac{m_0}{MT_s}, & m_0 < M/2 \\ -\frac{M-m_0}{MT_s}, & m_0 \geq M/2 \end{cases} \tag{10.93}$$

For each grid point $\mathbf{p}$ in the position space, evaluate the K matrices $\{\mathbf{G}_k\}$ and use FFT to find the ν_k that maximizes the expression $\mathbf{c}_k^H \mathbf{G}_k \mathbf{c}_k$. The emitter position estimate $\hat{\mathbf{p}}_0$ is determined as the position $\mathbf{p}$ that maximizes

$$L_{ks}(\mathbf{p}) = \sum_{k=1}^{K} \max_{\nu_k} \left\{\mathbf{c}_k^H \mathbf{G}_k \mathbf{c}_k\right\} \tag{10.94}$$

This concludes the derivation of the proposed method for unknown and known signals.

10.3.4 Computation Load

We assess the computation load of the DD and DPD method, assuming for simplicity two receivers.

Differential Doppler Approach

Consider the MLE in Equation (10.74). An efficient implementation can be obtained with a first step of computing the FFT of $\{r_{1,k}[n] r^*_{2,k}[n]\}$. This involves N complex multiplications for obtaining $\{r_{1,k}[n] r^*_{2,k}[n]\}$, $N \log_2 N$ multiplications for computing the FFT, and finally N complex multiplications for obtaining the squared absolute value. Thus, along the track the total number of complex multiplications is $KN(2 + \log_2 N) \cong KN \log_2 N$. This number will be doubled if the signals are known and the frequency is estimated for each of the two receivers separately. In the second step, the cost function in Equation (10.77) is evaluated. For each point in the grid, $14K$ real multiplications (or $7K$ complex multiplications) are required. Thus, for N_g grid points, the total number of operations is dominated by $7N_gK+KN\log_2 N$ if the signals are unknown, and therefore cross-correlation is used, or $7N_gK+2KN\log_2 N$ if the signals are known.

Proposed Approach

For known signals, the estimated position in our approach is determined by the maximum of Equation (10.94) over all grid points. Note that $\mathbf{G}_k=\mathbf{B}_k^H\mathbf{B}_k$, where $\mathbf{B}_k$ is an $L\times N$ matrix defined as $\mathbf{B}_k=\mathbf{V}_k^H\mathbf{S}_k$. The number of multiplications required to evaluate $\mathbf{B}_k$ is $2NL$ since $\mathbf{A}_{k,\ell}$ and $\mathbf{S}_k$ are diagonal matrices. Since $\mathbf{B}_k$ is an $L\times N$ matrix, the number of multiplications required to evaluate the Hermitian matrix $\mathbf{G}_k$ is approximately $0.5LN^2$. Therefore, we need a total of $2NL+0.5LN^2$ multiplications to evaluate $\mathbf{G}_k$.

The scalar $\mathbf{c}_k^H\mathbf{G}_k\mathbf{c}_k$ is evaluated by FFT as previously described. The FFT requires approximately $N\log_2 N$ multiplications. The total number of operations for N_g grid points is therefore $N_gK(2NL+0.5LN^2+N\log_2 N)\cong 0.5N_gKLN^2$. For unknown signals the evaluation of $\mathbf{V}_k$ in Equation (10.83) requires NL multiplications, and the evaluation of $\bar{\mathbf{Q}}_k$ requires an approximately additional $0.5NL^2$ multiplications. Finding the eigenvalues in Equation (10.86) requires L^3 multiplications. The total number of multiplications is $N_gK(L^3+0.5NL^2+NL)\cong N_gKLN(0.5L+1)$. Thus, in most cases the proposed approach requires considerably more computation than the DD, even if $L=2$.

10.3.5 Numerical Examples

We examined the performance of the proposed method and compared it with the DD method and with the CRLB (Amar and Weiss, 2008) using MC computer simulations. We focused on the position RMSE defined in Equation (10.38). To obtain statistical results, we used 100 MC trials.

The simulated signal was a 10-Kbps quadrature phase shift keying (QPSK) communication signal, sampled at 10^3 samples per second. Unless stated otherwise, we used 100 samples at each interception interval. The QPSK symbols were selected at random. The same signals were used for the known and unknown signal cases. The simulated nominal signal carrier frequency was $f_c=0.1$ GHz. The propagation speed was assumed to be $c=3\times10^8$ m/sec. The emitter's position was chosen at random within a square area of 10×10 Km. The unknown transmitted frequency shifts, $\{\nu_k\}$, were selected at random from the interval $[-100, 100]$ Hz. The channel attenuation was selected at random from a normal distribution with mean 1 and standard deviation 0.1, and the channel phase was selected at random from a uniform distribution over $[-\pi,\pi)$. These parameters were then used for all trials.

Consider two receivers ($L=2$), where one is moving leftward and one is moving rightward. Unless stated otherwise, the receivers' speed is $v=300$ m/sec.

Each receiver intercepts the signal at ten different intervals ($K = 10$) along its trajectory. The first receiver intercepts the signal every 1 Km starting at $[1, 0]$ Km and finishing at $[10, 0]$ Km; the second receiver intercepts the signal with the same spacing starting at $[10, 10]$ Km and finishing at $[1, 10]$ Km. As a second configuration, consider that three receivers ($L = 3$) are simulated. The trajectories, velocities, and interception points of the first two receivers are the same as previously described. The third receiver moves from $[1, -0.2]$ Km to $[10, -0.2]$ Km and intercepts the signal every 1 Km.

The localization performance of DPD is compared with the performance of the DD for known and unknown signals. The RMSE of position estimation versus SNR for known and unknown signals, and for two and three receivers, is shown in Figure 10.13. In all cases DPD outperforms DD at low SNR but at high SNR the methods are equivalent. (SNR is defined as the ratio of the average transmitted signal power to the average noise power.)

We now compare DD and DPD when the SNR is 20 dB for the two receivers at all interception intervals except for the SNR of the second receiver at the last three interception intervals. The direction and speed of each receiver is as previously stated. The SNR at the last three intervals is changed from 20 dB to -24 dB with a step of 2 dB.

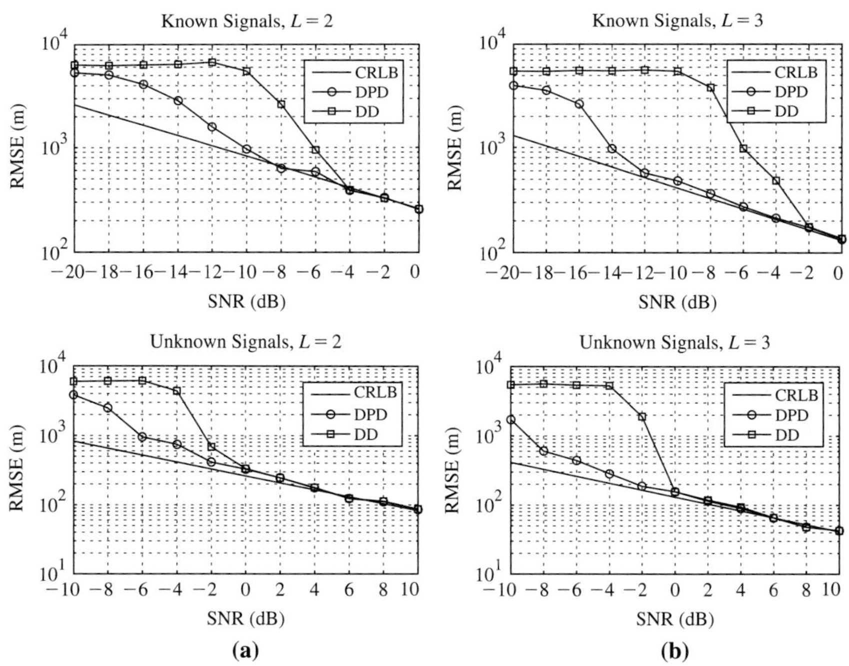

FIGURE 10.13 RMSE of DPD, DD, and CRLB versus SNR, with known and unknown transmitted signals, for two receivers (a) and three receivers (b).

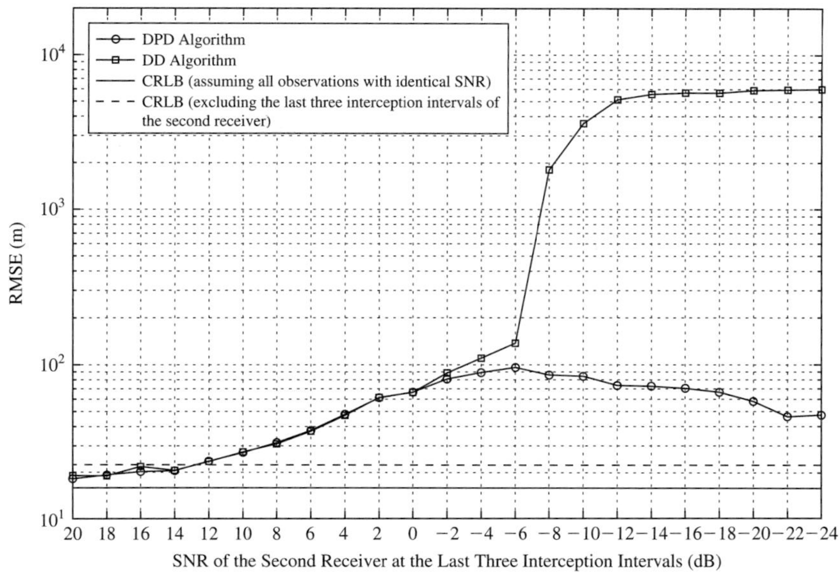

FIGURE 10.14 RMSE of DPD and DD versus SNR of the second receiver at the last three interception intervals.

Figure 10.14 shows the results. It can be seen that as the SNR at the last three intervals of the second receiver decreases, the performance of the two methods derogates. However, beyond a certain point the performance of DPD improves in contrast with conventional DD. DPD ignores the unreliable data and performs as if it does not exist. Practically, the DD outlier will probably be removed by a goodness-of-fit (chi-square) test. However, as demonstrated here DPD works well without such tests.

We examine the RMSE versus the number of samples, N, used in each interception interval. Consider two receivers ($L=2$). Assume that the first receiver moves from $[1, 0]$ Km to $[10, 0]$ Km and intercepts the signal every 1 Km. The second receiver moves from $[10, 1]$ Km to $[10, 10]$ Km and intercepts the signal with the same spacing. The signals are assumed unknown to the receivers. The SNR is -5 dB. The number of samples is changed from $N=100$ to $N=500$ with a step size of 20.

The position RMSE versus N for DD and DPD and the CRLB is shown in Figure 10.15. For a small number of samples, DPD outperforms DD, but as the number of samples increases the methods become equivalent. Note that as N increases, the number of unknowns increases since the signal samples are unknown. This explains the gap between the RMSE and the CRLB, which does not decrease with increasing N. Therefore, neither method is statistically efficient.

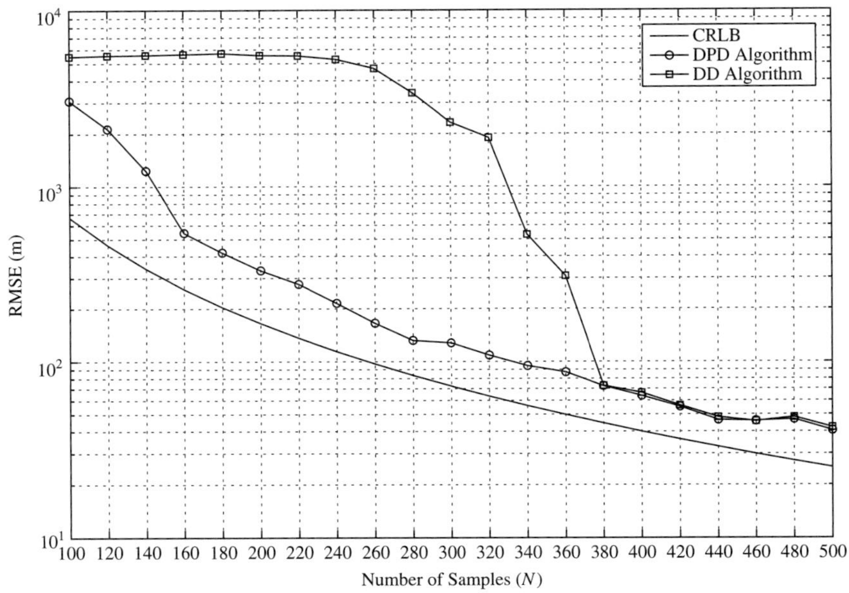

FIGURE 10.15 RMSE of DPD, DD, and CRLB versus the number of samples N.

10.4 TWO-STEP VERSUS ONE-STEP LOCALIZATION

As expected intuitively and demonstrated in this chapter, two-step procedures usually provide inferior performance with respect to single-step schemes. This property has been proven in the literature under asymptotic conditions, so it is of interest to identify indirect methods, appropriate for sensor data fusion, that are at least as asymptotically efficient as the direct methods. We prove here that if ML based on incomplete data is used in the first step and weighted least squares is used in the second step, indirect estimation achieves the performance of direct estimation for asymptotically large data sets.

10.4.1 Problem Formulation

Consider a parameter vector $\boldsymbol{\theta}$ in a space $\boldsymbol{\Theta}$, which is a subspace of the L-dimensional real Cartesian space. Similarly, consider a different parameter vector $\boldsymbol{\phi}$ in a space $\boldsymbol{\Phi}$, which is a subspace of the M-dimensional real Cartesian space where $M \leq L$. In short, $\boldsymbol{\theta} \in \boldsymbol{\Theta} \subset \mathbb{R}^L$, $\boldsymbol{\phi} \in \boldsymbol{\Phi} \subset \mathbb{R}^M$. Assume that for every $\boldsymbol{\phi} \in \boldsymbol{\Phi}$ there is a $\boldsymbol{\theta} \in \boldsymbol{\Theta}$; the mapping is known and is given by some injective function $\boldsymbol{\theta} = \mathbf{g}(\boldsymbol{\phi})$, $\forall \boldsymbol{\phi} \in \boldsymbol{\Phi}$. This mapping defines a subspace $\bar{\boldsymbol{\Theta}} \subseteq \boldsymbol{\Theta}$ given as $\bar{\boldsymbol{\Theta}} = \{\boldsymbol{\theta} : \boldsymbol{\theta} = \mathbf{g}(\boldsymbol{\phi}), \boldsymbol{\phi} \in \boldsymbol{\Phi}\}$.

Assume now that each of the elements of the vector $\boldsymbol{\theta}$ is a parameter that affects the p.d.f. $f(\mathbf{x}_j|\theta_j)$ of an observation vector denoted by $\mathbf{x}_j$. Denote by $\mathbf{x}_j(n)$ the nth realization of $\mathbf{x}_j$. We assume that each of the observation vectors

$\mathbf{x}_j, j=1,\ldots,L$ is collected at a different location and the locations are spatially separated. In other words, the estimation of θ_j must be based on $\{\mathbf{x}_j(n)\}_{n=1}^N$, not on $\{\mathbf{x}_i(n)\}_{n=1}^N$, $i \neq j$.

Based on a collection of N independent realizations $\{\mathbf{x}_j(n)\}_{j=1}^L$, we are interested in estimating the primary parameter of interest $\boldsymbol{\phi}$. The parameter $\boldsymbol{\phi}$ can be estimated directly from all observation vectors (the complete data) using a single step, or it can be estimated in two steps by first estimating $\boldsymbol{\theta}$ and then using the estimate $\hat{\boldsymbol{\theta}}$ for estimating $\boldsymbol{\phi}$. Since the jth element of $\boldsymbol{\theta}$ is estimated using only $\{\mathbf{x}_j(n)\}_{n=1}^N$, it is clear that each element of $\boldsymbol{\theta}$ is estimated using incomplete data.

Define the column vector $\mathbf{x}(n)$ consisting of a concatenation of the observation vectors $\mathbf{x}(n) \triangleq \left[\mathbf{x}_1^T(n), \mathbf{x}_2^T(n)\ldots\mathbf{x}_L^T(n)\right]^T$ and the matrix $\mathbf{X} \triangleq [\mathbf{x}(1), \mathbf{x}(2)\ldots\mathbf{x}(N)]$. Under the assumption that the vectors $\{\mathbf{x}(n)\}_{n=1}^N$ are independent, the log likelihood of $\mathbf{X}$ is given by $\tilde{\ell}(\mathbf{X}|\boldsymbol{\theta}) \triangleq \sum_{n=1}^N \ell(\mathbf{x}(n)|\boldsymbol{\theta})$, where $\ell(\mathbf{x}(n)|\boldsymbol{\theta}) \triangleq \sum_{j=1}^L \bar{\ell}(\mathbf{x}_j(n)|\theta_j)$ and $\bar{\ell}(\mathbf{x}_j(n)|\theta_j) \triangleq \ln(f(\mathbf{x}_j(n)|\theta_j))$.

Since $\boldsymbol{\theta} = \mathbf{g}(\boldsymbol{\phi})$, the exact ML estimator of $\boldsymbol{\phi}$ is defined by

$$\hat{\boldsymbol{\phi}}_1 = \underset{\boldsymbol{\phi}\in\boldsymbol{\Phi}}{\operatorname{argmax}}\, \tilde{\ell}(\mathbf{X}|\mathbf{g}(\boldsymbol{\phi})) \tag{10.95}$$

This is a single-step scheme for estimating $\boldsymbol{\phi}$ from all observations. Note that the search is performed only within the known space $\boldsymbol{\Phi}$ of the vector $\boldsymbol{\phi}$.

Although Equation (10.95) is well known, intuitive, and conceptually simple, it requires the collection of all observation vectors before $\hat{\boldsymbol{\phi}}_1$ can be evaluated. This may be a costly process in terms of power and transmission bandwidth and perhaps not a practical option in some cases.

In the two-step scheme (or indirect method) θ_j is estimated using the incomplete data $\{\mathbf{x}_j(n)\}_{n=1}^N$. Invoking the ML principle, we can write

$$\hat{\theta}_j = \underset{\theta_j\in\mathbb{R}}{\operatorname{argmax}} \sum_{n=1}^{N} \bar{\ell}(\mathbf{x}_j(n)|\theta_j)), \quad j=1,\ldots,L \tag{10.96}$$

The L estimates $\{\hat{\theta}_j\}_{j=1}^L$ are independent random variables. Also, since the estimate $\hat{\theta}_j$ is obtained by the jth sensor separately and without knowing what the other values of $\{\hat{\theta}_k\}_{k\neq j}$ are, there is no guarantee that $\hat{\boldsymbol{\theta}} \in \bar{\boldsymbol{\Theta}}$. Moreover, while $\forall\boldsymbol{\theta} \in \bar{\boldsymbol{\Theta}}$ we have $\boldsymbol{\phi} \in \boldsymbol{\Phi}$ by the injective property of $\mathbf{g}(\cdot)$, we do not necessarily have inverse mapping for any $\hat{\boldsymbol{\theta}}$ not in $\bar{\boldsymbol{\Theta}}$. Thus, the ML invariance theorem (Scharf, 1990, p. 218) is not applicable in this case.

We now use $\hat{\boldsymbol{\theta}}$ for estimating $\boldsymbol{\phi}$. Recall that $\boldsymbol{\theta}$ is estimated using the ML principle, and it is well known that under certain regularity conditions and $N \to \infty$, $\hat{\boldsymbol{\theta}}$ is normally distributed with mean $\boldsymbol{\theta}_0$ (the true value) and covariance $\mathbf{F}_0^{-1}$, where $\mathbf{F}_0$ is the FIM (Cramer, 1946). In other words, $\sqrt{N}(\hat{\boldsymbol{\theta}} - \boldsymbol{\theta}_0) \sim \mathcal{N}\left(0, N\mathbf{F}_0^{-1}\right)$. Here the FIM is a diagonal matrix and the nonzero elements are given by

$$[\mathbf{F}(\boldsymbol{\theta})]_{i,i} = N\mathbf{E}\left\{\left(\frac{\partial \bar{\ell}(\mathbf{x}_i(n)|\theta_i)}{\partial \theta_i}\right)^2\right\} \tag{10.97}$$

and $\mathbf{F}_0 \triangleq \mathbf{F}(\boldsymbol{\theta}_0)$.

Therefore, a reasonable estimate of the primary parameter vector $\boldsymbol{\phi}$ can obtained by minimizing a cost function of the form $[\hat{\boldsymbol{\theta}} - \mathbf{g}(\boldsymbol{\phi})]^T \mathbf{W} [\hat{\boldsymbol{\theta}} - \mathbf{g}(\boldsymbol{\phi})]$, where $\mathbf{W}$ is a diagonal weighting matrix. The element $\mathbf{W}_{j,j}$ should reflect the relative confidence in estimating θ_j, so the choice of $\mathbf{W}_{j,j} = [\mathbf{F}_0]_{j,j}$ is natural. Unfortunately, $\mathbf{F}_0$ may depend on the true value $\boldsymbol{\theta}_0$, which is not known. Thus, we propose the use of $\mathbf{W} = \mathbf{F}(\hat{\boldsymbol{\theta}})$, which is a random diagonal matrix. The second step in estimating $\boldsymbol{\phi}$ is therefore given by

$$\hat{\boldsymbol{\phi}}_2 = \underset{\boldsymbol{\phi} \in \boldsymbol{\Phi}}{\operatorname{argmin}}\ C(\boldsymbol{\phi}|\hat{\boldsymbol{\theta}}) \tag{10.98}$$

$$C(\boldsymbol{\phi}|\hat{\boldsymbol{\theta}}) \triangleq [\hat{\boldsymbol{\theta}} - \mathbf{g}(\boldsymbol{\phi})]^T \mathbf{F}(\hat{\boldsymbol{\theta}}) [\hat{\boldsymbol{\theta}} - \mathbf{g}(\boldsymbol{\phi})] \tag{10.99}$$

This concludes the description of the single-step (or direct) estimation procedure and the proposed two-step (indirect) estimation procedure.

10.4.2 Asymptotic Performance

We now evaluate the asymptotic covariance of both estimators. First note that the single-step estimator in Equation (10.95) is the ML estimator that under regularity conditions converges in probability to the true parameter value as $N \to \infty$. Further, $\hat{\phi}_1$ is asymptotically normal and efficient. Thus (Cramer, 1946),

$$\sqrt{N}(\hat{\boldsymbol{\phi}}_1 - \boldsymbol{\phi}_0) \sim \mathcal{N}(0, \boldsymbol{\Sigma}_1) \tag{10.100}$$

where ϕ_0 is the true parameter value and the covariance $\boldsymbol{\Sigma}_1$ is given by

$$\boldsymbol{\Sigma}_1^{-1} = \mathbf{E}\left\{\left(\frac{\partial \ell(\mathbf{x}(n)|\mathbf{g}(\boldsymbol{\phi}))}{\partial \boldsymbol{\phi}}\right)\left(\frac{\partial \ell(\mathbf{x}(n)|\mathbf{g}(\boldsymbol{\phi}))}{\partial \boldsymbol{\phi}}\right)^T\right\} \tag{10.101}$$

Note that $\partial \ell(\mathbf{x}(n)|\mathbf{g}(\boldsymbol{\phi}))/\partial \phi$ is a column vector evaluated at the true parameter value $\boldsymbol{\phi}_0$. Using the chain rule, we have

$$\frac{\partial \ell(\mathbf{x}(n)|\mathbf{g}(\phi))}{\partial \boldsymbol{\phi}} = \frac{\partial \ell(\mathbf{x}(n)|\boldsymbol{\theta})}{\partial \boldsymbol{\phi}} = \frac{\partial \boldsymbol{\theta}^T}{\partial \boldsymbol{\phi}} \frac{\partial \ell(\mathbf{x}(n)|\boldsymbol{\theta})}{\partial \boldsymbol{\theta}} = \mathbf{G} \frac{\partial \ell(\mathbf{x}(n)|\boldsymbol{\theta})}{\partial \boldsymbol{\theta}} \tag{10.102}$$

where $\mathbf{G}$ is a $M \times L$ matrix defined by

$$\mathbf{G} \triangleq \frac{\partial \boldsymbol{\theta}^T}{\partial \boldsymbol{\phi}} = \frac{\partial \mathbf{g}^T(\boldsymbol{\phi})}{\partial \boldsymbol{\phi}} \tag{10.103}$$

and evaluated at ϕ_0. By substituting Equation (10.102) in Equation (10.101) we obtain

$$\mathbf{\Sigma}_1^{-1} = \mathbf{G}\mathbf{E}\left\{\left(\frac{\partial \ell(\mathbf{x}(n)|\boldsymbol{\theta})}{\partial \boldsymbol{\theta}}\right)\left(\frac{\partial \ell(\mathbf{x}(n)|\boldsymbol{\theta})}{\partial \boldsymbol{\theta}}\right)^T\right\}\mathbf{G}^T = (1/N)\mathbf{G}\mathbf{F}_0\mathbf{G}^T \tag{10.104}$$

Thus, the asymptotic covariance of $\hat{\boldsymbol{\phi}}_1$ is given by $(\mathbf{G}\mathbf{F}_0\mathbf{G}^T)^{-1}$. This is also recognized as the CRLB on the covariance of any unbiased estimator of ϕ.

We turn to the two-step estimator in Equation (10.98). Because the estimate $\hat{\boldsymbol{\phi}}_2$ is the minimizer of $C(\boldsymbol{\phi}|\hat{\boldsymbol{\theta}})$, $(\partial C/\partial \boldsymbol{\phi})_{\hat{\boldsymbol{\phi}}_2,\hat{\boldsymbol{\theta}}} = \mathbf{0}$. Also, the true parameter ϕ_0 is the minimizer of $C(\boldsymbol{\phi}|\boldsymbol{\theta}_0)$ and therefore $(\partial C/\partial \boldsymbol{\phi})_0 = \mathbf{0}$, where $(\cdot)_0$ denotes an expression evaluated at $\boldsymbol{\phi} = \boldsymbol{\phi}_0$, $\hat{\boldsymbol{\theta}} = \boldsymbol{\theta}_0$.

Consider the first-order Taylor series expansion,

$$\begin{aligned}\left(\frac{\partial C}{\partial \boldsymbol{\phi}}\right)_{\hat{\boldsymbol{\phi}}_2,\hat{\boldsymbol{\theta}}} &= \left(\frac{\partial C}{\partial \boldsymbol{\phi}}\right)_0 + \left(\frac{\partial^2 C}{\partial \boldsymbol{\phi}^2}\right)_0 (\hat{\boldsymbol{\phi}}_2 - \boldsymbol{\phi}_0) \\ &\quad + \left(\frac{\partial^2 C}{\partial \boldsymbol{\phi}\partial \hat{\boldsymbol{\theta}}}\right)_0 (\hat{\boldsymbol{\theta}} - \boldsymbol{\theta}_0) + O(1/N)\end{aligned} \tag{10.105}$$

As indicated, the left expression is 0, as is the first expression on the right side. Thus,

$$-\left(\frac{\partial^2 C}{\partial \boldsymbol{\phi}^2}\right)_0 (\hat{\boldsymbol{\phi}}_2 - \phi_0) = \left(\frac{\partial^2 C}{\partial \boldsymbol{\phi}\partial \hat{\boldsymbol{\theta}}}\right)_0 (\hat{\boldsymbol{\theta}} - \boldsymbol{\theta}_0) + O(1/N) \tag{10.106}$$

Observe that

$$\left(\frac{\partial^2 C}{\partial \boldsymbol{\phi}^2}\right)_0 = 2\mathbf{G}\mathbf{F}_0\mathbf{G}^T \tag{10.107}$$

$$\left(\frac{\partial^2 C}{\partial \boldsymbol{\phi}\partial \hat{\boldsymbol{\theta}}}\right)_0 = -2\mathbf{G}\mathbf{F}_0 \tag{10.108}$$

Substituting Equation (10.107) and (10.108) in Equation (10.106) yields

$$\mathbf{G}\mathbf{F}_0\mathbf{G}^T(\hat{\boldsymbol{\phi}}_2 - \boldsymbol{\phi}_0) = \mathbf{G}\mathbf{F}_0(\hat{\boldsymbol{\theta}} - \boldsymbol{\theta}_0) + O(1/N) \tag{10.109}$$

Thus, noting that $\mathbf{E}\left\{(\hat{\boldsymbol{\theta}} - \boldsymbol{\theta}_0)(\hat{\boldsymbol{\theta}} - \boldsymbol{\theta}_0)^T\right\} = \mathbf{F}_0^{-1}$, for large enough N we get

$$\begin{aligned}\mathbf{\Sigma}_2 &\triangleq \mathbf{E}\left\{(\hat{\boldsymbol{\phi}} - \boldsymbol{\phi}_0)(\hat{\boldsymbol{\phi}} - \boldsymbol{\phi}_0)^T\right\} \\ &= (\mathbf{G}\mathbf{F}_0\mathbf{G}^T)^{-1}\mathbf{G}\mathbf{F}_0\mathbf{F}_0^{-1}\mathbf{F}_0\mathbf{G}^T(\mathbf{G}\mathbf{F}_0\mathbf{G}^T)^{-1} \\ &= (\mathbf{G}\mathbf{F}_0\mathbf{G}^T)^{-1}\end{aligned} \tag{10.110}$$

We see that, although a one-step estimator is expected to perform better than the two-step estimator, asymptotically (for large N) they are both unbiased and achieve the CRLB.

10.4.3 Illustrative Example

An example that illustrates the principles just discussed may be of interest. Consider the observation model

$$x_j(n)=\sqrt{\theta_j}+e_j(n)=\sqrt{\alpha_j\phi}+e_j(n); \quad j=1,\ldots,L; \; n=1,\ldots,N \qquad (10.111)$$

where $\{\alpha_j\}$ are real positive known scalars, $\phi>0$ is the primary scalar parameter, $\theta_j=\alpha_j\phi$, and $\{e_j(n)\}$ are i.i.d. Gaussian variables, $e_j(n)\sim\mathcal{N}(0,\sigma^2)$. Notice that $\boldsymbol{\theta}=\mathbf{g}(\phi)=[\alpha_1,\ldots\alpha_L]^T\phi$.

The true ML estimate of ϕ is given by maximizing the LLF of the observations in Equation (10.111). The single-step true ML estimate is thus

$$\hat{\phi}_1=\underset{\phi}{\operatorname{argmin}}\sum_{j=1}^{L}\sum_{n=1}^{N}\left[x_j(n)-\sqrt{\alpha_j\phi}\right]^2=\left(\frac{\sum_{j=1}^{L}\bar{x}_j\sqrt{\alpha_j}}{\sum_{j=1}^{L}\alpha_j}\right)^2 \qquad (10.112)$$

where $\bar{x}_j\triangleq\frac{1}{N}\sum_{n=1}^{N}x_j(n)$. The two-step estimator first finds an estimate for θ_j by maximizing the LLF function of $\{x_j(n)\}$:

$$\hat{\theta}_j=\underset{\theta_j}{\operatorname{argmin}}\sum_{n=1}^{N}\left[x_j(n)-\sqrt{\theta_j}\right]^2=\bar{x}_j^2 \qquad (10.113)$$

The diagonal elements of the FIM of $\boldsymbol{\theta}$ are given by

$$[\mathbf{F}_0]_{j,j}=\frac{N}{4\sigma^2\theta_j} \qquad (10.114)$$

Substituting $\hat{\theta}_j$ for θ_j in Equation (10.114) and then substituting the result in Equation (10.98) yields

$$\hat{\phi}_2=\frac{\sum_{j=1}^{L}\alpha_j}{\sum_{j=1}^{L}\alpha_j^2\big/\hat{\theta}_j}=\frac{\sum_{j=1}^{L}\alpha_j}{\sum_{j=1}^{L}\alpha_j^2\big/\bar{x}_j^2} \qquad (10.115)$$

It can be easily verified that in general $\hat{\phi}_1\neq\hat{\phi}_2$. However, for $L=1$ the two estimators yield the same result.

To check the performance of the estimators we performed MC computer simulations. The following parameter values were selected: $\phi_0=1$, $\sigma^2=3$, $\alpha_j=j$, $j=1,2,\ldots,10$. In Figures 10.16 and 10.17 we plot the empirical bias and empirical variance of each estimator. Each point on the plot is based on 1000 independent trials. As expected, for large N both estimators are unbiased and have the same variance. For smaller values of N the direct estimator is better.

In conclusion, two-step estimation schemes in multisensor applications are inferior to single-step ML estimation. We proved here that by using proper weighting in the second step the two-step estimator becomes asymptotically unbiased and efficient.

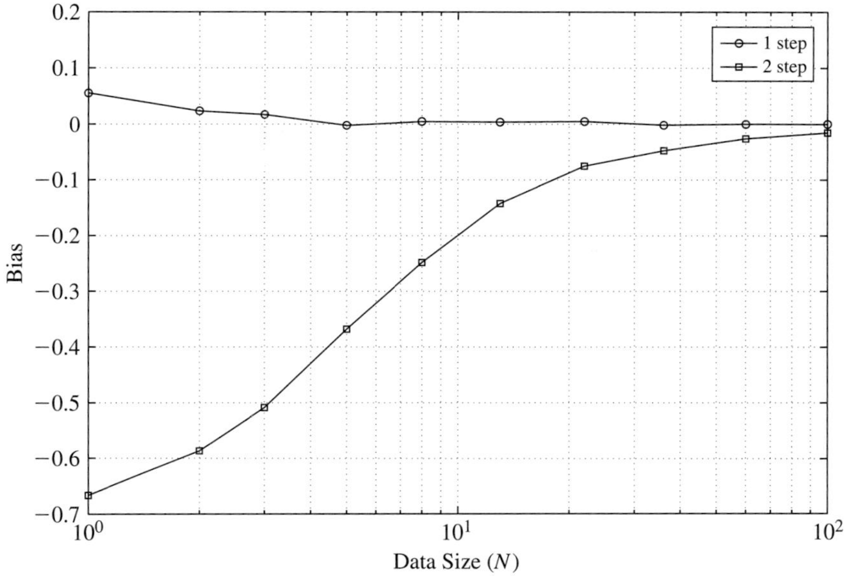

FIGURE 10.16 Bias of each of the two estimators versus data size.

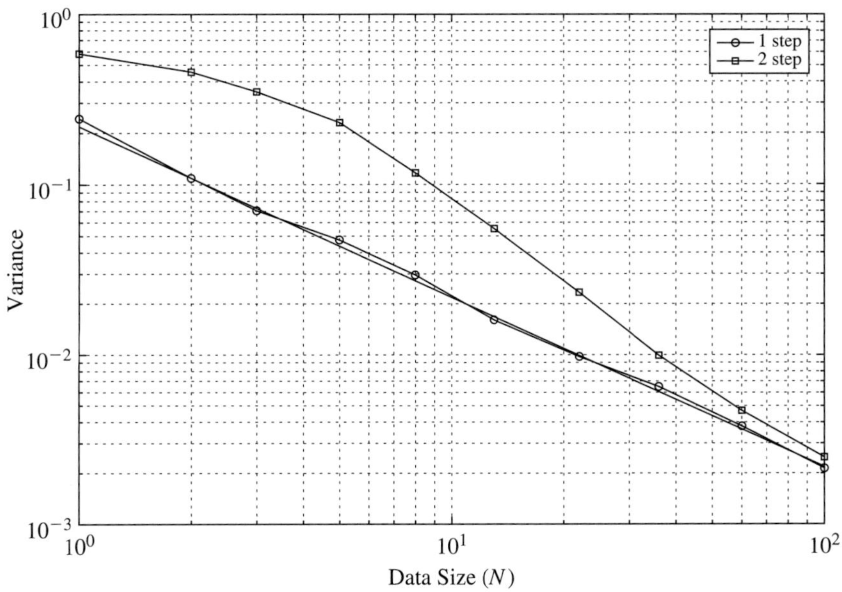

FIGURE 10.17 Variance of each of the two estimators versus data size.

10.5 CONCLUSION

DPD is a promising approach that yields superior results in difficult conditions such as low SNR, model errors, and NLOS. It requires good synchronization between base stations, both in frequency and time, which can be easily achieved by exploiting the global positioning system (GPS). It also requires the transfer of raw data between stations and therefore larger communication bandwidth.

Despite these drawbacks, the advantages of DPD, in precision, model error mitigation, and complexity, should make it the preferred method in many applications.

REFERENCES

Amar, A., Weiss, A.J., 2006. Direct position determination in the presence of model errors—known waveforms. Digital Signal Process. 16 (1), 52–83.

Amar, A., Weiss, A.J., 2007. Fundamental limitations on the number of resolvable emitters using geolocation system. IEEE Trans. Signal Process. 55 (5), 2193–2202.

Amar, A., Weiss, A.J., 2008. Localization of radio emitters based on Doppler frequency shifts. IEEE Trans. Signal Process. 56 (11), 5500–5508.

Becker, K., 1992. An efficient method of passive emitter location. IEEE Trans. Aerosp. Electron. Syst. 28 (4), 1091–1104.

Becker, K., 1999. Passive localization of frequency-agile radars from angle and frequency measurements. IEEE Trans. Aerosp. Electron. Syst. 35 (4), 1129–1144.

Bessis, J.L., 1981. Operational data collection and platform location by satellite. Remote Sens. Environ. 11 (5), 93–111.

Carter, G.C., 1993. Coherence and time delay estimation. IEEE Press. 3 (4), 67–94.

Chan, Y.T., Jardine, F.L., 1990. Target localization and tracking from Doppler shift measurements. IEEE J. Oceanic Eng. 15 (3), 251–257.

Chan, Y.T., Towers, J.J., 1992a. Sequential localization of a radiating source by Doppler shifted frequency measurements. IEEE Trans. Aerosp. Electron. Syst. 28 (4), 1084–1090.

Chan, Y.T., Towers, J.J., 1992b. Passive localization from Doppler shifted measurements. IEEE Trans. Signal Process. 40 (10), 2594–2598.

Chestnut, P.C., 1982. Emitter location accuracy using TDOA and differential Doppler. IEEE Trans. Aerosp. Electron. Syst. 18 (2), 214–218.

Cramer, H., 1946. Mathematical Methods of Statistics. Princeton University Press.

Fowler, M.L., 2000. Coarse quantization for data compression in coherent location systems. IEEE Trans. Aerosp. Electron. Syst. 36 (4), 1269–1278.

Fowler, M.L., 2001. Analysis of single platform passive emitter location with terrain data. IEEE Trans. Aerosp. Electron. Syst. 37 (2), 495–507.

Gezici, S., Tian, Z., Giannakis, G.B., Kobayashi, H., Molisch, A.F., Poor, H.V., Sahinoglu, Z., 2005. Localization via ultra-wideband radios: A look at positioning aspects for future sensor networks. IEEE Signal Process. Mag. 22 (4), 70–84.

Gustafsson, F., Gunnarsson, F., 2005. Mobile positioning using wireless networks: Possibilities and fundamental limitations based on available wireless network measurements. IEEE Signal Process. Mag. 22 (4), 41–53.

Haworth, D.P., Smith, N.G., Bardelli, R., Clement, T., 1997. Interference localization for EUTELSAT satellites—the first European transmitter location system. Int. J. Satell. Commun. 15 (4), 155–183.

Ho, K.C., Chan, Y.T., 1997. Geolocation of a known altitude object from TDOA and FDOA measurements. IEEE Trans. Aerosp. Electron. Syst. 33 (3), 770–783.

Knapp, C.H., Carter, G.C., 1977. Estimation of time delay in the presence of source or receiver motion. J. Acoust. Soc. Am. 61 (6), 1545–1549.

Kozick, R.J., Sadler, B.M., 2004. Source localization with distributed sensor arrays and partial spatial coherence. IEEE Trans. Signal Process. 52 (3), 601–616.

Krim, H., Viberg, M., 1996. Two decades of array signal processing research: The parametric approach. IEEE Signal Process. Mag. 13 (4), 67–94.

Levanon, N., 1980. Theoretical bounds on random errors in satellite Doppler navigation. IEEE Trans. Aerosp. Electron. Syst. 20 (6), 810–816.

Levanon, N., 1989. Interferometry against differential Doppler: performance comparison of two emitter location airborne systems. IEE Proc. F: Radar Signal Process. 136 (2), 70–74.

Levanon, N., Zaken, M.B., 1985. Random error in ARGOS and SARSAT satellite positioning systems. IEEE Trans. Aerosp. Electron. Syst. 21 (6), 783–790.

Liberti, J.C., Rappaport, T.S., 1999. Smart Antennas for Wireless Communications: IS-95 and Third Generation CDMA Applications. Prentice-Hall.

Pattison, T., Chou, S.I., 2000. Sensitivity analysis of dual satellite geolocation. IEEE Trans. Aerosp. Electron. Syst. 36 (1), 56–71.

Patwari, N., Ash, J.N., Kyperountas, S., Hero, A.O., Moses, R.L., Correal, N.S., 2005. Locating the nodes: Cooperative localization in wireless sensor networks. IEEE Signal Process. Mag. 22 (4), 54–69.

Rao, C.R., 2002. Linear Statistical Inference and Its Applications, Third ed. John Wiley & Sons.

Sayed, A.H., Tarighat, A., Khajehnouri, N., 2005. Network-based wireless location: Challenges faced in developing techniques for accurate wireless location information. IEEE Signal Process. Mag. 22 (4), 24–40.

Scales, W.C., Swanson, R., 1984. Air and sea rescue via satellite systems. IEEE Spectr. 21 (3), 48–52.

Scharf, L.L., 1990. Statistical Signal Processing: Detection, Estimation, and Time Series Analysis. Addison-Wesley.

Schmidt, R.O., 1986. Multiple emitter location and signal parameter estimation. IEEE Trans. Antennas Propag. 34 (3), 276–280.

Schultheiss, P.M., Weinstein, E., 1979. Estimation of differential Doppler shifts. J. Acoust. Soc. Am. 66 (5), 1412–1419.

Skolnik, M.I., 1980. Introduction to Radar Systems, Second ed. McGraw-Hill.

Stansfield, R.G., 1947. Statistical theory of DF fixing. J. IEE. 94 (15), 762–770.

Stein, S., 1993. Differential delay/Doppler ML estimation with unknown signals. IEEE Trans. Signal Process. 41 (8), 2717–2719.

Stoica, P., Nehorai, A., Sodersrom, T., 1995. Decentralized array processing using the MODE algorithm. Circuits Syst. Signal Process. 14 (1), 17–38.

Sun, G., Chen, J., Gou, W., Liu, K.J.R., 2005. Signal processing techniques in network-aided positioning: A survey of state-of-the-art positioning designs. IEEE Signal Process. Mag. 22 (4), 12–23.

Trees, H.L.V., 2002. Optimum Array Processing: Part IV—Detection, Estimation, and Modulation Theory. John Wiley & Sons.

Torrieri, D.J., 1984. Statistical theory of passive location systems. IEEE Trans. Aerosp. Electron. Syst. 20 (2), 183–197.

Wax, M., 1996. Model based processing in sensor arrays. Adv. Spectr. Anal. Array Process. 3 (4), 67–94.

Wax, M., Kailath, T., 1983. Optimum localization of multiple sources by passive arrays. IEEE Trans. Acoust. Speech Signal Process. 31 (5), 1210–1217.

Wax, M., Kailath, T., 1985. Decentralized processing in sensor arrays. IEEE Trans. Acoust. Speech Signal Process. 33 (5), 1123–1129.

Wax, M., Ziskind, I., 1989. On unique localization of multiple sources by passive sensor arrays. IEEE Trans. Acoust. Speech Signal Process. 37 (7), 996–1000.

Weinstein, E., 1981. Decentralization of the Gaussian maximum likelihood estimator and its applications to passive array processing. IEEE Trans. Acoust. Speech Signal Process. 29 (4), 945–951.

Weinstein, E., 1982. Measurement of the differential Doppler shift. IEEE Trans. Acoust. Speech Signal Process. 30 (1), 112–117.

Weinstein, E., Levanon, N., 1980. Passive array tracking of a continuous wave transmitting projectile. IEEE Trans. Aerosp. Electron. Syst. 16 (5), 721–726.

Weiss, A.J., Amar, A., 2005. Direct position determination of multiple radio signals. EURASIP J. Appl. Signal Process. 2005 (1), 37–49.

Weiss, A.J., Friedlander, B., 1993. Performance analysis of spatial smoothing with interpolated arrays. IEEE Trans. Signal Process. 41 (5), 1881–1892.

Weiss, A.J., Willsky, A.S., Levy, B.C., 1988. Maximum likelihood array processing for the estimation of superimposed signals. Proc. IEEE 76 (2), 203–205.

Zagami, J.M., Parl, S.A., Bussgang, J.J., Melillo, K.D., 1998. Providing universal location services using a wireless E911 location network. IEEE Commun. Mag. 36 (4), 66–71.

Index

M

N

O

P

U

V

W